A TEXTBOOK OF
ELECTRONIC DEVICES AND CIRCUITS

For Engineering Students

By

J.B. Gupta
B.Sc. (Elec. Engg.); C. Eng. (I);
F.I.E. (India); M.I.S.T.E.

S.K. KATARIA & SONS®
Educational Publishers
4885/109, Prakash Mahal, Dr. Subhash Bhargav Lane,
Opposite Delhi Medical Association, Daryaganj, New Delhi-110002 (INDIA)
Phone: +91-11-43551243; Mobile +91-9871775858, 7678425209
e-mail: katariabook@gmail.com; katariabooks@yahoo.com
Website: www.skkatariaandsons.com
(e-books, subscriptions and bulk buying also available on our website)

Published By:

S.K. Kataria & Sons®

4885/109, Prakash Mahal, Dr. Subhash Bhargav Lane,
Opposite Delhi Medical Association, Daryaganj, New Delhi-110002 (INDIA)
Phone: +91-11-43551243, *Mobile:* +91-9871775858, 7678425209
e-mail: katariabook@gmail.com, katariabooks@yahoo.com
Website: www.skkatariaandsons.com

First Edition: July, 2010; Reprint: 2025

Price: ₹ 495/-

Laser Typesetting by:
AISA ENTERPRISES, Delhi-110092

Printed at:
A.S. Offset Printers, Delhi-110051

PREFACE

The primary aim of presenting this book is to meet the needs of Second Year (Third Semester) students of B.E. (E.C.E.; E.E.; E.E.E.; E.I.) studying in Engineering Colleges affiliated to Maharishi Dayanand University (MDU), Rohtak, Haryana. The book has been prepared keeping in view the latest revised syllabus (W.E.F. Session 2010-11). The subject matter has been presented in an easily comprehensible manner, logical sequence, using simple language and to provide detailed exposition to fundamentals. A large number of workedout examples, theoretical questions, short answer type question with answers and numerical problems included in the text are from latest questions papers of universities/ Deemed Universities/Technical Universities including MDU or are of equivalent standard.

The book consists of 8 chapters divided into 4 sections according to the latest syllabus (W.E.F. Session 2010-11) prescribed, by Maharishi Dayanand University, for the subject of Electronic Devices and Circuits.

The solution to illustrative examples covering almost all topics/subtopics of different chapters makes the entire presentation easy to follow. Selected exercises and numerical problems with answers given at the end of each chapter help the students in evaluating their understanding of the subject. Short answer type questions with answers help the students in preparing for viva-voce/practical examinations.

Every care has been taken to eliminate misprints, omissions and errors but it is too much to expect that no inaccuracy/misprint/error has crept in and the author would be grateful to the readers for bringing to his notice any such errors/misprints/omissions they may come across while going through the book.

The author shall feel satisfied if the book meets the needs of the students for whom it is meant. Suggestions and criticism for the improvement of the book shall be welcomed.

The author will like to appreciate the sincere efforts and contribution made by his sons Rajeev Manglik and Rohit Manglik in presentation of the book in this form.

The author will like to thank all the Scientists, Engineers, Authors and Publishers whose works and text have been the source of enlightenment, inspiration and guidance in presenting this new edition.

Lastly, but not the least, the author is thankful to his publisher, Sanjeev Kumar Kataria for bringing out the book in such a short time and for pricing it moderately despite heavy cost of paper and printing.

July 14, 2010 **J.B. GUPTA**

To My Loving Grand Daughter

RUSHIL

OTHER USEFUL BOOKS

CONTENTS

Conducting Materials

CONTENTS OF THIS SECTION

CHAPTER 1

Conducting Materials

1.1. Introduction

Electrical Engineering materials may be divided into four broad groups according to purposes they serve. These four groups are : (*i*) Conductors, (*ii*) Semiconductors, (*iii*) Magnetic materials and (*iv*) Electrical insulating materials or dielectrics.

Materials used commercially for conducting electricity can be classed as *conducting materials.* Conducting materials (such as silver, copper, aluminium etc.) are characterised by large electrical conductivity (or small resistivity) and are used to provide path for flow of electric current in all types of electrical machine windings, electrical apparatus, devices, instruments etc. Conducting materials are required for the manufacture of cables and wires by means of which electrical energy is transmitted over long distances and then distributed to consumers.

1.2. Structure of an Atom

In order to explain many phenomena associated with conduction in gases, metals and semiconductors and the electron emission from a metal surface, it is necessary to assume that the atom has loosely bound electrons which n be torn away from it.

On the basis of experiments on the scattering of par icles, combined with the discovery of neutron made dwick in 1932, Rutherford suggested a complete he atom which is known as *Rutherford's Nuclear omic Structure*. According to this model :

Inside This Chapter

(*i*) The centre of an atom consists of a positively charged body which is called *nucleus.* All the protons and neutrons are present in the nucleus. Although the nucleus is small, it is heavy due to the presence of all the protons and neutrons in it. Since the mass of electron is negligible, the mass of an atom is equal to the sum of the masses of protons and neutrons, *i.e., the entire mass of an atom resides in its nucleus.*

(*ii*) The electrons keep revolving in the orbits around the nucleus at extremely high speed and at large distance from the nucleus as the planets revolve round the Sun (Fig. 1.1). Thus Rutherford's model is comparable to solar system and it is for this reason that the electrons revolving round the nucleus are generally referred to as *planetary electrons.*

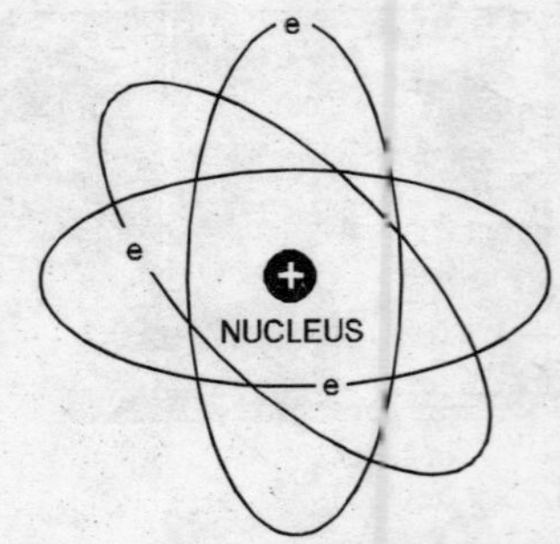

Fig. 1.1. ***Rutherford's Nuclear Model of Atomic Structure***

Since the atom is neutral, the number of electrons which are negatively charged particles is equal to the number of protons in the nucleus or the atomic number of the atom. The electrons revolve round the nucleus with such a speed that the centrifugal force (*i.e.*, repulsive force) acquired by the electrons due to their motion just balances the attractive force between the electrons and the nucleus.

The charge associated with each electron is 1.602×10^{-19} C. Compared to mass of nucleus, electrons are relatively tiny particles of almost negligible mass (each electron weighing 9.107×10^{-31} kg). Protons and neutrons each have mass about 1,800 times the mass of an electron.

The diameter of an atom is approximately 10^{-10} m and that of nucleus is about 10^{-15} m.

The ratio of e/m for an electron is equal to $\dfrac{1.602 \times 10^{-19}}{9.107 \times 10^{-31}} = 1.759 \times 10^{11}$ C/kg. Since it is not very easy to determine either the charge or the mass of an electron, electrons are usually identified by their high ratio of charge to mass. This is so high that a number of electrons making up a mass of 1×10^{-6} kg placed 100 km from a similar mass in air would experience a force of about 3×10^{9} kg. No other particle has so high a value of e/m, so that recognition by this means is easy and certain.

The number of protons (or electrons) in an atom is referred to as the *atomic number* (Z) of the atom. Now the term atomic number is often referred to as the *proton number*. The total number of protons and neutrons in the nucleus of an atom is called the *mass number* A, of the atom. The atomic mass (or atomic weight) is the actual weight of an atom and numerically slightly differs from the mass number of the atom. For example, aluminium has 79 protons and 118 neutrons. Thus its atomic number Z is equal to 79, being equal to the number of protons and the mass number A = 79 + 118 = 197.

The *force of attraction* between the electrons and nucleus follows Coulomb's law. It can be shown from classical mechanics that the resultant closed path will be a circle or an ellipse under the action of such a force.

Let us consider a hydrogen atom having a single electron, revolving around the nucleus which has a single proton, in a circular orbit with radius *r* as depicted is Fig. 1.2.

The force of attraction between the nucleus and the electron is $\dfrac{e^2}{4\pi \epsilon_0 r^2}$, where the electronic charge *e* is in coulombs,

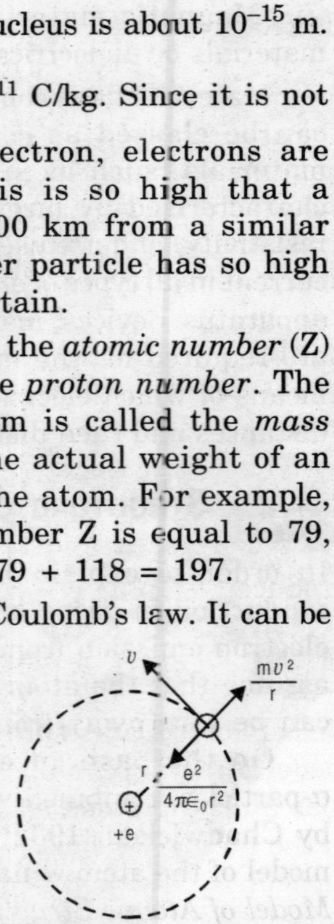

Fig. 1.2. ***Hydrogen Atom According To Rutherford's Hypothesis***

the separation r between the two particles is in metres, the force is in newtons, and ϵ_0 is the permittivity of free space and is equal to 8.854×10^{-12} F/m. By Newton's second law of motion, this must be equal to the product of electronic mass m in kg and the acceleration $\frac{v^2}{r}$ toward the nucleus, where v is the speed of the electron in the circular path, in m/s. Then

$$\frac{e^2}{4\pi \epsilon_0 r^2} = \frac{mv^2}{r} \qquad \text{...(1.1)}$$

Furthermore, the potential energy of the electron at a distance r from the nucleus is $\frac{-e^2}{4\pi \epsilon_0 r}$ and the kinetic energy is $\frac{1}{2} mv^2$. Then according to conservation of energy,

$$W = \frac{1}{2} mv^2 - \frac{e^2}{4\pi \epsilon_0 r} \text{ joules} \qquad \text{...(1.2)}$$

Combining Eqs. (2.1) and (2.2) we have

$$W = \frac{e^2}{8\pi \epsilon_0 r} - \frac{e^2}{4\pi \epsilon_0 r} = \frac{-e^2}{8\pi \epsilon_0 r} \qquad \text{...(1.3)}$$

The above Eq. (1.3) gives the desired relationship between the radius r and the total energy of the electron. This equation also shows that the total energy of the electron is always negative. The negative sign arises because the potential energy has been chosen to be zero when r is infinite. It is also evident from above equation that the energy of the electron becomes smaller (*i.e.*, more negative), as it approaches close to the nucleus.

1.2.1. Weakness of Rutherford's Atomic Model

The assumptions that electrons were orbiting about the nucleus was unfortunate. According to the classical electromagnetic theory if a charged particle accelerates around an oppositely charged particle, the former will radiate energy. If an electron radiates energy, its speed will decrease and it will go into spiral motion, finally falling into the nucleus, as shown in Fig. 1.3. This does not happen actually as then the atom would be unstable, which it is not. This was the chief weakness of Rutherford's atomic model.

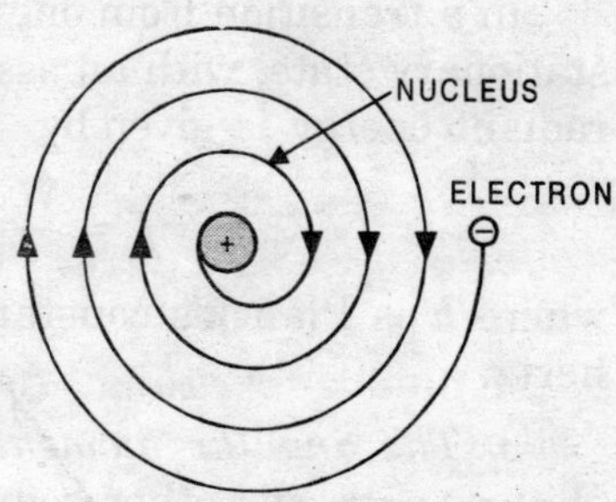

Fig. 1.3. *Spiral Path of Electron*

The other weakness of Rutherford's atomic model was that it fails to account for the following phenomenon.

When gases under low pressure are subjected to electric discharge *line spectra* are produced.

1.3. Bohr's Atomic Model

Neils Bohr, a Danish physicist, put forward his model of atom in 1913. He retained Rutherford's model of a very small positively charged nucleus at the centre. He also accepted the view that all the protons and neutrons (and hence most of the mass of the atom) are contained in the nucleus. He also agreed that the negatively charged electrons are revolving around the nucleus in the same way as the planets are revolving around the Sun. But he took a bold step forward by applying Planck's quantum theory to the electrons revolving around the nucleus. He postulated the following three fundamental laws :

1. *The electrons continue revolving in their respective orbits without losing energy. This is because energy in fraction of a quantum cannot be lost.*

In other words, the energy of an electron remains constant as long as it stays in the same orbit. This led to the idea that each orbit is associated with a definite energy, *i.e.*, with a definite whole number of quantas of energy. The orbits, therefore, are also known as *energy levels* or *energy shells*. Bohr gave numbers 1, 2, 3, 4, etc., (starting from the nucleus) to these energy levels, as shown in Fig. 1.4. These are now termed as *principal quantum numbers*. The various energy levels are also designated by letters K, L, M, N, etc. The farther the energy level from the nucleus, the greater is the energy associated with it. As long as the electron remains in a particular energy shell, it *neither emits nor absorbs energy*.

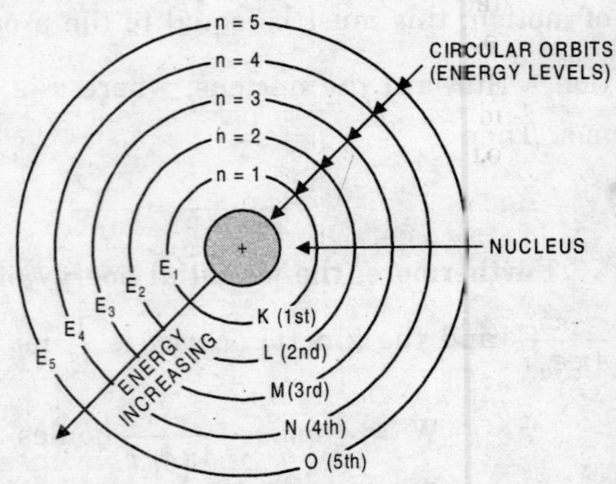

Fig. 1.4. ***Orbit-Like Representation of Various Energy Levels***

2. *Energy is emitted or absorbed by an atom only when an electron moves from one level to another*. The amount of energy emitted or absorbed is given by the difference of the energies associated with the two energy levels concerned. Thus the energy of an electron cannot change continuously. It changes only *abruptly* as the electron *jumps* from one energy level to another. The same idea has been expressed by the statement that *for a change of electronic energy, the electron has to jump and not to 'flow' from one energy level to another*.

In a transition from one stationary state corresponding to a definite energy W_2 to another stationary state, with an associated energy W_1 radiation will be emitted. The frequency of this radiant energy is given by

$$f = \frac{W_2 - W_1}{h} \qquad ...(1.4)$$

where h is Planck's constant in joule-seconds, the W's are expressed in joules, and f is in hertz.

3. *The angular momentum of an electron moving round the nucleus is quantized.* An electron, like any other body moving in a circular orbit, has an angular momentum given by mrv* where m is the mass, v is the velocity of the electron while r is the radius of the orbit. Bohr postulated that these orbits or energy levels are such that the angular momentum can have only *definite* or *discrete* values given by the expression

$$\text{Angular momentum} = mvr = \frac{nh}{2\pi} \qquad ...(1.5)$$ where n is an integer (1, 2, 3, etc.).

Thus the angular momentum of an electron may be $\frac{h}{2\pi}$ or simple whole number multiples of $\frac{h}{2\pi}$ such as $\frac{2h}{2\pi}, \frac{3h}{2\pi}, ... \frac{nh}{2\pi}$. This principle is known as *quantisation of angular momentum*.

* Angular momentum = $J\omega$ where J is the moment of inertia and ω is the angular velocity. Since $J = mr^2$ and $\omega = v/r$ where v is the linear velocity and r is the radius, therefore, angular momentum = $J\,\omega = mr^2 \times v/r = mrv$.

The energy level in joules of each state can be determined as follows :
From Eq. (1.5) we gave

$$mv = \frac{nh}{2\pi r}$$

$$\text{or } mv^2 = \frac{n^2h^2}{4\pi^2r^2m} \quad ...(1.6)$$

Again from Eq. (1.1) we have

$$mv^2 = \frac{e^2}{4\pi \in_0 r} \quad ...(1.7)$$

Comparing Eqs. (1.3) and (1.7) we have

$$\frac{e^2}{4\pi \in_0 r} = \frac{n^2h^2}{4\pi^2r^2m}$$

$$\text{or } r = \frac{\in_0 n^2h^2}{e^2m\pi} \quad ...(1.8)$$

The above Eq. (1.8) gives the expression for the radius of the oribit. Substituting the value of r from Eq. (1.8) in Eq. (1.3) we have

$$W = \frac{-e^2}{8\pi \in_0} \times \frac{e^2m\pi}{\in_0 n^2h^2} = \frac{-me^4}{8 \in_0^2 n^2h^2} \quad ...(1.9)$$

1.4. Electron Orbits and Energy Levels

As already discussed above electrons can occupy only certain rings or shells at fixed distances from the nucleus. These shells may be circular or elliptical and are numbered as K, L, M, N, O, P, and Q, K being nearest to the nucleus. Each shell can contain only a particular number of electrons. The formula $2n^2$ (n being the number of shell or level, the first shell or level being K) determines the number of electrons that can be accommodated in any level. For example Kth level can have maximum of 2 electrons. Similarly L = 2 × 2^2 = 8, M = 2 × 3^2 = 18, N = 2 × 4^2 = 32.

The limitation to the formula is that the number of electrons in the outermost shell of an atom cannot exceed 8. The electrons in the outer shell determine the electrical and chemical characteristics of each particular type of atom and are referred to as *valence electrons*. An atom may have its outer or *valence shell* completely filled or only partially filled.

From discussions made above it is clear that electrons can revolve only in the permitted orbits (*i.e.*, orbits of radii r_1, r_2 and r_3) and not in any intermediate orbit. Thus all the radii between r_1 and r_2 and between r_2 and r_3 are forbidden (for explanation refer to quantum physics). Each orbit has fixed amount of energy associated with it and the electrons moving in a particular orbit possess the energy of that orbit. Thus total energy of an electron can be identified with the size of its orbit—the larger the orbit, the greater is its energy *i.e.*, outer orbit electrons possess more energy than the inner orbit electrons.

The level of energy obtainable with different orbits can be conveniently determined from Eq. (1.9) and represented by horizontal lines, as illustrated in Fig. 1.5 (*b*). This is known as *energy level diagram*. Electrons in the smallest orbit are on the first energy level; electrons in the second orbit are on the second energy level and so on. Thus energy level is just another way of representing the orbital radius.

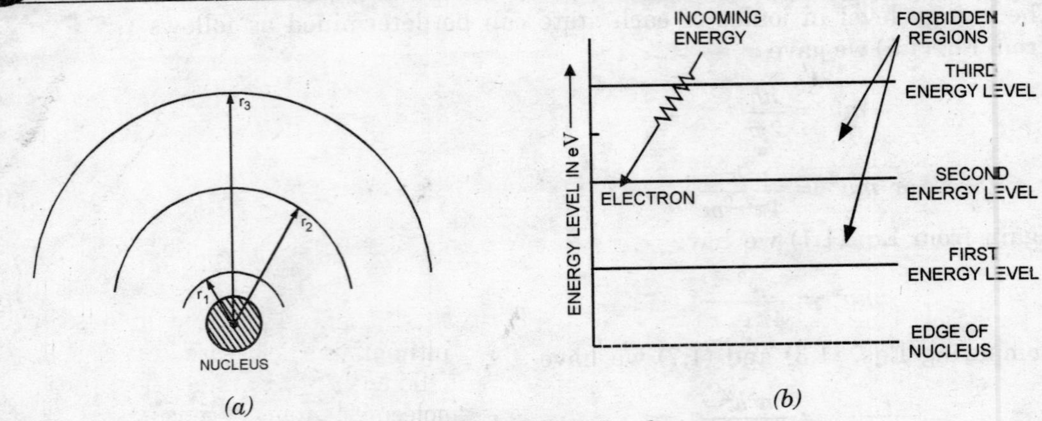

Fig. 1.5. *Energy Level*

Thus electrons closest to the nucleus need the greatest energy application to extract them from the atom. The electrons in inner orbits, thus, remain bound to the nucleus and are, therefore, called *bound electrons.*

The energy levels considered above are measured in electron volts (*e*V). An electron volt is defined as the amount of energy required to move an electron through a potential difference of 1 volt.

Between each energy level a *forbidden region exists.* Electrons cannot orbit in forbidden regions. However, they may quickly pass them. If electrons in a particular orbit acquire enough energy from some outside source (such as an external electric field or thermal energy), they will jump to a high energy level by passing quickly through the forbidden region. *A definite amount of energy is required to be spent in order to remove an electron from its orbit (level).*

Excited electrons are unstable, and they tend to fall back to low energy level by giving up their additional energy. The electrons that do leave their parent atom are termed *free electrons.*

1.5. Electron Volt

The joule (J) is the unit of energy in the SI system. In some engineering power problems this unit is very small, and a factor of 10^3 or 10^6 is introduced to convert from watts (1 W = 1 J/s) to kilowatts or megawatts, respectively. For a discussion of energies involved in electronic devices even the joule is too large a unit. This statement is not to be construed to mean that only minute amounts of energy can be obtained from electron devices. It is true that each electron possesses a tiny amount of energy, but an enormous number of electrons is involved even in a small current, so that considerable power may be represented.

A unit of work or energy, called the *electron volt* (*e*V), is given as

$$1\ eV = 1.602 \times 10^{-19}\ \text{J}$$

The name *electron volt* arises from the fact that, if an electron falls through a potential of 1 V, its kinetic energy will increase by the decrease in potential energy, or by one electron volt.

$$\text{i.e.}\quad 1\ \text{eV} = 1.602 \times 10^{-19}\ \text{C} \times 1\ \text{V} = 1.602 \times 10^{-19}\ \text{J}$$

However, the *e*V unit may be used for any type of energy, and is not restricted to problems involving electrons.

The abbreviations M*e*V and B*e*V are used to designate 1 million and billion electron volts respectively.

1.6. CRYSTALLINE STRUCTURE

Solids are characterized by *incompressibility*, *rigidity* and *mechanical strength*. This indicates that the molecules, atoms or ions that make up a solid are closely packed, *i.e.*, they are held together by strong forces and cannot move at random. On the contrary, **gases** are characterized by a high degree of compressibility. This is because in gases the molecules are having chaotic or random motion and exert little or no force on one another. Thus, while in gases we have '*molecular chaos*', in solids we have *well-ordered molecular arrangement*.

Some solids, like sodium chloride, sulphur and sugar, besides being incompressible and rigid, have also characteristic *geometrical forms*. Such substances are said to be *crystalline solids*. The X-ray diffraction studies reveal that their ultimate particles (viz., molecules, atoms or ions) are arranged in a definite pattern throughout the entire three-dimensional network of the crystal. This definite and orderly arrangement of molecules, atoms or ions (as the case may be) extends over a large distance. This is termed as *long-range order*.

Crystal Lattice and Unit Cell. The position of atoms, molecules or ions in a crystal relative to one another in space, are designated usually by points. Such a representation is called *space lattice* or *crystal lattice* or *crystalline structure*. A *crystal lattice* is an array of points showing how molecules, atoms or ions are arranged at different sites in three-dimensional space.

An example of array of points in a *three-dimensional* space lattice is shown in Fig. 1.6, to illustrate the point. Each point represents an identical atom or group of atoms and this gives a crystal structure.

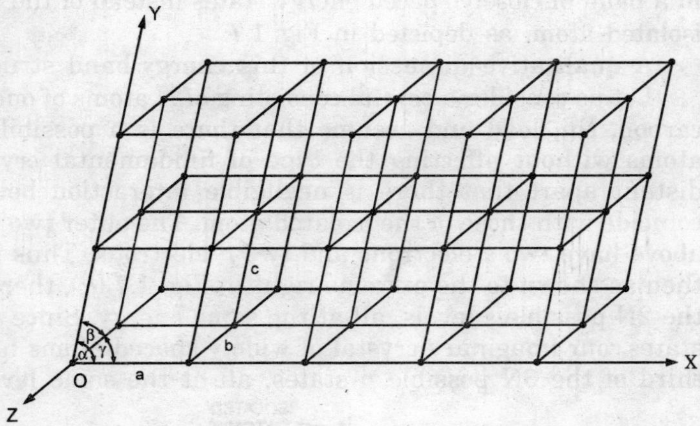

Fig. 1.6. *Space Lattice and a Unit Cell*

The lattice points can be broken up into a number of *unit cells*. This is done by connecting the points by a regular network of lines, as illustrated in the Fig. 1.6. A *unit cell* is the smallest repeating unit in space lattice which when repeated over and over again results in a crystal of the given substance. Thus, space lattice of a crystal has been likened to a wall paper on which single pattern is continuously repeated. Just as a pattern on the wall paper is repeated again and again, similarly a *unit cell* (representing a definite pattern) is repeated again and again to build up a crystal. The only difference is that while wall paper is in two dimensions, space lattice of a crystal is in three dimensions. The *unit cell*, in fact, is the smallest sample that represents the picture of the entire crystal. The crystal may be considered to consist of infinite number of unit cells. Each unit cell in a three-dimensional space has, evidently, three vectors, a, b and c, as shown in Fig. 1.6.

It should be clearly understood that these are the *points* and not the lines which constitute the *space lattice* or *crystal lattice*. The lines joining the points are drawn simply for representing three axes by means of which the relative positions of the points can be described. For example, in Fig. 1.6, three imaginary axes, OX, OY and OZ, which may be used to represent the unit cell, have been shown.

In order to describe a unit cell, we should know (*i*) the distances *a*, *b* and *c*, which give the lengths of the edges of the unit cell and (*ii*) the angles α, β and γ, which give the angle between the three imaginary axes OX, OY and OZ.

1.7. Energy Band Theory of Crystals

As discussed above it is revealed that most metals and semiconductors are crystalline in structure and a crystal consists of a space array of atoms or molecules (strictly speaking, ions) built up by regular repetition in three dimensions of some fundamental structural unit. The electronic energy levels discussed for a single free atom (as in a gas, where the atoms are sufficiently far apart and do not exert any influence on one another) do not apply to the same atom in a crystal. This is so because the potential characterizing the crystalline structure, is now a periodic function in space whose value at any point is the result of contributions from every atom. When atoms form crystals it is found that the presence of the neighbouring atoms do not affect appreciably the energy levels of the inner-shell electrons. However, the energy levels of the outer-shell electrons are changed considerably. This is because the outer-shell electrons are shared by more than one atom in a crystal. The new energy levels of the outer electrons can be determined by means of quantum mechanics, and it has been found that coupling between the outer-shell electrons of the atoms results in a band of closely spaced energy states instead of the widely separated energy levels of the isolated atom, as depicted in Fig. 1.7.

A qualitative discussion of this energy-band structure is given below :

Let us consider a crystal consisting of N atoms of one of the elements silicon, germanium, carbon, tin, lead and assume that there is a possibility of varying the spacing between atoms without affecting the type of fundamental crystal structure. If the atoms are so distant apart that there is negligible interaction between them, the energy levels will coincide with those of the isolated atom. The outer two subshells for each element indicated above have two *s* electrons and two *p* electrons. Thus if the inner-shell levels are ignored, then as shown to the extreme right in Fig. 1.7 (*a*), there are 2N electrons completely filling the 2N possible *s* levels, all at the same energy. Since the *p* atomic subshell has 6 possible states, our imaginary crystal of widely spaced atoms has 2N electrons, which fill only one-third of the 6N possible *p* states, all at the same level.

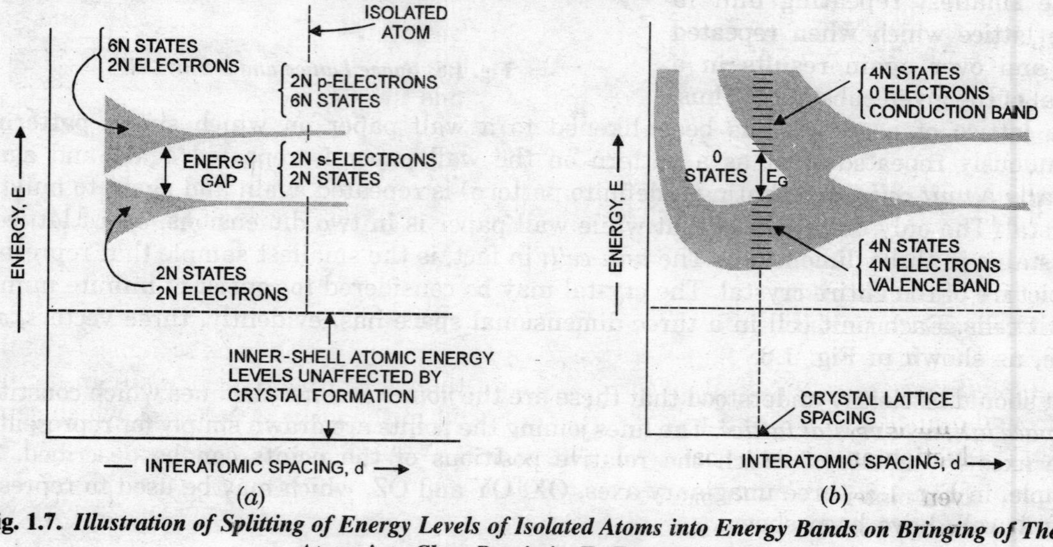

Fig. 1.7. ***Illustration of Splitting of Energy Levels of Isolated Atoms into Energy Bands on Bringing of These Atoms into Close Proximity To Form a Crystal***

Now, if the interatomic spacing of our imaginary crystal is reduced [moving from right to left in Fig. 1.7 (*a*)], an electric force will be exerted by an atom on its neighbours. Because of this coupling between atoms, the atomic-wave functions overlap and the crystal becomes an electronic system which must obey the Pauli exclusion principle.* Thus the 2N degenerate *s* states must spread out in energy. The separation between levels is small, but since N is very large ($\sim 10^{23}$ cm^{-3}), the total spread between the minimum and maximum energy may be of several eV if the interatomic distance is reduced sufficiently. This large number of discrete but closely spaced energy levels is known as *energy band*, and is depicted schematically by the lower shaded region in Fig. 1.7 (*a*). The 2N states in this band are completely filled with 2N electrons. In the same way, the upper shaded region in Fig. 1.7 (*a*) is a band of 6N states which has only 2N of its levels occupied by the electrons.

The worthnoting point is that there is an energy gap (a forbidden band) between the two bands discussed above and this gap reduces with the decrease in atomic spacing. For sufficiently small distances, these bands will overlap, as depicted in Fig. 1.7 (*b*). Under such circumstances the 6N upper states merge with the 2N lower states, providing a total of 8N levels, half of which are occupied by the 2N + 2N = 4N available electrons. At this spacing each atom has given up four electrons to the band; these electrons can no longer be said to orbit in *s* or *p* subshells of an isolated atom, but rather they belong to the crystal as a whole. The band occupied by these electrons is known as the *valence band*.

If the spacing between atoms is reduced below the distance at which the bands overlap, the interaction between atoms is indeed large. The energy-band structure in this condition depends upon the orientation of the atoms relative to one another in space (*i.e.* the crystal structure) and upon the atomic number, that determines the electrical constitution of each atom. At the crystal-lattice spacing (the dashed vertical line), we find the valence band filled with 4N electrons separated by a forbidden band of extent E_G from an empty band consisting of 4N additional states. This upper vacant band is known as the *conduction band*.

1.8. ENERGY BAND IN SOLIDS

As discussed earlier, the electrons of an isolated atom are acted upon only by the forces within that atom. However, when atoms are brought together, as in a solid, the electrons come under the influence of forces from other atoms. Under these conditions, the energy levels that may be occupied by electrons merge into bands of energy levels. Within any given material there are two *distinct energy* bands in which electrons may exist. These two energy bands are *valence energy band* (or *valence band*) and *conduction energy band* (or *conduction band*) and are separated by an energy gap in which no electron can normally exist, as illustrated in Fig. 1.8. This energy gap is termed as *forbidden energy gap* (or *forbidden gap*).

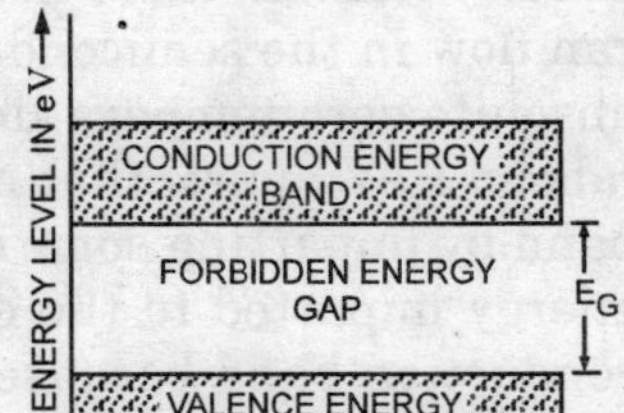

Fig. 1.8

The energy band of interest is the highest energy band or valence band. If a sufficient amount of energy is given to an electron in the valence band, the electron is freed from the atomic structure. Such an electron is said to possess enough energy to be in the conduction band, where it can take part in electric current flow. Free electrons (electrons in conduction band) can move readily under the influence of an external field. Electrons in the valence band are usually in normal orbit around a nucleus. Normally, electrons do not orbit in the forbidden region, which separates the conduction and valence bands, as illustrated in Fig. 1.8.

For any given material, the forbidden energy gap may be large, small or non-existent. The distinction between conductors, insulators and semiconductors is largely concerned with the

* The statement that no two electrons may occupy the same quantum state is known as the Pauli exclusion principle.

relative widths of the forbidden energy gaps. As illustrated in Figs. 1.9 (*a*), (*b*) and (*c*), there is a wide forbidden gap (more than 5 eV) for insulators, narrow forbidden gap (about 1 eV) in case of semiconductors and no forbidden gap in case of conductors.

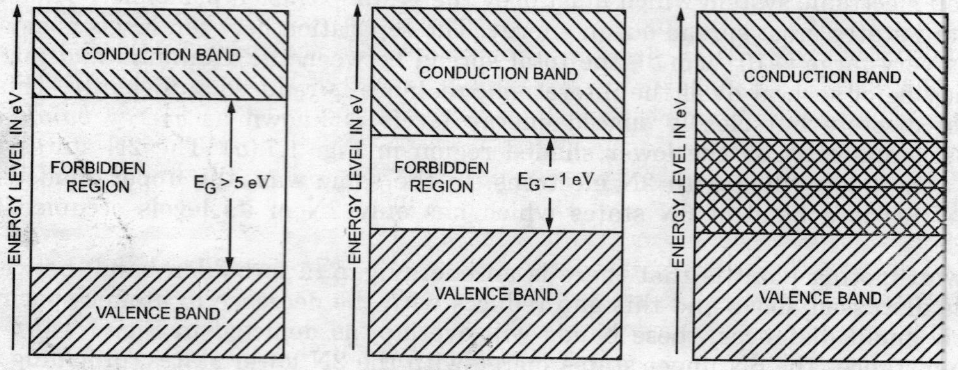

(*a*) *Insulator* (*b*) *Semiconductor* (*c*) *Metal*

Fig. 1.9. ***Energy-Band Structure***

In case of insulators, there is practically no electron in the conduction band, and the valence band is filled.* For an insulator, the valence and conduction bands are so apart (more than 5 eV) that at normal temperature (25°C) the energy which can be supplied to an electron from an applied field is too small to carry a particle from the filled band into the vacant band. Since the electron cannot acquire externally applied energy, conduction is impossible. Only at very high temperatures or under high external electric field, an electron can jump the gap. This is called the *breakdown of the insulating material*.

For semiconductors at a temperature of absolute zero (– 273·15°C) the valence band is usually full and there may be no electron in the conduction band. It means that no current can flow in the semiconductors and, therefore, the semiconductors behave as insulators at absolute zero temperature. However, in semiconductor materials, both bands are so close (about 1 eV apart) that an electron can be lifted from the valence band to the conduction band by imparting some energy to it. This energy must be more than energy gap E_G. If the energy imparted to the electron is less than E_G, it cannot be lifted from valence band to conduction band because no permissible energy level exists between the two bands. The energy imparted by the heat at room temperature is sufficient to lift electrons from the valence band to the conduction band. Some electrons do jump the gap and get into the conduction band. Thus, semiconductors are capable of conducting some electric current, even at normal ambient temperature.

In case of conducting materials there is no forbidden gap, and the valence and conduction bands overlap, as illustrated in Fig. 1.9 (*c*). The orbits in the conduction band are very large. An electron in the conduction band experiences almost negligible nuclear attraction. In fact, an electron in the conduction band does not belong to any particular atom. But it moves randomly throughout the solid. This is the reason that electrons in the conduction band are called the *free electrons*. Due to overlapping of valence and conduction bands, a slight potential difference across the conductor causes the free electrons to constitute electric current.

* Filled band means that all the permissible energy levels in the band are occupied with electrons. No electron can move in a filled band because there is no place to move. Thus an electron in a completely filled band cannot contribute to an electric current.

1.9. Conduction in Solids

Conduction occurs in any given material when an applied electric field causes electrons to move in a desired direction within the material. This may be due to one or both of two processes, *electron motion* and *hole transfer*. In case of former process, *free electrons* in the conduction band move under the influence of the applied electric field. Electrons being negatively charged are repelled from the negative terminal of the applied voltage and attracted towards the positive terminal. Hole transfer involves electrons which are still attached to the atoms *i.e.*, those in valence band.

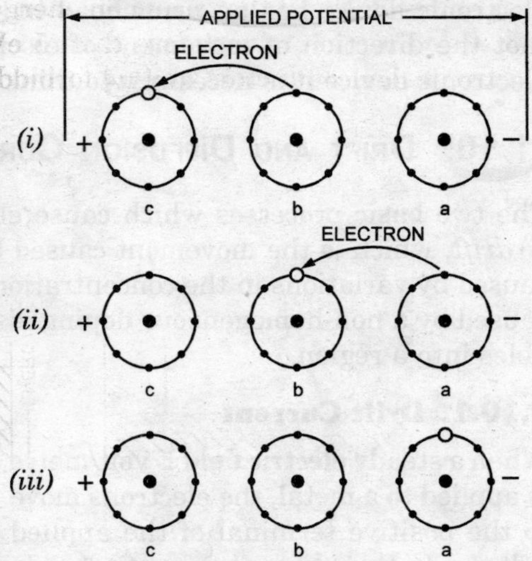

Fig. 1.10. ***Conduction By Transfer of Hole***

In case some of the energy levels in the valence band are not occupied by electrons, there are holes where electrons could exist. An electron may jump from one atom to fill the hole in another atom. When it jumps, the electron leaves a hole behind it, and thus the hole has moved in a direction opposite to that of the electron. The current flowing in such a way is said to be due to movement of holes.

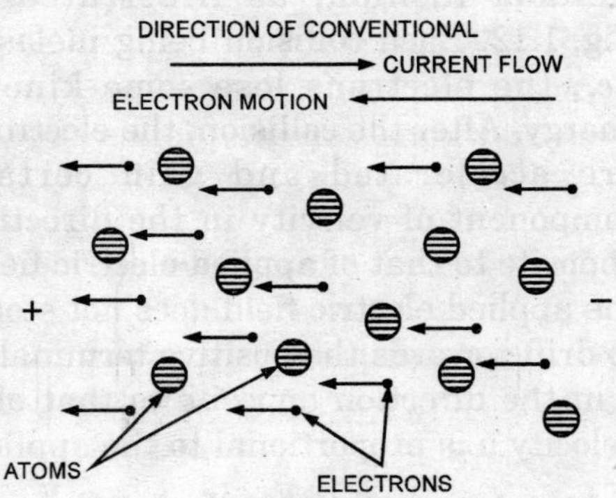

Fig. 1.11

In Fig. 1.10 there are no free electrons. However, those electrons in orbit around atoms experience a force of attraction to the positive terminal of the applied potential, and repulsion from the negative terminal. Consequently, an electron can be made to jump from one atom to fill the hole in another atom, so long as it is moving towards the positive terminal. In Fig. 1.10 (*i*), an electron is made to jump from atom *b* to atom *c*. The hole in the valence shell of atom *c* is filled but a hole is created in the valence shell of atom *b*, as illustrated in Fig. 1.10 (*ii*). If an electron now jumps from atom *a* to fill in the hole in atom *b*, a hole is left in the valence shell of atom *a*, as illustrated in Fig. 1.10 (*iii*). Thus the hole has moved from atom *c* to atom *b* and then to atom *a*.

Holes may be thought of as positive particles, and as such, they move through an electric field in a direction opposite to that of electrons; *i.e.*, positive particles are attracted towards the negative terminal of the applied voltage. In the condition illustrated above it is more convenient to think in terms of *hole movement* rather than in terms of electrons jumping from atom to atom.

Since the electric current flow is due to movement of electrons in the conduction band and that of holes in the valence band, electrons and holes are referred to as *charge carriers*. Each time a hole moves, an electron must be supplied with sufficient energy to enable it to escape from the atom. Free electrons need less energy to be applied for their movement in comparison to that required by holes. This is true due to the fact that the free electrons are already detached from their atoms. For this reason, electrons have *greater mobility* than holes.

Current flow from positive terminal to negative terminal is referred to as the *conventional current direction* whereas electron flow from negative terminal to positive terminal is referred to as *direction of electron flow*. It is to be noted that every graphic symbol used to represent an electronic device has an arrowhead indicating the direction of flow of conventional current (not the direction of movement of electrons). However, to understand the operation of an electronic device it is necessary to think in terms of movement of electrons.

1.10. Drift and Diffusion Currents

The two basic processes which cause electrons and holes to move in a semiconductor are : (*i*) *drift,* which is the movement caused by electric fields, and (*ii*) *diffusion,* which is the flow caused by variations in the concentration, *i.e.*, concentration gradients. Such gradients can be caused by a non-homogeneous doping distribution, or by injection of a quantity of electrons or holes into a region.

1.10.1. Drift Current

When a steady electric field E volt/metre is applied to a metal, the electrons move to the positive terminal of the applied voltage. In their way, they continuously collide with the atoms and rebound in a random fashion, as illustrated in Fig. 1.12. Each collision being inelastic *i.e.*, the electrons lose some kinetic energy. After the collision, the electrons are accelerated and gain certain component of velocity in the direction opposite to that of applied electric field (– E) and lose their energy at the next collision. Thus the applied electric field does not stop collisions and random motion but makes the electrons to drift towards the positive terminal. Consequently, the electrons gain average drift velocity v in the direction opposite to that of the applied electric field. It is observed that the drift velocity v is proportional to the applied electric field E, *i.e.*,

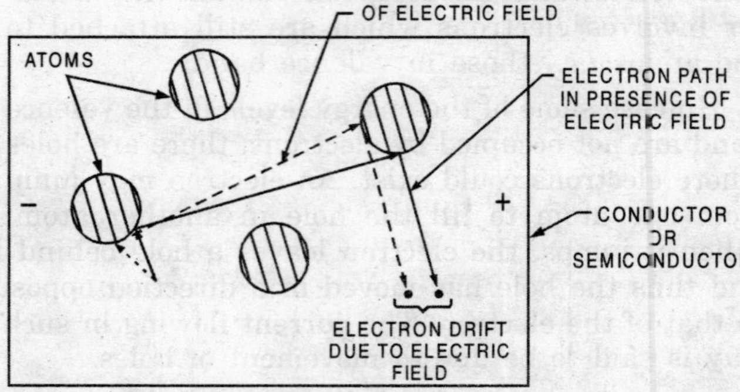

Fig. 1.12

$$v \propto E \quad \text{or} \quad v = \mu_e E \qquad ...(1.10)$$

where μ_e is called the electron *mobility** and is expressed in m^2/V-s.

This steady-state drift velocity is superimposed on the random motion of free electrons caused by thermal agitation. This steady flow of electrons in one direction caused by the applied electron field constitutes an electric current, called the *drift current*.

1.10.2. Diffusion Current

A second type of current, called the *diffusion current*, also exists in a semiconductor. The diffusion of charge carriers is as a result of a gradient of carrier concentration (*i.e.* the difference of carrier concentration from one region to another). In this case concentrations of charge carriers (either electrons or holes) tend to distribute themselves uniformly throughout the semiconductor crystal. This movement continues until all the carriers are evenly distributed throughout the material. This type of movement of charge carriers is called the *diffusion current*.

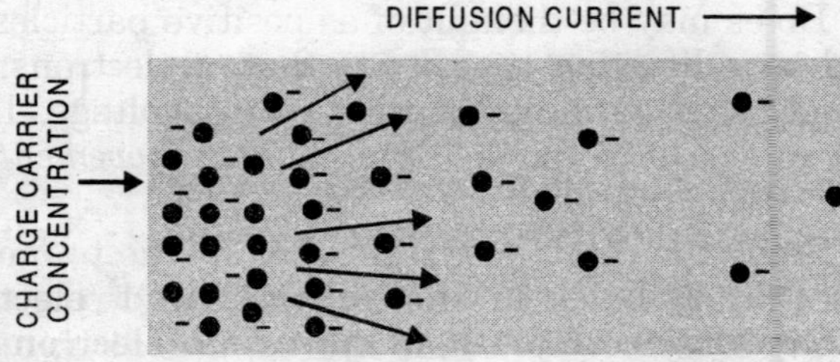

Fig. 1.13. *Flow of Particles From a Region of High Concentration To a Region of Low Concentration*

* Mobility is defined as the average particle drift velocity per unit electron.

Diffusion current occurs when charge carriers diffuse from a point of concentration, to spread uniformly throughout the volume of a piece of material, as illustrated in Fig. 1.13.

Both drift and diffusion occur simultaneously in a semiconductor device.

1.11. Fermi Dirac Energy Distribution

It is important to know what energies are possessed by a mobile carrier in a solid or semiconductor. This relationship is referred to as **energy distribution function.** We are concerned with the distribution of energy of the free electrons in a metal. The Fermi-Dirac statistics enables us to find the number of free electrons dn_E per unit volume, within energy range E to E + dE at temperature T.

The number of free electrons per cubic metre of a metal whose energies lie in the interval E to E + dE is given by

$$dn_E = \rho_E dE \qquad ...(1.11)$$

where dn_E represents the number of free electrons per cubic metre whose energies lie in the energy interval dE, and ρ_E represents the density of electrons in the energy interval dE.

In case of free electrons, it can be safely assumed that there is no potential variation within the metal. Accordingly, each cubic metre of the metal must have the same number of electrons. That is, the density in space (electrons/m^3) is a constant. However, within each unit volume of metal there will be electrons having all possible energies. It is this distribution in energy that is represented by ρ_E. ρ_E represents the number of electrons per eV per m^3 of the metal. The function ρ_E may be expressed as

$$\rho_E = N(E).f(E) \qquad ...(1.12)$$

where N(E) is the density of states (number of states per eV per cubic metre) in the conduction band, and f(E) is the probability that a quantum state with energy E is occupied by an electron.

The density of states, N(E) is found to be proportional to the half power of energy *i.e.*

$$N(E) \propto E^{1/2}$$

$$\text{or } N(E) = \gamma E^{1/2} \qquad ...(1.13)$$

where γ is a proportionality constant and is defined as

$$\gamma = 4\pi\left(\frac{2em}{h^2}\right)^{3/2} = 4\times\pi\times\left[\frac{2\times 1.602\times 10^{-19}\times 9.107\times 10^{-31}}{(6.626\times 10^{-34})^2}\right]^{3/2}$$

$$= 6.82 \times 10^{27} \qquad ...(1.14)$$

where m is the mass of electron in kg, and h is Planck's constant in joule-seconds.

The dimensions of γ are (m^{-3}) (eV)$^{-3/2}$.

The equation for f(E) is known as *Fermi-Dirac probability* function, and specifies the fraction of all states at energy E (electron volts) occupied under conditions of thermal equilibrium.

According to the principle of quantum mechanics, the Fermi-Dirac probability function is given by

$$f(E) = \frac{1}{1+e^{(E-E_f)/kT}} \qquad ...(1.15)$$

where k is Boltzmann's constant (k = 8.62 × 10^{-5} eV/K). The function f(E), the Fermi-Dirac distribution function, gives the probability that an available energy state at E will be occupied by an electron at absolute temperature T. The quantity E_F is called the *Fermi level,* and it represents an important quantity in the analysis of semiconductor behaviour. We notice that, for an energy E equal to the Fermi level energy E_F, the occupation probability is

$$f(E_F) = \frac{1}{1+e^{(E_F - E_F)/kT}} = \frac{1}{1+1} = \frac{1}{2} \qquad ...(1.16)$$

Thus an energy state at the Fermi level has a probability of 50 percent of being occupied by an electron.

A plot of f(E) versus $E - E_F$ is depicted in Fig. 1.14 (*a*) and of $E - E_F$ versus f(E) in Fig. 1.14 (*b*), both for T K and for larger values of temperature.

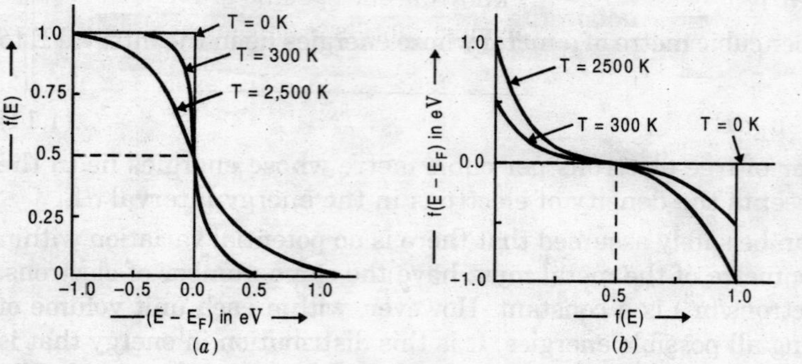

Fig. 1.14

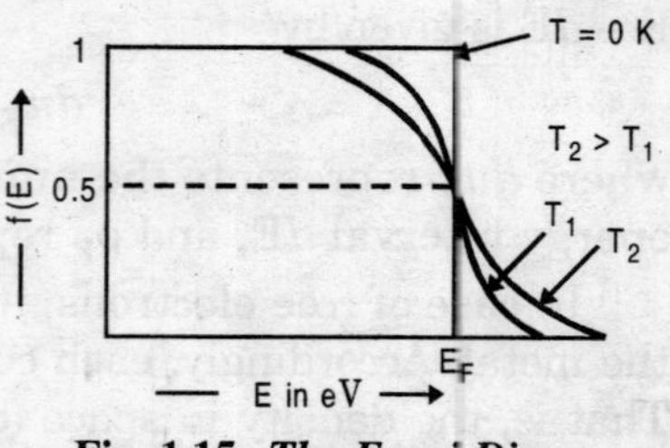

Fig. 1.15. The Fermi-Dirac Distribution Function

A closer examination of f(E) indicates that at 0 K the distribution takes the simple rectangular form depicted in Fig. 1.15. With T = 0 in the denominator of the exponent, f(E) is $\frac{1}{1+0} = 1$ when the exponent is negative ($E < E_F$), and is $\frac{1}{1+\infty} = 0$ when the exponent is positive ($E > E_F$). This rectangular distribution implies that at 0 K every available energy state up to E_F is filled up with electrons, and all states above E_F are empty.

From Eqs. (1.12), (1.13) and (1.15), we have at absolute zero temperature

$$\rho_E = \begin{cases} \gamma E^{1/2} & \text{for } E < E_F \\ 0 & \text{for } E > E_F \end{cases} \qquad ...(1.17)$$

At temperatures higher than 0 K, some probability exists for states above the Fermi level to be filled. For example, at $T = T_1$ in Fig. 1.15, there is some probability f(E) that states above E_F are filled, and there is a corresponding probability $[1 - f(E)]$ that states below E_F are empty. The Fermi function is symmetrical about E_F for all temperatures; that is, the probability $f(E_F + \Delta E)$ that a state ΔE above E_F is filled is the same as the probability $[1 - f(E_F - \Delta E)]$ that a state ΔE below E_F is empty. The symmetry of the distribution of empty and filled states above E_F makes the Fermi level a natural reference point in calculations of electron and hole concentrations in semiconductors.

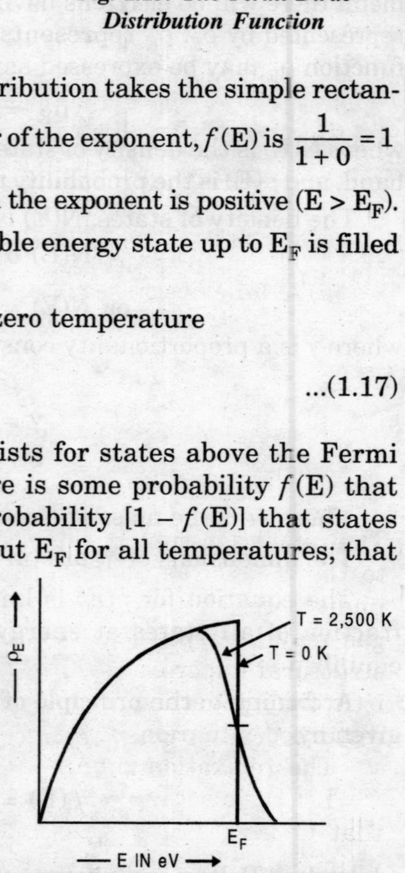

Fig. 1.16. Energy Distribution at 0 K and 2,500 K

The Fermi Level. An expression for E_F can be had from energy distribution curve depicted in Fig. 1.16 at 0 K. The area under the curve represents the total number of free electrons (per cubic metre of metal). Thus

$$n = \int_0^{E_F} \gamma E^{1/2}\, dE = \frac{2}{3}\gamma E_F^{3/2}$$

$$\text{or } E_F = \left(\frac{3n}{2\gamma}\right)^{2/3} \quad ...(1.18)$$

Substituting $\gamma = 6.82 \times 10^{27}$ from Eq. (1.14), we have

$$E_F = \left[\frac{3n}{2 \times 6.82 \times 10^{27}}\right]^{2/3} = 3.64 \times 10^{-19} (n)^{2/3} \quad ...(1.19)$$

The density n varies from metal to metal, therefore, E_F will vary from metal to metal. The density of electrons n and so E_F may be determined by knowing the specific gravity, the atomic weight, and the number of free electrons per atom. For most metals the numerical value of E_F is less than 10 eV.

Example 1.1. Find the probability for an electronic state to be occupied at room temperature, if the energy of this state lies, (*a*) at 0.1 eV above the Fermi level, (*b*) at 0.1 eV below the Fermi-level.

[*M.D. Univ. Third Semester May 2005*]

Solution : At any temperature T, the probability of state corresponding to the energy E being occupied by an electron is given by

$$f(E) = \frac{1}{1 + e^{(E - E_f)/kT}}$$

where k = Boltzmann's constant in eV/k = 8.61×10^{-5} eV/k

T = 300 k

$kT = 300 \times 8.61 \times 10^{-5} = 0.0258$ V

(*i*) At 0.1 eV above the Fermi level

i.e. $E - E_F = 0.1$ eV

$$\text{and } f(E) = \frac{1}{1 + e^{0.1/0.0258}} = 0.02 \quad \textbf{Ans.}$$

(*ii*) At 0.1 eV below the Fermi level

i.e. $E - E_F = -0.1$ eV

$$\text{and } f(E) = \frac{1}{1 + e^{-0.1/0.0258}} = 0.98 \quad \textbf{Ans.}$$

1.12. Relaxation Time, Collision Time and Mean Free Path

The relaxation time is closely related to the mean time of flight between collision and also to the mean free path of the conduction of electrons, which is the average distance of undisturbed motion between collision. All the collision processes occurring in the electron gas may be explained through relaxation time. The collisions are caused by thermal or structural imperfections in the lattice. The relaxation time is introduced as the characteristic time governing the establishment of equilibrium, through collision, from an initial disturbed situation in which $v_x \neq 0$.

The relaxation time is mathematically understood as follows :

Let the electrons in a metal have an average velocity $<v_x> 0$ at any instant $t = 0$; given that for $t > 0$ there is no external field, the electrons interact only with the lattice and ultimately the average velocity will go to zero. At any instant $t(> 0)$, let the velocity be given by the expression (Fig. 1.17)

$$\langle v_x \rangle_t = \langle v_x \rangle_0 \, e^{-t/\tau} \qquad ...(1.20)$$

where τ is the relaxation time of electrons.

The rate of change of velocity due to collision becomes

$$\left[\frac{d \langle v_x \rangle}{dt}\right]_{\text{coll}} = -\frac{1}{\tau} \langle v_x \rangle \qquad ...(1.21)$$

The subscript "coll" refers to collisions carried out by the electrons with the lattice.

Equation (1.20) is the solution of a modified equation of motion of electrons if we consider the frictional forces due to collisions.

Considering the relaxation time, the frictional force on the electrons due to collision may be expressed as $\frac{m \langle v_x \rangle}{\tau}$. Hence the modified equation of electrons becomes

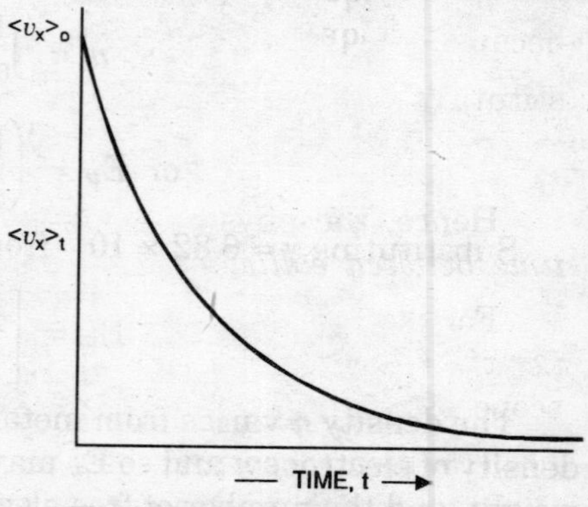

Fig. 1.17. ***Decay of Drift Velocity According to Eq. (1.20)***

$$m \frac{d \langle v_x \rangle}{dt} + \frac{m \langle v_x \rangle}{\tau} = -eE \qquad ...(1.22)$$

In the absence of the external forces $(-eE)$, the free motion satisfies

$$\frac{d \langle v_x \rangle}{dt} + \frac{1}{\tau} \langle v_x \rangle = 0 \qquad ...(1.23)$$

Or, if $\langle v_x \rangle$ 0 is the initial drift velocity in the non-equilibrium distribution, the approach to the equilibrium is described by the appropriate solution of Eq. (1.23)

$$\langle v_x \rangle_t = \langle v_x \rangle_0 \, e^{-t/\tau} \qquad ...(1.24)$$

The disturbances from equilibrium die out exponentially.

Let us now consider the collisions from the point of view of time. Assume that the probability for an electron to collide with an obstacle during a short time interval dt is given by $\frac{dt}{\tau'}$, where τ' is a constant with the dimensions of time. Let $F(t)$ represent the probability that an electron moves for a time t without suffering a collision, and let $F(t + dt)$ represent the same quantity for a period $(t + dt)$. We may now write

$$F(t + dt) = F(t) + \frac{dF}{dt} dt \qquad ...(1.25) \text{ (Taylor series expansion)}$$

and from probability theory

$$F(t + dt) = F(t)\, F(dt) = F(t) \left[1 - \frac{dt}{\tau}\right] \qquad ...(1.26)$$

Comparing Eqs. (1.25) and (1.26) we have

$$\frac{dF}{dt} = \frac{-F}{\tau'}$$

$$\text{or } F(t) = e^{-t/\tau'} \qquad ...(1.27)$$

Since $F(t) = 1$ for $t = 0$ which fixes the integration constant. The average time which elapses between collisions is given as

$$\langle t \rangle = -\int_0^{\infty} t \frac{dF}{dt} dt \qquad ...(1.28)$$

because $-\frac{dF}{dt}$ gives the probability that the electron collides between t and $(t + dt)$. Substituting F(t) from Eq. (1.27) in Eq. (1.28) we have

$$\langle t \rangle = \tau' \qquad ...(1.29)$$

Hence, the physical meaning of τ' introduced above is that it represents the *average time between collisions.*

For isotropic materials, the mean time of collision is the same as the relaxation time, *i.e.* $\tau' = \tau$. This is not true for non-isotropic materials, but such discussion is outside the scope of the book.

The mean free path of an electron, λ, will be defined as

$$\lambda = v\tau' \qquad ...(1.30)$$

where v is the total velocity of electron. In the case of metals, the conductivity always refers to a single total velocity, the one corresponding to *Fermi energy.* We know that free electrons in a metal can accept a series of discrete energy levels. At absolute zero, all energy levels below a certain value E_F are filled, and those above E_F being empty, E_F is called the Fermi level of electrons. The magnitude of E_F is of the order of several electron volts. At any temperature T, the probability of a state corresponding to the energy E being occupied by an electron is given as

$$f(E) = \frac{1}{1 + e^{(E - E_F)/kT}} \qquad ...(1.31)$$

where k is Boltzmann's constant.

The velocity of the electrons with the Fermi energy, v_F is given as

$$\frac{1}{2} m v_F^2 = E_F \qquad ...(1.32)$$

Thus mean free path, $\lambda = v_F\tau' \qquad ...(1.33)$

1.13. Conductivity of Metals

1.13.1. Electron Theory of Metals

This theory was developed by Drude and Lorentz and is also known as *Drude-Lorentz theory of metals.* The basic assumption of this theory is that a metal crystal consists of positive metal ions whose valence electrons are free to move between the ions as if they constituted an electron gas. The crystal is then pictured to be held together by electrostatic forces of attraction between the positively charged ions and the negatively charged electron gas. It is further assumed that the electrons are free to move throughout the entire crystal subject only to the laws of classical mechanics. The mutual repulsion between negative electrons is ignored in this theory and the potential field due to the positive ions is assumed to be completely uniform.

The free electrons in a metal move isotropically and move in a particular direction on an application of electric field. Because of elastic collisions, the magnitude of steady-state current is proportional to the voltage applied provided the temperature of metal remains constant. This leads to *ohm's law*. Moreover, as the free electrons can move easily, the metals exhibit high electrical and thermal conductivities. The ratio of the electrical conductivity 'σ' to the thermal conductivity 'K' should be constant for all metals at a constant temperature.

$$\text{So } \frac{\sigma}{K} = \text{Constant}$$

This is called *Wiedemann Franz law.*

This theory also explains the lusture and opaque properties of metals. When an electromagnetic radiation falls on the surface of metals, the free electrons are set into forced oscillation. The electrons return to their normal states by emitting the same amount of energy in all directions, and hence produces metallic lusture.

1.13.2. Electrical Conductivity

Consider a conductor of length l metres and cross-sectional area A square metres, as shown in Fig. 1.18. Let the number of free electrons available per m^3 of the conductor material *i.e.*, electron density be n and E be the applied electric field. Due to the applied electric field, let axial drift velocity of the electrons be v metres / second.

The force F on the particle of charge q when an electric field E is applied is given as $F = qE$, so the force acting on an electron having charge e on is given as

$$F = eE \qquad ...(1.34)$$

Also

$$F = ma \qquad ...(1.35)$$

where a is acceleration of electrons due to applied field E

Equating Eqs. (1.34) and (1.35) we have

$$\text{Acceleration, } a = \frac{eE}{m} \qquad ...(1.36)$$

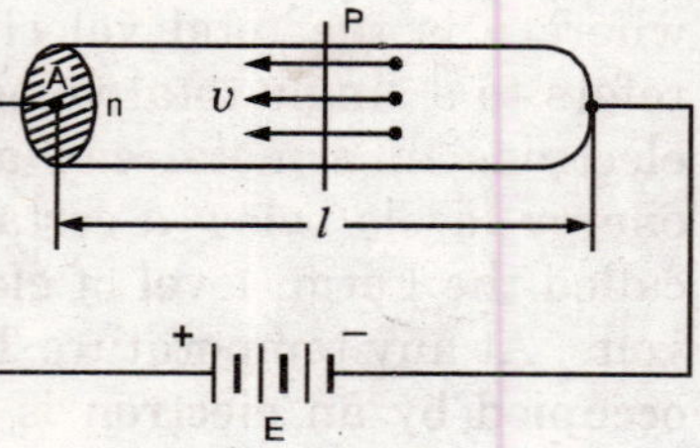

Fig. 1.18

Because of collisions of electrons during motion, the electrons will not get accelerated indefinitely. In fact, their velocity will fall to zero. If τ is the *relaxation* or *collision time*, the average velocity of electrons, known as *drift velocity* is given as

$$v = a \times \tau = \frac{eE}{m} \times \tau \qquad ...(1.37)$$

Let I be the current flowing through the conductor on application of electric field E corresponding to drift velocity v. In time dt, the electrons will travel a distance vdt and the number of electrons crossing the cross-sectional area A in time dt will be contained in volume $Avdt$. Thus, total charge flowing through the section in time dt

$$dq = envAdt \qquad ...(1.38)$$

Since the current in a conductor is defined as the rate of change of charge per unit time, current I is given as

$$I = \frac{dq}{dt} = envA \qquad ...(1.39)$$

The current density J*, defined as the current per unit area is given as

$$J = \frac{I}{A} = env \qquad ...(1.40)$$

The above equation may also be written as

$J = \rho v$ ...(1.41) where ρ is the charge density in coulombs/m^3 $= en$

Substituting the value of v from Eq. (1.37) in above Eq. (1.40), we have

$$J = en \times \frac{eE}{m} \times \tau = \frac{ne^2 E\tau}{m} \qquad ...(1.42)$$

---

* Current density is a vector quantity and is denoted by symbol $\bar{J}$ and, therefore, in vector notation, $I = \bar{J}\,\bar{A}$ where $\bar{A}$ is the vector notation for area '*a*'. For area of any shape we may write the expression for current I as $I = \int \bar{J}\,\bar{A}$.

For a particular material, the quantity $\frac{ne^2\tau}{m}$ in above equation is constant at particular temperature and is known as *electrical conductivity* 'σ' of the material and thus

Current density, $J = \sigma E$ ...(1.43)

So, the current density J is proportional to the conductivity of the material σ and the magnitude of applied electric field E. In terms of resistivity ρ, current density J is given as

$$J = \frac{E}{\rho} \quad ...(1.44) \qquad \because \sigma = \frac{1}{\rho}$$

We know that resistance R of a conductor is given as $R = \rho\frac{l}{A}$ where ρ is the resistivity of material and l is the length of conductor. Thus

$$I = JA = \frac{EA}{\rho} = \frac{El}{R} \quad ...(1.45) \qquad \because R = \rho\frac{l}{A}$$

Since $E = V/l$, the Eq. (1.45) may be written as

$$I = \frac{V}{l}\times\frac{l}{R} = \frac{V}{R}$$ ...(1.46) which is nothing but 'Ohms' law.

Electrical conductivity of the material is also given as

$$\sigma = \frac{ne^2\tau}{m} = ne\mu_e \quad ...(1.47)$$

where $\mu_e = \frac{e\tau}{m}$ is the mobility acquired by electrons due to presence of electric field.

Using Eq. (1.37), the mobility of electrons can also be expressed as

$$\mu_e = \frac{e\tau}{m} = \frac{v}{E} \quad ...(1.48)$$

Thus, the mobility of electron in the metal is defined as the steady-state drift velocity per unit electric field.

Let λ be the mean free path and v' be the root mean square velocity of electrons, then the *relaxation time* (commonly known as mean time between collisions) τ is given as

$$\tau = \frac{\lambda}{v'}$$

But $$v' = \sqrt{\frac{3k_B T}{m}}$$

$$\therefore \quad \tau = \lambda\sqrt{\frac{m}{3k_B T}} \quad ...(1.49)$$

Now the electrical conductivity σ can be expressed as

$$\sigma = \frac{ne^2\tau}{m} = \frac{ne^2\lambda}{m}\sqrt{\frac{m}{3k_B T}} = \frac{ne^2\lambda}{\sqrt{3mk_B T}} \quad ...(1.50)$$

Since electrical resistivity ρ is reciprocal of electrical conductivity σ,

$$\rho = \frac{\sqrt{3mk_B T}}{ne^2\lambda} \quad ...(1.51)$$

or $\rho \propto \sqrt{T}$

Above result shows that the resistivity varies as $\sqrt{T}$ whereas actually it is found to vary linearly with temperature.

Example 1.2. Calculate the conductivity and mobility of copper at 300 K. Collision time τ for electron scattering is 2×10^{-14} second at room temperature. Also calculate the average drift velocity of electrons when applied electric field is 10 V cm^{-1}. Given : Atomic weight = 63.5 and density = 8.92 gm/cc.

[*M.D. Univ. Third Semester December 2006*]

Solution :

$$\text{Molar volume} = \frac{\text{Atomic weight}}{\text{Density}} = \frac{63.5}{8.92} \text{ cc} = 7.12 \text{ cc or } 7.12 \times 10^{-6} \text{ m}^3$$

$$\text{Valence electrons per unit volume, } n = \frac{\text{Avogadro number}}{\text{Molar volume}} = \frac{6.025 \times 10^{23}}{7.12 \times 10^{-6}} = 8.46 \times 10^{28} \text{ m}^{-3}$$

$$\text{Electrical conductivity, } \sigma = \frac{ne^2\tau}{m} = \frac{8.46 \times 10^{28} (1.602 \times 10^{-19})^2 \times 2 \times 10^{-14}}{9.1 \times 10^{-31}}$$

$$= 4.77 \times 10^7 \text{ mho/metre } \textbf{Ans.}$$

$$\text{Mobility, } \mu_e = \frac{\tau e}{m} = \frac{2 \times 10^{-14} \times 1.602 \times 10^{-19}}{9.1 \times 10^{-31}} = 0.35 \times 10^{-2} \text{ m}^2\text{/V-s } \textbf{Ans.}$$

$$\text{Applied electric field, E} = 10 \text{ V cm}^{-1} = 1{,}000 \text{ Vm}^{-1}$$

$$\text{Average drift velocity, } v = \text{E} \times \mu_e = 1{,}000 \times 0.35 \times 10^{-2} = 3.5 \text{ m/s } \textbf{Ans.}$$

Example 1.3. In a certain copper conductor, the current density is 2.4 A/mm^2 and electron density is 5×10^{28} free electrons per m^3 of the copper. Determine the drift velocity of the electrons.

Solution : Current density, $J = 2.4 \text{ A/mm}^2 = 2.4 \times 10^6 \text{ A/m}^2$

Electron density, $n = 5 \times 10^{28}$

Electron charge, $e = 1.6 \times 10^{-19}$ coulomb

From Eq. (1.40) $J = e.n.v$

$$\therefore \text{ Drift velocity, } v = \frac{J}{en} = \frac{2.4 \times 10^6}{1.6 \times 10^{-19} \times 5 \times 10^{28}} = 0.3 \times 10^{-3} \text{ m/s or } 0.3 \text{ mm/s } \textbf{Ans.}$$

Example 1.4. The following data is known for a conductor,

Mobility of electrons = 7.04×10^{-3} m^2/volt-second

Number of conduction electrons per m^3 = 5.8×10^{28}

Calculate (*i*) relaxation time and (*ii*) resistivity of conductor. Given that $e = 1.6 \times 10^{-19}$ C, and $m = 9.1 \times 10^{-31}$ kg.

Solution : Mobility of electrons, $\mu_e = 7.04 \times 10^{-3} \text{ m}^2\text{/V-s}$

Number of conduction electrons, $n = 5.8 \times 10^{28}\text{/m}^3$

Charge on electron, $e = 1.6 \times 10^{-19}$ C

Mass of electron, $m = 9.1 \times 10^{-31}$ kg

From Eq. (1.48).

$$(i) \text{ Relaxation time, } \tau = \frac{\mu_e}{e} \times m = \frac{7.04 \times 10^{-3} \times 9.1 \times 10^{-31}}{1.6 \times 10^{-19}}$$

$$= 40.04 \times 10^{-15} \text{ second } \textbf{Ans.}$$

$$(ii) \text{ Resistivity of conductor, } \rho = \frac{1}{\sigma} = \frac{1}{ne\mu_e} = \frac{1}{5.8 \times 10^{28} \times 1.6 \times 10^{-19} \times 7.04 \times 10^{-3}}$$

$$= 0.01531 \times 10^{-6} \ \Omega\text{-m } \textbf{Ans.}$$

Example 1.5. A conduction wire has a resistivity of 1.54×10^{-8} Ω-m at room temperature. There are 5.8×10^{28} conduction electrons per m^3. Calculate the mobility and relaxation time of electrons.

[*I.E.S. 2000*]

Solution : From Eq. (1.47)

(*i*) Relaxation time, $\tau = \dfrac{\sigma \times m}{ne^2} = \dfrac{m}{ne^2 \times \rho} = \dfrac{9.1 \times 10^{-31}}{5.8 \times 10^{28} \times (1.6 \times 10^{-19})^2 \times 1.54 \times 10^{-8}}$

$= 3.979 \times 10^{-14}$ second **Ans.**

(*ii*) Mobility of electrons, $\mu_e = \dfrac{e \times \tau}{m} = \dfrac{1.6 \times 10^{-19}}{9.1 \times 10^{-31}} \times 3.979 \times 10^{-14} = 6.997 \times 10^{-3}\ \text{m}^2/\text{V-s}$ **Ans.**

Example 1.6. A conductor material has a free electron density of 10^{24} electrons per m^3. When a voltage is applied, a constant drift velocity of 1.5×10^{-2} m/s is attained by the electrons. If the x-sectional area of the material is 1 cm^2, calculate the magnitude of current. Electronic charge is 1.6×10^{-19} coulomb.

[*Aligarh Univ. Elec. Engineering 1981*]

Solution : Electron density, $n = 10^{24}$

Electron charge, $e = 1.6 \times 10^{-19}$ C

Drift velocity, $v = 1.5 \times 10^{-2}$ m/s $= 0.015$ m/s

Cross-sectional area, A $= 1$ cm^2 $= 1 \times 10^{-4}$ m^2

Magnitude of current, I $= envA = 1.6 \times 10^{-19} \times 10^{24} \times 0.015 \times 1 \times 10^{-4} = 0.24$ A **Ans.**

Example 1.7. The following data is known for a conductor

Fermi energy = 5.5 *e*V

Mobility of electrons = 7.04×10^{-3} m^2/V-s

Number of conduction electrons/m^3 = 5.8×10^{28}

Calculate (*i*) Relaxation time (*ii*) Resistivity of conductor (*iii*) Velocity of electron with the Fermi energy.

$e = 1.6 \times 10^{-19}$C, and $m = 9.1 \times 10^{-31}$ kg. [*U.P.S.C. I.E.S. Elec. Engineering-I, 1993*]

Solution : Mobility of electrons, $\mu_e = 7.04 \times 10^{-3}$ m^2/V-s

Number of conduction electrons, $n = 5.8 \times 10^{28}$/m^3

From Eq. (1.48)

(*i*) Relaxation time, $\tau = \dfrac{\mu_e}{e} \times m = \dfrac{7.04 \times 10^{-3} \times 9.1 \times 10^{-31}}{1.6 \times 10^{-19}} = 40.04 \times 10^{-15}$ second **Ans.**

(*ii*) Resistivity of conductor,

$$\rho = \frac{1}{\sigma} = \frac{1}{ne\mu_e} = \frac{1}{5.8 \times 10^{28} \times 1.6 \times 10^{-19} \times 7.04 \times 10^{-3}} = 1.531 \times 10^{-8}\ \Omega\text{-m}$$

(*iii*) Velocity of electrons with Fermi energy,

$$v_F = \sqrt{\frac{2E_F}{m}} = \sqrt{\frac{2 \times 5.5 \times 1.6 \times 10^{-19}}{9.1 \times 10^{-31}}} = 1.3907 \times 10^6 \text{ m/s}$$ **Ans.**

Example 1.8. What is meant by relaxation time, collision time and mean free path as applied to conduction phenomenon ?

The following data is known for copper :

Resistivity = 1.73×10^{-8} Ω-m;

Average time of collision of electrons = 2.42×10^{-14} second

Find (*i*) the valence electrons per unit volume, and the mobility of electrons.

$e = 1.6 \times 10^{-19}$ C; $m = 9.1 \times 10^{-31}$ kg. [*I.E.S. 1995*]

Solution : Resistivity, $\rho = 1.73 \times 10^{-8}$ Ω-m

Average time of collision of electrons, $\tau = 2.42 \times 10^{-14}$ second

From Eq. (1.47),

(*i*) Number of free electrons per m^3,

$$n = \frac{m\sigma}{e^2\tau} = \frac{m}{e^2\tau\rho} = \frac{9.1\times10^{-31}}{(1.6\times10^{-19})^2\times2.42\times10^{-14}\times1.73\times10^{-8}} = 0.849\times10^{29} \textbf{ Ans.}$$

(*ii*) Mobility of electrons,

$$\mu_e = \frac{e\tau}{m} = \frac{1.6\times10^{-19}}{9.1\times10^{-31}}\times2.42\times10^{-14} = 0.4255\times10^{-3}\ \text{m}^2/\text{V-s} \textbf{ Ans.}$$

Example 1.9. A conducting wire has a resistivity of 1.5 × 10⁻⁸ Ω-m at 300 K. Its Fermi energy is 5 eV. Its volume of 1 m³ contains 6 × 10²⁸ conducting electrons. Calculate the relaxation time of conducting electrons. Hence or otherwise obtain the value of drift velocity of electrons, when the conductor is subjected an electric field of 100 V/m. Write the significance of relaxation time and drift velocity. Given : Mass of electron = 9.107 × 10⁻³¹ kg; Electron charge = 1.601 × 10⁻¹⁹ C. [*I.E.S. 2001*]

Solution : Applied electric field, E = 100 V/m

Number of conducting electrons, $n = 6\times10^{28}/\text{m}^3$

From Eq. (1.47)

$$\text{Relaxation time, } \tau = \frac{\sigma\times m}{ne^2} = \frac{m}{ne^2\rho} = \frac{9.107\times10^{-31}}{6\times10^{28}\times(1.601\times10^{-19})^2\times1.5\times10^{-8}}$$

$$= 3.948\times10^{-14} \text{ second} \textbf{ Ans.}$$

$$\text{Drift velocity, } v = \frac{eE}{m}\times\tau = \frac{1.601\times10^{-19}\times100\times3.948\times10^{-14}}{9.107\times10^{-31}} = 0.694 \text{ m/s} \textbf{ Ans.}$$

Example 1.10. A copper wire of 2 mm diameter with conductivity of 5.8 × 10⁷ S/m and electron mobility of 0.0032 m²/V-s is subjected to an electric field of 20 mV/m. Find (*i*) the charge density of free electrons (*ii*) the current density (*iii*) the current flowing in the wire (*iv*) the electron drift velocity. [*I.E.S. 2002*]

Solution : Diameter of copper wire, d = 2 mm = 0.002 m

Conductivity of copper, $\sigma = 5.8\times10^7$ S/m

Electron mobility, $\mu_e = 0.0032\ \text{m}^2/\text{V-s}$

Applied electric field, E = 20 mV/m = 0.02 V/m

(*i*) From Eq. (1.47),

$$\text{Charge density, } n = \frac{\sigma}{e\mu_e} = \frac{5.8\times10^7}{1.6\times10^{-19}\times0.0032} = 1.133\times10^{29}/\text{m}^3 \textbf{ Ans.}$$

(*ii*) From Eq. (1.43),

$$\text{Current density, } J = \sigma E = 5.8\times10^7\times0.02 = 1.16\times10^6\ \text{A/m}^2 \textbf{ Ans.}$$

(*iii*) Current flowing in the wire, I = J × area of x-section of wire

$$= J\times\frac{\pi d^2}{4} = 1.16\times10^6\times\frac{\pi}{4}\times(0.002)^2 = 3.644 \text{ A} \textbf{ Ans.}$$

(*iv*) From Eq. (1.48)

$$\text{Electron drift velocity, } v = \mu_e\times E = 0.0032\times0.02 = 6.4\times10^{-5} \text{ m/s} \textbf{ Ans.}$$

Example 1.11. A *n*-type germanium crystal has a current density of 100 A/m². The crystal has a resistivity of 0.5 Ω-m and electron mobility of 0.4 m²/V-s. Calculate the drift velocity and the time taken by the electron to travel 10 micrometer in the crystal. Assume *e* = 1.6 × 10⁻¹⁹ C. [*I.E.S. 2004*]

Solution : Crystal resistivity, ρ = 0.5 Ω-m

Current density, J = 100 A/m²

Electron mobility, μ_e = 0.4 m²/V-s

Electron drift velocity, $v = \mu_e \times E = \mu_e \times \frac{J}{\sigma} = \mu_e \times J \times \rho = 0.4 \times 100 \times 0.5 = 20$ m/s **Ans.**

Time taken by the electron to travel 10×10^{-6} m in crystal $= \frac{10 \times 10^{-6}}{20} = 0.5 \times 10^{-6}$ second **Ans.**

1.14. CLASSIFICATION OF CONDUCTING MATERIALS

On the basis of resistivity of the materials, conducting materials are broadly classified into two categories namely, low resistivity materials and high resistivity materials.

1.14.1. Low Resistivity Materials

Low resistivity materials are used in house wiring, as conductors for power transmission and distribution, in the windings of transformer and machines like motors and generators. In fact low resistivity materials are used in all such applications where power loss and voltage drop should be low. For example, when power is transmitted from a generating station say Bhakra, to any place, say Delhi, care must be taken to select the conducting material for the transmission wire having such a value of resistivity that the voltage drop and the power loss in the wire along the way is as low as possible. Copper and aluminium are examples of commercially acceptable low resistivity materials. Silver has lower resistivity than copper but because of its prohibitive cost its use commercially as a conductor is not feasible. A low resistivity material, besides possessing very low value of resistivity, should also possess the following additional properties for reasons mentioned against each :

1. *Low Temperature Coefficient.* This means that the change of resistance with change in temperature should be low. This is necessary to avoid variation in voltage drop and power loss with variations in temperature. For example, the resistance of transmission lines, which are very long, will increase when exposed to hot summer sun. This will cause increase in voltage drop and power loss in the transmission line. The windings of electrical machines and apparatus become hot when loaded. This causes temperature rise and if the conducting material of the winding has high temperature coefficient of resistance, the voltage drop and power loss in the winding will be high.

2. *Sufficient Mechanical Strength.* The overhead line conductors used for transmission and distribution of electrical power are subject to stresses due to wind and their own weight. The conducting materials used for the windings of transformers, motors and generators develop mechanical forces when loaded, which can become very large if a high current flows due to a short circuit. Also, when the coils for the windings of such equipment are made on formers, the conducting material is subject to mechanical stresses. Therefore, to withstand the mechanical stresses, developed in the above mentioned applications, the conducting material should possess sufficient mechanical strength.

3. *Ductility.* Ductility is that property of a material which allows it to be drawn out into a wire shape. Conductors are required in different sizes and shapes. In some applications round wire section is used, while in others rectangular wire section is used. The conducting material should be ductile enough to enable itself being drawn into different sizes and shapes.

4. *Solderability.* Conductors have often to be jointed. The joint should offer minimum contact resistance. A simple joint would be to twist the conductors with the material to which it is to be jointed. But this gives high contact resistance. Minimum contact resistance

results, if the joint is soldered. All materials do not tend themselves to proper soldering. So while selecting a conducting material, this point should be kept in view.

5. *Resistance To Corrosion.* The conducting material should be such that it is not corroded when used in outdoor atmosphere.

Note : *The reader should not lead himself to believe that all conducting materials should possess all the above mentioned properties. Depending upon the applications and economic considerations an appropriate material should be chosen which may not have all the above properties but those which the particular applications call for.*

1.14.2. High Resistivity Materials

High resistivity materials are used for making resistance elements for heating devices, starters for electric motors, resistances used in precision measuring instruments, loading resistances and rheostats and filaments for incandescent lamps. In fact, high resistivity materials are used in all such applications where a large value of resistance is required. If low resistivity materials were used for such applications the length of the wire would be too large which would increase to a large extent the overall size of the equipment.

A high resistivity material besides possessing high value of resistivity should also possess the following additional properties for reasons mentioned against each :

1. *Low Temperature Coefficient.* High resistivity materials are often used as shunts in electrical measuring instruments, in making wire-wound precision resistances and resistance boxes. For such precision applications an important requirement is that the material of the element should have negligible temperature coefficient of resistance as otherwise the accuracy of measurements will be affected.

2. *High Melting Point.* In applications like loading rheostats and starters for electrical motors the material of the resistance element should be able to withstand high temperatures for a long time without melting. The consideration of the resistance temperature coefficient in these cases is also important but comparatively higher values than those mentioned in (1) above are permissible.

The consideration of high melting point is important also for resistance materials used in electrical heating devices like room heaters, furnaces etc.

3. *No Tendency For Oxidation.* Materials used as high resistance elements in heating appliances should be able to withstand high temperature for a long time without oxidation. This is because if an oxide layer is formed on the heating element the amount of heat radiation will reduce.

4. *Ductility.* High resistance materials are required to be made in the shape of very thin wires in the case of precision wire-wound resistors and in the shape of thick wires in the case of the elements used in ovens, heaters, starters etc. High resistance materials to be used for such applications should, therefore, be capable of being drawn into wires of different sizes and further be capable of being coiled.

5. *High Mechanical Strength.* High resistivity materials to be used for applications, where the wire must be very thin, are required to have high tensile strength as otherwise they may break during the drawing of the wire or during the assembly and subsequent operation.

1.15. Low Resistivity Materials and Their Applications

1. Copper. Amongst all the conducting materials, copper is the most widely used metal because of its high conductivity (*i.e.* low resistivity). Silver has the lowest resistivity but because of its high cost it is not used as a conducting material. Copper is a non-magnetic (non-ferrous) metal and has remarkable physical, chemical and electrical properties.

Copper is reddish in colour and can be made available in hard drawn or annealed form. Mechanical properties are different for hard drawn and annealed copper. Annealed copper is soft and has less tensile strength than hard drawn copper. Hard drawn copper becomes soft after annealing. Hard drawn copper is obtained by drawing copper bars or rods in cold condition. Annealing process involves heating at a specific temperature and then cooling. Hard drawn copper is spongy but annealed copper is flexible. Annealed copper has higher conductivity than hard drawn copper. The resistivity of annealed copper is 1.72×10^{-8} ohm-m and that of hard drawn copper is 1.77×10^{-8} ohm-m both at 20ºC. Copper can be drawn into very thin wires. Sheets and bars of various thicknesses can be made. Copper offers high resistance to corrosion. When exposed to atmosphere copper oxide layer is formed on it which acts as a protective layer and prevents corrosion. Density and melting point of annealed copper are 8.89 and 1,084 degrees centigrade and for hard drawn copper 8.93 and 1,084 degrees centigrade respectively. The tensile strength of copper varies from 8.15 to 4.72 tonnes/cm^2.

Copper can easily be soldered and welded which are necessary in electrical wiring. To remove the oxide layer formed on the surface of copper some solvent (flux) is necessary for good soldering. Copper joints offer low contact resistance as the oxide film formed on it also has conducting property. The most important applications of hard drawn copper are in overhead conductors, high voltage underground cables and bus-bars because of its high mechanical strength. Annealed copper is used for insulated conductors in low voltage power cables, winding wires for electrical machines and transformers, flexible wires and in making coils for many purposes.

Copper in pure form is not much used as an electrical contact material. When 10 to 30 % of nickel is mixed with it, then it becomes harder and cheaper. Due to its high electrical and thermal conductivity, it is commonly used as a contact material for control relays, motor starter switches and tap changers. It has poor resistance to oxidation which lowers its efficiency.

2. Aluminium. Aluminium is widely available in India and is used extensively in the field of electrical engineering. So far as electrical conductivity is concerned, it is next best to copper. Its resistivity is 2.8×10^{-8} ohm-m, *i.e.* about 1.6 times higher than copper. Its density is 2.68 which means that aluminium is much lighter than copper (Aluminium weighs $\frac{1}{3}$ times as much as copper). Keeping in view the resistivity, mechanical strength and density, for the same resistance and length of wire, an aluminium conductor should have a cross-sectional area 1.6 times that of copper conductor; the weight of the aluminium conductor will be equal to 0.48 times the weight of copper. Its melting point is 655 degrees centigrade. Like copper, it can be easily rolled and drawn hard. Its resistivity and density also change after the material is mechanically drawn. It is annealed (heat treated) after it is hard drawn. It can be drawn into thin wires. Aluminium is soft metal but when alloyed with some other material like magnesium, silicon or iron, it acquires higher mechanical strength and can be used for overhead transmission lines. Like copper, aluminium also forms an oxide layer over its surface when exposed to atmosphere and that layer prevents the material from further oxidation and acts as a resistive layer to corrosion. Unlike copper oxide layer, aluminium oxide layer when formed on the surface acts as an insulation because aluminium oxide has relatively higher resistivity. Thus, the contact resistance of aluminium wires is very high. Further, as it is a soft material, there is always a possibility of loose contact. Due to the insulating property of aluminium oxide formed on the surface, it is difficult to solder aluminium wires. This is a great disadvantage and the main objection against its use in house wiring and other places where joints and soldering have to be done. However, special flux is now being developed for soldering aluminium wires. As a substitute for copper, aluminium is quite extensively used for flexible wires, overhead transmission lines, bus-bars, squirrel cage induction motor rotor bars and in many other applications.

Use of aluminium as an electrical material, particularly in the aircraft industry, has considerable advantages because of the saving in weight involved. Again electrochemical plants are enormous users of aluminium busbars, because electrolytic cells operate with heavy currents at low voltages and to carry these currents, massive bars are required. Aluminium, because of its light weight, is being used more and more for such bars. The current carrying capacity of aluminium being 75% that of copper, and its density being approximately one-third that of copper, an aluminium busbar is only about half the weight of the copper busbar of equal current carrying capacity. Since aluminium costs much less than copper, an aluminium busbar will cost only about half as much as its counterpart copper.

However, for other important applications like winding of electrical machines and transformers, it is difficult to substitute aluminium for copper. This is because aluminium wires have tensile strength varying from 0.95 to 1.57 tonnes/cm^2, which is much lower than that of copper. This results in a tendency in the breaking of the wire under tension as well as developing kinks. Since its resistivity is higher compared to copper the wire has to have a thicker cross-section to keep the I^2R losses low. The winding thus occupies more space and the machine size increases. From example 1.12, given at page 31, it is obvious that space occupied by the aluminium winding will be more and hence the size of the transformer will be bigger. However, because of lower density of aluminium as compared to copper, aluminium wound machines have less weight. Efforts are being made to use aluminium as winding wires in transformers and such other applications.

For economic reasons aluminium has replaced copper in many fields. Unlike copper, it is available in abundance in India. Overhead transmission lines are now all made of aluminium conductors with steel reinforcement, called ACSR conductors (aluminium conductors steel reinforced). Steel reinforcement is made for giving higher strength to the overhead conductor. Figure 1.19 shows the cross-section of an ACSR conductor.

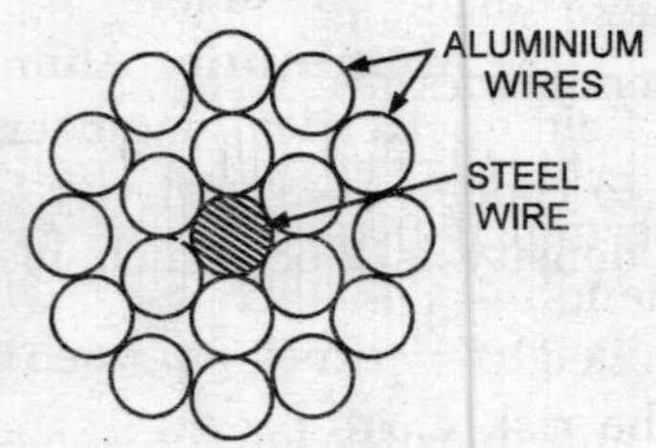

Fig. 1.19. *ACSR Conductor*

Copper and aluminium are low resistivity materials and if they are used for making the elements for electric heaters, the length of the wire would be too large, which would increase the overall size of the equipment to a large extent. Hence copper and aluminium are never used for this purpose.

3. Silver. Pure silver has high electrical conductivity and good corrosion resistance. It is used where high wear resistance is not required. In order to make it harder 15 % of copper is added into it. To make it more harder for use in commutator segments of small dc motors an alloy of silver-copper containing 40 % of copper is used. For brushes and collector rings of dc motors silver graphite alloy containing a small percentage of graphite is used because it provides sliding lubrication.

4. Gold. Gold is the best known electrical conductor. It is found all over the world but not in sufficient quantity to make it economical. Gold is generally found in veins among rocks and ores of other metals. It is also found in the form of dust in the beds of rivers. Gold has density of 19.3 times that of water at 20 °C. It melts at 1,063 °C and boils at 2,700 °C. It is malleable and ductile and can be easily beaten into translucent sheets as thin as 0.00001 mm. It is largely used as alloy to make coins and jewellery. Its good corrosion resistance property makes its alloy very much useful as contact materials in electrical fields. Its alloy is also used as corrosive resistant brazing material.

5. Steel. Steel contains iron and carbon. Iron itself is not exceptionally strong but it is in abundance. When carbon is added to iron the material that results, *steel*, possesses exceptional mechanical properties. About 92 per cent of steel produced is ordinary carbon steel. The rest is alloy steel, in which iron is alloyed not only with carbon but also with various amounts of manganese, nickel, chromium, molybdenum or vanadium.

The yield and tensile strength of the steel increases with the carbon content but the ductility decreases. Therefore, the carbon content may not be too high or else the steel will be brittle. Steels containing about 0.25 per cent carbon are classed as mild steels, those with 0.45 per cent carbon as medium steels, while those with about 0.7 per cent carbon and higher as high-carbon steels. High-carbon steels and also some of the low-carbon steels are usually heat-treated. Their tensile strength is greatly improved by heat treatment, wherein the steel is reheated to above critical temperature and then cooled at different rates; for rapid cooling it may be quenched in water.

The resistivity of steel is 8 to 9 times higher than that of copper. Due to this reason steel is not generally used as conducting material, although it has higher mechanical properties, availability in abundance, and low cost. When alternating current flows in steel wire, its resistivity increases and power losses are much higher than when direct current flows, due to magnetic properties of steel. Steel is easily corroded by moisture and heat. Overhead steel conductors are galvanised to prevent oxidation. In this method, steel is thoroughly cleaned and then it is dipped in a bath of molten zinc. The layer of zinc protects steel from rusting. As mentioned earlier, aluminium conductors are steel reinforced to increase their tensile strength for use as overhead transmission line conductors. Galvanised steel wires are used as overhead telephone wires and as earth wires. Steel overhead conductors are also used for transfer of small amounts of power.

6. Stranded Conductors. It is obvious that when a single conductor of large cross-section is used, it becomes rigid in construction, and is liable to kinks and breaks while handling. To avoid this conductors are made of a number of thin wires, bunched together, called strands. Stranding makes the conductor flexible and eliminates, to a large extent, the risk of its breaking through the insulation.

A stranded conductor is made by twisting the wire (strands) together to form layers. Generally, stranding is done in opposite directions for successive layers. This means, if the wires of one layer are twisted in left-hand direction, the next layer of wires will be twisted in the right hand direction and so on.

A standard stranding consists of 6 wires around 1 wire, then 12 wires around the previous 6, then 18 wires around the 12, then 24 wires around the 18 and so on. The number of layers to be provided will depend upon the number of wires to be provided. Note that the central wire is not counted as a layer. Instead of a single wire, three or four stranded wires may also be put in the centre and over them layers may be formed. If 3 stranded wires are put in the centre, 9 wires will be in the first layer then 15 wires in the second layer and so on. In each of the above cases, the increase in the number of wires in each successive layer is 6.

Formulae giving the number or wires for a particular layer, the total number of wires in a particular stranded conductor and the diameter of the stranded conductor is shown in Table 1.1.

TABLE. 1.1. ***Number of Wires in Different Layers, Total Number of Wires and Overall Diameter of Stranded Conductors***

Number of wires in the centre.	1 wire	3 wires	4 wires
Number of wires in the nth layer from centre	$6n$	$3 + 6n$	$4 + 6n$.
Total number of wires in a stranded conductor having n layers.	$1 + 3n(1 + n)$	$3(1 + n)^2$	$(4 + 3n)(1 + n)$
Diameter over the nth layer in centimetres, where d indicates the diameter of each wire in centimeters.	$(1 + 2n)d$	$(2.155 + 2n)d$	$(2.414 + 2n)d$

Stranded conductors are expressed as :

7/2.24, 19/2.50, 37/2.06, 61/2.59 etc.

The first number *i.e.,* 7, 19, 37, 61 etc. indicates the total number of wires in the stranded conductor, and the second number *i.e.,* 2.24, 2.50, 2.06, 2.59 etc. represents the diameter of each wire in millimetres.

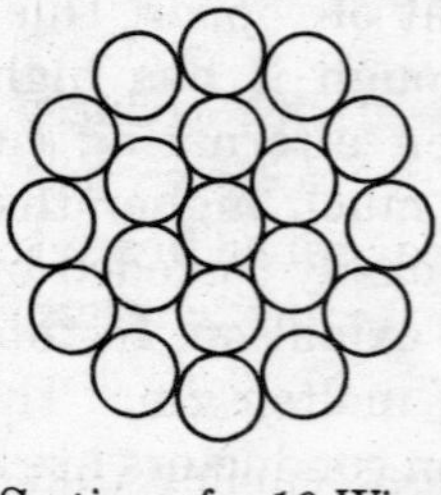

(a) Cross Section of a 19 Wire Stranded Circular Conductor Having 1 Wire at the Centre, 6 Wires in the Ist Layer and 12 Wires in the 2nd Layer.

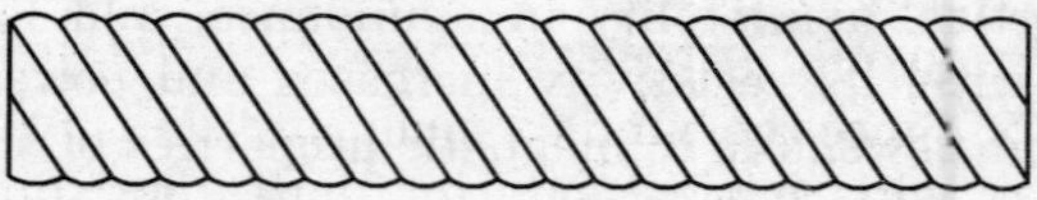

(b) A Stranded Circular Conductor Showing How Wires are Twisted Together.

Fig. 1.20

A 37/2.06 stranded conductor has 1 wire at the centre, 6 wires in first layer, 12 wires in second layer and 18 wires in the third layer. Therefore, total number of layers, $n = 3$.

$$\text{Total number of wires} = 1 + 3n(1 + n)$$

$$= 1 + 3 \times 3(1 + 3) = 37$$

$$\text{Overall diameter} = (1 + 2n)d = (1 + 2 \times 3) \times 2.06 \text{ mm} = 14.42 \text{ mm.}$$

Figure 1.21 shows how stranded conductors are formed from a number of wires.

Circular stranded conductors are normally used for single phase systems. The trend of recent years is to manufacture compact conductors both in circular and shaped form.

This results in overall economy by saving insulating and protective materials in the manufacture of cables. Compacting is carried out by pressing the circular stranded conductors through rollers or dies. How the overall dimensions of a compact circular stranded conductor are reduced can be seen from Fig. 1.21.

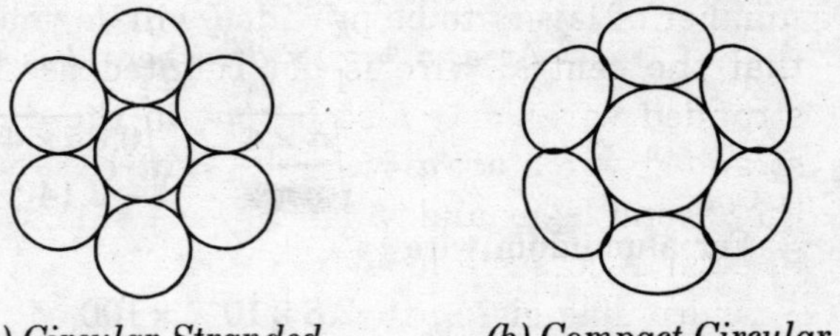

(a) Circular Stranded Conductor *(b) Compact Circular Stranded Conductor*

Fig. 1.21

7. Bundled Conductors. A bundled conductor is a conductor made up of two or more conductors, called the *subconductors*, per phase in close proximity compared with the spacing between phases (Fig. 1.22). The basic difference between a composite conductor and a bundled conductor is that the sub-conductors of a bundled conductor are separated from each other by a constant distance varying from 0.2 m to 0.6 m depending upon designed voltage and surrounding conditions throughout the length of the line with the help of spacers whereas the wires of a composite conductor touch each other. The bundled conductors have filler material or air space inside so that the overall diameter is increased.

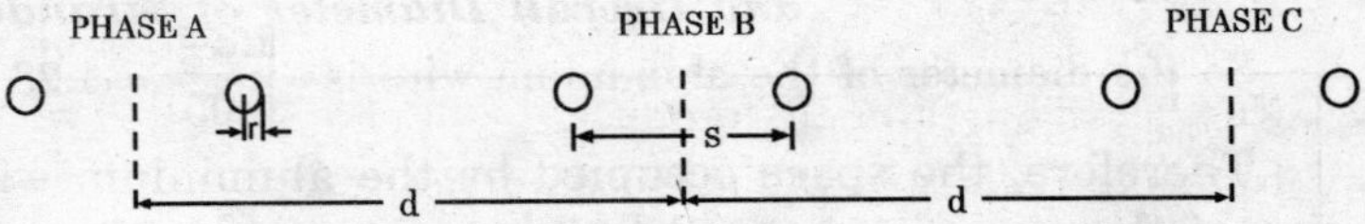

Fig. 1.22

The use of bundled conductors per phase reduces the voltage gradient in the vicinity of the line and thus, reduces the possibilities of the corona discharge.

Although the bundled conductors are used on EHV transmission lines primarily to reduce corona loss and radio interference, but they have several other advantages over single conductors such as given below :

1. The bundled conductor lines transmit bulk power with reduced losses, thereby giving increased transmission efficiency.
2. Since the bundled conductor lines have a higher capacitance to neutral in comparison with single conductor lines, therefore, they have higher charging current, which helps in improving the power factor.

8. Low Resistivity Copper Alloys. We have noticed earlier that copper becomes mechanically hard when it is drawn. However, hardening of copper can also be done by alloying with other metals.

Example 1.12. Calculate the diameter of copper wire of length 100 metres used as winding material in a transformer such that resistance of the whole winding is 2 ohms. Calculate the diameter of the wire if aluminium is to be used for the above winding, resistance remaining the same.

Solution : We know :

$$R = \rho \frac{l}{a}$$

Putting the values for copper wire we get :

$$2 = \frac{1.7 \times 10^{-8} \times 100}{a}$$

$$\text{or } a = \frac{1.7 \times 10^{-8} \times 100}{2} = 0.85 \text{ mm}^2$$

$$\text{Area, } a = \frac{\pi}{4} \times d^2, \text{ where } d \text{ is the diameter of the wire}$$

$$\text{or} \quad d = \sqrt{\frac{a \times 4}{\pi}} = \sqrt{\frac{0.85 \times 4}{3.14}} = 1.05 \text{ mm}$$

For aluminium wire :

$$a = \frac{2.8 \times 10^{-8} \times 100}{2} = 1.4 \text{ sq. mm}$$

$$\text{and} \quad d = \sqrt{\frac{1.4 \times 4}{3.14}} = 1.34 \text{ mm}$$

So the diameter of the aluminium wire is $\frac{1.34}{1.05} = 1.28$ times that of copper wire.

Therefore, the space occupied by the aluminium winding will be more and hence the size of the transformer will be bigger.

1.16. HIGH RESISTIVITY MATERIALS AND THEIR APPLICATION

Alloys of nickel, chromium, copper, iron, manganese are extensively used as resistance materials. These alloys have high specific resistance (about 25 times that of pure copper) and very low temperature coefficient of resistance (about one-twentieth of that of pure copper). In general the higher the specific resistance of the material, the shorter and stouter the resistance element with simple and stronger construction. The various materials used as resistance materials are described below :

1. *Platinum.* Though pure platinum is very good metal from the point of view of permanence but its use is limited due to its high cost, high temperature coefficient of resistance (0.003 Ω/Ω/per °C).

It is a heavy, greyish-white, non-corroding noble metal which is malleable and ductile, but poorly fusible and resistant to most chemicals. It has a very large specific weight 21,400 kg/m^3, a melting point of 1,775°C, a boiling point of 4,530°C, a thermal coefficient of expansion 9×10^{-4} per °C, and a resistivity of 0.1 μΩ-m. Platinum can be drawn into thin filaments and strips, and is used in different electrical devices, for example, as the heating elements in electrical laboratory ovens and furnaces (at extra high temperatures above 1,300°C it begins to atomise), in platinum-platinum-rhodium thermocouples working at temperatures up to 1,600°C etc. Multiple drawing of a bimetallic platinum-silver wire, after which the surface layer of silver is dissolved in nitric acid (that does not act on the platinum), is the way extra - thin elements of about one micron diameter are made for suspending movable parts in electric meters and in other sensitive instruments. Other applications of platinum are insoluble anodes, thermocouples, electrical contacts, corrosion-resistant laboratory accessories, electrical furnace windings, catalytic gas igniters, grids in special-purpose vacuum tubes etc.

2. *Tungsten.* Tungsten is a silver-white hard metal, obtained by a suitable chemical process from ore concentrates, having a specific weight of 20,000 kg/m^3. Its melting point is the highest of all metals comprising 3,410°C. It has a boiling point of about 5,900°C, a thermal coefficient of expansion of 4.4×10^{-6} per°C, a resistivity of 5.51×10^{-8} Ω-m and a temperature coefficient of resistance of 0.0045 Ω per Ω per °C. The thinner the tungsten wire, the greater is its tensile strength. The tensile strength of forged bars 6 by 3 mm amounts to tens of kg/mm^2, while that for thin filaments is as high as 300-400 kg/mm^2. Tungsten is one of the main materials used in electronics and vacuum engineering. It is used as the filament in incandescent lamps and as the electrodes, heaters, springs, catches and so on in electron X-ray and other kind of tubes.

3. *Molybdenum.* This metal is very much like tungsten as to its appearance and properties as well as the way it is processed. Molybdenum has a specific weight of 10,200 kg/m^3, a melting point of 2,620ºC, a boiling point of 3,700ºC, a thermal coefficient of expansion of 5.3×10^{-6} per ºC, a resistivity of 4.8×10^{-8} Ω-m, and a temperature coefficient of resistance of 0.0047 per ºC. Molybdenum is useful as the target in X-ray tubes and as structural members in high-vacuum electron tubes because of its ability to form a tight seal with glass. It is an important alloying element since it imparts hardness and corrosion resistance to ferrous and non-ferrous alloys containing it.

4. *Manganin.* It is an alloy of 84 per cent copper, 12 per cent manganese and 4 percent nickel of a reddish brown colour. Its properties are :

Specific resistance	0.42 – 0.74 μ Ω-m
Temperature coefficient of resistance	0.00003 Ω/Ω/ºC
Thermo-emf against copper	about 5μV/ ºC
Density	8,400 kg/m^3
Tensile strength	40 – 55 kg/mm^2
Maximum working temperature	Not over 60 ºC

It is most widely used material for the manufacture of resistance boxes, standard resistances and shunts.

5. *Constantan.* Constantan is an alloy of 60 percent copper and 40 percent nickel. Its properties are:

Resistivity	49×10^{-8} Ω-m
Temperature coefficient of resistance	Very small: may be either positive or negative.
Thermo-emf against copper	about 40 μV/ ºC
Density	8,900 kg/m^3
Tensile strength	40 – 50 kg/mm^2
Maximum working temp.	400 to 450 ºC

It is widely used in the manufacture of rheostats and heater resistors, provided their working temperature does not exceed 400 to 500 ºC. Due to generation of high thermo-emf, when in a junction with copper, it cannot be used in the manufacture of standard resistors for high precision instruments. The use of this property of constantan is made in thermo-couples for measuring temperature up to several hundred degrees.

6. *Nichrome.* Nichrome is an alloy of nickel and chromium. The properties of nichrome are:

Resistivity	about 1.0 μΩ-m
Temperature coefficient of resistance	about 0.00044 Ω/Ω/ºC
Density	8,400 kg/m^3
Tensile strength	70 kg/mm^2
Working temperature up to	1,000 ºC

Nichrome is available in the form of a wire or tape and is widely used as heating element in domestic appliances (such as heater, electric iron, toaster etc.) and furnaces with working temperatures up to 1,000 ºC.

1.17. Factors Affecting Electrical Conductivity (or Resistivity) of Metals

Resistivity of material is defined as the resistance between opposite faces of a unit cube of that material. Resistivity is measured in ohm- m or ohms per metre cube in SI system. It is expressed by ρ (rho).

Conductivity of material is reciprocal of resistivity and is expressed as σ

i.e.
$$\sigma = \frac{1}{\rho}$$

Conductivity may be defined as the conductance between the two opposite faces of a unit cube. It is measured in Siemens per metre.

The electrical conductivity $\sigma\left(=\frac{1}{\rho}\right)$ of a material is not constant. The factors which affect the value of conductivity σ of a material are:

(*i*) Temperature, (*ii*) Alloying, (*iii*) Mechanical stressing, (*iv*) Age hardening.

1.17.1. Resistance Variation With Temperature

The resistance of all pure metallic conductors increases with the increase in temperature but the resistance of the insulators and non-metallic materials generally decreases with the increase in temperature.

If the resistance of any pure metal is plotted on a temperature base, it is found that over the range of temperature from 0 to 100 °C the graph is practically a straight line, as illustrated in Fig. 1.23. If this straight line is extended, it cuts the temperature axis at some temperature, $-t_0$°C, known as *inferred zero resistance temperature.* This does not mean that the resistance of the metal is actually zero at that temperature, but $-t_0$ °C is the temperature at which the resistance would be zero if the rate of decrease between 100 and 0 °C were maintained constant at all temperatures. From the similarity of the triangles in Fig. 1.23

$$\frac{R_2}{R_1} = \frac{t_0 + t_2}{t_0 + t_1} \quad ...(1.52)$$

where R_1 and R_2 are the resistances at temperatures t_1 °C and t_2 °C respectively. Thus if the resistance R_1 for any temperature t_1 °C is known, then resistance for any other temperature t_2 can be computed from above expression provided that t_0 for that particular material is known. Inferred zero resistance temperatures, t_0 for various materials are given in Table 1.2.

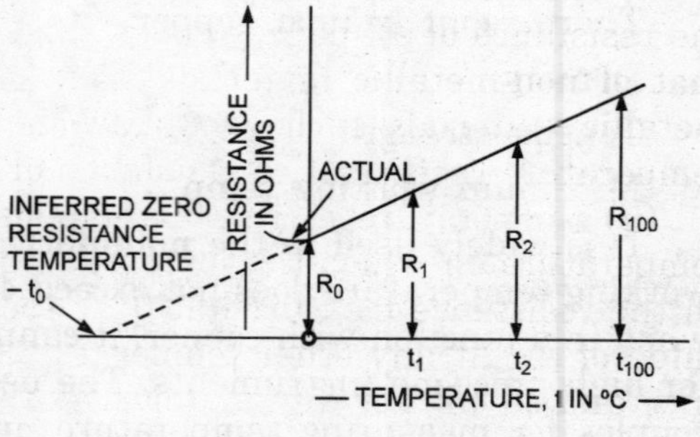

Fig. 1.23

The variation of resistance with temperature is often utilized in determining temperature variations. For example, in testing of an electric machine, the resistance of the coil is measured both before and after the test run, and the increase in resistance is a measure of the rise in temperature. For computation of temperature rise Eq. (1.52) may be transposed to the following form

$$t_2 - t_1 = \frac{R_2 - R_1}{R_1}(t_0 + t_1) \quad ...(1.53)$$

A temperature of 20 °C has been adopted as the *standard reference temperature* for measurement of resistance, and the handbooks give the resistance of the various materials at that temperature. Consequently, when a designer is computing the resistance of any conductor from its dimensions, the initial temperature t, at which the resistance is known, is generally 20 °C.

Temperature Coefficient of Resistance. Let a metallic conductor having a resistance of R_0 at 0 °C be heated to t °C and let its resistance at this temperature be R_t. From Eq. (1.52)

$$\frac{R_t}{R_0} = \frac{t_0 + t}{t_0 + 0}$$

$$\text{or} \quad R_t = R_0 + \frac{1}{t_0} R_0 t$$

$$\text{or change in resistance, } \Delta R = R_t - R_0 = \frac{1}{t_0} R_0 t = \alpha_0 R_0 t \qquad \text{...(1.54)}$$

where $\alpha_0 = \frac{1}{t_0}$ and is called the temperature coefficient of resistance of the material at 0 °C

From Eq. (1.54) it may be concluded that change in resistance due to change in temperature

(*a*) varies directly as its initial resistance,

(*b*) varies directly as rise in temperature and

(*c*) depends on the nature of the material of the conductor.

The Eq. (1.54) may be rewritten as:

$$\alpha_0 = \frac{\Delta R}{R_0 t} \qquad \text{...(1.55)}$$

So the temperature coefficient of resistance may be defined as the ratio of increase in resistance per degree rise of temperature to the original resistance.

If R_0 is the resistance of any conductor at 0°C and α_0 is the temperature coefficient of resistance at 0°C, then resistance at t °C is given as

$$R_t = \text{Original resistance + increase in resistance}$$

$$= R_0 + R_0 \alpha_0 t = R_0 (1 + \alpha_0 t) \qquad \text{...(1.56)}$$

The above expression holds good for both increase as well as decrease in temperature.

It is to be noted that

(*i*) temperature coefficient of resistance for all pure metallic conductors is positive *i.e.* the resistance of all pure metallic conductors increases with the increase in temperature, that of non-metallic materials such as of carbon is negative *i.e.* the resistance of non-metallic materials such as of carbon decreases with the increase in temperature. The temperature coefficient of resistance of alloys like constantan and manganin is negligible.

(*ii*) temperature coefficient of resistance is not constant but depends on the initial temperature on which the increment in resistance is based. When the increment is based on the resistance measured at 0 °C, then the temperature coefficient of resistance has the value of α_0. At any other temperature t °C, value of temperature coefficient of resistance is α_t and so on. For any material the temperature coefficient of resistance at 0 °C *i.e.* α_0 has the maximum value.

The temperature coefficient of resistance at any temperature t_1 is given as

$$\alpha_1 = \frac{1}{\frac{1}{\alpha_0} + t_1} \qquad \text{...(1.57) (Refer to Example 1.13)}$$

The temperature coefficient of resistance at temperature t_2 in terms of temperature coefficient of resistance at temperature t_1 is given as

$$\alpha_2 = \frac{1}{\frac{1}{\alpha_1} + (t_2 - t_1)} \qquad \text{...(1.58) (Refer to Example 1.14)}$$

If R_1 is the resistance of any conductor at t_1 °C and α_1 is the temperature coefficient of resistance at t_1 °C, then resistance of the conductor at t_2 °C is given as

$$R_2 = R_1 [1 + \alpha_1 (t_2 - t_1)] \qquad ...(1.59)$$

(*iii*) As the resistance of the material changes with the change in temperature so it is obvious that resistivity of the material depends on temperature.

Knowledge of temperature coefficient of resistance is utilized in determining the temperature rise of electrical machines.

Values of resistivity, temperature coefficient of resistance at 20 °C and inferred zero resistance temperature for various materials are given in Table 1.2.

TABLE 1.2. ***Resistivities, Temperature Coefficients and Inferred Zero Resistance Temperatures***

Material	Resistivity in Ω-m at 20 °C (× 10^{-8})	Temperature Coefficient of Resistance in ohms per ohm per °C rise in temperature at 20 °C	Inferred Zero Resistance Temperature in °C
Aluminium, annealed	2.82	0.0039	– 236
Aluminium, hard drawn	2.92	0.0038	– 243
Brass	6 – 8	0.002	– 480
Carbon	3,000 – 7,000	– 0.0005	—
Constantan or Eureka (60% copper and 40% nickel)	49	0.000008	– 1,25,000
Copper, annealed	1.724	0.00393	– 234.5
Copper, hard drawn	1.79	0.00382	– 242
German Silver (80% Cu, 12% Ni, 4% Zn)	33	0.0004	– 2,480
Gold	2.44	0.0034	– 274
Iron	12	0.005	– 180
Lead	22	0.0039	– 236
Manganin (84% Cu, 12%Mn, 4%Ni)	42 to 74	0.00003	– 33,000
Mercury	96	0.00089	– 1,100
Molybdenum	5.7	0.0033	– 280
Nichrome (60%Cu, 25%Fe, 15 %Cr)	100	0.00044	– 2,250
Nickel	7.8	0.006	– 147
Phosphor Bronze	5–10	0.0035	– 266
Platinum	10	0.003	– 310
Platinum–iridium	24.6	0.0012	– 814
Silver (99.98% pure)	1.6	0.0038	– 243
Tungsten	5.51	0.0045	– 200
Zinc	6.3	0.004	– 230

Note : *It has been found that the resistivity of the metallic conductors is significantly affected by variations in temperature but it is also affected by other factors like tension and pressure to some extent. For most metals except lithium and calcium resistivity decreases with the increase in pressure. However, resistivity increases with increase in tension.*

Example 1.13. Show that if α_0 be the resistance temperature coefficient of a conductor at 0 °C, then resistance temperature coefficient at t_1 °C is given as

$$\alpha_1 = \frac{1}{\frac{1}{\alpha_0} + t_1}$$

Solution : Let R_0, R_1 and R_2 be the resistances of a conductor at 0 °C, t_1 °C and t_2 °C temperatures respectively, then from Eq. (1.56)

$$\text{Resistance at } t_1\ ^\circ\text{C},\ R_1 = R_0\ (1 + \alpha_0 t_1) \qquad ...(i)$$

$$\text{Resistance at } t_2\ ^\circ\text{C},\ R_2 = R_0\ (1 + \alpha_0 t_2) \qquad ...(ii)$$

$$\text{Also } R_2 = R_1\ [1 + \alpha_1 (t_2 - t_1)] \qquad \text{(Refer to Eq. (1.59))}$$

$$\text{or } \frac{R_2}{R_1} = 1 + \alpha_1 (t_2 - t_1) \qquad ...(iii)$$

Dividing Eq. (*ii*) by Eq. (*i*), we get

$$\frac{R_2}{R_1} = \frac{1+\alpha_0 t_2}{1+\alpha_0 t_1} = \frac{1+\alpha_0 t_2 + \alpha_0 t_1 - \alpha_0 t_1}{1+\alpha_0 t_1}$$

$$= \frac{1+\alpha_0 t_1}{1+\alpha_0 t_1} + \frac{\alpha_0 t_2 - \alpha_0 t_1}{1+\alpha_0 t_1}$$

$$= 1 + \frac{\alpha_0 (t_2 - t_1)}{1+\alpha_0 t_1} = 1 + \frac{\alpha_0}{1+\alpha_0 t_1}(t_2 - t_1) \qquad ...(iv)$$

Comparing Eqs. (*iii*) and (*iv*) we get

$$1 + \alpha_1 (t_2 - t_1) = 1 + \frac{\alpha_0}{1+\alpha_0 t_1}(t_2 - t_1)$$

$$\text{or } \alpha_1 = \frac{\alpha_0}{1+\alpha_0 t_1} = \frac{1}{\frac{1}{\alpha_0} + t_1} \qquad \text{(Proved)}$$

Example 1.14. Show that if α_1 is the resistance temperature coefficient of a conductor at t_1 °C then resistance temperature coefficient at t_2 °C is given by

$$\frac{1}{\frac{1}{\alpha_0} + (t_2 - t_1)}$$

Solution : Let R_1, R_2 and R_3 be the resistances of a conductor at temperatures t_1°, t_2° and t_3°C respectively.

$$\text{Then } R_2 = R_1\ [1 + \alpha_1\ (t_2 - t_1)] \qquad ...(i)$$

$$\text{and } R_3 = R_1\ [1 + \alpha_1\ (t_3 - t_1)] \qquad ...(ii)$$

$$\text{Also } R_3 = R_2\ [1 + \alpha_2\ (t_3 - t_2)]$$

$$\text{or } \frac{R_3}{R_2} = 1 + \alpha_2\ (t_3 - t_2) \qquad ...(iii)$$

Dividing Eq. (*ii*) by Eq. (*i*) we get

$$\frac{R_3}{R_2} = \frac{1+\alpha_1 (t_3 - t_1)}{1+\alpha_1 (t_2 - t_1)} = \frac{1+\alpha_1 (t_2 - t_1) + \alpha_1 (t_3 - t_1) - \alpha_1 (t_2 - t_1)}{1+\alpha_1 (t_2 - t_1)}$$

$$= 1 + \frac{\alpha_1 (t_3 - t_2)}{1+\alpha_1 (t_2 - t_1)} \qquad ...(iv)$$

Comparing Eqs. (*iii*) and (*iv*) we get

$$1 + \alpha_2\ (t_3 - t_2) = 1 + \frac{\alpha_1 (t_3 - t_2)}{1+\alpha_1 (t_2 - t_1)}$$

$$\text{or } \alpha_2 = \frac{\alpha_1}{1+\alpha_1 (t_2 - t_1)} = \frac{1}{\frac{1}{\alpha_1} + (t_2 - t_1)} \text{ Proved.}$$

Example 1.15. A coil of relay is made of copper wire. At a temperature of 20 °C, the resistance of the coil is 400 Ω. Calculate the resistance of the coil at temperature of 80 °C. The temperature coefficient of copper is 0.0038 Ω/Ω/°C at 0 °C.

Solution : Temperature coefficient of resistance at 0 °C, $\alpha_0 = 0.0038\ \Omega/\Omega/^\circ\text{C}$

$$\text{Temperature coefficient of resistance at 20 °C, } \alpha_{20} = \frac{1}{\frac{1}{\alpha_0} + t} = \frac{1}{\frac{1}{0.0038} + 20} = 0.003532$$

Resistance of wire at 20 °C, $R_1 = 400\ \Omega$

Resistance of wire at 80 °C, $R_2 = R_1\ [1 + \alpha_{20}(t_2 - t_1)] = 400\ [1 + 0.003532\ (80 - 20)]$

$= 484.8\ \Omega$ **Ans.**

Example 1.16. A platinum coil has a resistance of 3.146 Ω at 40 °C and 3.767 Ω at 100 °C. Find the resistance at 0 °C and the temperature coefficient of resistance at 40 °C. *[Allahabad Univ. 1993]*

Solution : Let the temperature coefficient of resistance of platinum at 0 °C be α_0 and resistance of platinum coil at 0 °C be R_0, then

Resistance at 40 °C, $R_1 = R_0\ (1 + 40\ \alpha_0)$ ...(*i*)

Resistance at 100 °C, $R_2 = (1 + 100\ \alpha_0)$ ...(*ii*)

Dividing Eq. (*ii*) by Eq. (*i*) we have

$$\frac{R_2}{R_1} = \frac{1 + 100\,\alpha_0}{1 + 40\,\alpha_0}$$

$$\text{or} \quad \frac{3.767}{3.146} = \frac{1 + 100\,\alpha_0}{1 + 40\,\alpha_0}$$

$$\text{or} \quad \alpha_0 = 0.00379$$

Temperature coefficient of resistance at 40 °C,

$$\alpha_{40} = \frac{1}{\frac{1}{\alpha_0} + 40} = \frac{1}{\frac{1}{0.00379} + 40} = 0.00329\ \Omega/\Omega/\ ^\circ\text{C} \quad \textbf{Ans.}$$

Substituting $R_1 = 3.146$ and $\alpha_0 = 0.00379$ in Eq. (*i*) we have

$$3.146 = R_0\ (1 + 40 \times 0.00379)$$

$$\text{or} \quad R_0 = \frac{3.146}{1 + 40 \times 0.00379} = 2.732\ \Omega \quad \textbf{Ans.}$$

Example 1.17. A coil has resistance of 18 Ω when its mean temperature is 20 °C and of 20 Ω when its mean temperature is 50 °C. Find its mean temperature rise when its resistance is 21 Ω and the surrounding temperature is 15 °C. *[Allahabad Univ. 1992]*

Solution : Let R_0 be the resistance of the coil at 0 °C and α_0 be its temperature coefficient of resistance at 0 °C

Resistance at 20 °C, $18 = R_0\ (1 + 20\ \alpha_0)$ ...(*i*)

Resistance at 50 °C, $20 = (1 + 50\ \alpha_0)$ ...(*ii*)

Dividing Eq. (*ii*) by Eq. (*i*) we have

$$\frac{20}{18} = \frac{1 + 50\,\alpha_0}{1 + 20\,\alpha_0}$$

$$\text{or} \quad \alpha_0 = \frac{1}{250} = 0.004$$

If t °C is the temperature of coil when its resistance is 21 Ω, then

$21 = R_0\ (1 + 0.004\ t)$ ...(*iii*)

Dividing Eq. (*iii*) by Eq. (*ii*) we have

$$\frac{21}{20} = \frac{1 + 0.004t}{1 + 50 \times 0.004}$$

$$\text{or} \quad t = 65\ ^\circ\text{C}$$

Temperature rise $= t -$ surrounding temperature $= 65 - 15 = 50$ °C **Ans.**

Example 1.18. A copper conductor has its specific resistance of 1.6 × 10^{-6} Ω-cm at 0 °C and a resistance temperature coefficient of $\frac{1}{254.5}$ per °C at 20 °C. Find (*i*) the specific resistance and (*ii*) the resistance temperature coefficient at 60 °C. *[Pune Univ. F.Y. Engg. Nov, 1988]*

Solution : Temperature coefficient of resistance at 20 ºC,

$$\alpha_{20} = \frac{1}{254.5} \ \Omega/\Omega/^\circ C$$

Temperature coefficient of resistance at 60 ºC,

$$\alpha_{60} = \frac{1}{\frac{1}{\alpha_{20}} + (t_2 - t_1)} = \frac{1}{254.5 + (60-20)} = \frac{1}{294.5} \ \Omega/\Omega/^\circ C \ \textbf{Ans.}$$

If α_0 is the temperature coefficient of resistance at 0 ºC, then temperature coefficient of resistance at 20 ºC

$$\alpha_{20} = \frac{1}{\frac{1}{\alpha_0} + 20}$$

$$\text{or} \quad \alpha_0 = \frac{1}{\frac{1}{\alpha_{20}} - 20} = \frac{1}{254.5-20} = \frac{1}{234.5} \ \Omega/\Omega/^\circ C$$

Specific resistance at 60 ºC, $\rho_{60} = \rho_0 \, (1 + \alpha_0 \, t)$

$$= 1.6 \times 10^{-6} \left(1 + \frac{60}{234.5}\right) = 2.0094 \times 10^{-6} \ \Omega\text{-cm} \ \textbf{Ans.}$$

Note : *From the above example it is obvious that the specific resistance of a given material does not remain constant, but changes with the change in temperature.*

Example 1.19. The specific resistance of platinum at 0ºC is 10.3 μΩ-cm. How long must a wire of 32 SWG (diameter =0.0274 cm) platinum be to have a resistance of 4 Ω at 0ºC ? What will be the resistance of the wire at 100 ºC if the temperature coefficient of resistance is 0.0038 per ºC ? *[Allahabad Univ. 1974]*

Solution : Resistance of platinum wire, R = 4 Ω

Area of x-section of wire, $a = \frac{\pi}{4} d^2 = \frac{\pi}{4} (0.000274)^2$

Specific resistance of platinum, $\rho = 10.3 \ \mu\,\Omega\text{-cm} = 10.3 \times 10^{-8} \ \Omega\text{-m}$

Length of wire required, $l = \frac{R \times a}{\rho} = \frac{4 \times \frac{\pi}{4} (0.000274)^2}{10.3 \times 10^{-8}} = 2.29$ metres **Ans.**

Example 1.20. Find the current flowing at the instant of switching on a 100 watt metal filament lamp on to a 220 V circuit. Given that the incandescent filament temperature is 2,000 ºC and the resistance temperature coefficient at room temperature of 20 ºC is 0.005. *[Pb. Univ. 1973]*

Solution : Resistance of 100 watt metal filament, $R_{100} = \frac{V^2}{W} = \frac{(220)^2}{100} = 484 \ \Omega$

Let the resistance of 100 watt metal filament at room temperature be R_{20}

Temperature coefficient of resistance at 20 ºC, $\alpha_{20} = 0.005$

$$\because \quad R_{100} = R_{20} \, [1 + \alpha_{20} \, (t_2 - t_1)]$$

$$\therefore \quad R_{20} = \frac{R_{100}}{1 + \alpha_{20} \, (t_2 - t_1)} = \frac{484}{1 + 0.005 \times (2{,}000 - 20)} = \frac{484}{10.9} = 44.4 \ \Omega$$

Current flowing at the instant of switching on a 100 W metal filament lamp,

$$I_{20} = \frac{V}{R_{20}} = \frac{220}{44.4} = 4.95 \text{ A} \quad \textbf{Ans.}$$

Example 1.21. Two coils connected in series have resistances of 600 Ω and 300 Ω and temperature coefficients of 0.1% and 0.4% per °C at 20 °C respectively. Find the resistance of combination at a temperature of 50 °C. What is the effective temperature coefficient of the combination at 50 °C?
[Pb. Univ. 1972 Supp.]

Solution : Resistance of 600 Ω resistance at 50 °C = 600 [1 + (50 – 20) × 0.001] = 618 Ω

Resistance of 300 Ω resistance at 50 °C, = 300 [1 + (50 – 20) × 0.004] = 336 Ω

Resistance of combination at 50 °C, R_{50} = 618 + 336 = 954 Ω **Ans.**

Resistance of combination at 20 °C, R_{20} = 600 + 300 = 900 Ω

Resistance temperature coefficient of combination at 20 °C,

$$\alpha_{20} = \frac{\frac{R_{50}}{R_{20}}-1}{t_2-t_1} = \frac{\frac{954}{900}-1}{50-20} = \frac{\frac{0.06}{30}}{} = 0.002 \quad \because R_{50} = R_{20}\,[1+\alpha_{20}(t_2-t_1)]$$

Effective temperature coefficient of combination at 50 °C

$$\alpha_{50} = \frac{1}{\frac{1}{\alpha_{20}}+(t_2-t_1)} = \frac{1}{\frac{1}{0.002}+50-20} = \frac{1}{530} \text{ per °C } \textbf{Ans.}$$

Example 1.22. A copper wire has a resistivity of 1.8 × 10⁻⁸ Ω-m at 300 K. Assuming that copper is pure, estimate the resistivity at 700°C and percentage change in resistivity from room temperature at 700°C.
[M.D. Univ. Third Semester May 2005]

Solution : Resistivity at room temperature of 300 K,

$\rho_1 = 1.8 \times 10^{-8}$ Ω-m

Initial temperature, t_1 = 300 K

Final temperature, t_2 = 700 + 273 = 973 K

Since resistivity is proportional to absolute temperature,

$$\frac{\rho_2}{\rho_1} = \frac{973}{300}$$

or $$\rho_2 = \frac{973}{300}\times 1.8\times 10^{-8} = 5.84 \ \Omega\text{-m} \quad \textbf{Ans.}$$

Percentage change in resistivity from room temperature to 700°C.

$$= \frac{5.84\times 10^{-8} - 1.8\times 10^{-8}}{1.8\times 10^{-8}}\times 100 = 224.44\% \quad \textbf{Ans.}$$

1.17.2. Effect of Alloying on Resistivity

Alloying is another factor which affects the resistivity of a material. By adding traces of some impurities to a metal, its resistivity increases. Alloys have higher resistivity than pure base metal *e.g.* pure copper has $\rho = 1.7 \times 10^{-8}$ Ω-m while its alloy brass (60% Cu + 40% Zn) has $\rho = 7 \times 10^{-8}$ Ω-m at room temperature. Alloying elements greatly increase the lattice imperfections that in turn increases the scattering of electrons and phonons and hence the resistivity of metal increases with alloying.

The effect of impurity of alloying on resistivity may be determined by Nordion equation as

$$\rho_A = \rho_m + C\rho_l \qquad \text{...(1.60)}$$

where ρ_A, ρ_m and ρ_l are resistivities of alloy, base metal and alloying element. C is the concentration of alloying element in atomic percent. The above equation is applicable only for few binary solid solutions because it does not take into consideration the changes in the density of states with composition.

The increase in resistivity with alloying elements can be explained as :

(*a*) Atoms of different sizes cause a variation of lattice parameters which result into scattering of electrons.

(*b*) The valence difference between atoms of base metal and alloying metal introduces a local charge difference that increases the scattering.

(*c*) Alloying element having different electron concentration than base element changes the position of Fermi energy and hence the resistivity of the material, according to the following equation

$$\frac{1}{\rho} = \sigma = \frac{1}{3} e^2 V_f^2 \tau n(E_f) \quad ...(1.61)$$

1.17.3. Effect of Mechanical Stressing on Resistivity

The resistivity of a material also changes under the influence of mechanical treatment. The fabrication of conductor from the ingot to the final stage comprises initial hot working and final cold drawing. Cold-working operation (stressing) distorts the crystal structure of the metal. This generally tends to harden the material, increase its tensile strength and increase slightly its resistivity. The increase in tensile strength is very useful for many purposes such as overhead conductor. That is why many types of conductors are finally drawn in cold stage in which case they are identified as hard drawn. Although mechanical stressing increases the resistivity i.e., decreases the conductivity, annealing (heat treatment process) restores the electrical conductivity by establishing regularity in crystal structure. When a metallic conductor is strained, then there is a change in its resistance due to change in the length and cross-sectional area of the conductor.

The use of this effect of strain on the resistance of metallic conductors is utilised in manufacturing *strain gauges*. There are numerous applications of strain gauges such as measurement of stress, force, tension, torque in structures.

1.17.4. Effect of Age Hardening on Resistivity

The electrical resistivity of an alloy increases with age hardening. During this process crystal lattice undergoes distortion due to which electron mobility is decreased which in turn decreases the mean free path.

Example 1.23. The resistivity of copper rod as measured by Kelvin double bridge is 1.74 microhm-cm at 20ºC. If the resistivity of the copper is 1.73 microhm-cm at 20ºC, calculate the impurity (%) content in the rod. *[I.E.S. 2004]*

Solution : The percent impurity is given as

$$\% \text{ impurity} = \frac{\sigma - \sigma'}{\sigma} \times 100 = \frac{\frac{1}{\rho} - \frac{1}{\rho'}}{1/\rho}$$

where σ and σ′ are conductivities of pure metal and metal with impurity respectively.

$$\text{So } \% \text{ impurity} = \frac{\frac{1}{1.73} - \frac{1}{1.74}}{\frac{1}{1.73}} \times 100 = 0.5747\% \quad \textbf{Ans.}$$

1.18. ELECTRICAL CONTACT MATERIALS

The term electrical contact means a releasable junction between two conductors which is apt to carry current. Materials used for making contacts operate under the severest condition when the contacts serve to make and break electric circuits very frequently.

Depending on the power level of the circuits, contact materials are classified as *lightly* and *moderately* loaded when the current does not exceed 1 A and the voltage causing arc is of the order of 10 to 20 V. At higher power levels, the materials are classed as heavily-loaded. An important requirement for the materials is *stability* of contact resistance.

A peculiarity of circuit breaking contacts is that they have to withstand arcing or spark-overs whenever separated or brought together. Hence they deteriorate with time due to (*i*) corrosion resulting from oxidation and other chemical reactions owing to contact with aggressive constituents in the surrounding medium; (*ii*) errosion from fusing, evaporation and wear of the working surfaces during service. Erosion is usually associated with a certain carry-over of the material from one contact to the other, especially in dc circuits. Because of corrosion, contact surfaces usually acquire a film of oxide which has low conductivity and reduces the effectiveness of electrical contact. However, *poor contacting* can be avoided by using a contact-to-contact pressure high enough to break up the film each time contact is established. Under the influence of erosion and carry-over, growths can sometimes appear on one contact and craters on the other. Contacts designed to carry heavy currents may become welded to one another under unfavourable conditions. This may occur even when they are closed if an extra heavy current surges through the circuit. In low-power contacts, the surface may undergo sintering. Contacts operate more satisfactorily in a vacuum than in a gaseous medium as vacuum eliminates any possible arcs between them.

Lightly-loaded contacts are often made of pure noble metals such as platinum, palladium, silver and gold. Tungsten and molybdenum are also employed. Platinum does not oxidise in air and has no tendency to arc, but it may form bridges and 'needles' at low currents. It is often alloyed with iridium for making light contacts of extra-high reliability. Palladium has properties similar to those of platinum. Being much cheaper, it is often used to substitute platinum, but it is less resistant to cathodic erosion and oxidation in air. Alloys of palladium with silver are extensively employed. Gold readily supports arcing and is subject to considerable carry-over erosion. It is mainly employed in alloys with platinum, silver and nickel. Rhodium is an excellent contact material in light and precise devices. When pure silver is employed, its tendency to arc should be taken into consideration. Carry-over between silver contacts is less than between platinum or gold contacts due to oxidation of silver in air under the action of electric arcs. As silver oxides readily dissociate at a relatively low temperature (about 200ºC), the stability of contact resistance is only slightly affected by them. Nevertheless it is not recommended to use silver for contacts in light and precise devices with low contact pressure. Otherwise silver is extensively employed as a contact material, either alone or alloyed with copper. Silver readily combines with sulphur, therefore, measures should be taken to keep it from coming into contact with sulphur-bearing substances, rubber for instance.

Tungsten is one of the most widely used contact materials. Like platinum, it shows no tendency to arc, is less susceptible to erosion and carry-over than platinum, is very hard, has a high melting point (3,380ºC), and will not weld during service. The main drawback with tungsten is that atmospheric oxygen oxidizes it. Because of this drawback, a certain minimum pressure is required to be provided between contacts made of it. Tungsten contacts operating in an atmosphere of high relative humidity and temperature typical of moist tropical regions should not be located near phenol-bearing materials such as bakelized paper or laminates bonded with phenolic resins, because the gaseous products they evolve strongly corrode the tungsten. Forged bars of tungsten have a longitudinal fibrous structure and contacts cut from them will have larger wear resistance than those die-pressed from sheet tungsten. Tungsten contacts are used up to 200 V (ac and dc both) and up to 15 A. Its typical applications are: battery ignition systems, magneto-ignition systems, vibrators, electric razors, automatic horn contacts.

Molybdenum erodes faster than tungsten and corrodes in atmospheric conditions, forming a loosely packed oxide which may hinder close electrical contact.

Lightly-loaded non-arcing contacts can be made by electro-deposition of certain noble and high-cost metals on springs and contact surfaces, thus reducing the expenditure of these metals without any sacrifice in efficiency. Moreover, the metals deposited in this way have greater hardness and wear resistance than when in solid form. The metals employed for such purpose are silver, platinum, gold, palladium and rhodium.

Apart from metals, contacts for all power levels are also made from metal ceramics called the *cermet*. As compared to metal contacts, cermet contacts have the advantages of greater resistance to fusion, welding and wear. For example, for dc and ac from 0.5 to 4 A and voltages from 2 to 100 V better results were shown by cermet contacts of silver and nickel and of silver and tungsten than by silver and its alloys.

Moderately-loaded contacts are made from many of the materials used for lightly-loaded contacts, except platinum and palladium, various alloys of gold and electro-deposited metals.

Heavily loaded contacts are made from the metallic materials listed in Table 1.3, which also indicates their main fields of application.

TABLE 1.3. ***Highly-Loaded Electrical Contact Materials***

Contact Material	Contact Operating Conditions Main Application	Types of Electrical Apparatus
Silver; silver-palladium, Silver copper; silver cadmium	Voltages up to 500 V; Arcing during opening, High contact pressure	Current breakers of medium current rating
Silver; Copper cadmium	Main contacts	Air-break circuit breakers
Silver; Copper	Main contacts	Oil-break circuit breakers
Silver; Silver-copper; Copper; Copper-cadmium Copper silver	Medium and heavy alternating currents and voltages up to 500 V. Main contacts.	Industrial-control contactors and magnetic starters
Copper-cadmium	Arcing contacts	Same as above
Silver	Low-voltage, dc	DC contactors.

Circuit-breaking contacts are usually made of bimetallic materials in which the working layer consists of the main contact material and the support layer of copper, nickel, steel, etc.

Apart from the metals and alloys listed in Table 1.3, heavily loaded contacts are made from following cermets: Silver-cadmium oxide; silver-nickel; silver-graphite; silver-tungsten; silver-molybdenum; silver-tungsten carbide; silver-lead oxide; silver-cadmium-nickel; copper-tungsten; copper-graphite.

The characteristics of materials serving as circuit-breaking contacts are given below in tabular form (Table 1.4).

TABLE 1.4. ***Characteristics of Materials Used For Circuit-breaking Contacts***

Material	Type	Melting point, °C	Density in Kg/m^3	Resistivity in μ Ω-m	Hardness by Ball Test after Annealing in kg/mm^2
Platinum	Metal	1,770	21,400	0.105	50
Palladium	Metal	1,555	12,000	0.107	60
Rhodium	Metal	1,966	12,400	0.045	100
Tungsten	Metal	3,380	19,300	0.065	350
Silver	Metal	961	10,500	0.016	25
Platinum-iridium	Alloy	1,790 to 1,890	21,600 to 21,800	0.27 to 0.35	190 to 244
Silver-cadmium (20% cadmium)	Alloy	850	10,300	0.057	55
Silver-tungsten	Cermet	—	12,100 to 17,000	0.021 to 0.041	75 to 240
Silver - nickel	cermet	—	10,200 to 9,250	0.021 to 0.031	50 to 70

1.19. Fuse and Fuse Materials

Fuse is perhaps the simplest and cheapest device used for interrupting an electrical circuit under short-circuit, or excessive overload. As such, it is used for overload and (or short-circuit protection in high voltage (up to 66 kV) and low voltage (up to 400V) installations/circuits. In high voltage circuits their use is confined to those applications where their performance characteristics are particularly suitable for current interruption.

The action of a fuse is based upon the heating effect of the electric current. In normal operating conditions, when the current flowing through the circuit is within safe limits, the heat developed in the fuse element carrying this current is readily dissipated into the surrounding air and, therefore, fuse element remains at a temperature below its melting point. However, when some fault, such as short-circuit occurs or when load connected in a circuit exceeds its capacity, the current exceeds the limiting value. The heat generated due to this excessive current cannot be dissipated fast enough and the fusible element gets heated, melts and breaks the circuit. It thus protects a machine or apparatus or an installation from damage due to excessive current.

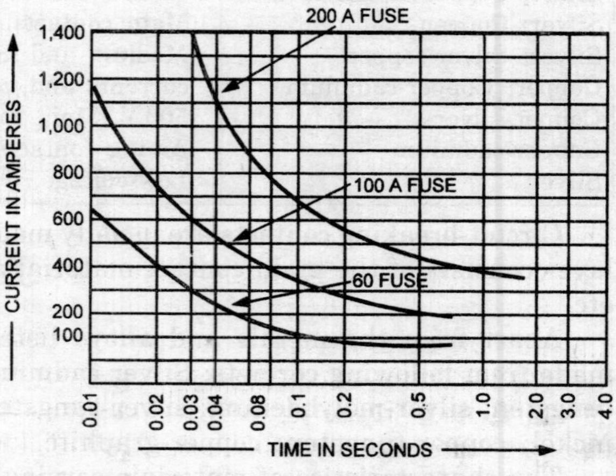

Fig. 1.24. *Time Current Characteristic*

The time for blowing out of fuse depends upon the magnitude of the excessive current. Larger the current, the more rapidly the fuse will blow i.e., the fuse has inverse time-current characteristic, as shown in Fig. 1.24. Such a characteristic is desirable for protective gear.

Essentially, a fuse consists of a fusible element in the form of a metal conductor of specially selected small cross-sectional area, a case or cartridge to hold the fusible element, and in some cases, provided with a means to aid arc extinction. The part which actually melts and opens the circuit is known as the *fuse element*. It forms a series part of the circuit to be protected against short-circuit or excessive overloads.

The function of fuse wire is (*i*) to carry the normal working current safely without heating and (*ii*) to break the circuit when the current exceeds the limiting current.

The materials used for fuse elements must be of low melting point, low ohmic loss, high conductivity (or low resistivity), low cost and free from deterioration. Experience has shown that the most generally suitable material for the fuse element is a low melting point material such as tin, lead or zinc. A low melting point is, however, available with a high specific resistance metal (refer Table 1.5).

A high specific resistance metal entails, for a given rating of fuse element, a greater mass metal than that would be required if the elements were made from high melting point metal of low specific resistance. Thus the use of low melting point metals introduces the problems of handling large mass of vaporised metal set free on fusion.

Oxidation of the fuse element tends to generate an outer layer which may be strong enough to support the metallic core of the fuse element when in a molten state. The effect is more pronounced with certain high melting point metals such as copper and iron than it is with low melting point metals such as tin or lead.

The above discussion reveals that no material possesses all the desirable characteristics. For instance lead has low melting point but it has high specific resistance and it is liable to oxidation. Similarly, copper has high conductivity and low cost but oxidises rapidly. Thus a compromise is required to be made in the selection of material for a fuse element.

The materials commonly used for fuse elements are tin, lead, silver, copper, zinc, aluminium and alloy of lead and tin. An alloy of lead and tin (lead 37% and tin 63%) is used for small current rating fuses (say not beyond 15 A). For current exceeding 15 A this alloy is not used as the diameter of the wire will be larger and after fusing the metal released will be excessive. Beyond 15 A rating circuits copper wire fuses are employed. Copper wire suffers from the drawback that it operates at a rather high temperature if a reasonably low fusing factor is desired. There is thus a tendency for the wire to overheat with the result that its cross-sectional area and fusing currents are gradually reduced and premature melting of the wire may occur. It is usual practice to tin the copper for its protection against oxidation. Zinc (in strip form only) is good if a fuse with considerable time-delay is required *i.e.*, one which does not melt very quickly with a small overload.

The present trend is to use silver as fuse element material despite its higher cost owing to its following advantages.

(*i*) It does not get oxidised and its oxide is unstable.

(*ii*) It does not deteriorate when used in dry air and it remains bright but when the air is moist and contains hydrogen sulphide, the silver surface is attacked and a layer of silver sulphide is formed at the top which shields the metal from further attack.

(*iii*) Owing to its very small coefficient of expansion no critical fatigue occurs and thus continuous full normal current capacity rating is assured for a long time.

(*iv*) The conductivity of silver does not deteriorate with oxidation.

(*v*) Owing to its high conductivity the mass of molten metal to be handled is minimum and thus operating speed is fast.

(*vi*) Vaporization temperature of silver is very much below the one at which the vapour will readily ionize. This means that on the formation of an arc through the vaporized portion of the element, the arc path has comparatively high resistance and the short-circuit current is quickly reduced.

(*vii*) Owing to its low specific heat silver fusible elements can be raised from normal operating temperature to vaporization very much quicker than other fusible elements. Also as the temperature of the silver reaches the melting point its resistance increases abruptly and thus the transition from melting to vaporization becomes nearly instantaneous. As a result, operation becomes very much faster at higher currents.

Either copper or lead-tin alloy is mostly used as an ordinary fuse wire.

TABLE 1.5

Metal	Melting Point in °C	Specific Resistance in μΩ-mm	Value of Fuse Constant *k* For *d* in m
Silver	980	16	—
Tin	240	112	12.8
Zinc	419	60	—
Lead	328	210	10.8
Copper	1,090	17	80
Aluminium	665	28	59

1.20. BIMETALS

A bimetal is made of two metallic strips of unlike metal alloys with different coefficients of thermal expansion. At a certain temperature the strip will bend and actuate a switch or a lever of a switch. The bimetal can be heated directly or indirectly. When heated the element bends so that metal having larger coefficient of expansion is on the outerside of the arc formed while that with lower coefficient is on the inner side. When cooled the element bends in the opposite direction. Alloys of iron and nickel with low coefficients of thermal expansion are employed as one element of the bimetallic strip. The other element of the bimetallic strip consists of materials having high value of coefficient of thermal expansion like iron, nickel, constantan, brass etc.

Bimetallic strips are employed in electrical apparatus and in devices such as regulators and relays. For example, a simple bimetallic regulator may be used in a heater to maintain a constant temperature. A bimetal relay or release can be employed for overload protection of electric motors or any circuit. In such a case, the circuit current is made to flow through the bimetallic strip. When the current exceeds its setting (many bimetal devices are adjustable for a particular current), the strip will get heated enough to bend and break the circuit directly or through an intermediate relay.

1.21. SOLDER MATERIALS

Solder is an alloy used to join two or more pieces of materials. The melting point of a solder is lower than that of the materials to be joined. The molten solder joins the pieces of metals and the process is known as *soldering.* Solders used for electrical purposes can be divided into two general classes: *soft solders* (with a melting point not exceeding 400°C), and *hard solders* (with a melting point exceeding 400°C). In selecting a solder, an important characteristic is the tensile strength. It is up to 5-7 kgf/mm^2 for soft solders, and can be as high as 50 kgf/mm^2 for hard solders.

A soft solder is an alloy of tin and lead. The most popular composition is 50% tin and 50% lead. Its melting point is about 185°C. Its tensile strength is 3.85 kgf/mm^2 and electrical conductivity is about 10% of that of copper. The tin-lead solders serve to join copper, bronze, brass and other metals.

A hard solder is an alloy of copper and zinc. It melts at a very high temperature and is used for joining brass, copper, iron and steel. There are two types of hard solders namely, brazing solder and silver solder. Brazing is a soldering at high temperature. The non-ferrous filler metal is used and has a melting point less than the base metals. The filler metal is distributed in the joint by capillary action.

The most important application of soft solders is in electronic devices and of hard solders in power apparatus for making permanent connections.

1.22. Electrode Materials

When in an electrical circuit, a metal rod is dipped in a solution of its salt, the metal rod is called *electrode*. The widespread use of these electrodes is in the emf cells and in electroplating industry. The electrode at which electrons leave the cell is called the *anode* and the one at which electrons enter the cell is called the *cathode*. Alkali and alkali earth elements can give up electrons more easily than the transition series elements and are said to be more anodic than the transition series elements. Zn and Mg are generally used as anode materials. A more widely used material is magnesium alloys because of low cost. Lithium is the most active and base metal and will go into solution readily. Gold is the most noble metal and it will not dissolve easily.

1.23. Electrical Carbon Materials

Electrical carbon materials are manufactured from graphite and other forms of carbon (coal, etc.). These all consist of practically pure carbon, which is a semiconductor and consequently possesses a negative temperature coefficient of resistivity. However, the carbon used has a conductivity less than that of the metals and their alloys. It is pressure sensitive (*i.e.* electrical resistance of the carbon contact decreases with the increase in pressure) and has low surface friction.

Carbon is used in applications like brushes for electrical machines and apparatus, electrodes for electric arc furnaces, and electrolytic baths, carbon pile resistances, non-wire resistors, membranes and other components for telecommunication equipment, battery cell elements, arc lamps, arc welding, components for vacuum valves and tubes (anodes, grids) etc. An important application of carbon is as carbon brushes for electrical machines and apparatus. Carbon brushes should possess mirror-smooth surface and adequate conductivity, should not wear out rapidly and should not cause mechanical damage due to friction to the surface on which they are mounted.

There are carbon-graphite, graphite, electrographitised (thermally processed in special electric ovens with the brush carbon transformed into crystalline graphite), copper-graphite (with copper added) and bronze-graphite brushes. The last two types have an especially low resistance and result in a small contact voltage drop between the brush and the commutator. The proper type of brush must be selected for reliable operation of the machine.

Another important application of carbon is as carbon resistors.

1.24. Thermal Conductivity of Metals

We know that in a solid, atoms vibrate about their equilibrium position at ordinary temperature and the amplitude of vibration decreases to a minimum at absolute zero of temperature. This vibration of atoms brings about elastic distortion in the lattice and, therefore, the waves associated with the vibrations are the elastic waves of varying frequencies. The waves superimpose to form a wave packet that moves through the solid. The quantum of elastic energy is known as *phonon*, whose energy is given by $E = h\nu$; ν being frequency of vibration. The increasing number of vibration of atoms with temperature is represented by increasing number of phonons passing through the crystal. It is the phonons together with valence electrons which are responsible for the transfer of thermal energy from high temperature region to low temperature region. The property that characterizes the ability of a material to transfer thermal energy (or hence the heat) is known as *thermal conductivity*. The thermal conductivity of a solid is proportional to its specific heat and to the mean free path of phonons and electrons. The mean free path of electrons is very large in comparison to the mean free path of phonons(10 to 100 Å), as a result, the electrons undergo less number of collisions. Since in metals the free electrons are sufficiently mobile, metals have shown high thermal conductivity due to the mobility of electrons and phonons as well. On the other hand, in insulators, phonons are the only particles to conduct heat, as the valence electrons are not present (ionic solids or covalent solids). In case of alloys, electrons remain scattered due to the presence of impurity elements, the thermal conductivity of alloys is reduced.

Conclusions: It has been observed that

(*i*) The thermal conductivity of metals decreases at higher temperatures.

(*ii*) The thermal conductivity of polycrystalline metals is lower than that of a single crystal due to scattering of electrons and phonons along grain boundaries.

(*iii*) The thermal conductivity of metals is lower than their alloys due to scattering of electrons and phonons with the atoms of alloying element.

(*iv*) The thermal conductivity of semiconducting materials increases with increasing temperature due to flow of excited electrons.

(*v*) The thermal conductivity of metals, alloys, semiconductors and dielectrics is in the descending order.

1.24.1. Wiedemann Franz Lorentz Relation

Suppose a homogeneous isotropic material is subjected to a temperature gradient $\frac{dT}{dx}$. The flow of heat will result in the direction opposite to the temperature gradient through the conducting medium. The heat flux Q (heat flow per unit time per unit area) will be proportional to the temperature gradient *i.e.* $Q \propto \frac{dT}{dx}$

$$\text{or } Q = -K\frac{dT}{dx} \quad \text{...(1.62)}$$

where K is the proportionality constant and is known as coefficient of thermal conductivity. If Q is expressed in W/m^2 and $\frac{dT}{dx}$ in K/m, the units of K will be W/mK.

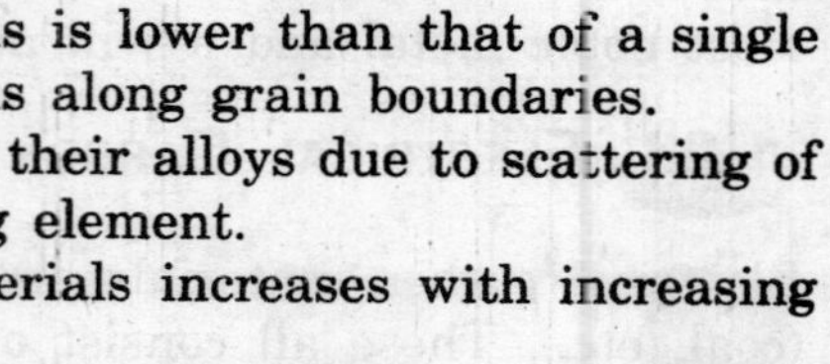

Fig. 1.25

As discussed earlier, the transformation of heat in solids is due to phonons and free electrons. Hence, the coefficient of thermal conductivity K can be written as

$$K = K_{phonon} + K_{electron}$$

In order to derive the expression for K, let us consider the heat flow from high temperature to low temperature in a metal slab having temperature gradient $\frac{dT}{dx}$.

Let C_v be the heat capacity, the heat transfer per unit area per second will be

$$Q = \frac{mnv}{3} C_v \frac{\lambda dT}{dx} \quad \text{...(1.63)}$$

where v is the velocity of electrons and λ being mean free path of collisions.

Also, heat flux $Q = K\frac{dT}{dx}$

$$\therefore K\frac{dT}{dx} = \frac{mnv}{3} C_v \lambda \frac{dT}{dx}$$

$$\text{or } K = \frac{mnv}{3} C_v \lambda \quad \text{...(1.64)}$$

The energy of free electron is given by

$$mC_vT = \frac{3}{2} K_B T, \text{ where } K_B \text{ is Boltzmann constant}$$

$$\therefore \text{ Thermal conductivity } K = \frac{3}{2} K_B . \frac{nv}{3} . \lambda$$

$$\text{or } K = K_B \left(\frac{nv\lambda}{2} \right) \quad \text{...(1.65)}$$

Specific heat at constant volume for an ideal gas is

$$C_v = \frac{3}{2} n K_B$$

$$\therefore K = \frac{1}{3} C_v \lambda v \quad \text{...(1.66)}$$

Above expression implies that the thermal conductivity of solids depends upon (*i*) specific heat, (*ii*) mean free path of collisions (*iii*) velocity of electrons.

Now consider the electrical conductivity σ.

$$\sigma = \frac{ne^2\tau}{m} \quad \text{...Refer to Eq. (1.47)}$$

and relaxation time (collision time) τ

$$\tau = \frac{\lambda}{v_d}$$

$$\text{Also } \frac{1}{2} m v_d^2 = \frac{3}{2} K_B T$$

$$\therefore \sigma = \frac{ne^2\lambda}{mv_d} = \frac{ne^2\lambda v_d}{3K_B T} \quad \text{...(1.67)}$$

Therefore, the ratio of thermal conductivity K to electrical conductivity σ is

$$\frac{K}{\sigma} = \frac{K_B n v \lambda}{2} \times \frac{3K_B T}{ne^2 \lambda v_d} = \frac{3}{2} \frac{K_B^2}{e^2} . T, \text{ if we assume } v = v_d \quad \text{...(1.68)}$$

$$\text{or } \frac{K}{\sigma T} = 5.838 \times 10^{-9} \ \Omega \text{ cal/K-sec} = 2.44 \times 10^{-8} \ \Omega \text{ W/°K}^2$$

which indicates that the ratio $\frac{K}{\sigma}$ is same for all metals and is a function of temperature only. This empirical law is known as Wiedemann-Franz Lorentz relation. Hence, we can say that best electrical conductor will be a best thermal conductor.

1.24.2. Heat Developed in Current Carrying Conductor

Joule observed that the heat developed in a conducting wire is given by I^2R; where I is current flowing through the wire having resistance R. If ρ be the electrical resistivity of wire, l being length of wire and area of cross-section is A, then

$$I^2R = \frac{V^2}{R} = \frac{(El)^2}{\rho\frac{l}{A}} = \sigma E^2.l\,A \qquad ...(1.69)$$

where V is the applied potential and E is the electric field developed across the wire of length l. Therefore, the heat developed per unit volume (lA) per second is

$$W = \sigma E^2 = JE \qquad (\because J = \sigma E) \qquad ...(1.70)$$

where J is the current density. If J is in amperes per m^2 and E is in volts per m; the units of W will be Watts per m^3.

As discussed earlier, the electrons in a conductor exert an acceleration 'a' on application of an electric field E and is given by

$$a = \frac{e}{m}E$$

The acceleration increases the velocity of the electron by $\frac{l}{m}Et$. Let v_x, v_y and v_z are components of velocity v along x, y and z directions at $t = 0$ *i.e.* electrons do not collide. At $t > 0$, the velocity components will be $v_x + \frac{e}{E}Et$, v_y and v_z if the electric field is applied along x-axis.

The increase in energy of the electron over the period t is

$$\Delta W = \frac{1}{2}m\left(v_x + \frac{e}{m}Et\right)^2 + \frac{1}{2}mv_y^2 + \frac{1}{2}mv_z^2 - \frac{1}{2}mv_x^2 - \frac{1}{2}mv_y^2 - \frac{1}{2}mv_z^2$$

$$= \frac{1}{2}m\left[\frac{2e}{m}Ev_x t + \left(\frac{e}{m}Et\right)^2\right] \qquad ...(1.71)$$

When averaged over a large number of electrons, the above expression becomes

$$\Delta W = \frac{1}{2}\frac{e^2E^2t^2}{m} \qquad ...(1.72)$$

Because v_x is zero for such a group of electrons. Here we have considered no collision between electrons.

The average energy increase of the electrons during the period between two collisions is given by

$$\Delta W = \int_0^{\infty} \frac{e^2E^2t^2}{2m}\frac{dt}{\tau}; \; \tau \text{ being collision time} \qquad ...(1.73)$$

The term $\frac{dt}{\tau}$ represents the probability that an electron will suffer a collision during a time dt.

$$\therefore \qquad \Delta W = \frac{e^2E^2\tau^2}{m}, \text{ for isotropic conductor } t = \tau \qquad ...(1.74)$$

Let n be the density of electrons and assuming that these electrons transfer their excess energy to the lattice; the total energy dissipated per unit volume per second will be

$$W = \frac{n}{\tau}\Delta W = \frac{ne^2\tau}{m}E^2 = \sigma E^2 \qquad ...(1.75)$$

This is the energy which is converted into heat during collision process.

1.24.3. Thermionic Emission From Metals

We know that moving electrons in a metal exert pressure similar to that of a gas and these moving electrons cannot escape to the outer space because of the attractive forces at the surface that tend to keep the electrons within the metal. The attractive forces are greater than the electron gas pressure. In order to escape from the surface of metal, an electron has to do a certain amount of work called *work function* ϕ of the surface and this is different for different metals. At ordinary temperature, the kinetic energy of the electrons is much less than the work function and hence the electrons cannot escape from the surface. With the increase in temperature of the metal surface, the kinetic energy of the electrons increases and if kinetic energy exceeds the work function, electrons can escape from the surface of the metal. This phenomenon is called *thermionic emission* and electrons are called *thermo-electrons* or *thermions*. The thermionic current I, is given by

$$I = AT^2 \exp\left(-\frac{\phi_0}{K_B T}\right) \quad \text{...(1.76)}$$

where A is a constant, T is the absolute temperature of the metal, ϕ_0 is the work function of the surface at absolute zero and K_B is Boltzmann constant. I measures the thermionic current in amperes per square metre of the emitting surface. The most commonly used emitters are (*i*) tungsten, (*ii*) thoriated tungsten and (*iii*) oxide coated cathodes containing barium or strontium.

The above equation is known as Richardson's equation for thermionic emission. The electron emitters are heated directly or indirectly. In case of direct heating, the current is passed through the filament which itself serves as the cathode. The indirectly heated electron emitters consist of a heater wire surrounded by a metal sleeve in the form of a cylinder whose surface is coated with electron emitting materials. The operating temperature for tungsten, thoriated tungsten and oxide coated emitters are respectively 2,200–3,000 K, 1,900 K and 1,000–1,150 K.

TABLE 1.6. ***Values of the Emission Constants***

Material	Work function ϕ_0 (eV)	A (A/m^2K)	Melting point (K)
Nickel	5.0	2.68×10^5	1,725
Platinum	5.32	3.20×10^5	2,047
Tantalum	4.1	6.02×10^5	3,123
Tungsten	4.52	6.02×10^5	3,655

1.25. Electron Scattering in Conductors

The resistivity of a perfect, pure single crystal of a metal goes to zero as temperature T approaches zero Kelvin (–273 °C). In fact, it can be shown theoretically that when the wave nature of the electrons is considered, an electron moving in a perfectly periodic lattice would not be scattered at all. In other words, the waves representing a moving electron in a perfect crystal resemble the electromagnetic waves corresponding to visible light which passes undeviated through a crystal such as sodium chloride. Thus, the present interpretation of the electrical resistivity of metals is based on the notion that *scattering of electrons* is due to deviations from perfect periodicity of the lattice. In pure metals, such deviations may be caused, for example, by the vibrations of the atoms in the lattice, they may also be caused by atoms which occupy wrong positions in the lattice, or by lattice sites which are unoccupied, or by impurity atoms. Thus, there are various kinds of obstacles which we have to consider in the scattering of electrons in metals, each of them giving rise separately to ascertain relaxation time τ. Since the probability for scattering by a particular kind of obstacle per unit time is equal to $1/\tau$ (for isotropic scattering) the resultant relaxation time may be obtained from the relation

$$\frac{1}{f} = \sum \frac{1}{\tau} \qquad \text{...(1.77)}$$

since scattering probabilities are additive.

At temperatures above so called Debye temperature of the material, the energy of the lattice vibration increases linearly with temperature T, and constantly, the resistivity increases linearly with T in the region, obeying the relation

$$\rho = \rho_i + \alpha T \qquad \text{...(1.78)}$$

where ρ_i and α are constants.

The separation of the resistivity into a temperature-independent component and a temperature-dependent component is referred to as *Matthiessen's rule*. The Debye temperature may be considered as a measure for the maximum frequency of the lattice vibrations, hard materials have higher Debye temperatures than soft materials, as may be seen from the values given in table below:

Metal	Ag	Al	Au	Cu	Fe	Pb
θ_D	215	398	180	315	420	88

The Debye temperature plays an important role in specific heat of a material, in that above the Debye temperature the specific heat drops rapidly, approaching zero as $T \to 0$.

1.26. SUPERCONDUCTIVITY

The resistivity of most metals increases with increase in temperature and *vice-versa*. There are some metals and chemical compounds whose resistivity becomes zero when their temperature is brought near 0 Kelvin (– 273°C). At this stage such metals or compounds are said to have attained *superconductivity*. For example, Mercury becomes superconducting at approximately 4.5 Kelvin (– 268.5°C). Superconductivity was discovered by Heike Kamerlingh Onnes at the University of Leiden in the Netherlands in 1911. The transition from normal conductivity to superconductivity takes place almost suddenly; it occurs over a very narrow range of temperature–about 0.05 K. The temperature at which the transition takes place from the state of normal conductivity to that of superconductivity is called *transition temperature.*

There are two types of superconductors commonly known as Type I and Type II superconductors. Type I superconductors are soft superconductors. They are usually pure specimens of some elements *i.e.* metals. They have very little use in technical applications. Type II superconductors are hard superconductors. They are usually alloys of metals with high value of resistivity in normal state. These are very useful as compared to Type I materials.

Experiments have shown that if a current is induced in a mercury ring at a temperature of 4.5 K, it will continue to flow for years without taking any power from the source of supply. Similarly a lead ring carried a current of several hundred amperes over a year with no change. Lead becomes superconducting at 7.22 K.

1.26.1. Superconducting Materials

Many metals and compounds have superconducting property at very low temperatures. Superconductivity has been observed to occur in poorer metallic conductors such as tin, lead and tantalum rather than in good conductors such as gold, silver and copper.

It has been found that superconductors may not only be pure metals but various alloys and chemical compounds as well. At present about 30 superconductor metals and more than 600 superconductor alloys are already known. The highest temperature at which, until now, superconductivity has been observed to occur is 20 K (–253°C) for a compound consisting of Niobium, Aluminium and Germanium.

1.26.2. Effect of Magnetic Field on Superconductivity

Application of a sufficiently strong magnetic field to superconductors causes the destruction of their superconductivity, *i.e.,* the restoration of their normal conducting state. The critical value of the magnetic field for the destruction of superconductivity is denoted by H_c and is functionally related to temperature as

$$H_c = H_c(0)\left[1 - \frac{T^2}{T_c^2}\right] \quad ...(1.79)$$

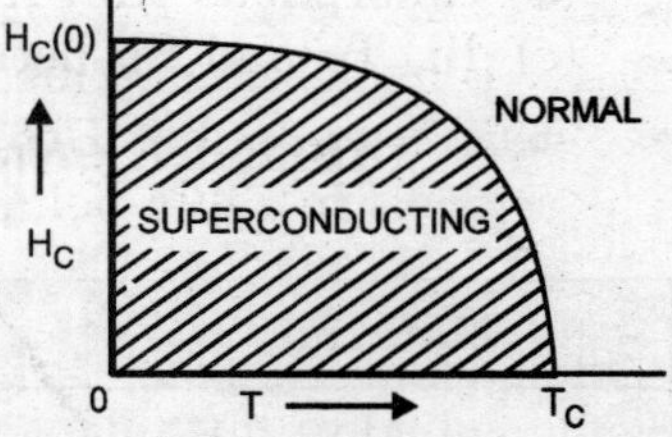

Fig. 1.26. ***Variation of Critical Field H_c With Temperature***

where H_c (0) is the critical field at 0 K, and has a specific value for each material. Note that the lower the temperature, the higher the value of H_c and the highest critical temperature occurs when there is no magnetic field, *i.e.,* $T = T_c$ when $H_c = 0$. A typical plot of critical magnetic field versus temperature for lead is shown in Fig. 1.26. Thus we find that the superconducting state is stable only in some definite ranges of magnetic fields and temperatures. For higher fields and temperatures, the normal state is more stable.

Some of the important properties of superconductors are as follows:

(*a*) The current in the superconductors persists for a very long time. This is demonstrated by placing a loop of the superconductor in a magnetic field, lowering its temperature below transition temperature T_c and then removing the field. The current which is set up is found to persist over a period longer than a year without any attenuation.

(*b*) The magnetic field does not penetrate into the body of the superconductor. The property known as the *Meissner effect,* is the fundamental characterization of superconductivity. However, when the magnetic field H is greater than a critical field H_c, the superconductor becomes a normal conductor.

(*c*) When a current through the superconductor is increased beyond critical value I_c (T), the superconductor again becomes a normal conductor, *i.e.,* the magnetic field which causes a superconductor to become normal from a superconducting state is not necessarily an external magnetic field, it may arise as a result of electric current flow in the conductor. The superconductivity may be destroyed when the current exceeds the critical value I_c, which at the surface of the wire will produce a critical field H_c given by

$$I_c = 2\pi r H_c \quad ...(1.80)$$

This is known as **Silsbee's rule.**

(*d*) The specific heat of the material shows an abrupt change at $T = T_c$ jumping to a large value for $T < T_c$.

(*e*) In all cases involving transition metals, the variation of T_c with number of valence electrons shows sharp maxima for $Z = 3$, 5 and 7.

(*f*) A rather striking correlation exists between T_c and Z^2 for elements Hg, La, Pb, Nb, Zn, Tl, In, Sn, V, Tc, Cd, Ga and Al.

(*g*) For a given value of Z, certain crystal structures seem more favourable than others, *e.g.,* β-tungsten and α-Mn structures are conducive to the phenomenon of superconductivity.

(*h*) T_c increases with a high power of the atomic volume and inversely as the atomic mass and is known as ***isotope effect.***

(*i*) Superconductivity occurs in materials having high normal resistivities. The condition $n\rho > 10^6$ is a good criterion for the existence of superconductivity, where n is number of valence electrons per cc and ρ is the resistivity in use at 20°C.

If one observes the total list of superconducting materials (some of them are listed in Table 1.7), the general features to be noted are :

(*a*) Monovalent metals are generally not superconductors.

(*b*) Ferromagnetic and antiferromagnetic metals are not superconductors.

(*c*) Good conductors at room temperature are not superconductors and superconducting metals are not good conductors at room temperature as the normal metals.

(*d*) Amorphous thin films of Be, Bi and Fe show superconductivity.

(*e*) Bi, Te and Sb become superconductor under high pressure.

TABLE 1.7. ***Critical Temperature T_c and Critical Field H_c (at 0 K) of Some Superconducting Materials***

Material	T_c (K)	H_c (Am^{-1})	Type
Al	1.196	7.9×10^3	I
Ga	1.09	4×10^3	I
Zn	0.79	3.8×10^3	I
Sn	3.72	24.5×10^3	I
Hg	4.12	33×10^3	I
Pb	7.175	65×10^3	I
V	5.1	105×10^3	I
Ta	4.48	67×10^3	II
Nb	9.25	156×10^3	II
Th	1.37	13×10^3	I
Nb-44% Ti	10.5	9.5×10^6	II
NbN	16.0	8×10^6	II
Nb-25% Zr	10.8	5.6×10^6	II
V_3Si	17.0	1.3×10^7	II
Nb_3Sn	18.5	1.6×10^7	II
V_3Ga	16.4	2.8×10^7	II
Nb_3Al	18.0	—	II
$BaBi_3$	5.69	59×10^3	II
Bi_2Pt	0.16	0.8×10^3	II
$CoSi_2$	1.40	8.4×10^3	II
Cr_3Ru	3.3	—	II

1.26.3. Meissner Effect

Meissner and Ochsenfeld in 1933 found that if a long superconductor is cooled in a longitudinal magnetic field to below the value of critical temperature corresponding to that field, then the lines of induction B are pushed out of the body of the superconductor at the transition (Fig. 1.27). This phenomenon is called the *Meissner effect***.** Such flux exclusion is also observed if the superconductor is first cooled below T_c and then placed in the magnetic field. The effect is of fundamental importance as it shows that a bulk superconductor *behaves in a external magnetic field H as if inside the specimen* $B = \mu_0 (H + M) = 0$ *or* $X = -1$; that is, a superconductor exhibits perfect diamagnetism. We shall now show that the perfect diamagnetism of superconductors is an independent property, not at all related to zero resistivity. Let us try to relate the two properties together and see what happens. From Ohm's Law $E = \rho j$, we see that if the resistivity ρ goes to zero while j is held infinite, then E must be zero. Using the Maxwell's equation

$$\nabla \times E = -\frac{\partial B}{\partial t} \qquad \text{...(1.81)}$$

we obtain B = constant. This concludes that the flux through the specimen cannot change on cooling through the transition. This means that when a perfect conductor ($\rho = 0$) is cooled in the magnetic field until its resistance becomes zero, the magnetic field in the material gets frozen in and cannot change subsequently irrespective of the applied field. This is obviously in contradiction to the Meissner effect. Thus, perfect diamagnetism and zero resistivity are the two independent essential properties of the superconducting state (Fig. 1.28).

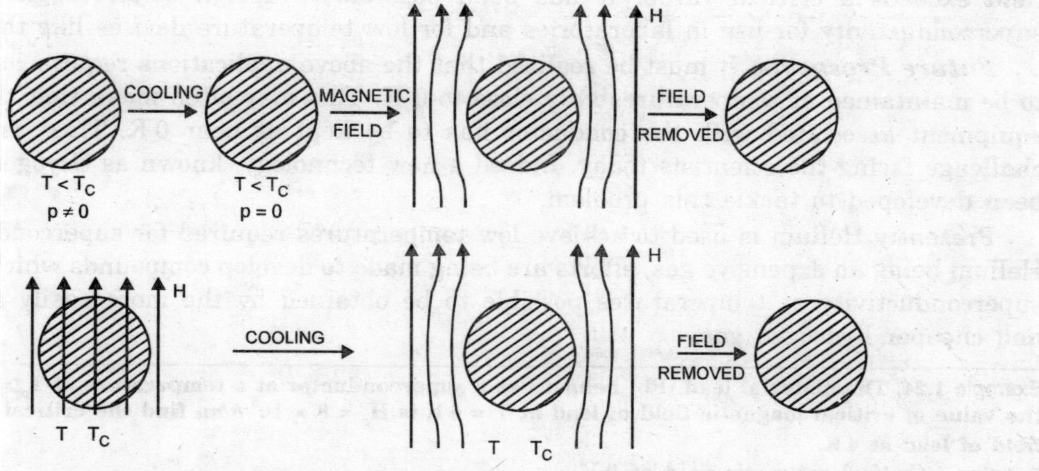

Fig. 1.27. ***An Applied Magnetic Field can Exist Inside a Superconductor Above T_C. When the Superconductor is Cooled Below T_c, Surface Currents Appear, Whose Effect is to Expel the Magnetic field from the interior of the superconductor***

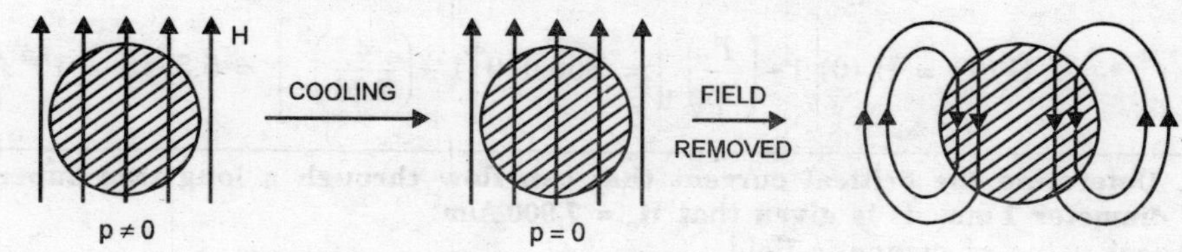

Fig. 1.28. ***Magnetic Behaviour of a Perfect Conductor***

1.26.4. Applications of Superconductors

1. Superconductors are used for producing very strong magnetic field of about 20–30 T which is much larger than the field obtained from an electromagnet and such high magnetic fields are required in power generators.

   Efforts are being made at present to develop electrical machines and transformers utilizing superconductivity. Calculations show that if we could use superconductors as conducting material, in addition to superconducting magnets, which are already being produced, it is possible to manufacture electrical generators and transformers in exceptionally small size, having an efficiency as high as 99.99 %.

2. Magnetic energy can be stored in large superconductors and drawn as required to counter the voltage fluctuations during peak loading.
3. The superconductors can be used to perform logic and storage functions in computers.
4. A superconductor material can be suspended in air against repulsive force from permanent magnet. The levitation can be used in transportation.
5. As there is no heat loss in superconductors (*i.e.* I^2R loss is zero), so power can be transmitted through the superconducting cables.

Superconducting materials if used for power cables enable transmission of power over very long distances using a diameter of a few centimetres without any significant power loss or drop in voltage.

Superconducting solenoids which do not produce any heat during operations have been produced. However, it must be noted that superconductivity can be destroyed if the magnetic field exceeds a critical value. It has been possible to design electromagnets using superconductivity for use in laboratories and for low temperature devices like the *maser*.

Future Prospects. It must be realized that the above applications require conductors to be maintained at temperatures very close to 0 K. This may often mean that the whole equipment associated with the conductor has to be kept at near 0 K. This is a great challenge facing the scientists today. Indeed a new technology, known as Cryogenics, has been developed to tackle this problem.

Presently Helium is used to achieve low temperatures required for superconductivity. Helium being an expensive gas, efforts are being made to develop compounds which exhibit superconductivity at temperatures possible to be obtained by the more easily available and cheaper hydrogen gas.

Example 1.24. The material lead (Pb) behaves as a superconductor at a temperature of T_c = 7.26 K. If the value of critical magnetic field of lead at T = 0 K is $H_c = 8 \times 10^5$ A/m, find the critical magnetic field of lead at 4 K.

Solution : Critical magnetic field at 0 K,

$$H_c(0) = 8 \times 10^5 = 800{,}000$$

Critical temperature, $T_c = 7.26$ K

From Eq. (1.79), the critical value of magnetic field at T = 4 K is given as

$$H_c(T) = H_c(0)\left[1-\left(\frac{T}{T_c}\right)^2\right] = 800{,}000\left[1-\left(\frac{4}{7.26}\right)^2\right] = 5.5715 \times 10^5 \text{ A/m} \quad \textbf{Ans.}$$

Example 1.25. Determine the critical current that can flow through a long thin superconducting conductor of diameter 1 mm. It is given that H_c = 7,900 A/m.

Solution : Critical value of magnetic field,

$$H_c = 7{,}900 \text{ A/m}$$

$$\text{Radius of conductor, } r = \frac{d}{2} = \frac{1}{2} \text{ mm} = 0.0005 \text{ m}$$

According to Eq. (1.80), Sibsbee's rule,

$$\text{Critical current, } I_c = 2\pi r H_c = 2\pi \times 0.0005 \times 7{,}900 = 24.818 \text{ A} \quad \textbf{Ans.}$$

Example 1.26. Determine the critical current density for 1 mm diameter wire of lead at 4.5 K. Given that for lead T_c = 7.2 K and $H_c = 8 \times 10^4$ A/m.

Solution : For lead, critical field, $H_c(0) = 8 \times 10^4$ A/m

For lead, critical temperature, T_c = 7.2 K

Critical field at T = 4.5 K,

$$H_c = H_c(0)\left[1-\left(\frac{T}{T_c}\right)^2\right] = 8\times10^4\left[1-\left(\frac{4.5}{7.2}\right)^2\right] = 4.875\times10^4 \text{ A/m} \quad \textbf{Ans.}$$

Critical current, $I_c = 2\pi r H_c = 2\pi \times 0.5 \times 10^{-3} \times 4.875 \times 10^4 = 153.15$ A **Ans.**

Highlights

1. Neils Bohr put forward his model of atom in 1913. He postulated the following three fundamental laws.
   (*i*) The electrons continue revolving in their respective orbits without losing energy.
   (*ii*) Energy is emitted or absorbed by an atom only when an electron moves from one level to another.

   In a transition from one stationary state corresponding to a definite energy W_2 to another stationary state, with an associated energy W_1 radiation will be emitted. The frequency of this radiant energy is given by

   $$f = \frac{W_2 - W_1}{h}.$$

   (*iii*) The angular momentum of an electron moving round the nucleus is quantized.
   The energy level in joules of each state is given by

   $$W = \frac{-me^4}{8\epsilon_0^2 n^2 h^2}$$

2. Electron volt is the unit of work or energy and
   $$1 \text{ eV} = 1.602 \times 10^{-9} \text{ joule}$$
3. Within any given material there are two distinct energy bands in which electrons may exist. These two energy bands are valence energy band (or valence band) and conduction energy band (or conduction band) and are separated by an energy gap in which no electron can normally exist. This energy gap is termed as forbidden energy gap (or forbidden gap). The distinction between conductors, insulators and semiconductors is largely concerned with the relative widths of the forbidden energy gaps (no forbidden gap in case of conductors, about 1 eV in case of semiconductors and more than 5 eV in case of insulators).
4. The two basic charge transport mechanisms are drift, due to an applied electric field, and diffusion, due to a density gradient.
   The steady flow of electrons in one direction caused by the applied electric field constitutes an electric current, called the drift current. The process whereby particles flow from a region of high concentration to a region of low concentration is known as diffusion process and the current that results from the diffusion of charged particles is termed as diffusion current.
5. The number of free electrons per cubic metre of a metal whose energies lie in the interval E to E + *d*E is given by

   $$dn_E = \rho_E \, dE$$

   where ρ_E represents the number of electrons/eV/m^3 of metal and is expressed as

   $$\rho_E = N(E)\, f(E)$$

   where N(E) is the density of states (number of states/eV/m^3 in the conduction band is given as

   $$N(E) = \gamma E^{1/2} \qquad \text{where } \gamma = 6.82\times10^{27}$$

The Fermi-Dirac probability function is given by

$$f(\text{E}) = \frac{1}{1+e^{(\text{E}-\text{E}_\text{F})/k\text{T}}} \quad \text{where k, the Boltzmann's constant} = 8.62 \times 10^{-5} \text{ eV/K}$$

6. Carrier mobility is the ratio of the average drift velocity and applied electric field *i.e.*

$$\text{Drift velocity, } v = \mu_e \text{E}$$

7. If τ is the relaxation or collision time, the average velocity of electrons, known as drift velocity is given by

$$v = \frac{e\text{E}}{m} \times \tau$$

The current density J, defined as the current per unit area, is given as

$$\text{J} = env = \frac{ne^2\text{E}\tau}{m} = \sigma\text{E}$$

$$\text{Electrical conductivity, } \sigma = \frac{ne^2\tau}{m} = ne\,\mu_e$$

8. On the basis of resistivity of the materials, conducting materials are broadly classified into two categories namely low resistivity materials and high resistivity materials.

A low resistivity material, besides possessing very low value of resistivity, should possess additional properties of low temperature coefficient, sufficient mechanical strength, ductility, solderability and resistiveness to corrosion.

A high resistivity material besides having high value of resistivity should also possess additional properties of low temperature coefficient, high melting point, ductility, high mechanical strength and no tendency for oxidation.

Copper, aluminium, silver etc. fall under the category of low resistivity materials whereas platinum, tungsten, molybdenum, manganin, constantan, nichrome etc. fall under the category of high resistivity materials.

9. Factors affecting electrical conductivity (or resistivity) of metals are : (*i*) Temperature, (*ii*) Alloying, (*iii*) Mechanical stressing, and (*iv*) Age hardening.

The resistance of all pure metallic conductors increases with the increase in temperature. Alloying is another factor which affects the resistivity of a material. By adding traces of some impurities to a metal, its resistivity increases. Alloys have higher resistivity than pure base metal *e.g.* pure copper has $\rho = 1.7 \times 10^{-8}$ Ω-m while its alloy brass (60% Cu + 40% Zn) has $\rho = 7 \times 10^{-8}$ Ω-m at room temperature. Alloying elements greatly increase the lattice imperfections that in turn increases the scattering of electrons and phonons and hence the resistivity of metal increases with alloying.

The resistivity of a material also changes under the influence of mechanical treatment. The fabrication of conductor from the ingot to the final stage comprises initial hot working and final cold drawing. Cold-working operation (stressing) distorts the crystal structure of the metal. This generally tends to harden the material, increase its tensile strength and increase slightly its resistivity. The increase in tensile strength is very useful for many purposes such as overhead conductor. That is why many types of conductors are finally drawn in cold stage in which case they are identified as hard drawn. Although mechanical stressing increases the resistivity i.e., decreases the conductivity, annealing (heat treatment process) restores the electrical conductivity by establishing regularity in crystal structure. When a metallic conductor is strained, then there is a change in its resistance due to change in the length and cross-sectional area of the conductor.

The electrical resistivity of an alloy increases with age hardening. During this process crystal lattice undergoes distortion due to which electron mobility is decreased which in turn decreases the mean free path.

10. The property that characterizes the ability of material to transfer thermal energy (or the heat) is known as thermal conductivity.

It has been observed that

(*i*) The thermal conductivity of metals decreases at higher temperatures.

(*ii*) The thermal conductivity of polycrystalline metals is lower than that of a single crystal due to scattering of electrons and phonons along grain boundaries.

(*iii*) The thermal conductivity of metals is lower than their alloys due to scattering of electrons and phonons with the atoms of alloying element.

(*iv*) The thermal conductivity of semiconducting materials increases with increasing temperature due to flow of excited electrons.

(*v*) The thermal conductivity of metals, alloys, semiconductors and dielectrics is in the descending order.

11. The phenomenon that electrical resistivity of some material e.g. mercury disappears completely at very low temperatures (below 4 K) is known as superconductivity.

Many metals and compounds have superconducting property at very low temperatures. Superconductivity has been observed to occur in poorer metallic conductors such as tin, lead and tantalum rather than in good conductors such as gold, silver and copper.

It has been found that superconductors may not only be pure metals but various alloys and chemical compounds as well. At present about 30 superconductor metals and more than 600 superconductor alloys are already known. The highest temperature at which, until now, superconductivity has been observed to occur is 20 K (–253°C) for a compound consisting of Niobium, Aluminium and Germanium.

Exercises

1. What is Bohr's atomic model ?
2. Discuss the formation of energy bands in solids. In what respects these bonds are different for insulators, conductors and semiconductors ?
3. Explain the existence of the various electron energy bands in solids. Based on these bands distinguish between insulators, conductors and semiconductors.
4. Compare and explain electrical conductivity of metals, semiconductors and insulators using energy band diagrams. **[M.D. Univ. Third Semester May 2007]**
5. Define and explain relaxation time, collision time, and mean free path. Derive mathematical relation for each term. **[M.D. Univ. Third Semester May 2006]**
6. Define relaxation time, collision time and mean free path. Derive mathematical relation for each term. **[M.D. Univ. Third Semester December 2006]**
7. Explain following terms :

   (*i*) Drift velocity, (*ii*) Mobility, (*iii*) Collision time. **[M.D. Univ. Third Semester May 2007]**
8. Classify the various conductor materials. Also discuss their properties and applications. **[M.D. Univ. Third Semester December 2007]**
9. Explain the properties and uses in electrical industry of the following conducting materials:

   (*i*) High conductivity materials and (*ii*) high resistivity materials.
10. Mention the characteristic features of low resistivity materials. Discuss some uses of low resistivity materials.
11. What are the properties that high conducting materials should possess ?

    What are the materials used for (*i*) electrical contacts (*ii*) rheostats and (*iii*) fuses ? Give reasons. **[I.E.S. 1996]**
12. State the relative advantages and disadvantages of aluminium as compared to copper as a conductor of electricity.

13. Prove that for the same volume of conducting materials the rating of the electrical apparatus with aluminium conductor is reduced by 22% as compared with copper conductors.

14. Discuss the important electrical and mechanical properties of copper. How do the electrical and mechanical properties of copper compare with that of aluminium for use as conductor ?

15. Discuss the properties and uses of the following materials
(*i*) Tungsten (*ii*) Nichrome (*iii*) Constantan (*iv*) Chromel.

16. What are the advantages of ACSR conductors ? What is a solder ? Explain the different uses of solder.
Explain the various uses of high resistivity materials used in electrical industry.

17. Discuss the factors affecting the conductivity of a conducting material.
[M.D. Univ. Third Semester December 2005]

18. Give various factors affecting conductivity of materials.
[M.D. Univ. Third Semester May 2007]

19. Explain the various factors affecting the resistivity of conducting materials.
[M.D. Univ. Third Semester May 2008]

20. What do you mean by temperature coefficient of resistance ?
Show that the temperature coefficient α_2 of a metallic conductor at temperature t_2 is related to the temperature coefficient α_1 at temperature t_1 by relation

$$\frac{\alpha_1 - \alpha_2}{\alpha_1 \alpha_2} = (t_2 - t_1).$$

21. On adding a higher conductivity material copper in small quantity to nickel of lower conductivity why does the conductivity of the alloy decrease ? Explain the phenomenon involved. **[I.E.S. 1998]**

22. Explain the different factors which affect the resistivity of conducting materials. Explain the requirements of materials used for electrical heating elements in heaters and electric furnace.

23. Explain the term specific resistance. What is an ACSR conductor ? Discuss the electrical and mechanical characteristics of pure copper.

24. Explain the properties, uses of electrical contact materials.

25. Write the properties and applications of any two noble metals used in electrical contacts.
[I.E.S. 2001]

26. Name some of the commonly used materials as fuse element. Why silver is preferred as fuse element inspite of its higher cost ?

27. Write short notes on soft and hard solders.

28. Explain thermal conductivity and how is it related to Wiedemann Franz law.
[M.D. Univ. Third Semester May 2006, December 2006]

29. State and explain Wiedemann Franz law. **[M.D. Univ. Third Semester December 2007]**

30. Prove that the heat developed per m^3 per second in a conductor carrying a current of density J as a result of an applied electric field E is given by $W = JE = \sigma E^2$ where σ is conductivity of conductor. **[M.D. Univ. Third Semester May 2008]**

31. What do you understand by electron scattering in conductors ? How is the resistivity dependent on it ? Hence explain Debye temperature and Mathiessen's rule. **[I.E.S. 1997]**

32. What is superconductivity ? How is it affected by (*i*) frequency (*ii*) magnetic field ? Give some applications of superconductivity. **[I.E.S. 1993]**

33. Explain the term superconductivity. State various applications of superconductors.
[M.D. Univ. Third Semester December 2005]

34. Explain the phenomenon of superconductivity and also discuss its behaviour in magnetic field. Give some applications of this phenomenon. **[M.D. Univ. Third Semester May 2008]**

35. What is superconductivity ? What are its uses ? Name some such materials.

36. What do you understand by the term "supeconductivity" ? Mention the possible applications of superconductors.

37. Explain the term superconductivity. Name two soft superconducting materials with their transition temperatures. Explain the effect of magnetic field on superconductivity. Describe some application areas of superconducting elements.

Short Answer Type Questions With Answers

Q. 1. What is a crystal ?

Ans. A crystal may be considered as a regular array of units — smallest groupings that repeat themselves in all directions in space. The repeating unit obeys the translation operation.

Q. 2. What is crystal lattice ?

Ans. A lattice is a regular periodic arrangement of imaginary points in space. The structure of all crystals is described in terms of lattice with a group of atoms attached to each lattice point. The groups of atoms is called basis, it is repeated in space to form the crystal structure.

Q. 3. What are lattice points ?

Ans. Lattice points are an array of points generated by the translation operations.

Q. 4. What is basis ?

Ans. The repeating unit assembly — atom, ion, molecules, radicals, etc. that is located at each lattice point is called the basis.

The crystal structure = Lattice + basis.

Q. 5. What is free electron theory ?

Ans. The first reasonable approach to an explanation of electrical conduction in solids was proposed by Drude. Drude imagined conductive metal in the form of a gas of free electrons. An applied electric field accelerates the electrons along the field, because the ions they encounter in the crystal lattice deflect them and they interpose resistance, the electrons settle into a constant drift velocity that is proportional to the strength of the applied electric field. Drude assumed that the mean free path i.e. average distance travelled between two collisions with the ions in the lattice, is of the order of interatomic distance.

Q. 6. How does free electron gas in metal differ from ordinary gas in some respects ?

Ans. Firstly the free electron gas is charged while the ordinary gas molecules are generally neutral. Secondly the concentration of electrons in metals is larger than that in the ordinary gas.

Q. 7. Classify materials into conductors, semiconductors and insulators.

Ans. According to the free electron theory, the total energy which is equal to the kinetic as well as the potential energy is assumed to be zero, is related to the wave number by the relation

$$E = \frac{h^2k^2}{2m} \qquad \text{where } k = \frac{2\pi}{\lambda}$$

In the free electron theory, when one considers the interaction of the electron with the period potential of the lattice, one finds that band gaps occur at Brillouin zone boundaries. In case the lower valence band is half filled, the material is said to be a good conductor. There is overlapping of zones/bands. This also refers to case where the material behaves like a metal or good conductor. At higher temperature ($K_BT \geq E_g$) an appreciable number of electrons get excited from the top of the valence band into the bottom of the conduction band. The material is then said to exhibit intrinsic semiconductivity; the magnitude of conductivity depends on the density and mobility of electrons in the conduction band and holes in the valence band. If the band gap is very large so that thermal excitation of electrons from valence to conduction band is negligible even at high temperatures, the material is deemed to be as an insulator.

Q. 8. What is the basis for classifying a material as a conductor, semiconductor or a dielectric ? What is the conductivity of a perfect dielectric ? **[I.E.S. 2002]**

Ans. Conductors possess high conductivity whereas the characteristic property of insulating materials (or dielectrics) is poor conductivity. Semiconductors occupy an intermediate position between conductors and insulators. Though there is no rigid line separating the conductors from semiconductors and semiconductors from insulators, but still according to resistivity the materials of resistivity of the order from 10^{-8} to 10^{-3}, 10^{-3} to 10^{6} and 10^{6} to 10^{18} ohm-metres may be classified as conductors, semiconductors and dielectrics respectively.

Another classification is based on temperature coefficient of resistivity. Metals have positive temperature coefficient of resistivity; semiconductors have small negative temperature coefficient of resistivity and insulators have large negative temperature coefficient of resistivity.

Conductivity of a perfect dielectric is zero.

Q. 9. What do you mean by the drift velocity and mobility of a free electron ?

Ans. The average velocity of a free electron is known as drift velocity whereas mobility of an electron is defined as the drift velocity per unit electric field.

Q. 10. What are the materials used for (*i*) electrical contacts (*ii*) rheostats and (*ii*) fuses ? **[I.E.S. 1996]**

Ans. High conductivity materials with small contact resistance should be used for electrical contacts.

Rheostat is a variable linear or circular wire wound resistor with a sliding contact and is usually connected in series with the circuit to control the current. Materials generally used for rheostats are constantan and nichrome.

Materials generally used for fuse wire are copper, aluminium, zinc, tin, lead and silver.

Q. 11. On adding a high conductivity material copper in small quantity to nickel of lower conductivity why does the conductivity of the alloy decrease ?

Ans. When one adds to nickel a small amount of copper, the conductivity decreases. The reason is that in replacing some of the nickel atoms in the lattice, the copper atoms destroy the lattice periodicity in their intermediate neighbourhood and thus acts as new scattering centres for the conduction electrons, resulting in decrease in relaxation time, hence resistivity increases.

Q. 12. What is an ACSR conductor ?

Ans. Aluminium conductors with steel reinforcement are called ACSR conductors (aluminium conductors steel reinforced).

Q. 13. What is meant by a fuse and on what principle does it operate ?

Ans. Fuse is a device used in an electrical circuit for protecting electrical equipment against overloads and/or short-circuits.

The action of a fuse is based upon the heating effect of electric current. During overloads or short-circuits the fuse gets heated up to melting point due to excessive current and interrupts the circuit.

Q. 14. What function does the fuse perform ?

Ans. The function of a fuse wire is (*i*) to carry the normal working current safely without heating and (*ii*) to break the circuit when current exceeds the limiting current.

Q. 15. Explain the silver contacts. **[I.E.S. 2003]**

Ans. Silver contacts are alloys, silver plus refractory constituent (tungsten), and silver plus a semirefractory constituent (cadmium oxide). Some of the compositions of alloys are: pure silver, cadmium silver, palladium.

Q. 16. Describe the factors on which the conductivity of a meterial depend. Give your answer discussing scattering phenomenon. **[I.E.S. 2004]**

Ans. The electrical resistivity of metals is based on the *notion* that scattering of electrons is due to deviations from perfect periodicity of the lattice. The resistivity is given by expression

$$\rho = \frac{m}{ne^2}\left[\frac{1}{\tau_i} + \frac{1}{\tau_T}\right]$$

where τ_i is the relaxation time associated with imperfections and τ_T is the relaxation time resulting from the thermal vibrations of the lattice.

Now with the increase in temperature T, the amplitude of the atomic vibration increases, hence the deviations from the periodicity become more severe and the scattering by lattice vibration increases. Consequently with the increase in T, the mean free path of an electron, λ decreases and since the velocity of electrons with the Fermi energy, v_F is constant, τ_T decreases. Thus with the increase in temperature resistivity increases.

Q. 17. What is superconductivity ?

Ans. The phenomenon that electrical resistivity of some material e.g. mercury disappears completely at very low temperatures (below 4 K) is known as superconductivity.

Q. 18. Is it possible to make a pure metal a superconductor ? Why ?

Ans. Not all the metals can be converted into superconductors because superconducting elements occur between well defined limits of atomic volume.

Q. 19. Give electric, magnetic and thermal properties of superconductors.

Ans. Electrical, magnetic and thermal properties of superconductors are given below :

1. Resistivity is almost zero.
2. If a magnetic field H is applied, the material remains superconducting until a critical field H_c is reached such that $H > H_c$, the material reverts from the superconducting state to the normal state.
3. For every superconductor there is a transition temperature T_c above which it ceases to be a superconductor.
4. Following relation holds between transition temperature and isotopic mass M of a superconductor.

$$M^{1/2}\, T_c = \text{Constant}$$

Q. 20. How is superconductivity affected by frequency ?

Ans. The electric and magnetic behaviour of superconductors at higher frequencies differs from zero frequency behaviour. In the radio-frequency (up to about 10^8 Hz) and microwave-frequency (from 10^8 to about 10^{11} Hz) regions of the electromagnetic spectrum, it is found that superconductors do not have zero resistance to the flow of current. The resistance and the accompanying electrical energy loss are still much smaller than in normal state, but they are not zero and they increase with the increase in frequency. On the other hand, in the optical region of the spectrum (about 10^{15} Hz) the electromagnetic response in the superconducting state is indistinguishable from that in the normal state.

Q. 21. What is critical field in a superconductor ? How is it affected by temperature ?

Ans. If we have a superconducting wire and a magnetic field is applied parallel to its length, then the resistance of the wire is suddenly restored at a definite field strength. This of course depends upon temperature. The magnetic field at which the resistance is restored is the characteristic of the particular metal and is called the critical field H_c. The H_c is functionally related to temperature as

$$H_c(T) = H_c(0)\left[1 - \frac{T^2}{T_c^2}\right]$$

where $H_c(0)$ is the critical field at 0 K.

Q. 22. What is energy gap in superconductors ?

Ans. In the superconducting state an energy gap $E_g = 3.5\, K_B T_c$ separates superconducting electrons below from normal electrons above the gap. This gap is detected in experiments on heat capacity, infrared absorption and tunneling.

Q. 23. Define critical magnetic field of a superconductor.

Ans. Critical magnetic field is the maximum field that can be applied to a superconductor without destroying the superconducting behaviour. It decreases from its maximum value at absolute zero to zero at the critical temperature T_c.

Q. 24. Define critical temperature of a superconductor.

Ans. Critical temperature T_c is the temperature that separates superconducting state from the normal state. Above critical temperature, T_c, the material is in the normal state with a finite resistivity, but below T_c it is in the superconducting state with zero resistivity.

Q. 25. Why type-II superconductors are more important ?

Ans. Type-II superconductors are more important because they can carry very high current densities, as a result they have great technological importance. To achieve high critical current densities, a super-conducting material should be fine grain material or heavily cold-worked or recovery-anealed material.

Problems

1. A conductor has a cross-section of 10 cm^2 and specific resistance of 7.5 μΩ-cm at 0 °C. What will be its resistance in ohms per km when the temperature is 40 °C. Take temperature coefficient of the material 0.005 per °C. **[Ans.** 0.09 Ω]
2. Find the temperature coefficient of resistance of material used in a resistor if the resistance at 23°C is 48 Ω and at 75°C is 59 Ω. **[Ans.** 4.904×10^{-3} per °C at 0°C]
3. Calculate the resistance of a wire at 50°C which is 300 m long and has an area of x-section of 25 mm^2. The wire is made of aluminium. The resistivity of aluminium at 20°C is 2.78×10^{-8} Ω-m. Temperature coefficient of aluminium is 0.004 Ω/°C at 0°C. **[Ans.** 0.3707 Ω]
4. It is found that the resistance of a coil of wire increases from 40 Ω at 15 °C to 50 Ω at 60 °C. Calculate the resistance temperature coefficient of the material and the resistance of the coil at 0 °C. **[Ans.** $\frac{1}{165}$ per °C; 36.667Ω]
5. A coil has a resistance of 18 Ω when its mean temperature is 25 °C and of 20 Ω when its mean temperature is 55 °C. Find the mean temperature rise when its resistance is 21 Ω and the surrounding temperature is 20 °C. Assume that temperature coefficient of resistance is constant. **[Ans.** 50 °C]
6. A piece of copper wire has a specific resistance of 1.7×10^{-6} Ω-cm at 0 °C and has a temperature coefficient of 0.005 at 20 °C. Find the specific resistance and temperature coefficient at 50 °C. **[Ans.** 2.17×10^{-6} Ω-cm ; 0.00435 per °C]
7. A potential difference of 250 V is applied to a field winding at 15 °C and the current is 5 A. What will be the mean temperature of the winding when the current has fallen to 3.91 A, applied voltage being constant. Assume $\alpha_{15} = \frac{1}{254.5}$. **[Pune Univ. Elec. Engineering, 1991]** **[Ans.** 86 °C]
8. The electron and hole concentration in a P-type semiconductor is 10^{12}/m^3 and 10^{22}/m^3 respectively. If the mobility of electrons and holes are 0.04 and 0.06 m^2/Vs respectively, determine the conductivity of the material. **[Ans.** 96 S/m]
9. Determine the number of electrons per m^3 if the resistivity of conductor material is 0.01531×10^{-6} Ω-m and mobility is 7.04×10^{-3} m^2/V-s. Also determine the relaxation time. Given that $e = 1.6 \times 10^{-19}$ C and $m = 9.1 \times 10^{-31}$ kg. **[Ans.** 5.8×10^{28}; 4.004×10^{-14} second]
10. Calculate the critical field and the critical current density for 1 mm diameter wire of lead at 7 K. Given that for lead T_c = 7.18 K and $H_c = 6.5 \times 10^4$ A/m. **[Ans.** 0.322×10^4 A/m; 12.873×10^6 A/m^2]

❑❑❑

SECTION-B

Semiconductors, Construction and Characteristics of Devices

Contents of This Section

SEMICONDUCTORS : Review of Si and Ge as Semiconducting Materials; Continuation Equation; P-N Junction; Drift and Diffusion; Diffusion and Transition Capacitances of P-N Junction.

CONSTRUCTION AND CHARACTERISTICS OF DEVICES : Brief Introduction to Planar Technology for Device Fabrication; Metal Semiconductor Junctions (Ohmic and Non-ohmic); Breakdown Mechanism in P-N Junction; Zener Diode; Electrical and Optical Excitation in Diodes.

CHAPTER 2. Semiconductors.

CHAPTER 3. Construction and Characteristics of Devices.

CHAPTER 2

Semiconductors

Inside This Chapter

2.1. Introduction

We are familiar with conducting and insulating materials. *Conducting materials* (such as silver, copper, aluminium etc.) are good conductors of electricity and are characterized by a large electrical conductivity and small electrical resistivity. *Insulating materials* (such as porcelain, glass, quartz, rubber, bakelite etc.) are characterized by poor electrical conductivity and are used to block current from flowing where it is not to pass. There is another group of materials (such as germanium, silicon, boron, selenium) which are neither good conductors nor good insulators. At room temperature such materials have conductivities considerably lower than those of conductors and much higher than those of insulators. Such materials are called the *semiconductors*. The resistivity of various semiconductor materials lies in a very wide range from 10^{-4} to about 0.5 Ω-m whereas the resistivities of conductors and insulators are of the order of 10^{-8} Ω-m and 10^{12} Ω-m respectively. An even more important property of semiconductors is that *their resistance depends largely on various factors and, therefore, it can be controlled.*

The resistance of conducting materials except that of carbon, increases with the increase in temperature and resistance of insulating materials remains almost constant up to a certain temperature after which it suddenly falls off and insulating materials start conducting. Semiconductors behave in somewhat different way. The resistance of semiconductors decreases with the increase in temperature, as shown in Fig. 2.1, *i.e.* temperature coefficient of resistance of semiconductors is negative. Thus, the semiconductors behave like insulators at very low temperatures but act as conductors at high temperatures.

The resistivity of semiconductors depends on illumination and decreases in brighter surroundings (the "light" resistivity of a semiconductor is smaller than its "dark" resistivity).

The resistivity of semiconductors depends on the magnitude of the electric field in the semiconductor, so that the current in it is not proportional to the voltage, as with conductors that obey Ohm's law, but increases by far more than the voltage, *i.e.*, semiconductors are *nonlinear resistors*.

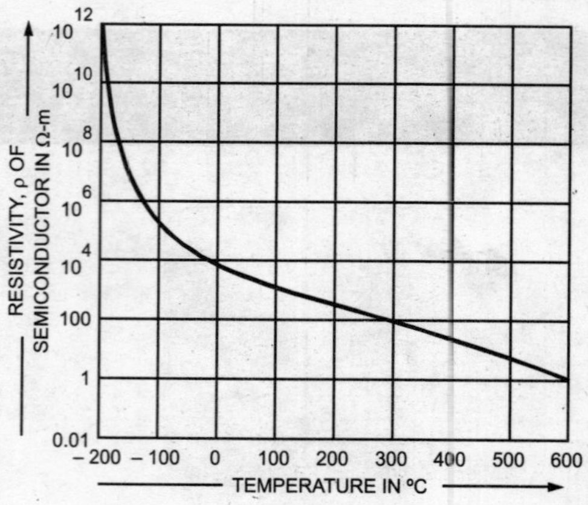

Fig. 2.1

The semiconductors have another very important property—*the resistivity (or conductivity) of semiconductors changes considerably when even minute amounts of certain other substances, called the impurities, are added to them.* Moreover, the introduction of certain impurities (such as arsenic, antimony, gallium, indium etc.) may even change the mechanism of electrical conductivity in the semiconductor, and certain physical phenomena will begin to occur at the junction between sections of the semiconductor having different impurities. Such phenomena make semiconductor devices important as *diodes, transistors, SCRs, UJTs, varistors* and *thermistors*.

The *elemental semiconductors* (*i.e.*, semiconductors comprising a single chemical element) are germanium (Ge), silicon (Si), grey (crystalline) tin, selenium (Se), tellurium (Te), boron (B) etc. *Germanium* and *silicon* are two of the best known and widely used semiconductors.

The devices (such as diodes, transistors etc.) manufactured from semiconductors have many advantages such as compact size, low cost, light weight, rugged construction, more resistive to shocks and vibrations, instantaneous operation (no heating required), low operating voltage, high operating efficiency (no heat loss) and long life with essentially no ageing effect if operated within permissible limits of temperature and frequency.

2.2. Atomic Bonds

Whether a material is a conductor, a semiconductor, or an insulator depends largely upon what happens to the outer shell electrons when the atoms bond themselves together to form a solid.

Atomic bonds are of two types viz. *primary* (or *chemical*) *bonds* and *secondary* (or *molecular*) bonds. Primary bonds are strongest bonds between atoms and can be further subdivided into (*i*) metallic (*ii*) covalent and (*iii*) ionic bonds. Secondary or molecular bonds are weaker than primary bonds. Attraction forces (also called *van der Waals forces*) exist between atoms or molecules.

In **metallic bonds**, atoms give up their valence electrons to form clouds of electrons which is shared by all atoms. Such a bond exists between elements having loosely held valence electrons so that they can be easily released to the common pool.

For example, in case of copper, the atoms give up easily detachable electrons and this creates a large mass of free electrons (or electron gas) drifting about through the spaces between the copper atoms. Each atom of copper becomes a positive ion because of losing an electron. The electron gas is, of course, negatively charged, consequently, an electrostatic force of attraction exists between the positive ions and the electron gas. This is the *bonding force* that holds the material together in a solid. In case of copper and other metals, the bonding force is termed *metallic bonding* or *sometimes electron gas bonding*. Such a bonding is shown in Fig. 2.2 (*a*).

Covalent bond is formed by sharing of electrons by two or more atoms. It can be formed between atoms of same or different elements. This arrangement is illustrated in Fig. 2.2 (*b*). In covalent bonding every valence shell of every atom appears to be filled, and consequently there are no holes and no free electrons drifting about within the material. The same covalent bonding arrangement exists in silicon and germanium.

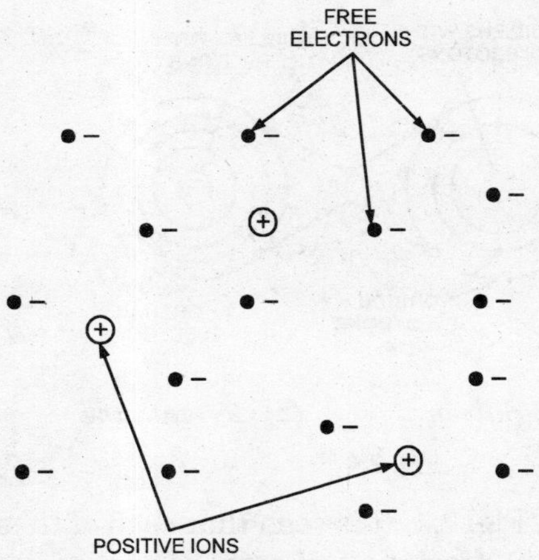

(*a*) *Metallic Bonding*

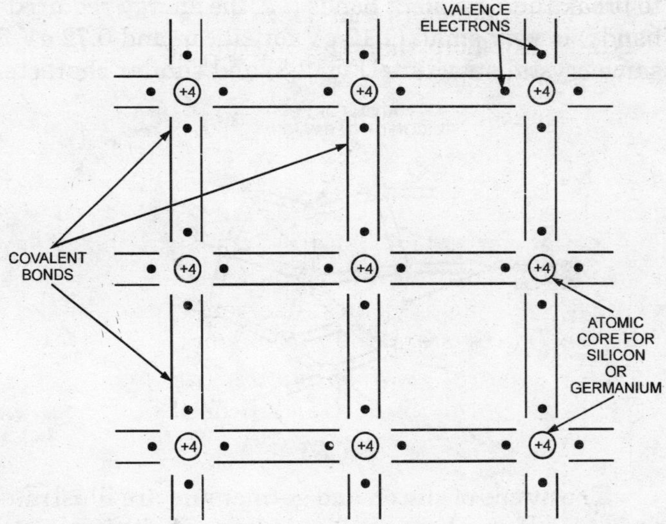

(*b*) *Covalent Bonding*

The Si (or Ge) atom at the centre share one of its valence electrons with each of the valence electrons of the four Si (or Ge) atoms surrounding it, as shown in Fig. 2.2 (*b*). This sharing of valence electrons produces four pairs of electrons that are covalently bonded thus completing its octlet.

In ionic bonds, atoms of one element transfer electrons to the other so that both have stable configurations.

In this arrangement, some atoms part with outer-shell electrons, but these are accepted into the orbit of other atoms. Thus, the atoms are ionized—those which give up electrons become *positive ions*, and those which accept the electrons become negative ions. This creates an electrostatic bonding force between the atoms, called the *ionic bonding* [Fig. 2.2 (*c*)]. Such bonding occurs in some of the insulating materials.

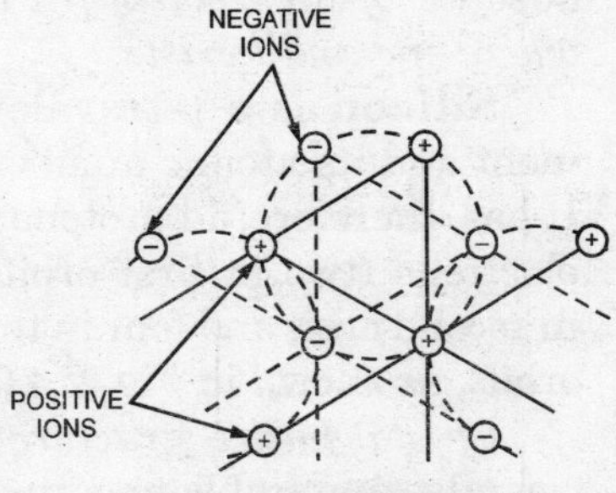

(*c*) *Ionic Bonding*

Fig. 2.2. *Bonding Between Atoms in Solids*

2.3. INTRINSIC SEMICONDUCTORS

Semiconductor materials may be classified into two classes viz. *intrinsic* (or *pure*) *semiconductors* and *extrinsic* (or *impure*) *semiconductors*.

An *intrinsic semiconductor* is one which is made of the semiconductor material in its extremely pure form. A semiconductor is not truely intrinsic unless its impurity content is less than one part in 100 million parts of semiconductor. In fact, recent advances have reduced impurity levels in the pure material to 1 part in 10 billion (1:10,000,000,000). One might ask if such low impurity levels are really required. They are certainly required. It is to be noted here that the addition of one part impurity (of the proper type) per million in a wafer of silicon material can change that material from a relatively poor conductor to a good conductor of electricity.

There are many semiconductor materials such as germanium, silicon, grey crystalline tin, selenium, tellurium, boron etc. available but silicon (Si) and germanium (Ge) are the two most widely used semiconductor materials in electronics. The reason for this is that energy required to break their covalent bonds (*i.e.* the energy required to release an electron from their valence bands) is very small (1.12 eV for silicon and 0.72 eV for germanium). Both elements have the same crystal structure (Fig. 2.3) and similar characteristics.

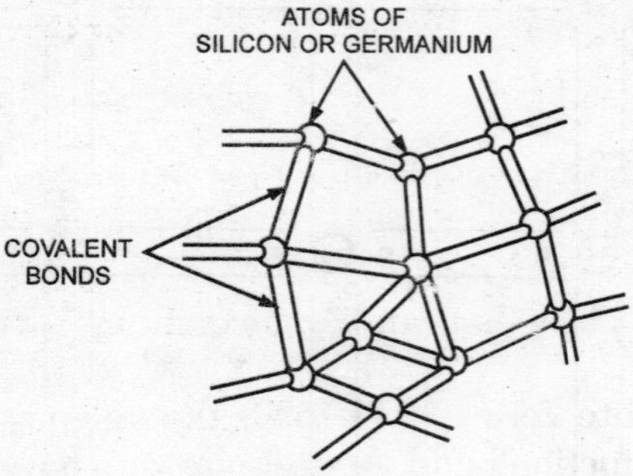

Fig. 2.3

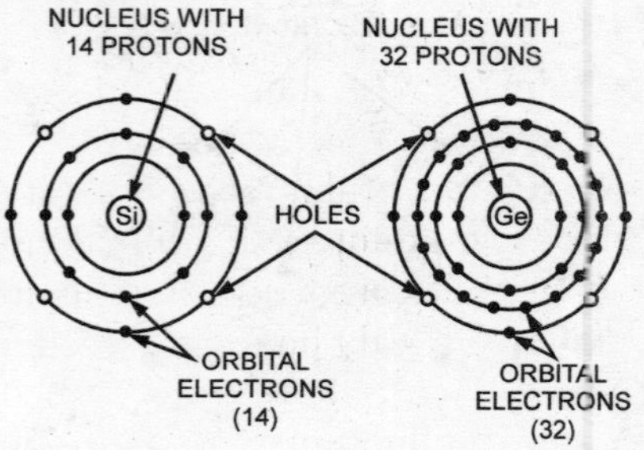

(a) Silicon *(b) Germanium*

Fig. 2.4

The atoms of silicon and germanium are illustrated in Fig. 2.4. It is seen that each of these atoms has four electrons in a valence shell that can have a maximum of eight. Thus we say, that their valence shells have four electrons and four holes. *A hole is defined simply as an absence of an electron in a shell where one could exist.*

Silicon is a tetravalent element and its atomic number is 14. It has, therefore, 14 protons and 14 electrons (two in first orbit, eight in second orbit and four in the third orbit), as shown in Fig. 2.4 (*a*).

Crystal lattice structure of silicon (also applicable to germanium) is shown in Fig. 2.3. Each atom is equidistant from its four neighbouring atoms. In a crystalline structure each atom shares its four valence electrons with the four adjacent atoms. This occurs because each tetravalent semiconductor atom needs a total of eight outer-shell electrons. This sharing of electron pairs is

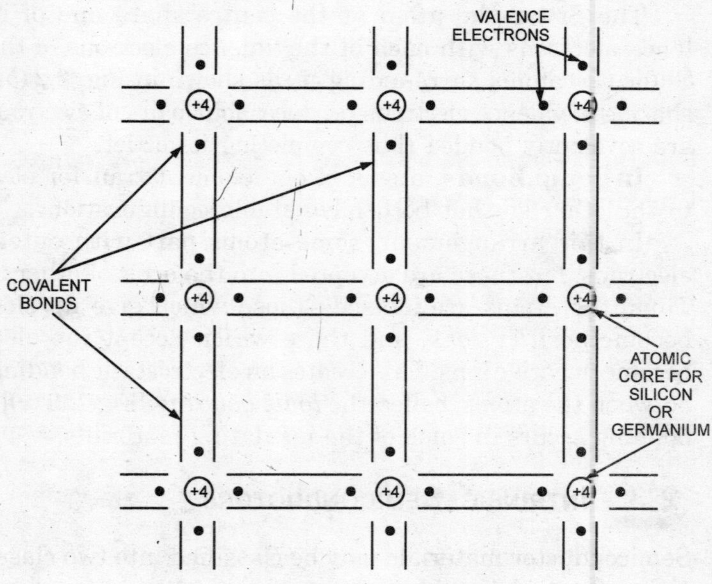

Fig. 2.5. ***Covalent Boding***

called *covalent bonding*. The covalent bonds are illustrated in Fig. 2.3 by means of rods connecting the adjacent atoms. For simplified discussion, the crystal structure may be reduced to two dimensional, as illustrated in Fig. 2.5. The core represents the nucleus and all the orbiting electrons except the valence electrons. Since there are as many protons in the nucleus as there are electrons orbiting around it and there are four valence electrons, the core will have an excess + 4 charge. The core of silicon contains 14 protons and 10 electrons (whereas germanium core has 32 protons and 28 electrons).

The energy band diagram for silicon is shown in Fig. 2.6 (*a*).

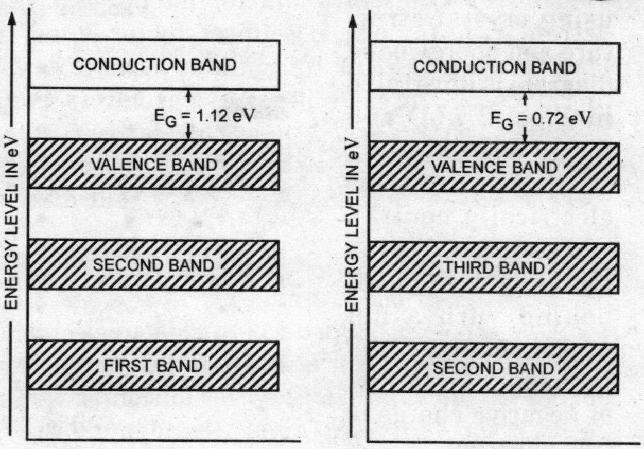

(a) For Silicon *(b) For Germanium*

Fig. 2.6. ***Energy Band Diagrams***

Germanium is also a tetravalent element *i.e.* it has 4 valence electrons. Its atomic number is 32. It has, therefore, 32, protons and 32 electrons [two in first orbit, eight in second orbit, eighteen in third orbit and four in fourth orbit], as shown in Fig. 2.4 (*b*). Its crystal lattice structure has already been discussed above. Energy band diagram for germanium is shown in Fig. 2.6 (*b*).

For semiconductors at a temperature of absolute zero (– 273·15°C) the valence band is usually full and there may be no electron in the conduction band. At absolute zero temperature E_G for silicon and germanium are 1.21 eV and 0·785 eV respectively and it is difficult to get additional energy of this magnitude from applied electric field. Hence the *conductivity of intrinsic semiconductors at absolute zero temperature is zero*. However, at room temperature (300 K), the energy provided by the heat is sufficient to lift electrons from the valence band to the conduction band. Some electrons do jump the gap and go into the conduction band. So at room temperature, semiconductors are capable of conducting some electric current.

Since energy required in transferring electrons from valence band to conduction band is more in case of silicon than that in case of germanium, the conductivity of silicon will be less than that of germanium at room temperature.

2.3.1. Production of Holes and Free Electrons

Breaking of covalent bond is equivalent to raising an electron from valence band to the conduction band. So the energy required to break a covalent bond is the same as the forbidden energy E_G and equals 0·72 eV in case of germanium and 1·12 eV in case of silicon. This energy may be supplied either (*i*) by raising the temperature and thereby imparting thermal energy (*ii*) by incidence of photons in the visible range or (*iii*) by bombardment of the semiconductor materials by the α-particles, electrons and other particles and X-rays.

When an electron breaks a covalent bond of an intrinsic semiconductor and moves away, a vacancy is created in the broken covalent bond. The vacancy so caused constitutes a hole and is represented by a small circle. Whenever a free electron is generated, a hole is simultaneously created. Thus, free electrons and holes are always generated in pairs. So concentration of free electrons and holes will always be equal in an intrinsic semiconductor. Such generation of free electron-hole pairs* is referred to as *thermal generation*. The formation of electron-hole pairs

* The liberated electron and the hole left behind is called an electron-hole pair (EHP).

using crystal structure and energy band diagram is illustrated in Figs. 2.7 (*a*) and 2.7 (*b*) respectively.

Since a crystal is electrically neutral, whenever an electron moves, a hole is left behind with a net positive charge equal in magnitude to that of negative charge of free electron.

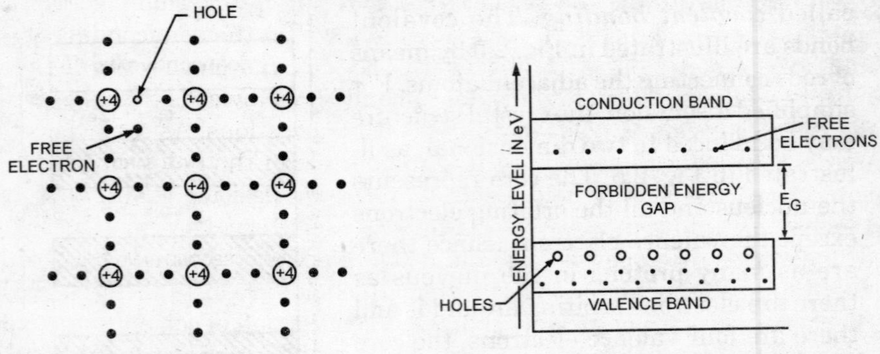

(a) Crystal Structure (b) Energy Band Diagram

Fig. 2.7. ***A Metallic Crystal Structure***

Although, strictly speaking, a hole is not a particle; for all practical purposes it is viewed as a positively charged particle capable of conducting current. This concept of a hole as a positively charged particle merely helps in simplifying the explanation of conduction in semiconductors.

2.3.2. Carrier Generation and Recombination

In an intrinsic semiconductor crystal equal number of free electrons and holes are created by thermal (heat) energy. The reason being that a hole is created only by removing an electron from the covalent bond. Thus the electrons and holes are created in pairs. Since the free electrons and holes move randomly within the crystal lattice, there is a possibility of an electron meeting a hole. Although this possibility is small, even then when a free electron approaches the hole, it gets attracted and falls into the hole. This merging of a free electron and a hole is called *recombination*. In this process of recombination, both the free electron and hole disappear and energy* is released as a quantum heat or light. This quantum is absorbed by another electron to break away from its valence band and creates a new electron-hole pair (EHP).

The recombination rate is proportional to the product of the concentrations or densities of free electrons and holes. Thus the processes of breaking of covalent bonds and recombination of electrons and holes take place simultaneously. The amount of time between the creation and disappearance of a free electron is called the *lifetime*. It varies from a few nanoseconds to several microseconds, depending how perfect the crystal is and other factors.

The rate of production of electron-hole pairs increases with the rise in temperature; the rate of recombination is proportional to the product of concentrations of free electrons and holes and at a given temperature, the concentrations or densities of free electrons and holes remain fixed and equal. So the rate of generation of electron-hole pairs must be equal to the rate of their recombination. With the increase in temperature, the rate of generation of electron-hole pairs increases. The concentration of electrons and holes goes on increasing until rate of recombination equals the increased rate of generation. Thus with the increase in temperature, concentrations of free electrons and holes increase.

2.3.3. Conduction in Intrinsic Semiconductors

When the electrons are liberated on breaking of covalent bonds, they move randomly through the crystal lattice. These free electrons are neither attracted by the nuclei of the atoms nor repelled by the electrons bound by covalent bonds because their electrical effects are fully en-

* When an electron is freed from valence band, certain amount of energy is absorbed. The same amount of energy is released when an electron falls into a hole.

gaged in maintaining the covalent bond. When an external electric field is applied to a pure (intrinsic) semiconductor, the conduction through the semiconductor is by both free electrons and holes. The free electrons in the conduction band move towards the positive terminal of the battery while the holes in the valence band move towards the negative terminal of the battery *i.e.* the electrons and holes move in opposite directions. As the holes reach the negative terminal B, the electrons reaching there combine with the holes which are thus destroyed. At the same time, the loosely held electrons near the positive terminal A are attracted away from their atoms into the positive terminal of the battery. Now new holes are created which again drift towards the negative terminal B. The electric current flows through the intrinsic semiconductor in the same direction as in which the holes are moving (the holes having positive charge). Since electrons are negatively charged, the direction of conventional current is opposite to that of movement of electrons. The total current inside the conductor is thus the sum of currents owing to free electrons and holes (though the two types of charge carriers move in the opposite directions) but current in the external wires is only because of electrons, as illustrated in Fig. 2.8.

The mobility of electrons is more than that of holes because the probability of an electron having the energy required to move to an empty state in the conduction band is much greater than the probability of an electron having the energy required to move to the empty state in valence band. Thus the current due to movement of electrons is greater than that due to the hole drift in the semiconductor. *The mobility of hole is about half that of an electron*. The net current is very small.

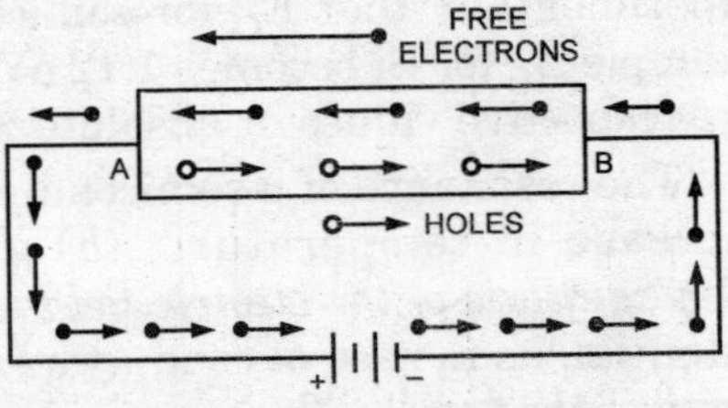

Fig. 2.8. *Conduction of Current in an Intrinsic Semiconductor*

2.3.4. Movement of Charge Carriers

There are two mechanisms in which electrons and holes move through a semiconductor. One of these is termed *drift* and other is *diffusion*. When the movement of charge carriers is under the influence of an applied electric field, as explained above (Art. 2.3.3), the resultant current is termed as *drift current*.

A second type of current, called the *diffusion current*, also exists in a semiconductor. The diffusion of charge carriers is as a result of a gradient of carrier concentration (*i.e.* the difference of carrier concentration from one region to another). In this case concentrations of charge carriers (either electrons or holes) tend to distribute themselves uniformly throughout the semiconductor crystal. This movement continues until all the carriers are evenly distributed throughout the material. This type of movement of charge carriers is called the *diffusion current*.

Diffusion current occurs when charge carriers diffuse from a point of concentration, to spread uniformly throughout the volume of a piece of material, as illustrated in Fig. 2.9.

Both drift and diffusion occur simultaneously in a semiconductor device.

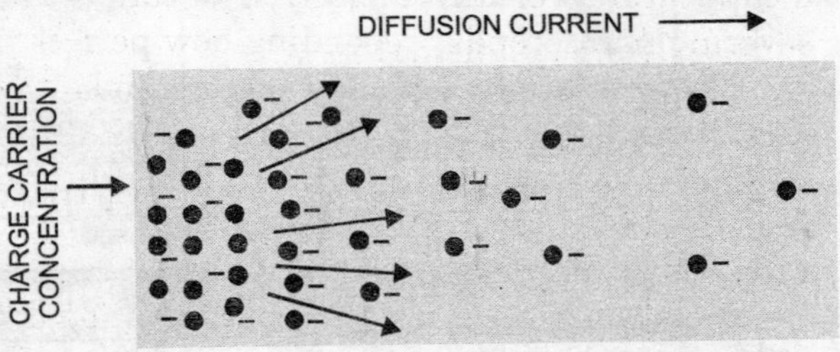

Fig. 2.9. *Flow of Particles From a Region of High Concentration To a Region of Low Concentration*

2.3.5. Effects of Heat and Light on Conductivity of Intrinsic Semiconductors

At absolute zero temperature (– 273°C), all the electrons of intrinsic semiconductors (germanium or silicon) are tightly held by their atoms. The inner orbit electrons are bound to nucleus whereas, the valence electrons are bound by the forces of covalent bonds. Thus, at absolute zero temperature, no free electron is available in the intrinsic semiconductor and so it behaves like a perfect insulator.

When the material is heated, electrons breakaway from their atoms and move from the valence band to conduction band. This produces holes in the valence band and free electrons in the conduction band. Conduction can then occur by electron movement and by hole transfer. With the increase in temperature, the rate of generation of electron-hole pairs is increased. This in turn increases the rate of recombination. Thus with the increase in temperature, the concentration of charge carriers (electrons and holes) increases. As more charge carriers are made available, the conductivity of a pure semiconductor increases with the increase in temperature. In other words resistivity of a pure semiconductor decreases with the increase in temperature *i.e.* the semiconductors have *negative temperature coefficient of resistance*.

When the semiconductor is illuminated, its resistance decreases in the same way as in case of increase in temperature.

The forbidden energy gap E_G also depends somewhat on temperature. It has been determined experimentally that E_G for semiconductors decreases with the increase in temperature. For example E_G for Si becomes 1.12 eV and that for Ge becomes 0.72 eV at room temperature while the respective values at absolute zero temperature are 1.21 eV and 0.785 eV.

The resistance of a semiconductor depends largely on (*a*) *temperature*—decreases with the increase in temperature, (*b*) *illumination*—decreases in brighter surroundings and (*c*) *magnitude of the electric field* — the current does not increase in proportion to the applied potential, as in case of conductors obeying Ohm's law but increases by far more than the voltage *i.e.*, *the semiconductors are non-linear*.

2.3.6. Fermi Level in an Intrinsic Semiconductor

The *Fermi level* is simply a reference energy level. It is the energy level at which the probability of finding an electron *n* energy units above it in the conduction band is equal to the probability of finding a hole (electron absence) *n* energy units below it in the valence band. Very simply, it can be considered to be the average energy level of the electrons, as illustrated in Fig. 2.10 (*a*).

Let at any temperature T K

Number of electrons in the conduction band be n_C

Number of electrons in the valence band be n_V

Total number of electrons in both bands, $n = n_C + n_V$

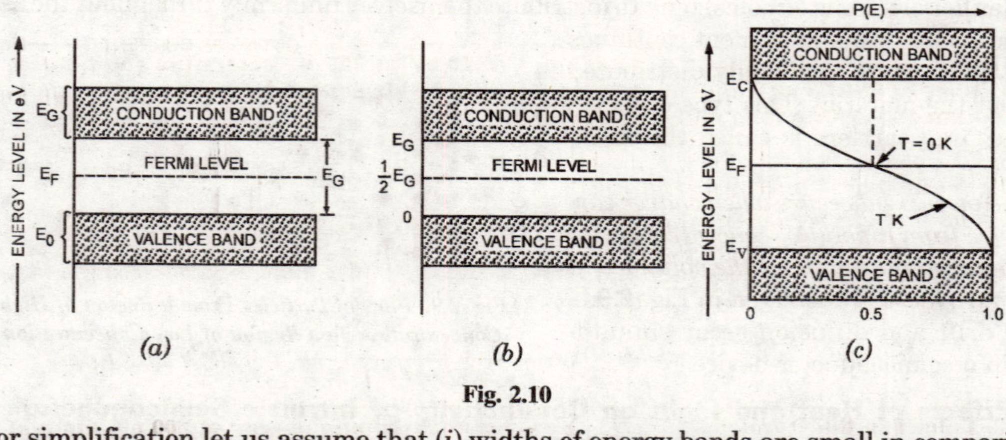

Fig. 2.10

For simplification let us assume that (*i*) widths of energy bands are small in comparison to forbidden energy gap between them (*ii*) all levels in a band have the same energy, bandwidths being assumed to be small (*iii*) energies of all levels in valence band are E_0, as shown in Fig. 2.10 (*a*) and energies of all levels in conduction band are E_G.

Let the zero-energy reference level be taken arbitrarily at the top of the valence band, as shown in Fig. 2.10 (*b*).

Now number of electrons in conduction band, $n_C = n.\,P\,(E_G)$ ...(2.1)

where $P\,(E_G)$ represents the probability of an electron having energy E_G. Its value may be determined from Fermi-Dirac probability distribution function given as

$$P\,(E) = \frac{1}{1+e^{(E-E_F)/kT}} \quad ...(2.2)$$

where P (E) is the probability of finding an electron having any particular value of energy E.

$$\text{So } P\,(E_G) = \frac{1}{1+e^{(E_G-E_F)/kT}} \quad ...(2.3)$$

where E_F is the Fermi level.

Substituting the value of $P\,(E_G)$ from Eq. (2.3) in Eq. (2.1) we have

$$n_C = \frac{n}{1+e^{(E_G-E_F)/kT}} \quad ...(2.4)$$

Now number of electrons in valence band, $n_V = n\,P\,(0)$ ...(2.5)

The probability P (0) of an electron being found in the valence band with zero energy can be determined by substituting E = 0 in Eq. (2.2)

$$P\,(0) = \frac{1}{1+e^{(0-E_F)/kT}} = \frac{1}{1+e^{-E_F/kT}} \quad ...(2.6)$$

$$\text{So } n_V = \frac{n}{1+e^{-E_F/kT}} \quad ...(2.7)$$

Now total number of electrons in both bands,

$$n = n_C + n_V = \frac{n}{1+e^{(E_G-E_F)/kT}} + \frac{n}{1+e^{-E_F/kT}}$$

$$\text{or } 1 - \frac{1}{1+e^{-E_F/kT}} = \frac{1}{1+e^{(E_G-E_F)/kT}}$$

$$\text{or } E_F = \frac{1}{2}E_G \quad ...(2.8)$$

i.e. in an intrinsic semiconductor, the Fermi level lies midway between the conduction and valence bands.

Example 2.1. The energy band gap of germanium is 0.72 eV at 300 K. Determine the fraction of the total number of electrons (conduction band as well as valence band) in the conduction band at 300 K. Boltzmann constant is 8.61 × 10^{-5} eV/K.

Solution : From Eq. (2.4)

$$\frac{\text{Number of electrons in conduction band, } n_C}{\text{Total number of electrons in both bands, } n} = \frac{1}{1+e^{(E_G-E_F)/kT}}$$

Substituting $E_F = \frac{1}{2}E_G = \frac{1}{2} \times 0.72 = 0.36$ eV, $k = 8.61 \times 10^{-5}$ eV/K and T = 300 K in above equation we have

$$\frac{n_C}{n} = \frac{1}{1+e^{0.36/300 \times 8.61 \times 10^{-5}}} = 8.85 \times 10^{-7} \text{ Ans.}$$

Example 2.2. Calculate the wavelength λ of an electron with kinetic energy of 300 eV. Mass of an electron = 9.108 × 10^{-31} kg, charge = 1.602 × 10^{-19} C. Planck's constant, *h* = 6.626 × 10^{-34} J-s.

[U.P.S.C. I.E.S. Electrical Engineering-I, 1999]

Solution : Energy associated with electron

$$E = 300 \text{ eV} \quad \text{or } 300 \times 1.602 \times 10^{-19} \text{ J}$$

Mass of electron, $m = 9.108 \times 10^{-31}$ kg

Velocity of electron, $v = \sqrt{\dfrac{2E}{m}} = \sqrt{\dfrac{2 \times 300 \times 1.602 \times 10^{-19}}{9.108 \times 10^{-31}}} = 10.273 \times 10^{6}$ m/s

Wavelength of an electron with KE of 300 eV,

$$\lambda = \frac{hv}{E} = \frac{6.626 \times 10^{-34} \times 10.273 \times 10^{6}}{300 \times 1.602 \times 10^{-19}} = 1.416 \times 10^{-10} \text{ m} = 1.416 \text{ Å} \quad \textbf{Ans.}$$

2.4. Extrinsic Semiconductors

Intrinsic (or pure) semiconductor by itself is of little significance as it has little current conduction capability at ordinary room temperature. However, the electrical conductivity of intrinsic semiconductor can be increased many times by adding very small amount of impurity (of the order of one atom per million atoms of pure semiconductor) to it in the process of crystallization. This process is called *doping* and the doped material is called the *impurity* or *extrinsic* semiconductor.

Germanium and silicon are tetravalent. So the impurity or doping material may be either pentavalent or trivalent. Accordingly the impurity introduced may be of two types viz. (*i*) donor type or N-type and (*ii*) acceptor type or P-type. Depending on the type of impurity added, the extrinsic conductors can be divided into two classes namely (*i*) N-type semiconductors and (*ii*) P-type semiconductors.

1. N-Type Extrinsic Semiconductors. When a small amount of pentavalent impurity, such as *arsenic, antimony, bismuth* or *phosphorus*, is added to a pure semiconductor crystal during the crystal growth, the resulting crystal is called the *N-type extrinsic semiconductor*, where N stands for negative.

When a pentavalent or a donor impurity is added to silicon (or germanium), the impurity atoms form covalent bonds with the silicon (or germanium) atoms; but since intrinsic semiconductor atoms have only four electrons and four holes in their valence shells, one spare valence-shell electron is produced for each impurity atom added. Each spare electron so produced enters the conduction band of pure semiconductor as a free electron. The pentavalent impurity is called the *donor type impurity* as it donates one electron to the conduction band of a pure semiconductor. Though each impurity provides only one free electron yet an extremely small amount of impurity provides enough atoms to supply millions of free electrons.

The effect of adding donor impurity in a semiconductor is illustrated in Fig. 2.11 (*a*). As shown in the figure, the unbound electron remains near its parent atom. This occurs because the impurity atom has a proton in its nucleus which attracts the unbound electron. However, due to the geometry involved, the electron is partially shielded from its parent atom and the attractive force between them is somewhat reduced. The energy required to detach this fifth electron from the atom is of the order of only 0.05 eV for silicon (0.01 eV for germanium). This energy is so small that at room temperature practically all such electrons become free. In other words, at room temperature each impurity atom donates one electron to the conduction band. These donated electrons are called *excess electrons*, because they are in excess to the electrons which are thermally generated by breaking covalent bonds. However, the number of thermally generated electron-hole pairs will be very small in comparison to the number of free electrons due to the impurity atoms. The effect of this doping process on the relative conductivity can best be explained by energy band diagram shown in Fig. 2.11 (*b*). When donor impurities are added to an intrinsic semiconductor, allowable energy levels are introduced at a very small distance below the conduction band, as illustrated in Fig. 2.11 (*b*). These new allowable levels are essentially a discrete level because the added impurity atoms are far apart in the crystal structure and hence their interaction is small. In the case of silicon, the distance of the new discrete allowable energy level is only 0.05 eV (0.01 eV in germanium) below the conduction band and, therefore, at room temperature almost all of the "fifth" electrons of the donor impurity are raised into the conduction band and the conductivity of the material increases considerably. At room temperature in an intrinsic silicon there is about 1 free electron for every 10^{12} atoms (1 to 10^9 for germanium). If the dosage level were 1 in 10 million (10^7), the ratio $10^{12}/10^7$ *i.e.* 10^5 would indicate that the carrier concentration has increased by a ratio $10^5 : 1$.

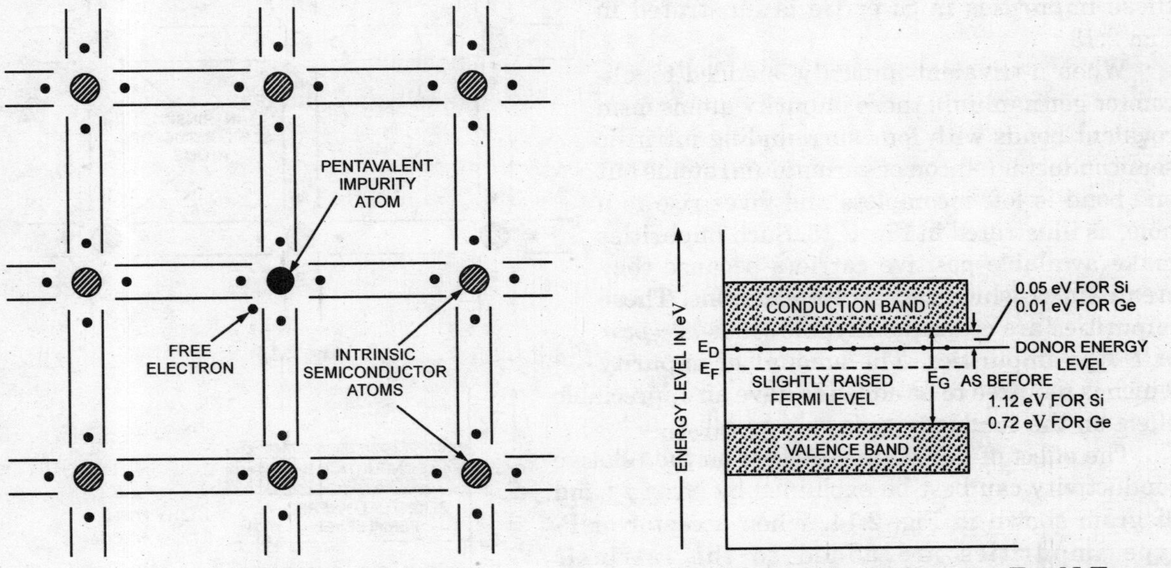

(*a*) (*b*) *Energy Band diagram For N-Type Semiconductor*

Fig. 2.11

Negative or N-Type Conductivity. When an electric field is applied, the excess electrons donated by impurity atoms move towards the positive terminal of the battery, as shown in Fig. 2.12. This constitutes the electric current. This type of conductivity is called the *negative or N-type conductivity* because the current flows through the crystal due to free electrons (negatively charged particles).

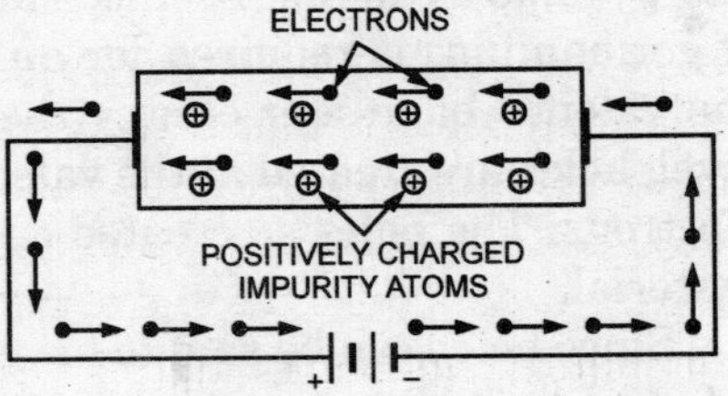

Fig. 2.12

Although N-type semiconductor has excess of electrons but it is electrically neutral. This is due to the fact that electrons are created by the addition of neutral pentavalent impurity atoms to the pure semiconductor *i.e.* there is no addition of either negative charges or positive charges.

Since pentavalent impurity donates electrons, which have negative charge, to the pure semiconductor, such an impurity is also called the *N-type impurity*.

Since current flowing through the crystal is primarily due to free electrons (negatively charged particles), such a conductor is called the *N-type semiconductor*.

In an N-type semiconductor material, the number of holes is small in comparison to parent intrinsic semiconductor because the larger number of electrons present increase the rate of recombination of electrons with the holes. Thus, the number of free electrons becomes far greater than the number of holes. That is why it is said that an N-type semiconductor has electrons as *majority* carriers, and the holes as *minority* carriers.

2. P-Type Extrinsic Semiconductors. When a small amount of trivalent impurity, such as *boron, gallium, indium* or *aluminium,* is added to a pure semiconductor crystal during the crystal growth, the resulting crystal is called the *P-type extrinsic semiconductor*, where P stands for positive. The effect of adding one of these impurities in Si or Ge is illustrated in Fig. 2.13.

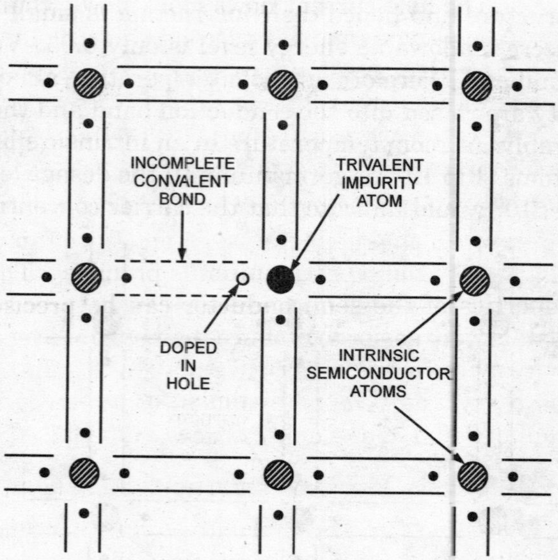

Fig. 2.13

When a trivalent impurity is added to silicon (or germanium), these impurity atoms form covalent bonds with four surrounding intrinsic semiconductor (silicon or germanium) atoms but one bond is left incomplete and gives rise to a hole, as illustrated in Fig. 2.13. Such impurities make available positive carriers because they create holes which can accept electrons. These impurities are consequently known as *acceptor* or *P-type* impurities. The amount of impurity which is required to be added to have an appreciable effect on the conductivity is very small.

The effect of this doping process on the relative conductivity can best be explained by energy band diagram shown in Fig. 2.14. When acceptor or P-type impurities are added to the intrinsic semiconductor, they produce an allowable discrete energy level which is just above the valence band, as shown in Fig. 2.14. Since a very small amount of energy (0.08 eV in case of silicon and 0.01 eV in case of germanium) is required for an electron to leave the valence band and occupy the acceptor energy level, holes are created in the valence band by these electrons. The holes so created constitute the larger number of carriers in the semiconductor material.

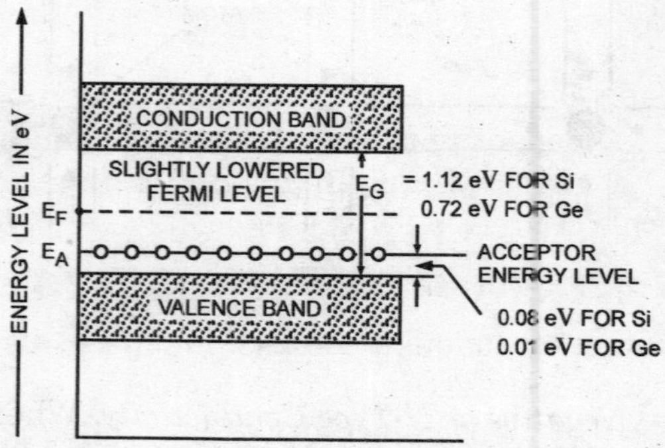

Fig. 2.14. ***Energy Band Diagram For P-Type Semiconductor***

Since holes can be said to have a positive charge, acceptor-doped semiconductor material is referred to as *P-type semiconductor*. The resulting P-type material is electrically neutral, for the same reasons as for the N-type material.

The worthnoting points about P-type semiconductors are:

(*i*) In P-type semiconductors, the majority carriers are holes while the minority carriers are electrons.

(*ii*) P-type semiconductor remains electrically neutral as the number of mobile holes under all conditions remains equal to the number of acceptors.

(*iii*) When an electric field is applied across a P-type semiconductor, as shown in Fig. 2.15, the current conduction is primarily due to holes. Here the holes

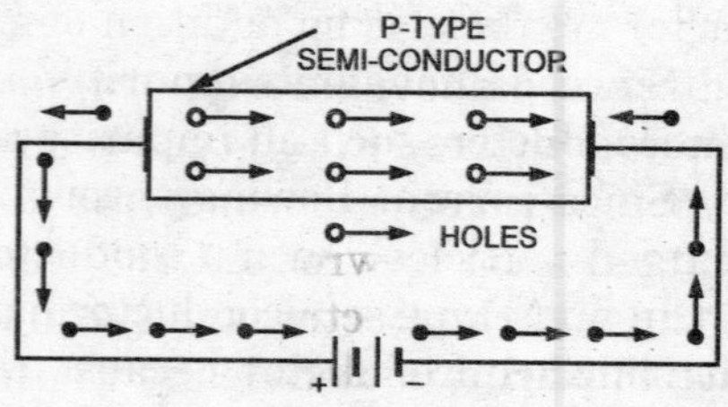

Fig. 2.15

are shifted from one covalent bond to another covalent bond. As the holes are positively charged, they are directed towards the negative terminal and constitute the hole current. The hole current flows more slowly than electron current in N-type semiconductor.

(*iv*) In P-type conductivity, the valence electrons move from one covalent bond to another covalent bond unlike N-type where conduction is by electrons.

Since the current flowing through the crystal is primarily due to holes, which have positive charge, such a semiconductor is called the P-type semiconductor and the conductivity is called the P-type conductivity. The impurity making P-type semiconductor is also called the P-type impurity.

Note: In the manufacturing process of an extrinsic semiconductor, before doping a pure crystal of silicon or germanium is produced. Then by controlling the amount of impurity, the properties of the semiconductor can be precisely controlled. Historically, pure germanium crystals were easier to produce than pure silicon crystals. This is why the earliest semiconductor devices were made of germanium. Eventually, manufacturing techniques improved and pure silicon crystals became available. Because of its advantages, silicon has become the most popular and useful semiconductor material.

2.4.1. Terms Used in Extrinsic Semiconductors

Various terms used in reference to extrinsic semiconductors are explained below.

1. *Doping.* Addition of impurity atoms to intrinsic semiconductor crystal is called the *doping* and the impurity used for doping is called the *dopant*.
2. *Diffusion.* The process of doping a semiconductor with impurities is known as *diffusion*.
3. *Donor.* Arsenic, antimony or phosphorus or any other pentavalent impurity used as dopant to produce an N-type semiconductor is called the *donor type impurity*.
4. *Acceptor.* Boron, gallium, indium or any other trivalent impurity used as dopant to produce a P-type semiconductor is called the *acceptor type impurity*.

2.4.2. Effect of Temperature on Extrinsic Semiconductors

Addition of a small amount of impurity (donor or acceptor) produces a large number of charge carriers in an extrinsic semiconductor, as already discussed above. In fact, this number is so large that the conductivity of an extrinsic semiconductor is many times that of an intrinsic semiconductor at room temperature.

If the temperature of an extrinsic semiconductor (say N-type) is raised, the additional thermal energy increases the thermally generated carriers. As a result, the concentration of minority carriers increases. Eventually, a temperature is reached, called the *critical temperature*, (85°C in case of Ge and 200°C in case of Si) when the number of covalent bonds, that are broken is very large, so that the number of holes becomes approximately the same as the number of electrons. The extrinsic semiconductor now behaves like an intrinsic semiconductor but with higher conductivity.

2.4.3. Charge Carriers

As discussed above, current conduction in an N-type semiconductor is due to excess of free electrons whereas in a P-type semiconductor it is due to excess of holes. One may think that N-type semiconductor has a net negative charge and P-type has a net positive charge. But this conclusion is wrong. Even though the N-type semiconductor has excess free electrons but these extra electrons were supplied by the atoms of donor impurity and each atom of donor impurity is electrically neutral. When the donor impurity is added, the term *"excess electrons"* refers to the excess with reference to the number of electrons required to fill the covalent bonds in the semiconductor crystal. The extra free electrons are there to increase the conductivity of the semiconductor, however, N-type semiconductor is electrically neutral. Similar situation occurs in P-type semiconductor which is also electrically neutral. Free electrons in N-type semiconductors and holes in P-type semiconductors are *charge carriers*.

2.4.4. Majority and Minority Charge Carriers

In the intrinsic state, the number of free electrons in silicon or germanium is only due to those few electrons in the valence band that have acquired sufficient energy from heat or light sources to break the covalent bond. The vacancies left behind in the covalent bonding structure represent very limited supply of holes. In an N-type material, the number of holes has not changed significantly from this intrinsic level and the number of electrons (both added and thermally generated) far outweighs the number of thermally generated holes. Hence in an N-type material electrons are majority carriers and responsible for flow of current.

The reverse is true for the P-type material *i.e.* P-type materials have holes as majority carriers and free electrons as minority carriers, as illustrated in Fig. 2.16 (*b*).

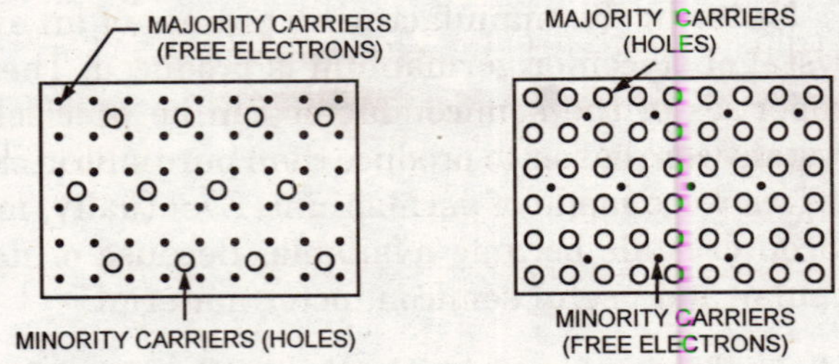

(*a*) *N-Type Material* (*b*) *P-Type Material*

Fig. 2.16

2.5. Mass Action Law

As already mentioned in Art. 2.3.1, concentration of free electrons and holes is always equal in an intrinsic semiconductor. When N-type impurity is added to an intrinsic semiconductor, the concentration of free electrons is increased (or the concentration of holes is reduced below the intrinsic value). Similarly addition of P-type impurity results in reduction of concentration of free electrons below the intrinsic value. Theoretical analysis reveals that under thermal equilibrium the product of concentration of free electrons and concentration of holes is constant and is independent of the amount of doping by donor and acceptor impurities. This is known as *mass action law*. Thus,

$$n\,p = n_i^2 \qquad \text{...(2.9)}$$

where n_i is the intrinsic concentration and is a function of temperature. For an intrinsic semiconductor $n = p = n_i$

The above discussion reveals that the doping of impurity of either type in an intrinsic semiconductor causes increase in the conductivity and results in either free electrons or holes to dominate the situation.

This law is very important from the point of view that it enables us to determine minority carriers concentration. According to this law the addition of impurities to an intrinsic semiconductor increases the concentration of one type of carriers, which consequently become majority carriers and simultaneously decreases the concentration of the other carriers, which as a result become minority carriers.

2.6. Charge Densities in an Extrinsic Semiconductor

Equation (2.9) gives relationship between the electron and hole concentrations n and p respectively.

If N_D is the concentration of donor atoms, there will be N_D immobile positive charges per unit volume contributed by donor ions as the donor atoms are ionized. The positive charge density will be $N_D + p$. Similarly as the acceptor atoms are ionized at room temperature there will be N_A immobile negative charges per unit volume and total negative charge density will be $N_A + n$, where N_A is the concentration of acceptor atoms. Since the semiconductor is neutral, the magnitude of positive charge density will be equal to the magnitude of negative charge density.

$$i.e.\ N_D + p = N_A + n \qquad ...(2.10)$$

In N-type semiconductors there is no acceptor doping *i.e.* $N_A = 0$. Also the number of electrons is much greater than the number of holes *i.e.* $n >> p$

$$\text{So } n = N_D \qquad ...(2.11)$$

i.e. in N-type semiconductors, the free electron concentration is approximately equal to the density of donor atoms.

Now the concentration of holes in N-type semiconductors will be given by the expression

$$p = \frac{n_i^2}{n} = \frac{n_i^2}{N_D} \qquad ...(2.12)$$

Similarly in case of P-type semiconductors $p = N_A$ ...(2.13)

$$\text{and } n = \frac{n_i^2}{N_A} \qquad ...(2.14)$$

Example 2.3. A specimen of germanium at 300 K for which the density of carriers is 2.5 × 10¹³/cm³, is doped with impurity atoms such that there is one impurity atom for 10⁶ germanium atoms. All the impurity atoms may be assumed ionized. The resistivity of doped material is 0.039 Ω-cm. Carrier mobility for germanium at 300 K is 3,600 cm²/V-s. For the doped material, find the electron and hole densities. e = 1.602 × 10⁻¹⁹ C. [*U.P.S.C. I.E.S. 1992*]

Solution : Resistivity of doped material, $\rho = 0.039\ \Omega\text{-cm}$

Conductivity of doped material, $\sigma_n = \frac{1}{\rho} = \frac{1}{0.039} = 25.64$ mho/cm

Electron charge, $e = 1.602 \times 10^{-19}$ C

Carrier mobility, $\mu_e = 3{,}600\ \text{cm}^2/\text{V-s}$

Since $\sigma_n = n\, e\, \mu_e = N_D\, e\, \mu_e$

So concentration of donor atoms, $N_D = \frac{\sigma_n}{e\,\mu_e} = \frac{25.64}{1.602 \times 10^{-19} \times 3{,}600} = 4.45 \times 10^{16}/\text{cm}^3$

Concentration of electrons $n \simeq N_D = 4.45 \times 10^{16}/\text{cm}^3$ **Ans.**

Concentration of holes, $p = \frac{n_i^2}{n} = \frac{(2.5 \times 10^{13})^2}{4.45 \times 10^{16}} = 1.4 \times 10^{10}/\text{cm}^3$ **Ans.**

Example 2.4. A donor type impurity is added to the extent of 1 atom per 10⁶ atoms of an intrinsic semiconductor (silicon). Calculate:

(*i*) Resulting donor atom concentration.

(*ii*) Resulting mobile electron concentration.

(*iii*) Resulting hole concentration.

(*iv*) Conductivity of doped silicon sample.

(*v*) If silicon bar is 0.5 cm long, cross-sectional area of (50 × 10⁻⁴)² cm². Find its resistivity. The concentration of silicon atoms = 5 × 10²² cm⁻³ and n_i = 1.45 × 10¹⁰ cm⁻³

Solution : (*i*) Donor atom concentration,

$$N_D = \text{Number of silicon atoms/cm}^3 \times \text{donor impurity}$$
$$= 5 \times 10^{22} \times \frac{1}{10^6} = 5 \times 10^{16} \text{ per cm}^3 \textbf{ Ans.}$$

(*ii*) Mobile electron concentration, $n \simeq N_D = 5 \times 10^{16}/\text{cm}^3$ **Ans.**

(*iii*) Hole concentration, $p = \dfrac{n_i^2}{N_D} = \dfrac{(1.45 \times 10^{10})^2}{5 \times 10^{16}} = 4.205 \times 10^3/\text{cm}^3$ **Ans.**

(*iv*) Conductivity of doped silicon sample,

$$\sigma = ne\mu_e = 5 \times 10^{16} \times 1.602 \times 10^{-19} \times 1{,}300 = 10.413 \text{ S/cm} \textbf{ Ans.}$$

Taking μ_e for Si as 1,300 cm²/V-s

$$\text{Resistivity, } \rho = \frac{1}{\sigma} = \frac{1}{10.413} = 0.096 \ \Omega\text{-cm}$$

$$\text{Resistance of given semiconductor, } R = \frac{\rho l}{a} = \frac{0.096 \times 0.5}{(50 \times 10^{-4})^2} = 1{,}920 \ \Omega \textbf{ Ans.}$$

Example 2.5. What is the concentration of holes in Si crystals having donor concentration of 1.4 × 10^{24}/m³ when the intrinsic carrier concentration is 1.4 × 10^{18}/m³ ? Find the ratio of electron to hole concentration.

Solution : Intrinsic carrier concentration, $n_i = 1.4 \times 10^{18}/\text{m}^3$

Donor concentration, $N_D = 1.4 \times 10^{24}/\text{m}^3$

Concentration of electrons, $n \simeq N_D = 1.4 \times 10^{24}/\text{m}^3$

$$\text{Concentration of holes, } p = \frac{n_i^2}{n} = \frac{(1.4 \times 10^{18})^2}{1.4 \times 10^{24}} = 1.4 \times 10^{12}/\text{m}^3$$

$$\text{Ratio of electron to hole concentration} = \frac{n}{p} = \frac{1.4 \times 10^{24}}{1.4 \times 10^{12}} = 1 \times 10^{12} \textbf{ Ans.}$$

Example 2.6. Find the conductivity of *n*-type Ge at room temperature, assuming one donor atom in each 10^8 atoms. The density of Ge is 5.32 × 10^3 kg/m³ and the atomic weight is 72.6 kg/k-mol. Comment on the result [e = 1.6 × 10^{-19} C, μ_e = 0.38 m²/V-s, μ_h = 0.18 m²/V-s]. [*I.E.S. 2002*]

Solution :

$$\text{Number of germanium atoms per cm}^3 = 6.023 \times 10^{23} \times \frac{\text{atoms}}{\text{mole}} \times \frac{1 \text{ mole}}{72.6 \text{g}} \times \frac{5.32 \text{g}}{\text{cm}^3}$$
$$= 4.41 \times 10^{22}$$

$$\text{Donor atom concentration, } N_D = \text{Number of germanium atoms/cm}^3 \times \text{donor impurity}$$
$$= \frac{4.41 \times 10^{22}}{10^8} = 4.41 \times 10^{14} \text{ atoms/cm}^3$$

Concentration of electrons, $n = 4.41 \times 10^{14}$ atom/cm³

$$\text{Concentration of holes, } p = \frac{n_i^2}{n} = \frac{(2.5 \times 10^{13})^2}{4.41 \times 10^{14}} = 1.417 \times 10^{12} \text{ holes/cm}^3$$

$\because$ n_i at 300 K for Ge = 2.5 × 10^{13}/cm³

Since $n >> p$.

Conductivity of *n*-type germanium $\sigma = ne\mu_e = 4.41 \times 10^{14} \times 1.6 \times 10^{-19} \times 0.38 \times 10^4$

$= 0.268$/ohm-cm $= 26.8$/ohm-m **Ans.**

Comment : The addition of 1 donor atom in 10^8 germanium atoms has multiplied the conductivity by a factor of $\dfrac{26.8}{2.24} \approx 12$ (conductivity of intrinsic germanium is 2.24/ohm-m).

2.7. CONDUCTIVITY OF SEMICONDUCTOR

As already mentioned in Art. 2.3.4, conduction through a semiconductor is due to two mechanisms viz. drift and diffusion.

The movement of charge carriers (electrons and holes) under the influence of an applied electric field E is termed as the *drift current*.

The conductivity of σ_e of the semiconductor due to electrons in conduction band is given as

$$\sigma_e = n\, e\, \mu_e \quad ...(2.15)$$

where n is the number of electrons per unit volume of the conductor (*i.e.* electron density or concentration), e is the electron charge and μ_e is the electron mobility.

Similarly, the conductivity σ_h of the semiconductor due to holes is given as

$$\sigma_h = p\, e\, \mu_h \quad ...(2.16)$$

where p is the number of holes per unit volume and μ_h is the hole mobility.

Since in semiconductors the conduction is by free electrons as well as by holes, total conductivity σ is given as

$$\sigma = \sigma_e + \sigma_h = n\, e\, \mu_e + p\, e\, \mu_h = e\,(n\, \mu_e + p\, \mu_h) \quad ...(2.17)$$

Since in **intrinsic semiconductors** $n = p = n_i$, the intrinsic concentration of holes or free electrons in the semiconductor, **intrinsic semiconductor conductivity**

$$\sigma_i = \sigma_n + \sigma_h = e\,(n_i\, \mu_e + n_i\, \mu_h) = n_i\, e\,(\mu_e + \mu_h) \quad ...(2.18)$$

$$\text{Current density, } J = n_i\, e\,(\mu_e + \mu_h)\, E = \sigma\, E \quad ...(2.19)$$

$$\text{Current, } I = J\,A = n_i\, e\,(\mu_e + \mu_h)\, EA = n_i\, e\,(\mu_e + \mu_h)\, A\, V/l \quad ...(2.20)$$

where E is applied electric field, V is the applied potential difference across the two ends of the conductor and A is the cross-sectional area of the conductor.

In case of N-type semiconductor hole concentration p_h is negligible and electron concentration $n_n = N_D$. So conductivity in N-type semiconductor

$$\sigma_n = e\, N_D\, \mu_e \quad ...(2.21)$$

Similarly in case of P-type semiconductor electron concentration is negligible and hole concentration $p_h = N_A$. So conductivity in P-type semiconductor

$$\sigma_h = e\, N_A\, \mu_h \quad ...(2.22)$$

Total current density due to drift of electrons and holes on application of electric field E,

J = Current density due to hole drift J_h + current density due to electron drift, J_e

$$= E\, p\, e\, \mu_h + E\, n\, e\, \mu_e = E\, e\,(p\, \mu_h + n\, \mu_e). \quad ...(2.23)$$

Intrinsic Concentration. It has been established that the intrinsic concentration n_i varies with temperature according to the relation

$$n_i^2 = A_0\, T^3 e^{-E_{GO}/kT} \quad ...(2.24)$$

where A_0 is a constant and is independent of temperature, k is a Boltzmann constant in eV per K and E_{GO} is the energy gap in electron volts at zero absolute temperature, T is temperature in degrees absolute and n_i is concentration in number per cubic metre. A_0 is taken as 2.33×10^{43}.

Mobility. The mobility μ of charge carriers (electrons and holes) varies as T^{-m} over a temperature range of 100 K and 400 K. For silicon $m = 2.5$ for electrons and 2.7 for holes. For germanium $m = 1.66$ for electrons and 2.33 for holes.

The carrier currents are also due to concentration gradients in the doped material which leads to **diffusion of carriers** from high concentration region to low concentration region.

A concentration gradient exists if either the number of holes or electrons is greater in one region as compared to the rest of the region. As already mentioned in Art. 2.3.4, when a concentration gradient of carriers exists in a material, the charge carriers tend to move from the region of high concentration to the region of lower concentration. The process is called the *diffusion* and the current due to this process is called the *diffusion current*.

Let us assume that at reference plane x, the density of holes is p per m^3. Let the carriers have a mean free path of Δx and mean free time of Δt between collisions. Assuming uniform density gradient of holes as $\frac{-dp}{dx}$.

The average density of holes in the region $x - \Delta x$ and x

$$= \frac{1}{2}\left[p + p - \frac{dp}{dx}\cdot\Delta x\right]$$

$$= p - \frac{1}{2}\frac{dp}{dx}\cdot\Delta x$$

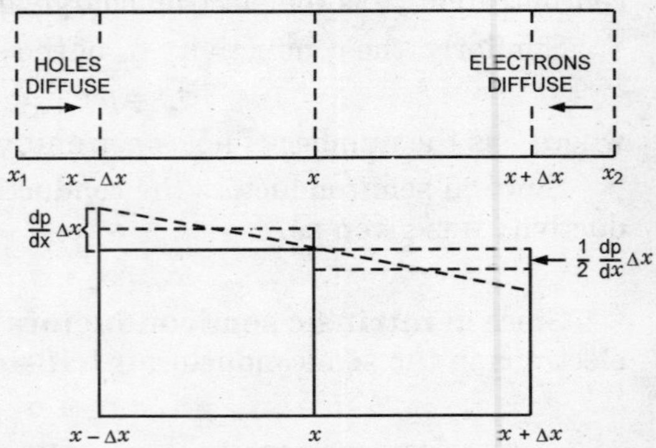

Fig. 2.17

Flow of holes to the right per unit time,

$$p_1 = \frac{1}{2}\left[p - \frac{1}{2}\frac{dp}{dx}\Delta x\right]\frac{\Delta x}{\Delta t}$$

Left directed hole flow across reference plane x per unit time,

$$p_2 = \frac{1}{2}\left[p + \frac{1}{2}\frac{dp}{dx}\Delta x\right]\frac{\Delta x}{\Delta t}$$

The net rate of flow of holes is to the right and is given as

$$p = p_1 - p_2 = \frac{1}{2}\left[p - \frac{1}{2}\frac{dp}{dx}\Delta x\right]\frac{\Delta x}{\Delta t} - \frac{1}{2}\left[p + \frac{1}{2}\frac{dp}{dx}\Delta x\right]\frac{\Delta x}{\Delta t}$$

$$= -\frac{1}{2}\Delta x\cdot\frac{\Delta x}{\Delta t}\cdot\frac{dp}{dx} = -D_h\frac{dp}{dx} \qquad \text{...(2.25)}$$

where $D_h = \frac{1}{2}\Delta x\cdot\frac{\Delta x}{\Delta t} = \frac{1}{2}\frac{\Delta x^2}{\Delta t}$ m^2/s, called the *diffusion constant*.

The current density due to diffusion of holes is given as

$$J_h = -e\,D_h\frac{dp}{dx} \qquad \text{...(2.26)}$$

Similarly current density due to diffusion of electrons is given as

$$J_e = e\,D_e\frac{dn}{dx} \qquad \text{...(2.27)}$$

Total Current in a Semiconductor. It is possible that a potential gradient and a concentration gradient may exist within semiconductor. In such a case, the total current is the sum of drift current due to potential gradient and the diffusion current due to charge carrier concentration gradient.

$$\text{Thus } J_h = E\,e\,p\,\mu_h - e\,D_h\frac{dp}{dx} \qquad \text{...(2.28)}$$

$$\text{and } J_e = E\,e\,n\,\mu_e + e\,D_e\frac{dn}{dx} \qquad \text{...(2.29)}$$

The constants E_{G0}, μ_e, μ_h, D_e, D_h and many other important physical quantities for silicon and germanium are given on next page in tabular form (Table 2.1).

Einstein Relationship. Diffusion constant D and mobility μ are statistical thermodynamic phenomena. Hence they are not independent of each other. The diffusion constants D_h and D_e and mobilities μ_h and μ_e are related by Einstein's equation given as

$$\frac{D_h}{\mu_h} = \frac{D_e}{\mu_e} = V_T \qquad ...(2.30)$$

where V_T is the *volt-equivalent of temperature* and is defined as

$$V_T = \frac{k'T}{e} = kT = \frac{T}{11,600} \qquad ...(2.31)$$

where k' is Boltzmann constant in joule/K and k is Boltzmann constant in eV/ K

TABLE 2.1. Properties of Silicon and Germanium

Property	Silicon	Germanium
Atomic number	14	32
Atomic weight	28.1	72.6
Density	2,330 kg/m^3	5,320 kg/m^3
Dielectric constant	12	16
Atoms/m^3	5.0×10^{28}	4.4×10^{28}
E_{G0} at 0 K	1.21 *eV*	0.785 *eV*
E_G at 300 K	1.12 *eV*	0.72 *eV*
Intrinsic concentration n_i at 300 K	1.5×10^{16}/m^3	2.5×10^{19}/m^3
Intrinsic resistivity ρ_i at 300 K	2,300 Ω-m	0.45 Ω–m
Electron mobility, μ_e	0.13 m^2/ V-s	0.38 m^2/ V–s
Hole mobility, μ_h	0.05 m^2/V-s	0.18 m^2/V–s
Diffusion constant, D_e	0.0034 m^2/s	0.0099 m^2/s
Diffusion constant, D_h	0.0013 m^2/s	0.0047 m^2/s

At room temperature (300 K)

$$V_T = \frac{300}{11,600} = 0.026 \text{ V or } 26 \text{ mV}$$

$$\text{and } \mu = \frac{11,600}{300} D = 39 D$$

Example 2.7. Consider intrinsic germanium at 300 K. By what percentage does the conductivity increase per degree rise in temperature ? [*M.D. Univ. Third Semester May 2005*]

Solution : The intrinsic concentration is given by

$$n_i = AT^{3/2} e^{-eV/2kT}$$

where A = Constant depending upon material and for germanium A = 9.64×10^{15}

T = Temperature in Kelvin

e = Charge of an electron = 1.6×10^{-19} C

k = Boltzmann constant = 1.38×10^{-23} J/k

V = 0.75 V for germanium

$$\text{So, } n_i \text{ at 300 K} = 9.64 \times 10^{1.5} \times (300)^{1.5} e^{\frac{-1.6\times10^{-19}\times0.75}{2\times1.38\times10^{-23}\times300}} = 2.54469\times10^{13}/\text{cm}^3$$

$$n_i \text{ at 301 K} = 9.64 \times 10^{15} \times (301)^{1.5} e^{\frac{-1.6\times10^{-19}\times0.75}{2\times1.38\times10^{-23}\times301}} = 2.68358\times10^{13}/\text{cm}^3$$

Intrinsic conductivity at 300 K

$$\sigma_i = n_i e(\mu_e + \mu_n) \text{ S/cm} = 2.54469 \times 10^{13} \times 1.6 \times 10^{-19}(3{,}800 + 1{,}800) = 0.0228 \text{ S/cm}$$

Intrinsic conductivity at 301 K

$$\sigma_1' = 2.68358 \times 10^{13} \times 1.6 \times 10^{-19} (3{,}800 + 1{,}800) = 0.02405 \text{ S/cm}$$

Percentage increase in conductivity per degree rise in temperature

$$= \frac{0.02405 - 0.0228}{0.0228} \times 100 = 5.48\% \textbf{ Ans.}$$

Example 2.8. Find the conductivity and resistivity of an intrinsic semiconductor at temperature of 300 K. It is given that $n_i = 2.5 \times 10^{13}/\text{cm}^3$, $\mu_n = 3{,}800 \text{ cm}^2/\text{V-s}$; $\mu_p = 1{,}800 \text{ cm}^2/\text{V-s}$, $e = 1.6 \times 10^{-19}$ C.

Solution : Intrinsic concentration, $n_i = 2.5 \times 10^{13}/\text{cm}^3$

Electron charge, $e = 1.6 \times 10^{-19}$ coulomb

Hole mobility, $\mu_h = 1{,}800 \text{ cm}^2/\text{V-s}$

Electron mobility, $\mu_e = 3{,}800 \text{ cm}^2/\text{V-s}$

Intrinsic conductivity, $\sigma_i = n_i e (\mu_e + \mu_h)$

$= 2.5 \times 10^{13} \times 1.6 \times 10^{-19} (3{,}800 + 1{,}800) = 0.0224 \text{ (ohm-cm)}^{-1}$ **Ans.**

Intrinsic resistivity, $\rho_i = \dfrac{1}{\sigma_i} = \dfrac{1}{0.0224} = 44.64\ \Omega\text{-cm}$ **Ans.**

Example 2.9. The intrinsic resistivity of germanium at 300 K is 0.47 ohm-m. The electron mobility (μ_e) at 300 K in germanium is 0.39 m²/V-s. The hole mobility μ_h at 300 K in germanium is 0.19 m²/V-s. The electronic charge (e) is 1.6×10^{-19} C. Calculate the density of electrons in the intrinsic material. Also calculate the drift velocity of holes and electrons for an electric field (E) = 10^4 V/m.

Solution : Intrinsic resistivity, $\rho_i = 0.47\ \Omega\text{-m}$

Intrinsic conductivity, $\sigma_i = \dfrac{1}{\rho_i} = \dfrac{1}{0.47} = 2.12766$ S/m

Electron mobility, $\mu_e = 0.39 \text{ m}^2/\text{V–s}$

Hole mobility, $\mu_h = 0.19 \text{ m}^2/\text{V–s}$

Electron charge, $e = 1.6 \times 10^{-19}$ C

Since $\sigma_i = n_i e (\mu_e + \mu_h)$,

Density of electrons = Intrinsic concentration n_i

$$= \frac{\sigma_i}{e(\mu_e + \mu_h)} = \frac{2.12766}{1.6 \times 10^{-19}(0.39 + 0.19)} = 2.293 \times 10^{19}/\text{m}^3 \textbf{ Ans.}$$

Electric field, E = 10^4 V/m

Drift velocity for electrons, $v_n = \mu_e \text{E} = 0.39 \times 10^4 = 3{,}900$ m/s **Ans.**

Drift velocity for holes, $v_h = \mu_h \text{E} = 0.19 \times 10^4 = 1{,}900$ m/s **Ans.**

Example 2.10. The intrinsic carrier concentration for silicon at room temperature (300 K) is $1.5 \times 10^{10}/\text{cm}^3$. If the mobilities of electrons and holes are 1,300 cm²/V–s and 450 cm²/V-s respectively, what is the conductivity of silicon (intrinsic) at 300 K? If silicon is doped with 10^{18} boron atoms per cm³, what is its conductivity?

Solution : Intrinsic carrier concentration of silicon,

$n_i = 1.5 \times 10^{10}/\text{cm}^3$

Hole mobility, $\mu_h = 450 \text{ cm}^2/\text{V-s}$

Electron mobility, $\mu_e = 1{,}300 \text{ cm}^2/\text{V-s}$

Conductivity of silicon (intrinsic), $\sigma_i = n_i e (\mu_e + \mu_h) = 1.5 \times 10^{10} \times 1.6 \times 10^{-19} \times (1{,}300 + 450)$

$= 4.2 \times 10^{-6} \text{ (ohm-cm)}^{-1}$ **Ans.**

Doping level, $N_A = 10^{18}/cm^3$

Conductivity of the resulting P-type silicon semiconductor,

$$\sigma_p = e\, N_A\, \mu_h = 1.6 \times 10^{-19} \times 10^{18} \times 450 = 72\ (\text{ohm-cm})^{-1}\ \textbf{Ans.}$$

Example 2.11. Find the conductivity of intrinsic germanium at 300 K. If donor type impurity is added to the extent of 1 impurity atom in 10^7 germanium atoms, find the conductivity. Given that n_i at 300 K is $2.5 \times 10^{13}/cm^3$ and μ_n and μ_p in germanium are 3,800 and 1,800 cm^2/V-s respectively.

Solution : Intrinsic conductivity, $\sigma_i = n_i\, e\, (\mu_e + \mu_h)$

$$= 2.5 \times 10^{13} \times 1.6 \times 10^{-19}\ (3{,}800 + 1{,}800) = 0.0224\ (\text{ohm-cm})^{-1}\ \textbf{Ans.}$$

Number of germanium atoms/cm^3 $= 4.41 \times 10^{22}$

Donor impurity = 1 donor atom/10^7 germanium atoms

Concentration of donor atoms, $N_D = 4.41 \times 10^{22} \times 10^{-7} = 4.41 \times 10^{15}/cm^3$

Concentration of electrons, n $N_D = 4.41 \times 10^{15}/cm^3$

$$\text{Concentration of holes, } p = \frac{n_i^2}{N_D} = \frac{(2.5 \times 10^{13})^2}{4.41 \times 10^{15}} = 1.417 \times 10^{11}\ \text{holes/cm}^3$$

So conductivity in N-type germanium semiconductor

$$\sigma_n = e\, N_D\, \mu_e = 1.6 \times 10^{-19} \times 4.41 \times 10^{15} \times 3{,}800 = 2.68\ (\text{ohm-cm})^{-1}\ \textbf{Ans.}$$

Example 2.12. A pd of 10 V is applied longitudinally to a rectangular specimen of intrinsic germanium of length 25 mm, width 4 mm and thickness 1.5 mm. Determine at room temperature (*i*) electron and hole drift velocities (*ii*) the conductivity of intrinsic germanium if intrinsic carrier density is $2.5 \times 10^{19}/m^3$ and (*iii*) the total current.

Given that $e = 1.6 \times 10^{-19}$ C, $\mu_e = 0.38\ m^2\ V^{-1}\ s^{-1}$. $\mu_h = 0.18\ m^2\ V^{-1}\ s^{-1}$.

Solution : Applied electric field, $E = V/l = \dfrac{10}{0.025} = 400$ V/m

(*i*) Electron drift velocity, $v_e = \mu_e \times E = 0.38 \times 400 = 152$ m/s **Ans.**

Hole drift velocity, $v_h = \mu_h \times E = 0.18 \times 400 = 72$ m/s **Ans.**

Intrinsic carrier concentration, $n_i = 2.5 \times 10^{19}/m^3$

(*ii*) Intrinsic conductivity of Ge, $\sigma_i = n_i\, e\, (\mu_e + \mu_h)$

$$= 2.5 \times 10^{19} \times 1.6 \times 10^{-19}\ (0.38 + 0.18) = 2.24\ (\text{ohm-m})^{-1}\ \textbf{Ans.}$$

(*iii*) Total current, $I = \sigma_i\, E\, a = 2.24 \times 400 \times 4 \times 10^{-3} \times 1.5 \times 10^{-3} = 5.376$ mA **Ans.**

Example 2.13. Find the diffusion coefficients of holes and electrons for germanium at 300 K. The carrier mobilities in cm^2/V-s at 300 K for electrons and holes are respectively 3,600 and 1,700. Density of carriers is $2.5 \times 10^{13}/cm^3$. Boltzmann constant $k = 1.38 \times 10^{-23}$ J/ K. $e = 1.602 \times 10^{-19}$ C.

[*U.P.S.C. I.E.S. Elec. Engineering-I, 1996*]

Solution : From Eqs. (2.30) and (2.31)

$$\text{Diffusion constant, } D_e = \mu_e \frac{k'T}{e} = \frac{3{,}600 \times 1.38 \times 10^{-23} \times 300}{1.602 \times 10^{-19}} = 93\ cm^2/s\ \textbf{Ans.}$$

$$\text{Diffusion constant, } D_h = \mu_n \frac{k'T}{e} = \frac{1{,}700 \times 1.38 \times 10^{-23} \times 300}{1.602 \times 10^{-19}} = 43.93\ cm^2/s\ \textbf{Ans.}$$

Example 2.14. Find the conductivity of intrinsic silicon at 300 K. It is given that n_i at 300 K in silicon is $1.5 \times 10^{10}/cm^3$ and mobility of electrons and holes in silicon are 1,300 cm^2/V-s and 500 cm^2/V-s respectively.

If donor type impurity is added to extent of 1 impurity atom in 10^8 silicon atoms, find the conductivity.

If acceptor type impurity is added to extent of 1 impurity atom in 10^8 silicon atoms, find the conductivity. [*M.D. Univ. Third Semester May 2006*]

Solution : Intrinsic carrier concentration of silicon,

$$n_i = 1.5 \times 10^{10}/\text{cm}^3$$

$$\text{Hole mobility, } \mu_h = 500 \text{ cm}^2/\text{V-s}$$

$$\text{Electron mobility, } \mu_e = 1{,}300 \text{ cm}^2/\text{V-s}$$

$$\text{Conductivity of silicon (intrinsic), } \sigma_i = n_i \times e \times (\mu_e + \mu_h) = 1.5 \times 10^{10} \times 1.6 \times 10^{-19} (1{,}300 + 500)$$

$$= 4.32\ (\Omega\text{-cm})^{-1} \textbf{ Ans.}$$

(*i*) Number of silicon atoms/cm^3 = 5.0×10^{22}

$$\text{Concentration of electrons, } n \simeq N_D = 5.0 \times 10^{22} \times \frac{1}{10^8} = 5 \times 10^{14} \text{ atoms/cm}^3$$

$$\text{Conductivity of doped material, } \sigma_n = N_D e \mu_e = 5.0 \times 10^{14} \times 1.6 \times 10^{-19} \times 1{,}300 = 0.104\ (\Omega\text{-cm})^{-1} \textbf{ Ans.}$$

(*ii*) Concentration of holes, $p \simeq N_A = 5.0 \times 10^{22} \times \frac{1}{10^8} = 5 \times 10^{14}$ atoms/cm^3

$$\text{Conductivity of doped material, } \sigma_p = N_A e \mu_h = 5 \times 10^{14} \times 1.6 \times 10^{-19} \times 500 = 0.04\ (\Omega\text{-cm})^{-1} \textbf{ Ans.}$$

2.8. FERMI LEVEL IN EXTRINSIC SEMICONDUCTORS

The equation for f (E) specifying the fraction of all states at energy E (electron volts) occupied under conditions of thermal equilibrium is called the *Fermi-Dirac probability function* and is given as

$$f(E) = \frac{1}{1 + e^{(E - E_F)/kT}} \qquad \text{...(2.32)}$$

where k is Boltzmann constant in eV/K, T is temperature in K and E_F is the *Fermi level* or *characteristic energy* for crystal in eV.

Concentrations of free electrons n and concentration of holes p are given as

$$n = n_C\, e^{-(E_C - E_F)/kT} \qquad \text{...(2.33)}$$

$$\text{and } p = n_V\, e^{-(E_F - E_V)/kT} \qquad \text{...(2.34)}$$

where n_C is the number of electrons in conduction band, n_V is the number of electrons in the valence band, E_C is the lowest energy in the conduction band in eV and E_V is the maximum energy of valence band in eV.

The only parameter that changes with the addition of impurity in Eqs. (2.33) and (2.34) is the Fermi level.

Fermi Level in N-type Extrinsic Semiconductor. In case of an intrinsic semiconductor, E_F lies in the middle of the energy gap, indicating equal concentrations of free electrons and holes (refer to Art. 2.3.6).

When a donor type impurity is added to the crystal, then at a given temperature and assuming all donor atoms, the first N_D states in the conduction band will be filled. Hence it will be more difficult for the electrons from the valence band to bridge the energy gap by the thermal agitation. Consequently the number of electron-hole pairs thermally generated for that temperature will be reduced. Since the Fermi level is a measure of the probability of occupancy of the allowed energy states, it is clear that E_F must be more closer to the conduction band to indicate that many of the energy states in that band are filled by the donor electrons and fewer holes exist in the valence band. This situation is illustrated in Fig. 2.18 (*a*), for an N-type extrinsic semiconductor.

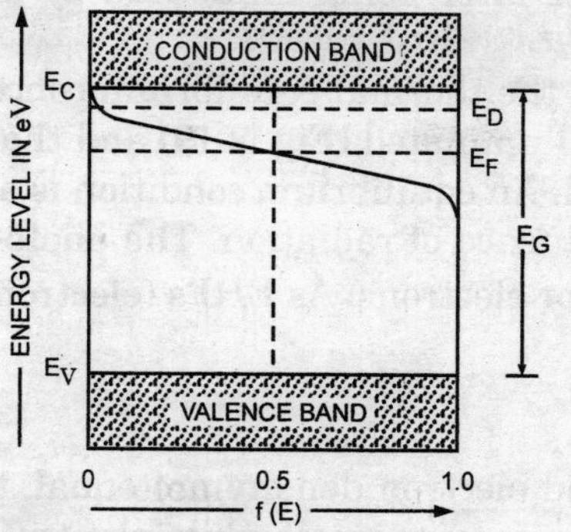

(a) Position of Fermi Level in an N-Type Semiconductor

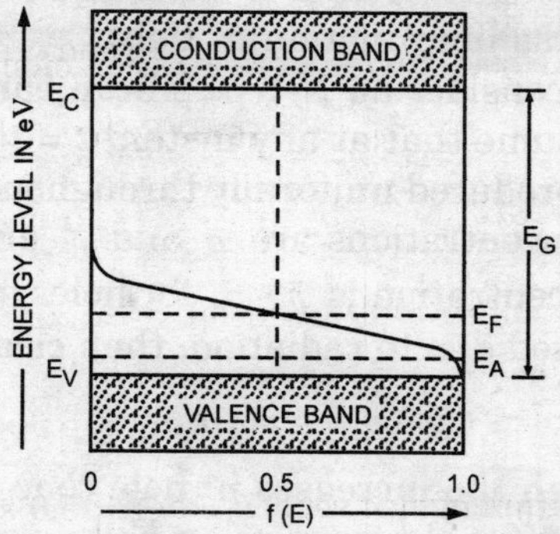

(b) Position of Fermi Level in a P-Type Semiconductor

Fig. 2.18

The same kind of argument leads to the conclusion that E_F must move from the centre of the forbidden gap closer to the valence band for a *P-type extrinsic semiconductor*.

Exact position of the Fermi level in an N-type extrinsic semiconductor can be determined by substituting $n = N_D$ in Eq. (2.33) and thus we have

$$N_D = n_C \, e^{-(E_C - E_F)/kT} \quad \text{...(2.35)}$$

Taking logarithm we have, $\log_e N_D = \log_e n_C - \dfrac{(E_C - E_F)}{kT}$

Solving the above equation we have, $E_F = E_C - kT \log_e \dfrac{n_C}{N_D}$ ...(2.36)

Similarly for the P-type extrinsic semiconductor,

$$E_F = E_V + kT \log_e \frac{n_V}{N_A} \quad \text{...(2.37)}$$

2.9. Carrier Lifetime

As already mentioned in Art. 2.3.1, in an extrinsic semiconductor the number of holes is equal to the number of free electrons. Due to thermal agitation new electron-hole pairs are continuously generated while other electron-hole pairs disappear as a result of recombination *i.e.* free electrons falling into empty covalent bonds. On an average, a hole exists for τ_h second known as *mean lifetime of hole* and similarly an electron exists for τ_e second known as *mean lifetime of electron.* These lifetimes vary from a few nanoseconds to several microseconds. The lifetimes of carriers (holes and electrons) form very important factors in semiconductor devices because they indicate the time required for hole and electron

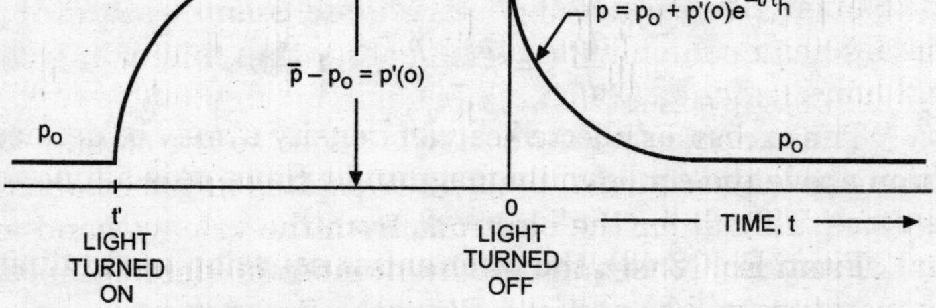

Fig. 2.19. ***Hole (Minority) Concentration in an N-Type Semiconductor Bar as a Function of Time Due to Generation and Recombination***

concentrations to return to their equilibrium values after being influenced by any external cause such as illumination by light.

Let us consider an N-type silicon bar containing the thermal-equilibrium concentration p_o and n_o. Assume that at any instant $t = t'$ the bar is illuminated (Fig. 2.19) and that additional EHPs are produced uniformly throughout the crystal. An equilibrium condition is attained and the new concentrations are $\bar{p}$ and $\bar{n}$ under the influence of radiation. The photo-injected, or excess, concentration is $\bar{p} - p_o$ for holes and $\bar{n} - n_0$, for electrons. As EHPs (electron-hole pairs) are generated due to radiation, then clearly

$$\bar{p} - p_o = \bar{n} - n_0 \quad \text{...(2.38)}$$

Although the increases in hole concentration and electron density are equal, the percentage increase for electrons in an N-type semiconductor is very small while percentage increase in holes may be tremendous. This is because electrons are plentiful and holes are scarce in an N-type crystal. In short it can be said that the majority carrier concentration is hardly affected due to radiation. So our discussion will be limited to the behaviour of the minority carriers.

When a steady-state condition is reached, at time $t = 0$ (Fig. 2.19), the radiation is removed. Now it is to be shown that the excess carrier density returns to zero exponentially with time.

From the definition of mean lifetime τ_h and assuming that τ_h is independent of magnitude of the hole concentration,

$$\frac{p}{\tau_h} = \text{Decrease in hole concentration per second due to recombination} \quad \text{...(2.39)}$$

From the definition of generation rate,

$$g = \text{Increase in hole concentration per second due to thermal generation} \quad \text{...(2.40)}$$

Since charge can neither be generated nor destroyed, the increase in hole concentration per second must be equal to $\frac{dp}{dt}$. This rate must at every instant of time, equal the algebraic sum of the rates given in Eqs. (2.39) and (2.40) *i.e.*

$$\frac{dp}{dt} = g - \frac{p}{\tau_h} \quad \text{...(2.41)}$$

Under steady-state conditions, the rate of increase in hole concentration $\frac{dp}{dt}$ must be equal to zero, and with no radiation falling on the sample, the hole concentration attains its thermal-equilibrium value p_o.

So $g = \frac{p_o}{\tau_h}$ and the above Eq. (2.41) becomes

$$\frac{dp}{dt} = \frac{p_o - p}{\tau_h} \quad \text{...(2.42)}$$

The excess, or injected carrier density p' may be defined as the increase in minority concentration above the equilibrium magnitude. Since p' is a function of time, then

$$p' = p - p_o = p'(t) \quad \text{...(2.43)}$$

From Eq. (2.43), the differential equation controlling p' is

$$\frac{dp'}{dt} = \frac{-p'}{\tau_h} \quad \text{...(2.44)}$$

The rate of change of excess concentration is proportional to this concentration—an intuitively correct result. The minus sign indicates that the change is a decrease in the case of recombination and an increase when the concentration is recovering from a temporary depletion.

Since the radiation causes an initial (at ≤ 0) excess concentration $p'(0) = \bar{p} - p_0$ and then this excitation is removed, the solution of Eq. (2.44) for $t \geq 0$ is

$$p'(t) = p'(0)\, e^{-t/\tau_h} = (\bar{p} - p_0)\, e^{-t/\tau_h} = p - p_o \qquad \text{...(2.45)}$$

The excess concentration falls exponentially to zero ($p' = 0$ or $p = p_o$) with a time constant equal to the mean lifetime τ_h, as shown in Fig. 2.19.

The most important mechanism through which recombination of holes and electrons occur is the mechanism involving recombination centres or traps which contribute electronic states in the energy gap of the semiconductor material. These new states are associated with imperfections in the crystal. Specifically, metallic impurities in the semiconductor are capable of introducing energy states in the forbidden gap. Recombination is affected not only by volume impurities, but also by surface imperfections in the crystal.

Gold is extensively employed as a recombination agent in semiconductor manufacture. Thus the designer of the semiconductor device can obtain desired carrier lifetimes by introducing gold into silicon under controlled conditions.

2.10. CONTINUATION EQUATION

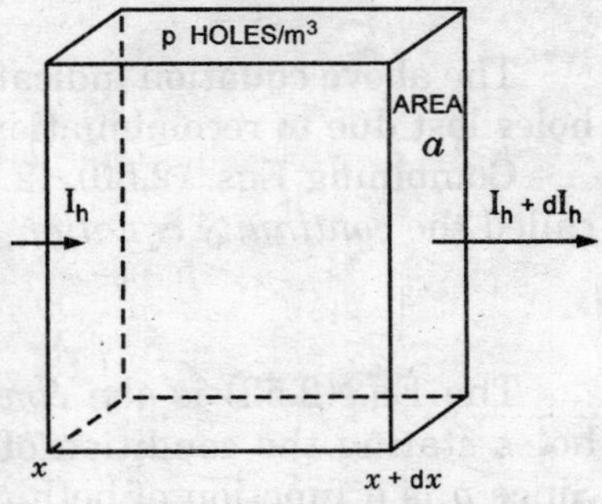

Fig. 2.20

The continuity equation states a condition of dynamic equilibrium for the concentration of mobile carriers in any elementary volume of the semiconductor.

In Art. 2.9 we have seen that on disturbing the equilibrium concentrations of carriers (electrons and holes) in a semiconductor, the concentrations of holes or electrons vary with time approaching the equilibrium value exponentially. In general, however, the carrier concentration in the body of a semiconductor is a function of both time and distance. The differential equation governing this functional relationship, called the *continuity equation*, is based upon the fact that charge can be neither created nor destroyed.

Consider an infinitesimal element of volume of area a and length dx, as shown in Fig. 2.20, within which the average hole concentration is p. If τ_h is the mean lifetime of holes, then p/τ_h equals the holes per second lost by recombination per unit volume. If e is the electronic charge, then, because of recombination, the number of coulombs per second decreases within the volume and

$$\text{Decrease within the volume} = e\, a\, dx\, \frac{p}{\tau_h} \qquad \text{...(2.46)}$$

If g is the thermal rate of generation of electron-hole pairs per unit volume, the number of coulombs per second increases within the volume and

$$\text{Increase within the volume} = e\, a\, dx\, g \qquad \text{...(2.47)}$$

In general, the current varies with distance within the semiconductor. If the current entering the volume at x is I_h and leaving at $x + dx$ is $I_h + d\,I_h$, as shown in Fig. 2.20, the number of coulombs per second decreases within the volume and

$$\text{Decrease within the volume} = d\, I_h \qquad \text{...(2.48)}$$

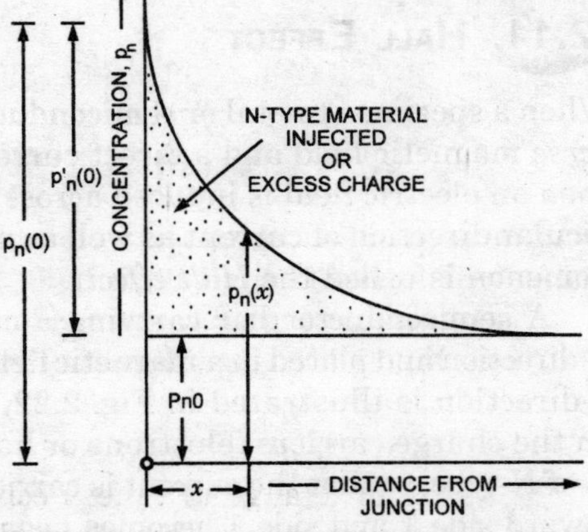

Fig. 2.21. ***Hole Concentration Components in N-Type Material***

Because of the three effects enumerated above, the hole concentration must change with time, and the total number of coulombs per second increases within the volume and

$$\text{Increase within the volume} = e\,a\,dx\,\frac{dp}{dt} \qquad ...(2.49)$$

Since the charge must be conserved, so

$$e\,a\,dx\,\frac{dp}{dt} = -e\,a\,dx\,\frac{p}{\tau_h} + e\,a\,dx\,g - d\,I_h \qquad ...(2.50)$$

The hole current I_h is the sum of drift current and diffusion current so

$$I_h = a\,E\,p\,e\,\mu_h - a\,e\,D_h\,\frac{dp}{dx} \qquad ...(2.51) \text{ [Refer to Eq. (2.28)]}$$

If the semiconductor is in thermal equilibrium with its surroundings and is subjected to no applied field, the hole density will attain a constant value p_0. Under these conditions $I_h = 0$ and $\frac{dp}{dt} = 0$. So from the Eq. (2.50) we have

$$g = \frac{p_0}{\tau_h} \qquad ...(2.52)$$

The above equation indicates that the rate of thermal generation of holes equals the rate of holes lost due to recombination, under equilibrium conditions.

Combining Eqs. (2.50), (2.51) and (2.52) we have the equation of *conservation of charge*, called the *continuity equation*

$$\frac{dp}{dt} = -\frac{p - p_0}{\tau_h} + D_h\,\frac{d^2 p}{dx^2} - \mu_h\,\frac{d(pE)}{dx} \qquad ...(2.53)$$

The Eq. (2.53) is the *continuation equation* or *equation of conservation* of charge for holes stating the condition of dynamic equilibrium for the density of mobile carrier holes. Since p is a function of both t and x, partial derivatives should be used and Eq. (2.53) may be modified as

$$\frac{\partial p}{\partial t} = -\frac{p - p_0}{\tau_h} + D_h\,\frac{\partial^2 p}{\partial x^2} - \mu_h\,\frac{\partial(pE)}{\partial x} \qquad ...(2.54)$$

Similarly, the continuity equation for electrons states the condition of dynamic equilibrium for the density of mobile carrier electrons and is given by

$$\frac{\partial n}{\partial t} = -\frac{n - n_0}{\tau_e} + D_e\,\frac{\partial^2 n}{\partial x^2} - \mu_e\,\frac{\partial(nE)}{\partial x} \qquad ...(2.55)$$

2.11. HALL EFFECT

When a specimen (metal or semiconductor) is placed in a *transverse* magnetic field and a direct current is passed through it, then an electric field is induced across its edges in the perpendicular direction of current as well as magnetic field. This phenomenon is called the *Hall effect*.

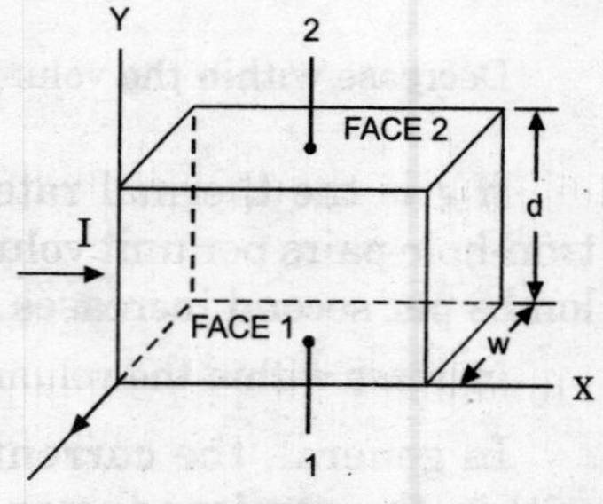

Fig. 2.22

A semiconductor bar carrying a current **I** in the positive X-direction and placed in a magnetic field **B** acting in the positive Z direction is illustrated in Fig. 2.22. Now a force is exerted on the charge carriers (electrons or holes) in the negative Y direction. If the semiconductor is of N-type, so that the current is carried by electrons, these electrons will be forced downward toward side 1 and side 1 becomes negatively charged with respect to side 2. Thus a potential difference V_H, called the *Hall voltage*, is developed between surfaces 1 and 2. The polarity of

Hall voltage enables us to determine whether the semiconductor specimen is N-type or P-type. In case of P-type semiconductor side 1 will become positively charged with respect to side 2.

In the equilibrium state the electric field intensity E due to Hall effect must exert a force on the charge carriers which just balances the magnetic force

$$i.e.\ e\,E_H = Bev \qquad ...(2.56)$$

where e is the magnitude of charge on electron or hole and v is the drift velocity.

Now electric field, $E_H = \dfrac{V_H}{d}$

$$\text{or } V_H = E_H d = Bvd \qquad ...(2.57) \quad \because \text{ from Eq. (2.56) } E_H = Bv$$

where d is the distance between surfaces 1 and 2.

The current density is given as $J = \dfrac{I}{A} = \dfrac{I}{wd} = \rho v$

where ρ is the charge density and w is the width of specimen.

$$\therefore v = \frac{I}{\rho w d} \qquad ...(2.58)$$

From Eqs. (2.57) and (2.58) we have

$$V_H = \frac{BdI}{\rho w d} = \frac{BI}{\rho w} \qquad ...(2.59)$$

Thus from the above equation charge density ρ can be determined if the quantities B, I, V_H and w are measured.

Hall field per unit current density per unit magnetic field is called the *Hall coefficient*. So Hall coefficient,

$$R_H = \frac{\text{Electric field due to Hall effect}}{\text{Current density} \times \text{magnetic field}}$$

$$= \frac{E_H}{J \times B} = \frac{V_H/d}{J \times B} \qquad \because E_H = \frac{V_H}{d}$$

$$= \frac{BI}{\rho w} \times \frac{1}{d \times J \times B} \qquad \because \text{ From Eq. (2.59) } V_H = \frac{BI}{\rho w}$$

$$= \frac{I}{\rho w d J} = \frac{1}{\rho} \qquad \because I = wdJ$$

$$= \frac{1}{ne} = \frac{V_H w}{BI} \qquad ...(2.60) \quad \because \rho = ne$$

In any extrinsic semiconductor, the conductivity is due to charge of one type. Hence the conductivity σ is given as

$$\sigma = \rho\mu \qquad ...(2.61)$$

If the conductivity σ is measured together with the Hall coefficient, the mobility μ can be determined from the relation

$$\mu = \sigma R_H \qquad ...(2.62)$$

The above equation is derived by considering that all the charge carriers move with the mean drift velocity v. Actually the charge carriers have a random thermal distribution in speed. If such distribution is taken into account, the Eq. (2.62) is modified as

$$\mu = \frac{8\sigma}{3\pi} R_H \qquad ...(2.63)$$

2.11.1. Hall Angle

The net electric field E in the specimen is a vector sum of electric field component in the X-direction because of flow of current, E_x and electric field due to Hall effect *i.e.* E_H, as shown in Fig. 2.23.

So resultant electric field E acts at an angle of θ_H to the X-axis and this angle is called *Hall angle*.

Thus, Hall angle, $\theta_H = \text{Tan}^{-1} \dfrac{E_H}{E_x}$

From Eq. (2.57), $E_H = \dfrac{V_H}{d}$

and electric field component in the X-direction

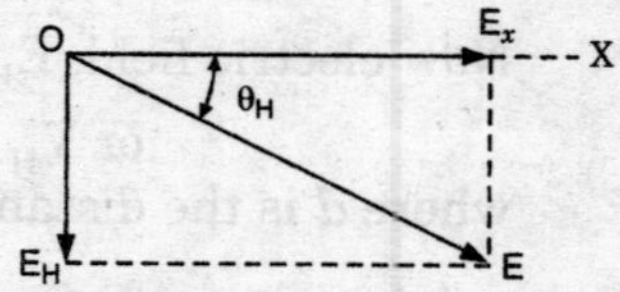

Fig. 2.23

$$E_x = \frac{\text{Voltage drop along the length}}{\text{Length of specimen}}$$

$$= \frac{IR}{l} = \frac{I \times \text{resistivity} \times l}{l \times a} = \frac{I}{a} \times \text{resistivity} = J \times \frac{1}{\sigma} = \frac{J}{\sigma} \quad \text{...(2.64)}$$

Substituting the values of $E_H = \dfrac{V_H}{d}$ and $E_x = \dfrac{J}{\sigma}$ in expression for Hall angle we have

$$\theta_H = \text{Tan}^{-1} \frac{E_H}{E_x} = \text{Tan}^{-1} \frac{V_H/d}{J/\sigma} = \text{Tan}^{-1} \frac{BI/\rho w d}{J/\sigma} \quad \because \text{From Eq. (2.59) } V_H = \frac{BI}{\rho w}$$

$$= \text{Tan}^{-1} \frac{BJ/\rho}{J/\sigma} \quad \because J = \frac{I}{wd}$$

$$= \text{Tan}^{-1} \frac{B\sigma}{\rho} = \text{Tan}^{-1} \sigma B R_H \quad \text{...(2.65)} \quad \because \text{From Eq. (2.60), } R_H = \frac{1}{\rho}$$

$$= \text{Tan}^{-1} \mu B \quad \text{...(2.66)} \quad \because \text{From Eq. (2.62), } \mu = \sigma R_H$$

2.11.2. Applications of Hall Effect

The Hall effect may be used to

(*i*) determine whether a semiconductor is of N-type or P-type.

(*ii*) determine the carrier (electron or hole) concentration.

(*iii*) measure the conductivity of a material and then compute mobility.

(*iv*) measure magnetic flux density because Hall voltage V_H, for a given current, is proportional to magnetic flux density B.

(*v*) measure power in an electromagnetic wave.

The Hall effect multiplier gives the output that is proportional to the product of two signals. If current I is proportional to one input signal and B is linearly related to the other signal then Hall voltage, V_H, which is proportional to the product of B and I, is proportional to the product of both signals.

Example 2.15. A sample of N-type semiconductor has a resistivity, of 0.1 Ω-cm and Hall coefficient of 100 cm³/Coulomb. Assuming only electrons as carriers, determine the electron density and mobility.

[*U.P.S.C. I.E.S., Electrical Engineering-I, 2006*]

Solution : Conductivity, $\sigma = \dfrac{1}{\text{Resistivity}} = \dfrac{1}{0.1} (\Omega\text{-cm})^{-1} = 10\ (\Omega\text{-cm})^{-1} = 0.1\ (\Omega\text{-m})^{-1}$

Charge density, $\rho = \dfrac{1}{R_H} = \dfrac{1}{100} = 0.01$ coulomb/cm^3 $= 10^4$ C/m^3

$$\text{Electron density} = \frac{\text{Charge density}}{e} = \frac{10^4}{1.602 \times 10^{-19}} = 6.25 \times 10^{22}/\text{m}^3 \quad \textbf{Ans.}$$

$$\text{Mobility, } \mu = \sigma R_H = 0.1 \times \frac{100}{10^6} = 10^{-5} \ \text{m}^2/\text{V-s} \quad \textbf{Ans.}$$

Example 2.16. Calculate the Hall voltage across the width of a semiconducting specimen from the following data :

Width of specimen = 0.1 m, thickness of the specimen = 0.01 m, field applied perpendicular to width and length = 0.6 T, current flowing lengthwise = 10 mA, Hall coefficient = 3.8 × 10^{-4} m^3/C.

[U.P. Technical Univ. Electrical Engineering Materials 2004-05]

Solution : Width of the specimen, $w = 0.1$ m

Flux density, $B_Z = 0.6$ T

Hall coefficient, $R_H = 3.8 \times 10^{-4}$ m^3 C^{-1}

Current, I = 10 mA

$$\text{Hall voltage, } V_H = \frac{R_H \times B_Z \times I}{w} = \frac{3.8 \times 10^{-4} \times 0.6 \times 10 \times 10^{-3}}{0.1} = 22.8 \ \mu\text{V} \quad \textbf{Ans.}$$

Example 2.17. Find the magnitude of Hall voltage V_H in an N-type germanium bar used in Fig. 2.22, having majority carrier concentration $N_D = 10^{17}$/cm^3. Assume $B_Z = 0.1$ Wb/m^2; $w = d = 4$ mm; E_x = 5V/cm and μ_{eGe} = 3,800 cm^2/V-s.

Solution : Applied electric field, E_x = 5 V/cm

Electron mobility for germanium, μ_e = 3,800 cm^2/V-s

Flux density, B_z = 0.1 Wb/m^2

$d = 4 \text{ mm} = 4 \times 10^{-3}$ m

Drift velocity $v = \mu_e \times E_x = 3{,}800 \times 5 = 19{,}000$ cm/second = 190 m/s

Hall voltage, $V_H = B_Z \times v \times d = 0.1 \times 190 \times 4 \times 10^{-3} = 76$ mV **Ans.**

Example 2.18. An N-type germanium sample has a donor density of 10^{21}/m^3. It is arranged in a Hall experiment having magnetic field of 0.2 T (Wb/m^2) and the current density is 600 A/m^2. Determine the Hall voltage if d = 4 mm.

Solution : Concentration of donor atoms, $N_D = 1 \times 10^{21}$/m^3

Magnetic field density, B_Z = 0.2 T

Current density, J = 600 A/m^2

Concentration of electrons, $n = N_D = 1 \times 10^{21}$/m^3

$d = 4 \text{ mm} = 4 \times 10^{-3}$ m

From Eq. (2.60) $$\frac{V_H \times w}{BI} = \frac{1}{ne}$$

$$\text{or Hall voltage } V_H = \frac{BI}{ne} \times \frac{1}{w} = \frac{BJ \times w \times d}{ne\, w} = \frac{BJd}{ne} = \frac{0.2 \times 600 \times 4 \times 10^{-3}}{1 \times 10^{21} \times 1.6 \times 10^{-19}}$$

$$= 3 \text{ mV} \quad \textbf{Ans.}$$

Example 2.19. The resistivity of semiconductor material was known to be 0.00912 Ω-m at room temperature. The flux density in the Hall model was 0.48 Wb/m^2. Calculate the Hall angle for a Hall coefficient of 3.55 × 10^{-4} m^3/coulomb.

[U.P. Technical Univ. Electrical and Electronics Engineering Materials 2005-06]

Solution : Flux density in Hall model, B = 0.48 Wb/m^2

Hall coefficient, $R_H = 3.55 \times 10^{-4}$ m^3/C

$$\text{Conductivity, } \sigma = \frac{1}{\text{Resistivity}} = \frac{1}{0.00912} = 109.65 \text{ (ohm-m)}^{-1}$$

From Eq. (2.65) Hall angle, $\theta_H = \text{Tan}^{-1} \sigma B R_H = \text{Tan}^{-1} 109.65 \times 0.48 \times 3.55 \times 10^{-4}$

$= \text{Tan}^{-1} 0.01868 = 1.0704°$ **Ans.**

2.12. Basic Structure of P-N Junction

Most semiconductor devices employ one or more P-N junctions. The P-N junction is the control element for the performance of all semiconductor devices such as rectifiers, amplifiers, switching devices, linear and digital integrated circuits. The P-N junction is produced* by placing a layer of P-type semiconductor next to the layer of N-type semiconductor. The interface separating the N and P regions is referred to as the metallurgical junction.

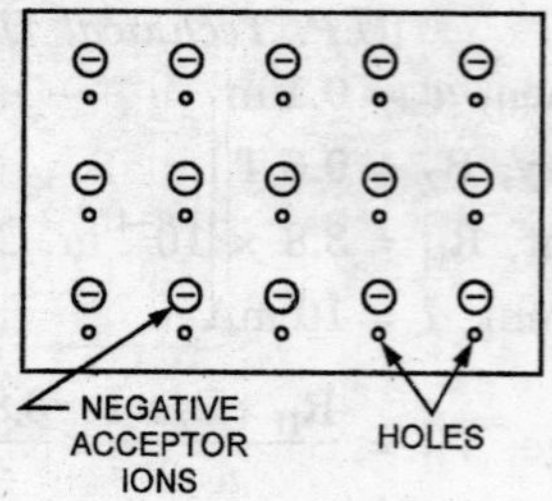

(a) P-Type Semiconductor Material

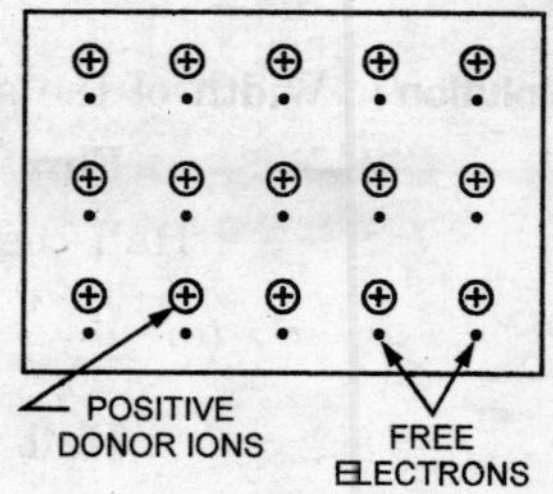

(b) N-Type Semiconductor Material

Fig. 2.24

Figure 2.24 represents two blocks of semiconductor material, one P-type, and the other N-type. The P type semiconductor block has mobile holes (shown by small circles) and the same number of fixed negative acceptor ions (shown by encircled minus sign). Similarly the N-type semiconductor block has mobile or free electrons (shown by dots) and the same number of fixed donor positive ions. Normally the holes, which are the majority charge carriers in P-type of material, are uniformly distributed through the volume of that material. Similarly the electrons, which are the majority charge carriers in N-type of material, are uniformly distributed through the volume of that material. Each region is electrically neutral because each of them carries equal positive and negative charges.

On the formation of P-N junction some of the holes from P-type material tend to diffuse across the boundary into N-type material and some of the free electrons similarly diffuse into the P-type material, as illustrated in Fig. 2.25. This happens due to density gradient (as concentration of holes is higher on P-side than that on N-side and concentration of electrons is higher on N-side than that on P-side). This process is known as *diffusion*.

The doping profile of an ideal uniformly doped P-N junction is depicted in Fig. 2.26.

As a result of the displacement of the charges, an electric field appears across the junction. Equilibrium is established when the field becomes large enough to restrain the process of diffusion. The general shape of the charge distribution may be, as illustrated in Fig. 2.25 (*b*). The electric charges are confined to the neighbourhood of the junction, and consists of immobile ions.

We see that the free electrons crossing the junction create negative ions on the P-side by giving some atoms one more electron than their total number of protons. The electrons also leave positive ions (atoms with one fewer electron than the number of protons) behind them on the N-side. As negative ions are created on the P-side of the junction, the P-side acquires a

* It is to be noted that a P-N junction is not formed by merging two separate P-type and N-type materials. If this were done, a polycrystalline structure would have resulted in, which would have not performed very well. The P-N junction found in the semiconductor devices is actually formed in a single monocrystalline structure by adding carefully controlled amounts of donor and acceptor impurities.

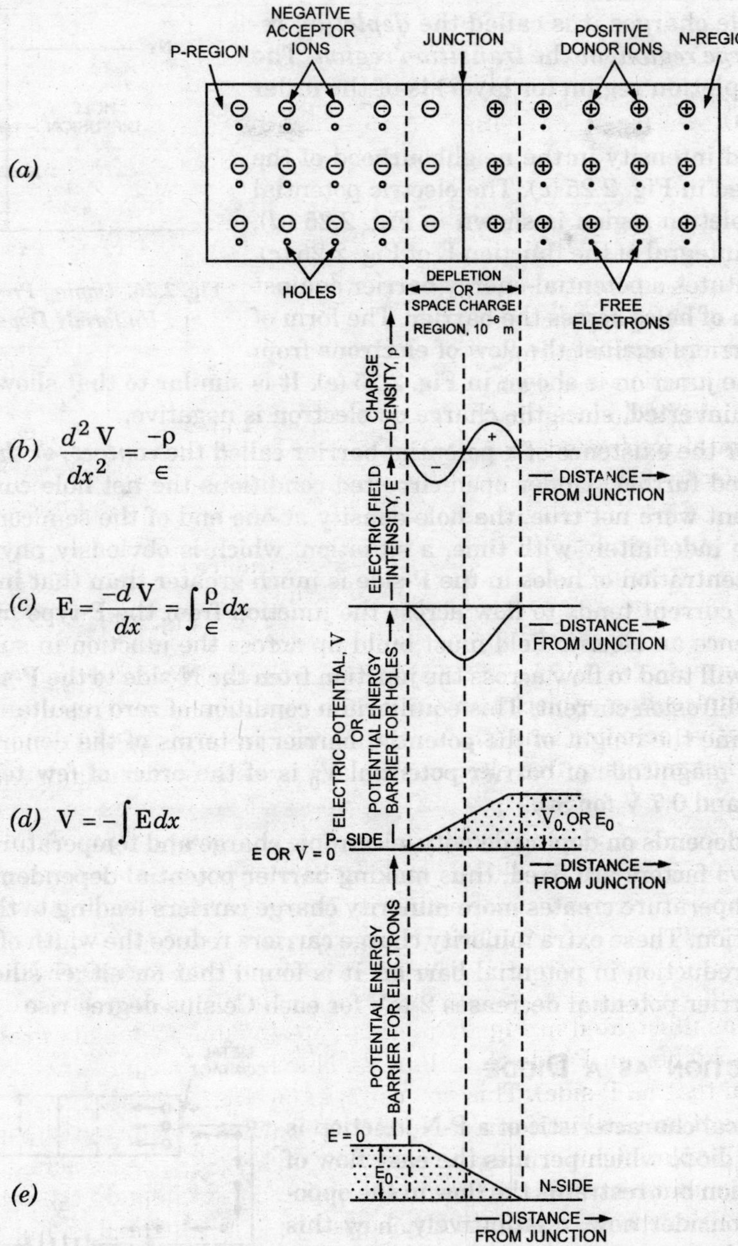

Fig. 2.25. ***A Schematic Diagram of a P-N Junction, Including The Charge Density, Electric Field Intensity, and Potential Energy Barriers at The Junction (Not To Scale)***

negative potential, as illustrated in Fig. 2.25 (*e*). Similarly the positive ions are created on the N-side and the N-side acquires a positive potential as illustrated in Fig. 2.25 (*d*). The negative potential on the P-side prevents the migration of any more electrons from the N-type material to the P-type material. Similarly the positive potential on the N-side prevents any further migration of holes across the boundary. Thus, the initial diffusion of charge carriers creates a *barrier potential* at the junction.

The region around the junction is completely ionised. As a result, there are no free electrons on the N-side, nor there are holes on the P-side. Since the region around the junction is depleted of mobile charges it is called the *depletion region*, the *space-charge region*, or the *transition region*. The thickness of the depletion region (or layer) is of the order of 1 micron (10^{-6} m).

The electric field intensity in the neighbourhood of the junction is illustrated in Fig. 2.25 (*c*). The electric potential variation in the depletion region is shown in Fig. 2.25 (*d*), and is the negative integral of the function E of Fig. 2.25 (*c*). The variation constitutes a potential-energy barrier against the further diffusion of holes across the barrier. The form of potential-energy barriers against the flow of electrons from the N-side across the junction is shown in Fig. 2.25 (*e*). It is similar to that shown in Fig. 2.25 (*d*), except that it is inverted, since the charge on electron is negative.

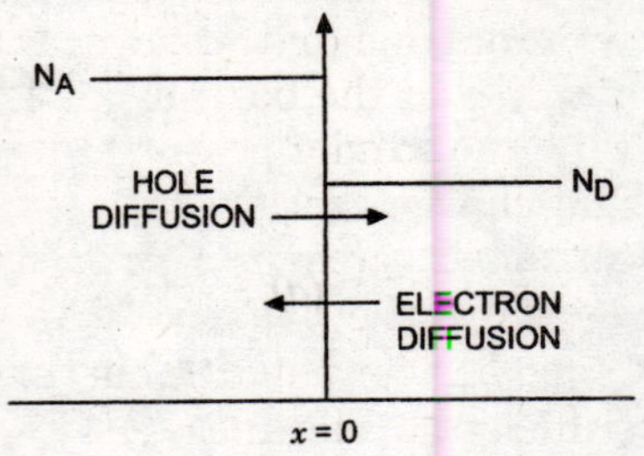

Fig. 2.26. ***Doping Profile of an Ideal Uniformly Doped P-N Junction***

The necessity for the existence of a potential barrier called the *contact*, or *diffusion potential* is now considered further. Under open-circuited conditions the net hole current must be zero. If this statement were not true, the hole density at one end of the semiconductor would continue to increase indefinitely with time, a situation, which is obviously physically impossible. Since the concentration of holes in the P-side is much greater than that in the N-side, a very large diffusion current tends to flow across the junction from the P-type material to the N-type material. Hence an electric field must build up across the junction in such a direction that a drift current will tend to flow across the junction from the N-side to the P-side in order to counterbalance the diffusion current. This equilibrium condition of zero resultant hole current allows us to determine the height of the potential barrier in terms of the donor and acceptor concentrations. The magnitude of barrier potential V_0 is of the order of few tenths of a volt (0.3 V in case of Ge and 0.7 V for Si).

Barrier voltage depends on doping density, electronic charge and temperature. For a given junction, the first two factors are fixed, thus making barrier potential dependent on temperature. Increase in temperature creates more minority charge carriers leading to their increased drift across the junction. These extra minority charge carriers reduce the width of the depletion layer equivalent to reduction in potential barrier. It is found that for either silicon or germanium diodes, the barrier potential decreases 2 mV for each Celsius degree rise.

2.13. P-N Junction as a Diode

The essential electrical characteristic of a P-N junction is that it constitutes a diode which permits the easy flow of current in one direction but restrains the flow in the opposite direction. We consider now, qualitatively, how this diode action comes about.

2.13.1. Forward Biasing

When an external field, with P-region connected to positive terminal and N-region connected to negative terminal of the battery, is applied across the junction, as shown in Fig. 2.27, the junction is said to be *forward biased*. In this circuit arrangement, the holes on the P-side being positively charged particles are repelled from the positive bias terminal and driven towards the junction. Similarly, the electrons on the N-side are repelled from the negative bias

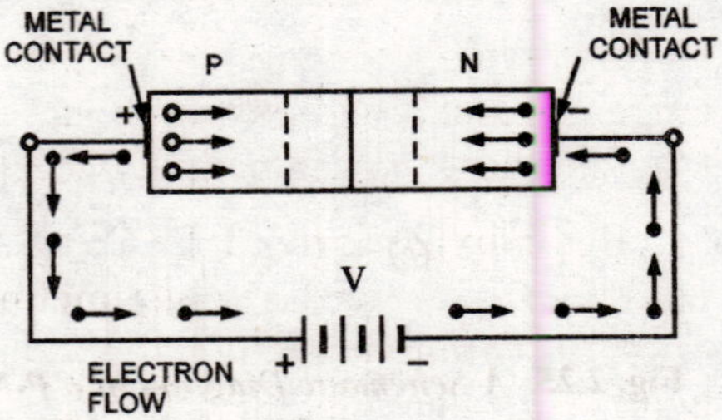

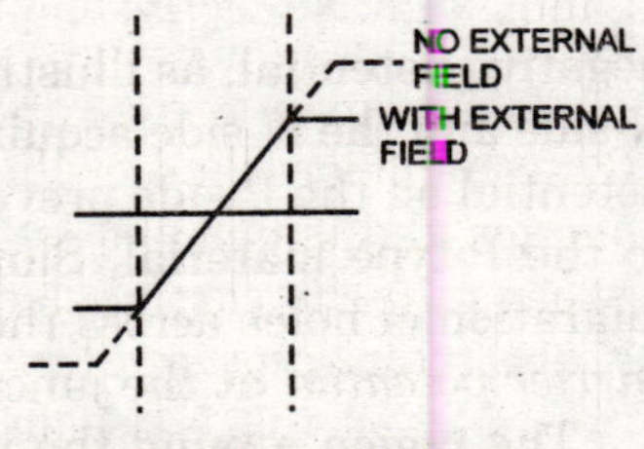

Fig. 2.27. ***Forward Biasing***

terminal and driven towards the junction. The result is that the depletion region is reduced in width, and the barrier potential is also reduced. If the applied bias voltage is increased from zero, the barrier potential gets progressively smaller and smaller until it effectively disappears and charge carriers can easily flow across the junction. Electrons from the N-side are then attracted across to the positive bias terminal on the P-side, and holes from the P-side flow across to the negative bias terminal on the N-side. Thus a majority carrier current flows. Since barrier potential is very small (0.3 V for Ge and 0.7 V for Si), a small forward voltage is sufficient to eliminate the barrier completely. Once the barrier is eliminated by the application of forward voltage, junction resistance becomes almost zero and a low resistance path is established in the entire circuit. The current, called the *forward current*, flows in the circuit.

In brief it can be said that when the junction is forward biased (P-region connected to the positive terminal and N-region connected to the negative terminal of the battery), (*i*) barrier is reduced and at a forward voltage of 0.3 V in case of germanium (0.7 V in case of silicon) it is eliminated altogether, (*ii*) the junction offers low resistance, called the *forward resistance* r_f, to the flow of current and (*iii*) current flows in the circuit due to establishment of low resistance path and the magnitude of current depends upon the magnitude of applied forward voltage.

The mechanism of current flow in a forward biased P-N junction is summed up as follows :

(*i*) The electrons from the negative bias terminal continue to arrive into the N-region while the free electrons in the N-region move towards the P-N junction.

(*ii*) The electrons travel through the N-region as free electrons *i.e.* the current in N-region is by free electrons.

(*iii*) These free electrons on reaching P-N junction, combine with holes and become valence electrons. Since a hole is in the covalent bond and, therefore, when a free electron combines with a hole, it becomes a valence electron.

(*iv*) The electrons travel through P-region as valence electrons and current in this P-region is, therefore, by holes.

(*v*) These valence electrons, on reaching the left end of the crystal, flow into the positive terminal of the battery.

Thus we see that the current in the N-type region is due to movement of electrons whereas in the P-type region it is carried by the holes. However, in the external circuit (*i.e.* in connecting wires) the current is carried by electrons.

2.13.2. Zero Applied Bias

Here we will examine the properties of the step junction in thermal equilibrium, where no currents exist and no external excitation is applied. We will determine built-in potential barrier through the depletion or space charge region, electric field and width of the space charge region.

Built-in Potential Barrier. When no voltage is applied across the P-N junction (*i.e.* with zero applied bias), the junction is in the thermal equilibrium—the Fermi energy level is constant throughout the entire specimen. The valence and conduction bands energies have to bend while moving through the space charge region because the relative position of these bands with respect to the Fermi energy changes between P and N regions.

Electrons in the conduction band of the N-region face a potential barrier when moving into the conduction band of the P-region. This potential barrier is known as the *built-in potential barrier* and is denoted by V_{bi} or V_0. The built-in potential barrier maintains equilibrium between majority carrier electrons in the N-region and minority carrier electrons in the P-region. It also maintains equilibrium between majority carrier holes in the P-region and minority carrier holes in the N-region. The potential difference across the junction cannot be measured with a voltmeter as new potential barriers are being formed between the semiconductor and

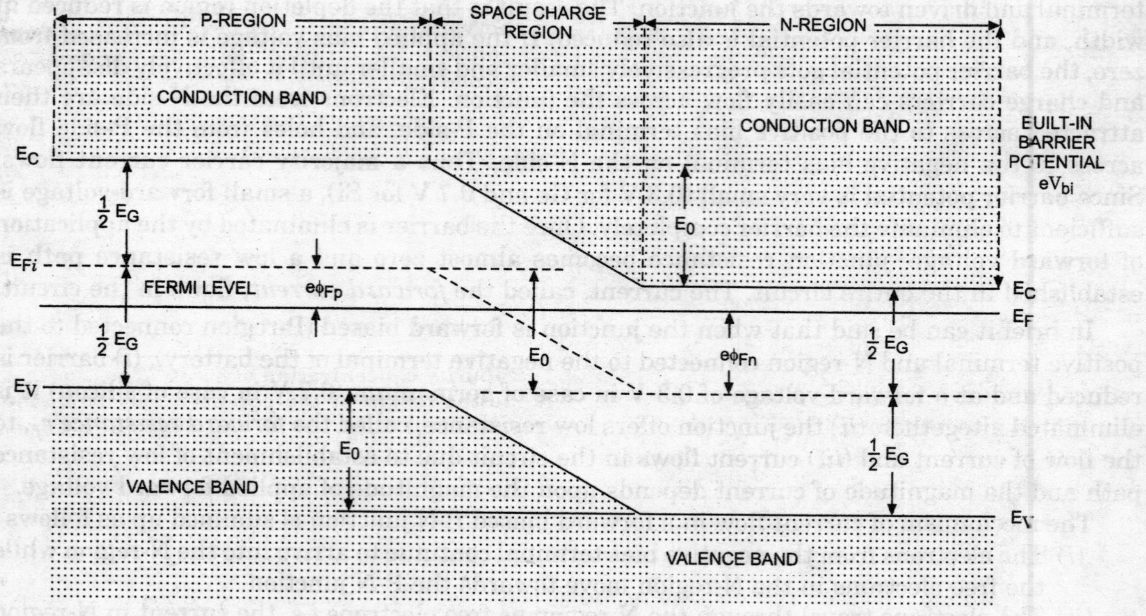

Fig. 2.28. ***Energy Band Diagram of a P-N Junction in Thermal Equilibrium***

the probes which will cancel built-in potential barrier V_{bi}. As the condition pertains to equilibrium so no current is caused by this voltage.

The intrinsic Fermi level is equidistant from the conduction band edge through the junction, as illustrated in figure. The built-in potential barrier V_{bi} is the difference between the intrinsic Fermi levels of the N and P-regions.

$$V_{bi} = \phi_{Fn} + \phi_{Fp} \quad \text{...(2.67)}$$

The electron concentration in the N-region conduction band,

$$n_0 = n_C e^{-(E_C - E_F)/kT} = n_i e^{(E_F - E_{Fi})/kT} = n_i e^{-e\phi_{Fn}/kT} \quad \text{...(2.68)}$$

where E_{Fi} and n_i are the intrinsic Fermi energy and intrinsic carrier concentration respectively.

Potential ϕ_{Fn} in the N-region is defined as

$$e\,\phi_{Fn} = E_{Fi} - E_F \quad \text{...(2.69)}$$

Taking natural log of both sides of Eq. (2.69), setting $n_o = N_D$ we have

$$\phi_{Fn} = \frac{-k'T}{e} \log_e \frac{N_D}{n_i} \quad \text{...(2.70)}$$

For the P-region the hole concentration

$$\phi_{Fp} = \frac{+k'T}{e} \log_e \frac{N_A}{n_i} \quad \text{...(2.71)}$$

The built-in potential barrier for the step junction is then

$$V_{bi} = \frac{k'T}{e} \log_e \frac{N_A N_D}{n_i^2} = V_T \log_e \frac{N_A N_D}{n_i^2} \quad \text{...(2.72)}$$

where V_T is defined as the thermal voltage and is equal to $\frac{k'T}{e}$.

Example 2.20. What is the contact potential ? Obtain the value of contact potential of an abrupt junction at room temperature of intrinsic concentration n_i = 1.5 × 10^{16}/m^3 with doping level of $N_D = N_A = 10^{21}$/m^3. *[U.P. Technical Univ. Semiconductor Materials and Devices 2005-06]*

Solution :

Room temperature, T = (273 + 27) = 300 K Assuming room temperature to be 27ºC

Instrinsic concentration, $n_i = 1.5 \times 10^{16}/\text{m}^3$

Doping level, $N_D = N_A = 10^{21}/\text{m}^3$

$$\text{Contact potential, } V_0 = V_T \log_e \frac{N_D N_A}{n_i^2} = \frac{T}{11{,}600} \log_e \frac{N_D N_A}{n_i^2} \qquad \because V_T = \frac{T}{11{,}600}$$

$$= \frac{300}{11{,}600} \log_e \frac{(1 \times 10^{21})^2}{(1.5 \times 10^{16})^2} = 0.5745 \text{ V } \textbf{Ans.}$$

Example 2.21. Consider a Ge diode with $N_D = N_A$ and with impurity concentration of 8 × 10^{14}/cm^3. Assume n_i = 2 × 10^{13}/cm^3. At room temperature of 300 K calculate the height of the potential barrier under open-circuit conditions. Assume Boltzmann's constant k = 8.61 × 10^{-5} eV/K and electron charge, e = 1.6 × 10^{-19} C.

Solution :

Impurity concentration, $N_D = N_A = 8 \times 10^{14}/\text{cm}^3$

Intrinsic carrier concentration, $n_i = 2 \times 10^{13}/\text{cm}^3$

Boltzmann's constant, k = 8.61 × 10^{-5} eV/K

Temperature, T = 300 K

$$\text{Potential barrier, } V_0 = kT \log_e \frac{N_D N_A}{n_i^2} = 300 \times 8.61 \times 10^{-5} \log_e \frac{(8 \times 10^{14})^2}{(2 \times 10^{13})^2}$$

$$= 0.19 \text{ V } \textbf{Ans.}$$

2.13.3. Reverse Applied Bias

If an external bias voltage is applied with positive terminal to N-side and negative terminal to the P-side of a P-N junction, as shown in Fig. 2.29 (*a*), the junction is said to *be reverse-biased*. In this arrangement electrons from the N-side are attracted to the positive bias terminal and holes from P-side are attracted to the negative bias terminal. Thus, as shown in Fig. 2.29 (*a*), holes from the impurity atoms on the P-side of the junction are attracted away from the junction, and the electrons are attracted from the atoms on the N-side of the junction. Thus the depletion region is widened, and the barrier potential is increased by the magnitude of the applied bias. With the increased barrier potential, there is no possibility of majority carrier current flow across the junction. Thus the P-N junction is in a non-conductive state.

In brief it can be said that when the P-N junction is reverse biased (*i*) barrier potential is increased and width of depletion is also increased (*ii*) the junction

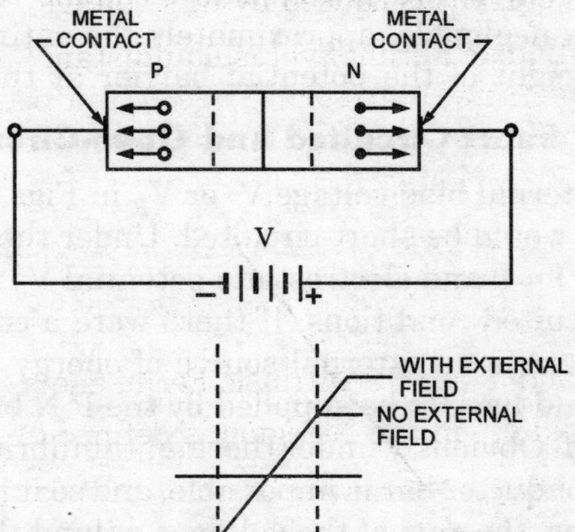

(a) Reverse Biasing

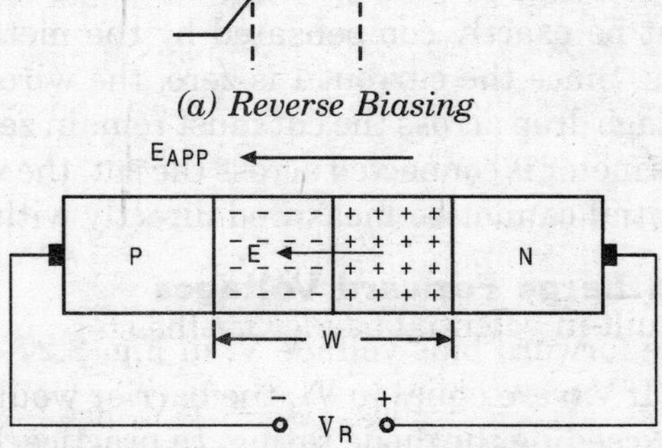

(b) A P-N Junction, With an Applied Reverse Bias, Indicating The Directions of The Electric Field Induced By Voltage VR and Space Charge Electric Field

Fig. 2.29

offers high resistance, called the reverse resistance r_r, to the flow of current (*iii*) owing to establishment of high resistance path, very small current, called the *leakage current* or *reverse current*, flows through the circuit. The reverse or leakage current is not affected by barrier and depends on temperature, material and construction of the junction.

Thus the P-N junction allows flow of large current only when forward biased and only small reverse (or leakage) current when reverse biased.

2.14. METAL-SEMICONDUCTOR JUNCTIONS

It has been assumed that the external bias voltage (forward or reverse) appears directly across the junction and has the effect of lowering or raising the electrostatic potential across the junction. In order to justify the above assumption it should be specified how electrical contact is made to the semiconductor from the external bias circuit. In Figs. 2.27 and 2.29 (*a*) we indicate metal contacts with which the homogeneous P-type and N-type materials are provided. Thus two metal-semiconductor junctions, one at each end of the diode, have been introduced. So a contact potential is expected to be developed across these additional junctions. However, it can be assumed that metal-semiconductor contacts shown in Figs. 2.27 and 2.29 (*a*) have been manufactured in such a way that they are non-rectifying *i.e.*, the contact potential across these junctions is almost independent of direction and magnitude of current. A contact of this type is referred to as an *ohmic contact.*

We are now in a position to justify our assumption that the entire applied voltage appears as a *change* in the *height* of the potential barrier. In as much as the metal-semiconductor contacts are low-resistance ohmic contacts and the voltage drop across bulk of the crystal is negligible, approximately the entire applied voltage will indeed appear as a change in the height of the potential barrier at the P-N junction.

2.14.1. Short-Circuited and Open-Circuited P-N Junction

If the external bias voltage V or V_R in Figs. 2.27 and 2.29 (*a*) were set equal to zero, the P-N junction would be short-circuited. Under these conditions, as explained below, no current can flow *i.e.*, I = 0 and electrostatic potential V_0 remains unchanged and equal to the value under open-circuited conditions. If there were a current *i.e.* I ≠ 0, the metal would become heated. Since there is no external source of energy available, the energy required to heat the metal wire would have to be supplied by the P-N bar. The semiconductor bar, therefore, would have to cool off. Obviously under thermal equilibrium the simultaneous heating of metal and cooling of semiconductor bar is impossible, and so it is concluded that I = 0. Since under short-circuited conditions, the sum of the voltages around the closed loop must be zero the junction potential V_0 must be exactly compensated by the metal-to-semiconductor contact potentials at ohmic contacts. Since the current I is zero, the wire can be cut without changing the situation, and the voltage drop across the cut must remain zero. If in an attempt to measure junction potential V_0 a voltmeter is connected across the cut, the voltmeter will indicate zero *i.e.*, contact difference of potential cannot be measured directly with a voltmeter.

2.14.2. Large Forward Voltages

Suppose forward bias voltage V, in Fig. 2.27, is increased until V approaches junction potential V_0. If V were equal to V_0, the barrier would disappear and the current could be arbitrarily large, exceeding the diode rating. In practice, barrier cannot be reduced to zero because, as the current increases without limit, the bulk resistance of the crystal and the resistance of the ohmic contacts, will limit the current. Thus it is no longer possible to assume that all the voltage V appears as a change across the P-N junction. It is concluded that, as the forward bias voltage V becomes comparable with V_0, the current through a P-N junction diode will be governed by the ohmic-contact resistance and the crystal bulk resistance. Thus the V-I characteristic becomes almost a straight line.

2.14.3. Bulk Resistance

After the barrier potential is overcome, all that impedes the current is the resistance of the P and N regions, if ohmic contact resistance is neglected. The sum of these resistances is called the *bulk resistance* r_B of the diode. The bulk resistance r_B depends on the doping level, and the size of the P and N regions. Typically the bulk resistance of rectifier diodes is less than 1 Ω.

2.14.4. Energy Band Structure of an Open-Circuited P-N Junction

Let us consider that a P-N junction is formed by placing P and N-type materials in intimate contact on an atomic scale. Under such conditions the Fermi level must be constant throughout the specimen at equilibrium. If this were not true, electrons on one side of the junction would have an average energy higher than those on the other side and there would be a transfer of electrons and energy until the Fermi level on the two sides equalize. In Art. 2.8 we have seen that Fermi level E_F is closer to the conduction band edge E_{CN} in the N-type material and closer to the valence band edge E_{VP} in the P-type material. Obviously the conduction band edge E_{CP} in the P-type material cannot be at the same level as E_{CN}, nor can the valence band edge E_{VN} in the N-side line up with E_{VP}. Thus on formation of a P-N junction, conduction band edge E_{CP} in P-type material is higher than E_{CN} in N-type material and valence band edge E_{VP} in P-type material is higher than E_{VN} in N-type material. This is illustrated in Fig. 2.30, where a shift in energy level E_0 is shown.

The total shift in the energy level E_0 is given as

$$E_0 = E_{CP} - E_{CN} + E_{VP} - E_{VN} = E_1 + E_2 \quad ...(2.73)$$

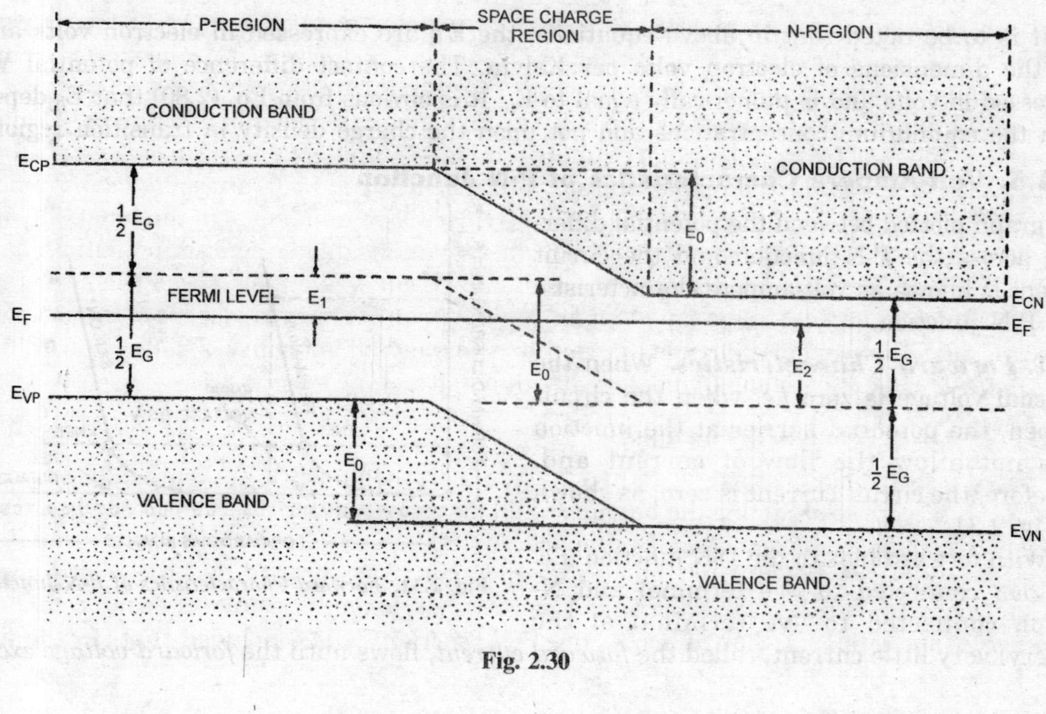

Fig. 2.30

where E_1 and E_2 represent the shifts in the Fermi level from the intrinsic condition in P- and N-type materials. The energy E_0 represents the potential energy of the electrons at the junction, as shown in Fig. 2.25 (*e*). From Fig. 2.30, we have

$$E_F - E_{VP} = \frac{1}{2} E_G - E_1 \quad ...(2.74)$$

$$\text{and } E_{CN} - E_F = \frac{1}{2} E_G - E_2 \quad ...(2.75)$$

Adding Eqs. (2.74) and (2.75) we have

$$E_1 + E_2 = E_G - (E_{CN} - E_F) - (E_F - E_{VP})$$

$$\text{or } E_0 = E_G - (E_{CN} - E_F) - (E_F - E_{VP}) \quad ...(2.76)$$

From Eqs. (2.33) and (2.34) we have

$$n_i^2 = np = n_C n_V e^{-[(E_C - E_F) + (E_F - E_V)]/kT}$$

$$= n_C n_V e^{-(E_C - E_V)/kT} = n_C n_V e^{-E_G/kT}$$

$$\text{or } E_G = kT \log_e \frac{n_C n_V}{n_i^2} \quad ...(2.77)$$

From Eqs. (2.36) and (2.37) we have

$$E_{CN} - E_F = kT \log_e \frac{n_C}{N_D} \quad ...(2.78)$$

$$\text{and } E_F - E_{VP} = kT \log_e \frac{n_V}{N_A} \quad ...(2.79)$$

Substituting values of E_G, $(E_{CN} - E_F)$ and $(E_F - E_{VP})$ from Eqs. (2.77), (2.78) and (2.79) in Eq. (2.76) we have

$$E_0 = kT \log_e \frac{n_C n_V}{n_i^2} - kT \log_e \frac{n_C}{N_D} - kT \log_e \frac{n_V}{N_A}$$

$$= kT \log_e \left[\frac{n_C n_V}{n_i^2} \cdot \frac{N_D}{n_C} \cdot \frac{N_A}{n_V}\right] = kT \log_e \frac{N_D N_A}{n_i^2} \quad ...(2.80)$$

It is to be noted that in above equations, the E's are expressed in electron volts and k has the dimensions of electron volts per Kelvin. The contact difference of potential V_0 is expressed in volts and is numerically equal to E_0. It is obvious from Eq. (2.80) that E_0 depends upon the equilibrium concentrations and not upon the charge density in transition region.

2.14.5. Volt-Ampere Characteristics of P-N Junction

The graph plotted between the potential difference across the P-N junction and the circuit current is known as volt-ampere characteristic of a P-N junction.

1. *Forward Characteristics.* When the external voltage is zero *i.e.* when the circuit is open, the potential barrier at the junction does not allow the flow of current and, therefore, the circuit current is zero, as shown in Fig. 2.31.

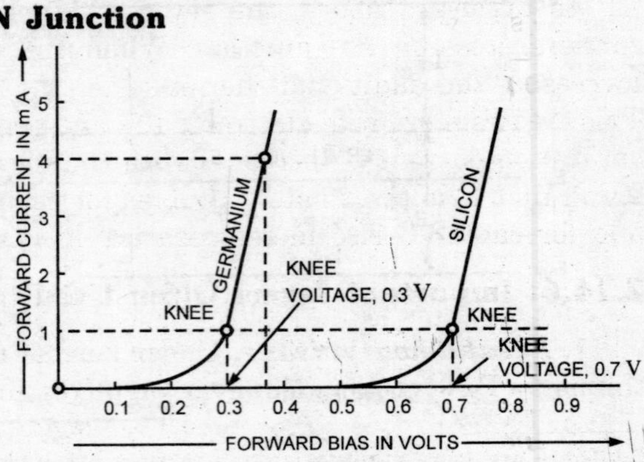

Fig. 2.31. *Forward Characteristics of P-N Junction*

With forward bias to the P-N junction (*i.e.* P-region connected to +ve terminal and N-region connected to –ve terminal of the battery) very little current, called the *forward current*, flows until the *forward voltage exceeds* the junction barrier potential (0.3 V for Ge and 0.7 V for Si). The characteristic follows an exponential law. As the forward voltage is increased to the knee of the characteristic, the barrier potential is progressively reduced to zero, allowing more and more majority charge carriers to flow across the junction. Beyond the knee of the characteristic, the potential barrier is completely eliminated, forward current increases almost linearly with the increase in forward voltage and the P-N junction starts behaving as a resistor. If the forward voltage is increased beyond a certain value, extremely large current will flow and the P-N junction may get destroyed due to overheating.

Because of the nonlinear knee of the curve, if a very small signal voltage is applied to the diode so that it only operates around the knee, the signal will be distorted. The signal must, therefore, be large enough to operate the diode over the linear part of the characteristic. Forward resistance of P-N junction is low (of the order of 75 Ω in case of Ge and 150 Ω in case of Si).

2. *Reverse Characteristics.* When a reverse bias is applied (*i.e.* when P-region is connected to the negative bias terminal and N-region is connected to the positive bias terminal), as shown in Fig. 2.29 (*a*), the potential barrier at junction is increased, therefore, the junction resistance becomes very high and there is no possibility of majority carriers flowing across a reverse biased junction. But still minority carriers generated on each side can cross the junction. Electrons on the P-side are attracted across the junction to the positive bias terminal on the N-side and holes on the N-side may travel across the junction to the negative bias terminal on the P-side. This results in a very small current, as illustrated in Fig. 2.32. This current is called *reverse current* and is due to minority carriers.

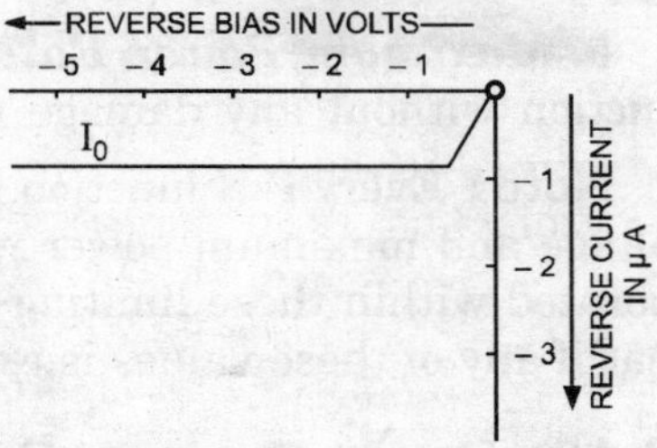

Fig. 2.32. ***Reverse Characteristics of P-N Junction***

For silicon the reverse current* is typically less than 1 μA and may be as low as 1 *n*A. In case of germanium, the reverse current may exceed 10 μA. Only a very small reverse voltage is necessary to direct all minority carriers across the junction, and when all minority carriers are flowing across, further increase in bias voltage will not increase the current. This current is referred to as a *reverse saturation current* and is designated as I_0. Reverse resistance is very high (of the order of MΩ).

As discussed above, the reverse saturation current I_0 is made up of minority charge carriers crossing the junction. When the temperature of the semiconductor material is increased, the additional thermal energy causes more electrons to breakaway from atoms. This creates more hole-electron pairs and generates more minority charge carriers. So reverse saturation current I_0 increases with the increase in the junction temperature.** It has been found that I_0 approximately doubles for each 10°C rise in temperature in case of germanium and for each 8°C rise in temperature in case of silicon.

2.14.6. Important Terms Often Used in P-N Junction

1. *Breakdown Voltage.* Under normal reverse voltage, a very little reverse current flows through a P-N junction. However, on increasing the reverse voltage, a point may reach at which

* There are more minority charge carriers available in germanium than in silicon, since charge carriers are more easily detached from germanium atoms.

** It can be shown that reverse current I_0 depends upon electronic charge, doping density and junction area, as well as temperature. With the exception of temperature, all other factors (*i.e.* electronic charge, doping density and junction area) are constant for a given junction, I_0 is affected only by variation in temperature.

the junction breaks down with a sudden rise in reverse current. Breakdown voltage is defined as the reverse voltage at which P-N junction breaks down with a sudden rise in reverse current.

2. *Knee Voltage.* As already explained above in Art. 2.14.5, with forward bias to P-N junction very little current flows until the forward voltage exceeds the junction barrier potential (0.3 V in case of Ge and 0.7 V in case of Si). The forward voltage, at which the current through the junction starts increasing rapidly, is called the *knee voltage* or the *cut-in voltage*.

3. *Maximum Forward Current.* It is the highest instantaneous forward current that a P-N junction can conduct without damage to the junction. In case the forward current exceeds this rating, the junction will get destroyed due to overheating.

4. *Peak Inverse Voltage.* It is the maximum reverse voltage that can be applied to the P-N junction without damage to the junction. If the reverse voltage across the junction exceeds its peak inverse voltage (PIV), the junction may get destroyed owing to excessive heat.

5. *Maximum Power Rating.* It is the maximum power that can be dissipated at the junction without any damage to it.

Note : Every P-N junction has limiting values of maximum forward current, peak inverse voltage and maximum power rating. The P-N junction gives satisfactory performance when operated within these limiting values. The P-N junction may get destroyed due to excessive heat if any of these values is exceeded.

2.15. SEMICONDUCTOR DIODE

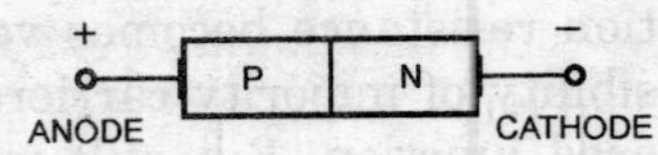

(a) Circuit Symbol

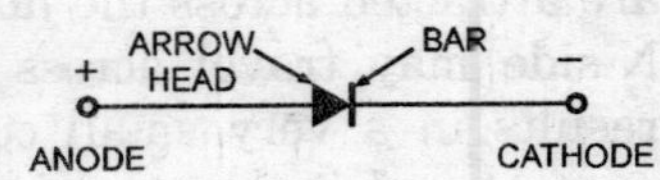

(b) Graphical Symbol

Fig. 2.33. *Semiconductor Diode*

The term *diode* denotes a two-electrode device. A *semiconductor diode* is simply a P-N junction with connecting leads or terminals on the two sides of the P-N junction. P-N junction is also called the crystal diode because the P-N junction is grown out of crystal. A diode is a unidirectional device permitting the easy flow of current in one direction but restraining the flow in opposite direction. A major application of diodes is in *rectification i.e.* in conversion of ac into dc. Semiconductor diode is becoming more and more popular these days due to its smaller size, cheapness, robustness and higher operating efficiency.

Circuit symbol of a semiconductor diode is given in Fig. 2.33. The symbol for a diode is shown in Fig. 2.33 (*b*). The arrow in the symbol indicates the direction of conventional current flow when the diode is forward biased *i.e.* from the positive terminal through the device to the negative terminal. The P-side of the diode is always the positive terminal for forward bias and is designated the *anode*. The N-side is called the *cathode* and is the negative terminal when the device is forward biased.

The semiconductor diode may be either of silicon or germanium. The silicon junction diode is similar in appearance to the germanium diode but differs in properties. Silicon diodes have, in general, higher peak inverse voltage (PIV) and current ratings and wider temperature ranges than germanium diode.

(*i*) The peak inverse voltage ratings for the silicon can be as high as 1,000 V whereas for germanium it is near about 400 V.

(*ii*) Silicon can be operated at temperatures up to about 200°C compared with about 75–80°C for germanium.

(*iii*) The forward voltage drop of silicon diode is approximately double of that of the corresponding germanium diode. It is typically of the order of 0.7 V for commercially available silicon diodes and 0.3 V for germanium diodes.

(*iv*) The reverse current at a given temperature and voltage is approximately a hundredth of that of the corresponding germanium diode, but there is little sign of current saturation as in case of germanium. In fact, the reverse current of a silicon diode is approximately proportional to the square root of the voltage until breakdown is approached.

(*v*) The reverse current of a silicon diode, for a given voltage, practically doubles for every 8ºC rise of temperature, compared with 10ºC for germanium.

2.16. FABRICATION TECHNIQUES

In practice, the P-N junction is formed from a single monocrystalline structure by adding carefully controlled amounts of donor and acceptor impurities. Here discussion is limited only to acquaintation to the basic techniques and terminology (not expertise in fabrication).

The first and foremost requirement is to obtain an extremely pure germanium or silicon. Impurity of less than one part in ten billion (10^{10}) is required for most semiconductor device fabrication today. For obtaining pure semiconductor material it is first purified chemically. For reducing the impurities further, and to ensure the formation of a monocrystalline structure, a technique known as *floating zone* is quite often employed. This is illustrated in Fig. 2.34.

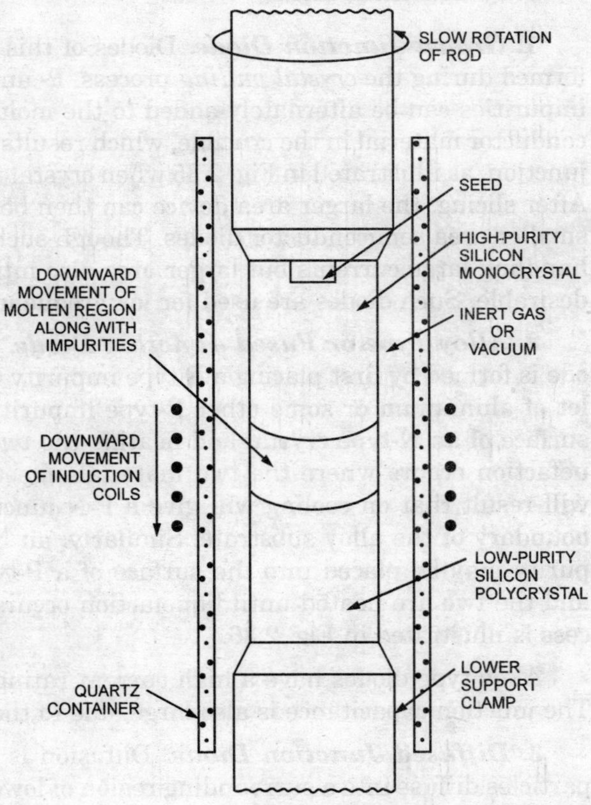

Fig. 2.34

The monocrystalline structure is formed through the use of a small seed of semiconductor (*e.g.*, silicon or germanium). The seed itself is a monocrystal that has been very carefully cut along the face of its cubic lattice. Support clamps are employed to hold the low-purity polycrystalline rod.

The rod, seed, and support clamps are placed in a quartz cylinder. The process may be carried out in vacuum or by surrounding the semiconductor with an inert gas. Great care is required to ensure that the semiconductor is not further contaminated during the floating zone process.

Induction coils encompass the quartz container, as illustrated in Fig. 2.34. The coils are excited by a radio-frequency voltage. Circulating currents are induced in the semiconductor due to the magnetic field set up by RF currents flowing through the induction coils. The heat so produced melts the zone which is exposed to it. As a result, a molten region is formed.

Slow rotation of rod makes the semiconductor atoms to align themselves with the atoms in the monocrystalline seed. Thus, as the induction coils are moved downward, the molten region follows, and a monocrystalline structure continues to grow from the seed. Purification of the semiconductor rod occurs simultaneously with the formation of the monocrystal. The impurities in the semiconductor rod tend to become 'more liquid' than the semiconductor. Thus as the induction coils are moved downward, the impurities tend to follow the molten region. Once the induction coils have traversed the length of the semiconductor rod, the impurities are collected in its lower end. The lower end of the semiconductor rod may then be cut-off, and the process is repeated until the desired impurity level is attained.

Once a pure monocrystalline semiconductor has been produced, carefully controlled amount of donor and acceptor impurities are added to the semiconductor without disturbing the orderly monocrystalline structure. Thus a P-N junction is formed.

Semiconductor diodes are normally one of the following types :

1. Grown junction diode.
2. Alloy type or fused junction diode.
3. Diffused junction diode.
4. Epitaxial grown or planar diffused diode.
5. Point contact diode.

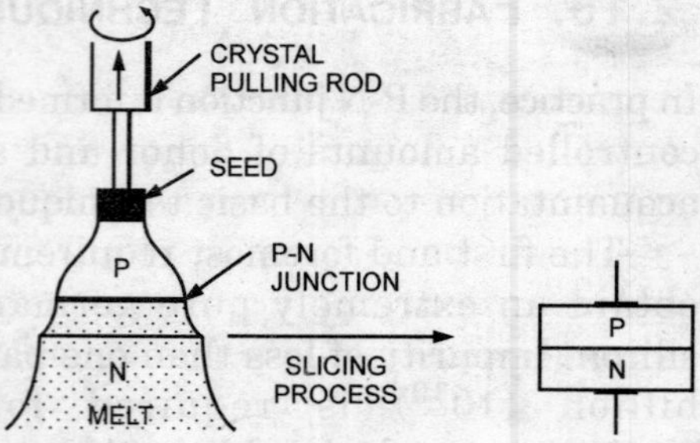

Fig. 2.35

1. *Grown Junction Diode.* Diodes of this type are formed during the *crystal pulling* process. P- and N-type impurities can be alternately added to the molten semiconductor material in the crucible, which results in a P-N junction, as illustrated in Fig. 2.35 when crystal is pulled. After slicing, the larger area device can then be cut into a large number (say in thousands) of smaller area semiconductor diodes. Though such diodes, because of larger area, are capable of handling large currents but larger area also introduces more capacitive effects, which are undesirable. Such diodes are used for low frequencies.

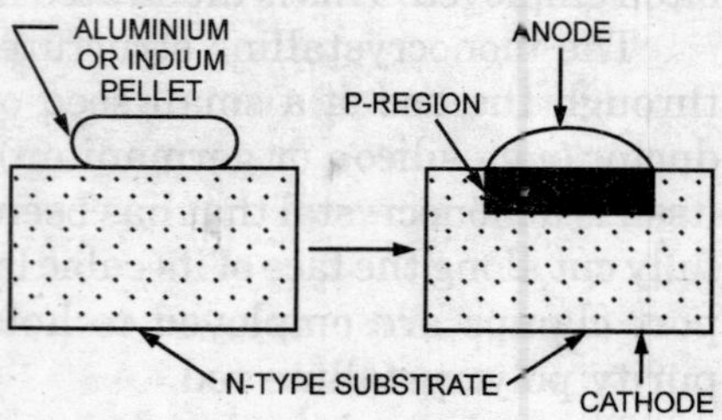

Fig. 2.36

2. *Alloy Type or Fused Junction Diode.* Such a diode is formed by first placing a P-type impurity (a tiny pellet of aluminium or some other P-type impurity) into the surface of an N-type crystal and heating the two until liquefaction occurs where the two materials meet. An alloy will result that on cooling will give a P-N junction at the boundary of the alloy substrate. Similarly, an N- type impurity may be placed into the surface of a P-type crystal and the two are heated until liquefaction occurs. The process is illustrated in Fig. 2.36.

Alloy type diodes have a high current rating and large PIV (peak inverse voltage) rating. The junction capacitance is also large, due to the large junction area.

3. *Diffused Junction Diode.* Diffusion is a process by which a heavy concentration of particles diffuse into a surrounding region of lower concentration. The main difference between the diffusion and alloy process is the fact that liquefaction is not reached in the diffusion process. In the diffusion process heat is applied only to increase the activity of elements involved.

For formation of such diodes, either solid or gaseous diffusion process can be employed.

The process of solid diffusion starts with formation of layer of an acceptor impurity on an N-type substrate and heating the two until the impurity diffuses into the substrate to form the P-type layer, as illustrated in Fig. 2.37 (*a*). A large P-N junction is divided into parts by cutting process. Metallic contacts are made for connecting anode and cathode leads.

In the process of gaseous diffusion instead of layer formation of an acceptor impurity, an N-type substrate is placed in a gaseous atmosphere of acceptor impurities and then heated. The impurity diffuses into the substrate to form P-type layer on the N-type substrate. The process is illustrated in Fig. 2.37 (*b*).

Though, the diffusion process requires more time than the alloy process but it is relatively inexpensive, and can be very accurately controlled. The diffusion technique leads itself to the simultaneous fabrication of many hundreds of diodes on one small disc of semiconductor material and is most commonly used in the manufacture of semiconductor diodes. This technique is also used in the production of transistors and ICs (integrated circuits).

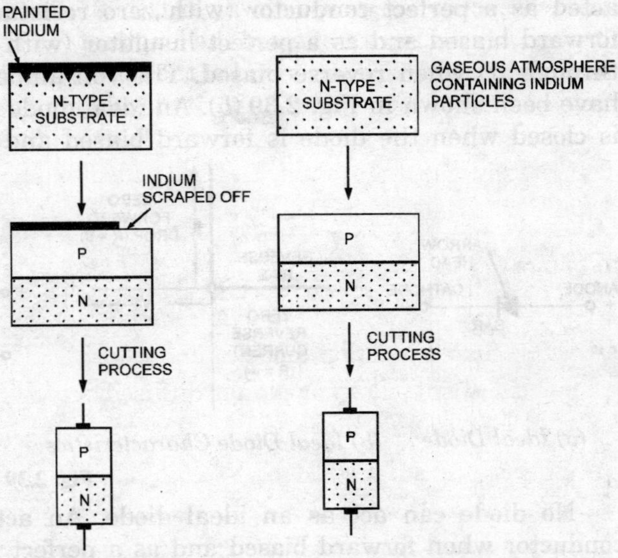

(*a*) *Solid Diffusion* (*b*) *Gaseous Diffusion*

Fig. 2.37. ***Diffusion Process***

4. *Epitaxial Growth or Planar Diffused Diode.* The term "epitaxial" is derived from the Latin terms *epi* meaning 'upon' and *taxis* meaning "arrangement".

To construct an epitaxially grown diode, a very thin (single crystal) layer of semiconductor material (silicon or germanium) is grown on a heavily doped substrate (base) of the same material. This complete structure then forms the N-region on which P-region is diffused. SiO_2 layer is thermally grown on the top surface, photo-etched and then aluminium contact is made to the P-region. A metallic layer at the bottom of the substrate forms the cathode to which lead is attached.

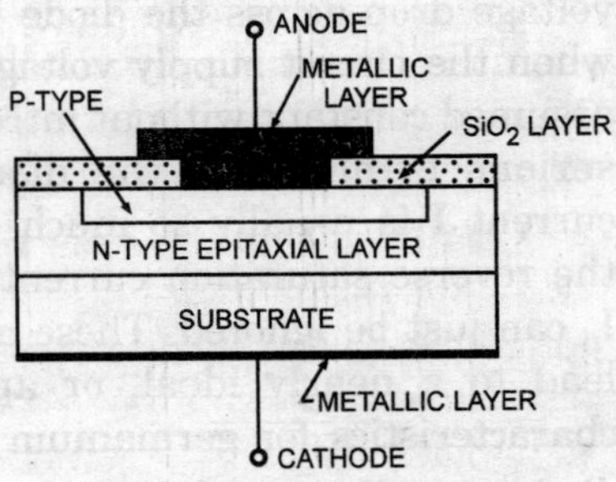

Fig. 2.38. (*a*) *Epitaxially Grown or Planar Diffused Diode*

This process is usually employed in the fabrication of IC chips.

5. *Point Contact Diode.* It consists of an N-type germanium or silicon wafer about 12.5 mm square by 0.5 mm thick, one face of which is soldered to a metal base by radio-frequency heating and the other face has a phosphor bronze or tungsten spring pressed against it. A barrier layer is formed round the point contact by a pulsating current forming process. This causes a P-region to be formed round the wire and since pure germanium is N-type, a very small P-N junction in the shape of a hemisphere is formed round the point contact. The forming process cannot be controlled with precision. Because of small area of the junction, point contact diode can be used to rectify only very small currents (of the order of mA). On the other hand, the shunting capacitance of point contact diodes are very valuable in equipment operating at super high frequencies (as high as 25,000 MHz).

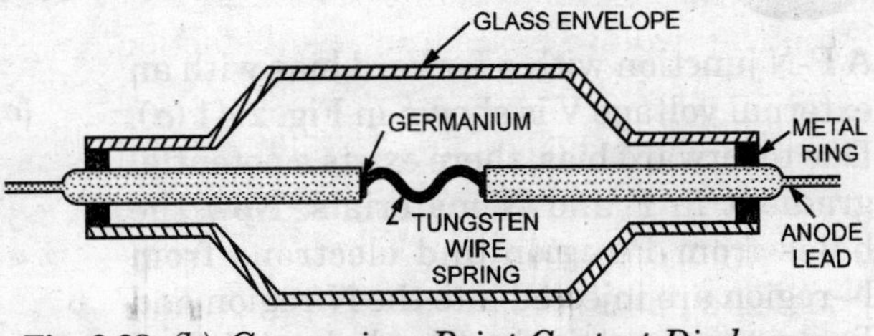

Fig. 2.38. (*b*) *Germanium Point Contact Diode*

2.17. IDEAL DIODE

It is a two terminal device and permits only *unidirectional conduction.* It conducts well in the forward direction and poorly in the reverse direction. It would have been ideal if it acted as a perfect conductor (with zero resistance or zero voltage drop across it) when forward biased and as a perfect insulator (with an infinite resistance or with no current through it) when reverse biased. The volt-ampere characteristics of such an ideal diode have been shown in Fig. 2.39 (*b*). An ideal diode acts like an *automatic switch.* The switch is closed when the diode is forward biased and is opened when reverse biased.

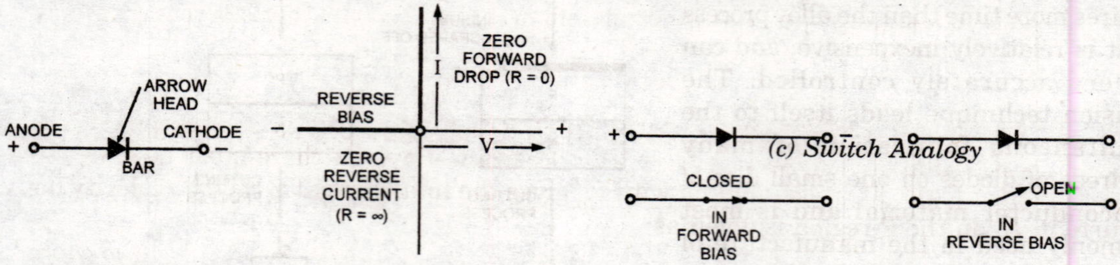

(a) Ideal Diode *(b) Ideal Diode Characteristics* *(c) Switch Analogy*

Fig. 2.39

No diode can act as an ideal diode. An actual diode does not behave as a perfect conductor when forward biased and as a perfect insulator when reverse biased. Neither it offers zero resistance when forward biased nor infinite resistance when reverse biased. However, there are many applications in which diodes can be assumed to be nearly ideal devices, if the voltage drop across the diode when it is forward biased *i.e.*, *v* is taken into account. In cases when the circuit supply voltage V_S is much larger than the forward voltage drop *v*, *v* can be assumed constant without introducing any serious error. Also, the diode forward current I is usually so much larger than the reverse saturation current I_0 that the I_0 can just be ignored. These assumptions lead to a nearly ideal, or approximate, characteristics for germanium and silicon diodes, as illustrated in Fig. 2.40.

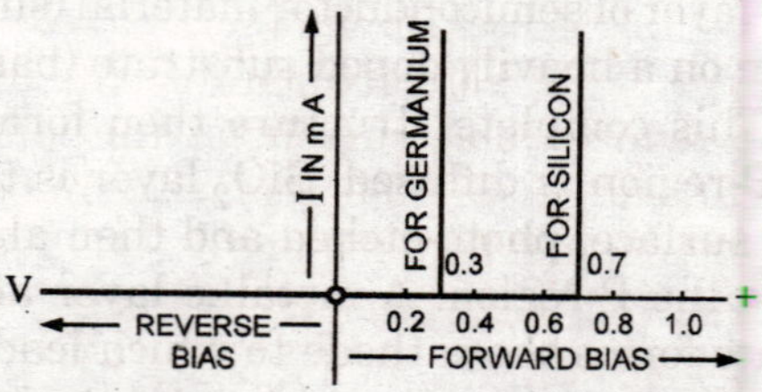

Fig. 2.40. ***Approximate Characteristics For Ge and Si Diodes***

2.18. Current Components in a P-N Diode

A P-N junction with a forward bias with an external voltage V is shown in Fig. 2.41 (*a*). Due to forward bias, there exists a potential gradient in P and N materials. Now the holes from P-region and electrons from N-region are injected into the N-region and P-region respectively. The holes injected into the N-region and electrons injected into the P-region are the minority carriers. The minority carriers diffuse away from the junction exponentially with the distance, as illustrated in Fig. 2.41 (*b*). Their concentration falls steadily due to recombination with electrons and holes respectively. Since the diffusion current due to minority carriers is proportional to the

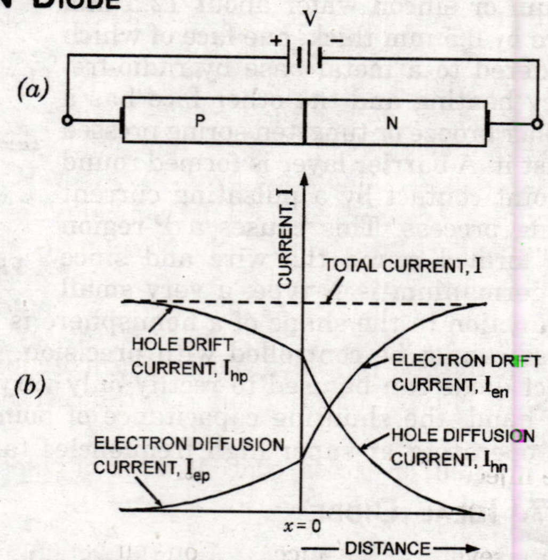

Fig. 2.41. ***Current Components in a Forward Biased P-N Junction Diode***

concentration gradient [refer to Eqs. (2.26) and (2.27)], this current must also vary exponentially with distance. There are two minority (diffusion) currents, I_{ep} and I_{hn} as shown in Fig. 2.41 (*b*). Let the symbols $I_{ep}(x)$ and $I_{hn}(x)$ represent the electron current in P-region and hole current in N-region respectively.

Electrons crossing the junction at $x = 0$ from right to left constitute a current in the same direction as the holes crossing the junction from left to right. Thus the total current I at junction (*i.e.*, $x = 0$) is given as

$$I = I_{hn}(0) + I_{ep}(0) \quad \text{...(2.81)}$$

Since the current is the same throughout a series circuit, total current I is independent of x, and is shown as a horizontal line in Fig. 2.41 (*b*). So, in the P-region there must be a second component of current I_{hp} which combining with I_{ep} gives the total current I. So the majority (or hole) current, I_{hp} in the P-region is given as

$$I_{hp}(x) = I - I_{ep}(x) \quad \text{...[2.82(a)]}$$

Similarly the majority (or electron) current, I_{en} in the N-region is given as

$$I_{en}(x) = I - I_{hn}(x) \quad \text{...[2.82(b)]}$$

The currents I_{hp} and I_{en} are plotted as a function of distance in Fig. 2.41 (*b*). The Fig. 2.41 (*b*) is drawn for unsymmetrically doped diode, so that $I_{ep} \neq I_{hn}$.

It may be noted that far away from the junction in P-region the current is drift current I_{hp} of holes sustained by small electric field in the semiconductor. As the holes approach the junction, some of them recombine with the electrons which have been injected into the P-region from the N-region. Thus part of the current I_{hp} becomes a negative current just equal in magnitude to the diffusion current I_{ep}. The drift current I_{hp}, therefore, decreases toward the junction. The rate of decrease is such that the total current remains constant, independent of distance. The reduced hole current at the junction now enters the N-region and becomes hole diffusion current I_{hn}. Similar remarks can be made with respect to electron drift current in N-region, I_{en}. Thus in a forward-biased P-N diode, the current enters the P-region as a hole current and leaves the N-region as an electron current of the same magnitude.

From the above discussion it is concluded that the current in a P-N diode is bipolar in character (because it is made up of both positive and negative charge carriers) and the total current is constant throughout the device, but the proportion of current due to holes and that due to electrons varies with the distance, as illustrated in Fig. 2.41 (*b*).

2.19. Quantitative Theory of P-N Diode Currents

Here an expression for the total current will be derived as a function of the applied voltage with the assumption that depletion layer thickness is negligible (*i.e.*, *barrier* width is zero).

When a P-N diode is forward biased, holes are injected from P-region into the N-region. The concentration p_n of the holes in the N-region is increased above its thermal equilibrium value p_{no}. The hole concentration in N-region is given as

$$p_n(x) = p_{no} + p'_n(0)\, e^{-x/L_h} \quad \text{...(2.83)}$$

where p_{no} is the hole concentration in thermal-equilibrium condition, L_h is the diffusion length for holes in N-type material and x is the distance from the junction where concentration is considered.

The injected, or excess, concentration of holes at $x = (0)$, $p'_n(0)$ is given as

$$p'_n(0) = p_n(0) - p_{no} \quad \text{...(2.84)}$$

These several hole concentration components are shown in Fig. 2.42, which shows that the concentration $p_n(x)$ falls exponentially with distance x into the N-type material.

The diffusion hole current in the N-region is given as

$$I_{hn} = -A\,e\,D_h \frac{d\,p_n}{dx} \quad ...(2.85)$$

Taking the derivative of Eq. (2.83) and substituting the value of $\frac{d\,p_n}{dx}$ in above Eq. (2.85) we have

$$I_{hn}(x) = \frac{A\,e\,D_h\,p_n'(0)}{L_h}\,e^{-x/L_h} \quad ...(2.86)$$

At junction *i.e.*, $x = 0$

$$I_{hn}(0) = \frac{A\,e\,D_h\,p_n'(0)}{L_h} = \frac{A\,e\,D_h}{L_h}[p_n(0) - p_{no}] \quad ...(2.87)$$

where A is area of material in m^2, D_h is diffusion constant for holes in m^2/s, e is the magnitude of charge on holes, L_h is the diffusion length of holes in metre, and $p'_n(0)$ is the excess of hole concentration at junction.

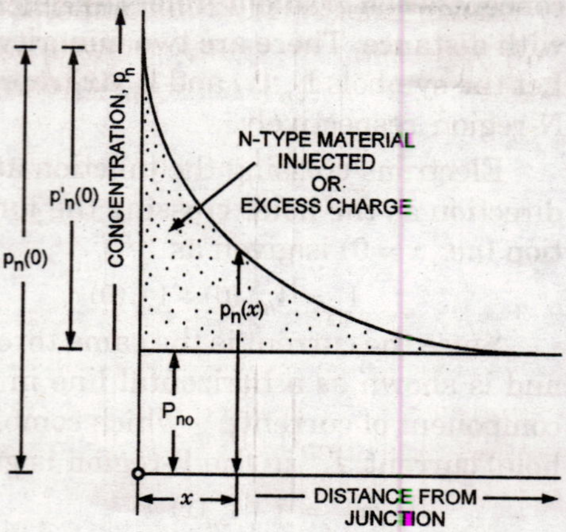

Fig. 2.42. *Hole Concentration Components in N-Type Material*

Using Boltzmann relationship of kinetic gas theory, we have

$$p_n(0) = p_{no}\,e^{V/V_T} \quad ...(2.88)$$

where V is the voltage applied across the P-N diode and V_T is the volt-equivalent of temperature and from Eq. (2.31) it is given as

$$V_T = \frac{kT}{e} = \frac{T}{11{,}600}$$

where k′ is the Boltzmann constant.

The Eq. (2.88) is known as the *law of junction.*

2.19.1. Diffusion Length

Diffusion length is defined as the distance travelled by free charge carriers (electrons or holes) before recombination. It may also be defined as the average distance covered by an excess charge carrier during its lifetime τ.

The diffusion length L_h and lifetime τ_h are related to each other as :

$$L_h = \sqrt{D_h \tau_h} \quad ...(2.89)$$

Similarly in case of P type semiconductor

$$L_e = \sqrt{D_e \tau_e} \quad ...(2.90)$$

2.19.2. Forward Currents

The total diode current I at $x = 0$ is given as

$$I = I_{hn}(0) + I_{ep}(0) \quad ...(2.91)$$

where $I_{hn}(0)$ is the current caused by holes entering the N-region and $I_{ep}(0)$ is the current caused by electrons entering the P-region.

From Eqs. (2.87) and (2.88)

$$I_{hn}(0) = \frac{A\,e\,D_h}{L_h}\left[p_{no}e^{V/V_T} - p_{no}\right] = \frac{A\,e\,D_h\,p_{no}}{L_h}\left[e^{V/V_T} - 1\right] \quad ...(2.92)$$

$$\text{Similarly } I_{ep}(0) = \frac{A\,e\,D_e\,n_{po}}{L_e}\left[e^{V/V_T} - 1\right] \quad ...(2.93)$$

From Eqs. (2.91), (2.92) and (2.93) we have total diode current

$$I = \left[\frac{Ae\,D_h\,p_{no}}{L_h} + \frac{Ae\,D_e\,n_{po}}{L_e}\right](e^{V/V_T} - 1)$$

$$\text{or } I = I_0\,(e^{V/V_T} - 1) \qquad \text{...(2.94)}$$

$$\text{where } I_0 = \frac{Ae\,D_h\,p_{no}}{L_h} + \frac{Ae\,D_e\,n_{po}}{L_e} \qquad \text{...(2.95)}$$

2.19.3. Reverse Saturation Current

In the foregoing discussion a positive value of V indicates a forward bias. Equation (2.94) is equally valid for reverse bias [*i.e.* for negative values of applied voltage V]. For a reverse bias voltage exceeding V_T (*i.e.* 26 mV) at room temperature (27 °C or 300 K), current $I \to -I_0$. Thus I_0 is called the *reverse saturation current.*

From Eq. (2.12), concentration of holes in N-type region,

$$p_{no} = \frac{n_i^2}{N_D}$$

and from Eq. (2.14), concentration of electrons in P-type' region,

$$n_{po} = \frac{n_i^2}{N_A}$$

where N_D and N_A are the concentrations of donor and acceptor atoms respectively.

Substituting $p_{no} = \frac{n_i^2}{N_D}$ and $n_{po} = \frac{n_i^2}{N_A}$ in Eq. (2.95) we have

$$I_0 = Ae\left[\frac{D_h}{L_h N_D} + \frac{D_e}{L_e N_A}\right] n_i^2 \qquad \text{...(2.96)}$$

From Eq. (2.24) $$n_i^2 = A_0\,T^3 e^{-E_{GO}/kT} = A_0\,T^3 e^{-V_{GO}/V_T} \qquad \text{...(2.97)}$$

where V_{GO} is a voltage numerically equal to the forbidden energy gap E_{GO} in electron volts, and V_T is the volt equivalent of temperature and is equal to T/11,600.

For germanium the diffusion constants D_h and D_e vary approximately inversely proportional to temperature T. Hence I_0 may be given as

$$I_0 = K_1\,T^2 e^{-V_{GO}/V_T} \qquad \text{...(2.98)}$$

where K_1 is a constant, independent of temperature T.

In the discussion made above, the effect of carrier generation and recombination in the space-charge region has been ignored. Such an assumption is valid for a germanium diode (not for a silicon diode). For the silicon diode, a diffusion current is negligible in comparison with the transition-layer charge-generation current, which is given approximately by

$$I = I_0\,(e^{V/\eta V_T} - 1) \qquad \text{...(2.99)}$$

where $\eta \approx 2$ for small (rated) currents and $\eta \approx 1$ for large currents. Also I_0 is found to be proportional to n_i instead of n_i^2 so

$$I_0 = K_2\,T^{1.5} e^{-V_{GO}/2V_T} \qquad \text{...(2.100)}$$

where K_2 is a constant, independent of temperature T.

2.20. Volt-Ampere Characteristics of Junction Diodes

The volt-ampere characteristic of a silicon junction diode is depicted in Fig. 2.43. The same characteristic is depicted in Fig. 2.44 with some scale expanded and other compressed to reveal details—the changes in scale have resulted in the apparent discontinuity at the origin.

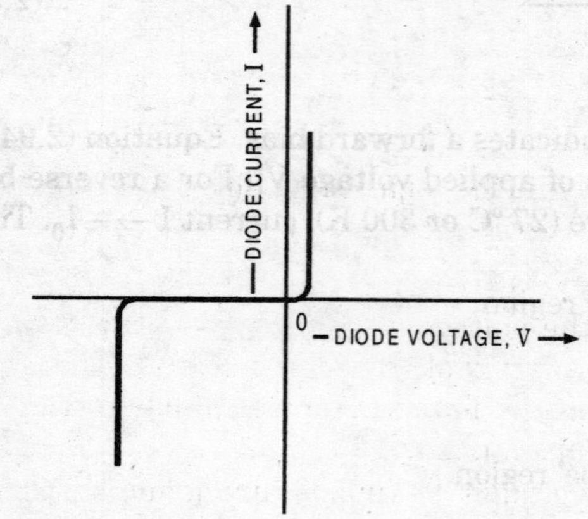

Fig. 2.43. *Volt-Ampere Characteristic of a Silicon Junction Diode*

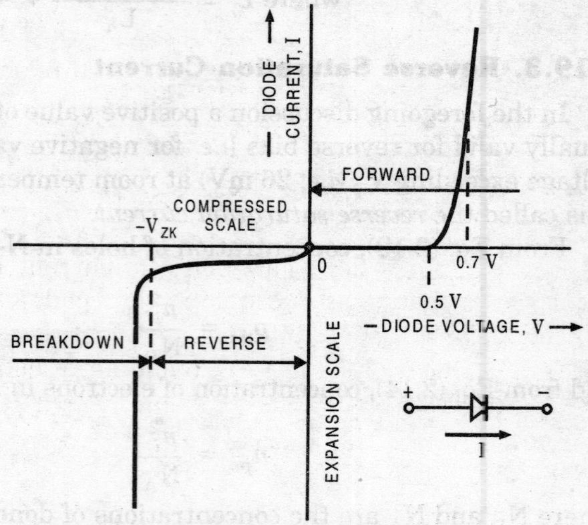

Fig. 2.44. *V-I Characteristics of the Diode With Some Scales Expanded and Others Compressed In Order To Reveal Details*

As shown in the figure, the characteristic curve consists of three distinct regions :

1. The forward-bias region, determined by $V > 0$.
2. The reverse-bias region, determined by $V < 0$.
3. The breakdown region, determined by $V < -V_{ZK}$.

The above three regions of operation are described below :

2.20.1. Forward-Bias Region

A positive value of current I means the current flowing from the P-region to the N-region. The diode is forward biased if the applied voltage V is positive *i.e.* P-side is connected to the positive terminal and N-side is connected to negative terminal as shown in Fig. 2.27.

The current flowing through the junction diode is given by Eq. (2.99).

In Eq. (2.99) I_0 is a constant for a given diode at a given temperature. A formula for I_0 in term of the temperature will be given in Art. 2.21. The current I_0 is usually called the saturation current, and sometimes denoted by I_s. Another name for the saturation current is the *scale current.* This name arises from the fact that I_0 is directly proportional to the cross-sectional area of the diode. Thus doubling of junction area results in a diode with double the value of I_0 and, as the diode equation indicates, double the value of current I for a given forward voltage V. For "small-signal" diodes (small size diodes intended for low-power applications), I_0 is of the order of 10^{-15} A. The value of I_0 is, however, a very strong function of temperature. As a thumb rule, I_0 approximately doubles for every 5°C rise in temperature.

Voltage V_T in Eq. (2.99) is the voltage equivalent of temperature and is equal to $\frac{T}{11,600}$. At room temperature V_T is taken equal to be 2.6 mV.*

In Eq. (2.99) the constant η has a value between 1 and 2, depending on the material and physical structure of diode. Diodes made using the standard IC fabrication process exhibits $\eta = 1$ when operated under normal conditions. Diodes available as discrete two-terminal components generally exhibit $\eta = 2$.

For appreciable current I in the forward direction, specifically for $I >> I_0$. Equation (2.99) can be approximated by the exponential relationship

$$I = I_0 e^{V/\eta V_T} \qquad \text{...(2.101)}$$

This relationship can be expressed alternatively in the logarithmic form

$$V = \eta V_T \log_e \frac{I}{I_0} \qquad \text{...(2.102)}$$

The exponential relationship of current I to the voltage V holds over many decades of current (a span of as many as seven decades—*i.e.*, a factor of 10^7—can be found) This is quite a remarkable property of junction diodes, one that is also found in bipolar junction transistors and that has been exploited in many interesting applications.

From Eq. (2.101), the currents I_1 and I_2 for voltages V_1 and V_2 are given respectively by equations

$$I_1 = I_0 e^{V_1/\eta V_T} \qquad \text{...(2.103)}$$

$$I_2 = I_0 e^{V_2/\eta V_T} \qquad \text{...(2.104)}$$

Dividing Eq. (2.104) by Eq. (2.103), we have

$$\frac{I_2}{I_1} = e^{(V_2 - V_1)/\eta V_T}$$

$$\text{or } V_2 - V_1 = \eta V_T \log_e \frac{I_2}{I_1}$$

$$\text{or } V_2 - V_1 = 2.3 \eta V_T \log_{10} \frac{I_2}{I_1} \qquad \text{...(2.105)}$$

The above equation reveals that for a decade (factor of 10) change in diode current, the diode voltage drop changes by $2.3\eta V_T$, which is about 60 mV for $\eta = 1$ and 120 mV for $\eta = 2$. This also suggests that the diode volt-ampere characteristic is most conveniently drawn on semilog paper. Using the vertical, linear axis for V and horizontal, log axis for current I, we have a straight line with a slope of $2.3\eta V_T$ per decade of current. Finally, it is worthmentioning that not knowing the exact value of η, circuit designer uses the convenient approximate number of 0.1 V/decade for the slope of the diode logarithmic characteristic.

Voltage-current characteristic in the forward region (Fig. 2.44) reveals that the current is negligibly small for voltage smaller than about 0.5 V. This value is usually referred to as the *cut-in* voltage. It needs to be emphasized, however, that this apparent threshold in the characteristic is simply a consequence of the exponential relationship. Another consequence of this exponential relationship is the rapid increase of current I. Thus, for a "fully conducting" diode, the average drop lies in a narrow range, approximately 0.6 V to 0.8 V. This gives rise to a simple "model" for the diode where it is assumed that a conducting diode has approximately

---

* A slightly higher ambient temperature (25°C or so) is usually assumed for electronic equipment operating inside a cabinet. At this temperature, V_T = 25.8 mV ≃ 26 mV. Nevertheless, for the sake of simplicity, and to promote rapid circuit analysis, sometimes more arithmetically convenient value of 25 mV is used.

0.7 V drop across the diode. Diodes with different current ratings (*i.e.* different areas and correspondingly different I_0) will exhibit the 0.7 V drop at different currents.

2.20.2. Reverse-Bias Region

When the diode is reverse-biased and V is several times V_T in magnitude, current, $I = -I_0$. The reverse current is, therefore, constant, independent of the applied reverse bias, as shown in Fig. 2.44. Consequently, I_0 is referred to as the *reverse saturation current.*

Real diodes exhibit currents, that, though quite small, are much larger than I_0. For instance, a small-signal diode whose current I_0 is of the order of 10^{-14} to 10^{-15} A could show a reverse current of the order of 1 nA. The reverse current also increases somewhat with the increase in magnitude of the reverse voltage. Because of the very small magnitude of the current, these details are not clearly evident on the diode V-I characteristic depicted in Fig. 2.44.

A large component of the reverse current is due to leakage effects. These leakage currents are proportional to junction area, just as I_0 is. Their dependence on temperature, however, is different from that of I_0. Thus, whereas I_0 doubles for every 5°C rise in temperature, the corresponding rule of thumb for the temperature dependence of the reverse current is that it doubles for every 10°C rise in temperature.

2.20.3. Breakdown Region

The third distinct region of diode operation is the breakdown region, which can be easily identified on the volt-ampere characteristic depicted in Fig. 2.44. The breakdown region is entered when the magnitude of the reverse voltage exceeds a threshold value that is specific to the particular diode, called the *breakdown voltage.* This is the voltage at the "knee" of the V-I characteristic curve in Fig. 2.44, and is denoted by V_{ZK}, where the subscript Z stands for zener (to be explained shortly) and K denotes knee.

As can be seen from Fig. 2.44, in the breakdown region the reverse current increases rapidly with the associated increase in voltage drop being very small. Diode breakdown is normally not destructive provided that the power dissipated in diode is limited by external circuitry to a "safe" level. This safe value is normally specified on the data sheet of the device. It is, therefore, necessary to limit the reverse current in the breakdown region to a value consistent with the permissible power dissipation.

The fact that the diode V-I characteristic in breakdown is almost a vertical line enables it to be used in voltage regulation.

Example 2.22. A germanium diode carries a current of 1 mA at room temperature when a forward bias of 0.15 V is applied. Estimate the reverse saturation current at room temperature.

Solution : Applied forward bias voltage, V = 0.15 V

Forward current, I = 1 mA

Volt-equivalent of room temperature, V = 26 mV = 0.026 V

If reverse saturation current = I_0

then, diode current is given as, $I = I_0 \left(e^{V/\eta V_T} - 1\right)$

$$\text{or } I_0 = \frac{I}{\left(e^{V/V_T} - 1\right)} = \frac{1 \times 10^{-3}}{\left(e^{0.15/0.026} - 1\right)} \qquad \because \text{ for Ge, } \eta = 1$$

$$= 3.12 \times 10^{-6} \text{ A} = \mathbf{3.12\ \mu A} \quad \textbf{Ans.}$$

Example 2.23. The current flowing through a certain germanium P-N junction at room temperature when reverse biased (bias voltage being very large in comparison to V_T) is 0.15 μA. Determine the current flowing through the germanium diode when the applied voltage is 0.12 V.

Solution : The diode current is given as $I = I_0 \, (e^{V/V_T} - 1)$

From given datas,

Large reverse bias current, $I \simeq I_0 = 0.15 \times 10^{-6}$ A

Applied voltage, $V = 0.12$ V

Volt-equivalent of temperature, $V_T = 26$ *m*V or 0.026 V at room temperature

$$\text{So } I = 0.15 \times 10^{-6} \, (e^{0.12/0.026} - 1) = 15 \ \mu\text{A} \quad \textbf{Ans.}$$

Example 2.24. A silicon diode has reverse saturation current of 2.5 μA at 300 K. Find forward voltage for a forward current of 10 mA.

Solution : From Eq. (2.99)

$$I = I_0 \, (e^{V/\eta V_T} - 1)$$

Substituting forward current I = 10 mA (or 0.01 A), reverse saturation current $I_0 = 2.5 \times 10^{-6}$ A in above equation, we have

$$0.01 = 2.5 \times 10^{-6} \, (e^{V/\eta V_T} - 1)$$

$$\text{or} \quad e^{V/\eta V_T} = \frac{0.01}{2.5 \times 10^{-6}} + 1 = 4 \times 10^3 + 1 = 4{,}001$$

$$\text{or} \quad \frac{V}{\eta V_T} = \log_e 4001$$

$$\text{or } V = \eta \times V_T \log_e 4001 = 2 \times 0.026 \times \log_e 4001 = 0.43 \text{ V} \quad \textbf{Ans.}$$

$\because$ $\eta = 2$ for silicon and V_T at 300 K = 26 mV or 0.026 V

Example 2.25. Find the reverse saturation current density in an abrupt silicon junction. Given the following data :

$N_D = 10^{21}$ m^{-3}; $N_A = 10^{22}$ m^{-3}; $D_e = 3.4 \times 10^{-3}$ m^2 s^{-1};
$D_h = 1.2 \times 10^{-3}$ m^2 s^{-1}; $L_e = 7.1 \times 10^{-4}$ m
$L_h = 3.5 \times 10^{-4}$ m; $n_i = 1.6 \times 10^{16}$ m^{-3}.

Solution : From Eq. (2.96) reverse saturation current.

$$I_0 = Ae \left[\frac{D_h}{L_h N_D} + \frac{D_e}{L_e N_A} \right] n_i^2$$

or reverse saturation current density

$$= \frac{I_0}{A} = \left[\frac{D_h}{L_h N_D} + \frac{D_e}{L_e N_A} \right] e n_i^2$$

$$= \left[\frac{1.2 \times 10^{-3}}{3.5 \times 10^{-4} \times 10^{21}} + \frac{3.4 \times 10^{-3}}{7.1 \times 10^{-4} \times 10^{22}} \right] \times 1.602 \times 10^{-19} \times (1.6 \times 10^{16})^2$$

$$= 1.6 \times 10^{-7} \text{ A} \quad \text{or} \quad 0.16 \ \mu\text{A} \quad \textbf{Ans.}$$

2.21. Temperature Dependence of P-N Diodes

Temperature is very important consideration in the analysis or design of electronic devices or systems. It affects virtually all of the characteristics of a semiconductor device.

The diode current at a given voltage depends upon the operating temperature. The variation may be determined from Eq. (2.99), where the temperature is contained implicitly in V_T and also in the reverse saturation current I_0.

From Eq. (2.99) $\quad I = I_0\,(e^{V/\eta V_T} - 1)$

At room temperature V_T = 26 mV = 0.026 V

For germanium η = 1 and for silicon η = 2

So substituting the above values in the expression for diode current we have

$$\text{For germanium, } I = I_0\,(e^{V/0.026} - 1) \quad I_0\,e^{40\,V} \qquad \text{...(2.106)}$$

$$\text{and for silicon, } I = I_0(e^{V/2 \times 0.026} - 1) \quad I_0\,e^{20\,V} \qquad \text{...(2.107)}$$

The dependence of I_0 on temperature T is, from Eqs. (2.98) and (2.100), given approximately as

$$I_0 = K T^m e^{-V_{GO}/\eta V_T} \qquad \text{...(2.108)}$$

where K is a constant and eV_{GO} is the forbidden energy gap in joules

For silicon : η = 2 $\quad m$ = 1.5 $\quad V_{GO}$ = 1.21 V

For germanium : η = 1 $\quad m$ = 2 $\quad V_{GO}$ = 0.785 V

Taking the logarithm on both sides of Eq. (2.108), we have

$$\log_e I_0 = mK \log_e T - \frac{V_{GO}}{\eta V_T}$$

Taking the derivative of the above equation we have

$$\frac{d(\log_e I_0)}{dT} = \frac{1}{I_0}\frac{dI_0}{dT} = \frac{m}{T} + \frac{V_{GO}}{\eta T V_T} \qquad \text{...(2.109)}$$

At room temperature (300 K), from Eq. (2.109)

$$\frac{d(\log_e I_0)}{dT} = 0.08 \text{ per °C for silicon and } 0.11 \text{ per °C for germanium.}$$

The performance of commercial diodes is only consistent with above result. The reason for the discrepancy is that, in a practical diode, there is a component of the reverse saturation current due to leakage over the surface that has not been taken into account in Eq. (2.108). Since leakage component is independent of temperature, it may be expected to find a smaller rate of change of I_0 with temperature than that predicted above.

From experimental data it is found that reverse saturation current I_0 increases 7 percent per °C for both silicon and germanium. Since $(1.07)^{10}$ $\;$ 2.0, the reverse saturation current approximately doubles for every 10°C rise in temperature.

Germanium is more temperature dependent than silicon because its reverse saturation current is approximately 1,000 times larger.

The change in V-I characteristics of a semiconductor diode due to variations in temperature above and below room temperature is shown in Fig. 2.45.

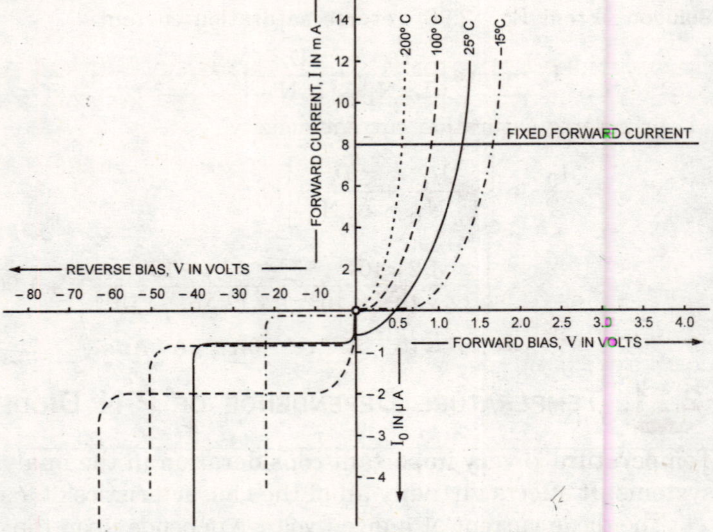

Fig. 2.45. *Variation in Diode Characteristics With Variations in Temperature*

From Fig. 2.45 we see that for a fixed level of forward voltage, forward current increases with the increase in temperature. If the forward current is measured, at the increased temperature, for several levels of forward voltage and the results are plotted it is seen that current-voltage characteristic is moved to the left. The horizontal line on Fig. 2.45 shows that for a fixed forward current, forward voltage drops as the junction temperature increases *i.e.* forward voltage has a negative temperature coefficient. It is found that the temperature coefficient for the forward voltage of a P-N junction is approximately – 1.8 m V/ °C for Si and – 2.02 mV/ °C for Ge. Zener potential is also experiencing a pronounced change in level.

2.22. DIODE RESISTANCE

As already discussed, a forward-biased diode conducts easily whereas a reverse-biased diode restrains the flow of current, so a diode has a low forward resistance as compared to high blocking or reverse resistance.

1. Forward Resistance. The resistance offered by a diode in the circuit, when forward biased, is known as the *forward resistance.* This resistance is not the same for dc as well as changing current. Thus, forward resistance is of two types viz. dc or *static resistance* and ac or *dynamic resistance*

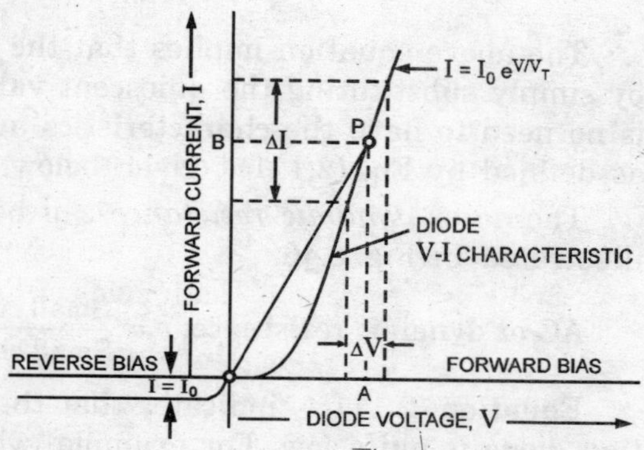

Fig. 2.46

DC or **static resistance**, R is the resistance offered by a diode to the direct current. It is simply the ratio of the dc voltage across the diode to the direct current flowing through it. It can be determined by determining the ratio of voltage drop across the diode to the current flowing through it. At any point P on the V-I characteristic of the diode (Fig. 2.46), the voltage across the diode is OA and the corresponding current is OB.

So dc or static resistance, $R = \frac{V}{I} = \frac{OA}{OB}$

Thus at any point on the V-I characteristic of the diode, the dc or static resistance R is equal to the reciprocal of the slope of the line joining the operating point to the origin.

DC or static resistance is not constant but depends on the operating point on the V-I characteristic of the diode. It is different at different points, as illustrated in Fig. 2.46. From Fig. 2.46 it is obvious that the dc resistance of a diode in the forward-bias region decreases as the region of higher voltages and currents is approached.

In the reverse-bias region, the current $I = I_0$, constant in magnitude, the resistance of the ideal diode is essentially infinite resistance (open-circuit equivalent).

Any way dc or static resistance does not form a useful parameter of the diode.

AC or **dynamic resistance** *r* is resistance offered by a diode to the changing forward current. It may also be defined as the reciprocal of the slope of the forward characteristic of the diode. It forms an important device parameter for small signal operation. However, it varies with the operating voltage.

From Eq. (2.99)
$$r = \frac{dV}{dI} = \frac{\eta V_T}{I_0\, e^{V/\eta V_T}} = \frac{\eta V_T}{I + I_0} \qquad \text{...(2.110)}$$

For a forward bias greater than a few tenths of a volt, $\left|\frac{V}{\eta V_T}\right|$ >> 1 and according to Eq. (2.99) $I >> I_0$, and dynamic resistance *r* is given approximately by

$$r = \frac{\eta V_T}{I} \qquad \text{...(2.111}a\text{)}$$

From above equation it is obvious that for forward bias r varies inversely with the current I. At room temperature (27°C or 300 K) V_T = 0.026 V or 26 mV.

Substituting η = 1 and V_T = 26 mV in above Eq. (2.111*a*), we have

$$r = \frac{26 \text{ mV}}{I \text{ (mA)}} \qquad \text{...(2.111}b\text{)}$$

The above equation implies that the dynamic resistance of a diode can be determined by simply substituting the quiescent value of the diode current into the equation. There is no need to have the characteristics available or to worry about sketching tangent line as defined by Eq. (2.111*c*), given below.

The *ac* or *dynamic resistance* can be determined from the forward characteristic, as illustrated in Fig. 2.46.

$$\text{AC or dynamic resistance, } r = \frac{\text{Small change in forward voltage}}{\text{Small change in forward current}} = \frac{\Delta V}{\Delta I} \qquad \text{...(2.111}c\text{)}$$

Equation (2.111*b*), indicates that the incremental forward resistance of the P-N junction diode is quite low. For example, when diode current is 26 mA, the diode resistance is 1 Ω.

Actually diode ac or dynamic resistance for forward bias is slightly different from that obtained from above equation. This is because of bulk and contact resistance of the semiconductor device. This additional resistance level can be included in Eq. (2.111*b*), by adding a factor denoted r_B as appearing in Eq. (2.111*d*)

$$r' = \frac{26 \text{ mV}}{I \text{ (mA)}} + r_B \qquad \text{...(2.111}d\text{)}$$

For large values of forward current the first term $\frac{26 \text{ mV}}{I \text{ (mA)}}$ giving apparent forward or junction resistance r is negligible so $r' = r_B$. For small values of forward current, r_B is negligible so $r' = \frac{26 \text{ mV}}{I \text{ (mA)}}$.

The incremental backward resistance (*i.e.*, resistance with reverse bias), is ideally infinite because of the zero slope of the V-I characteristic in this region.

This conclusion also follows from Eq. (2.110), upon inserting $I = -I_0$ for negative values of bias voltage. Thus diode resistance for reverse bias

$$= \frac{\eta V_T}{I + I_0} = \frac{\eta V_T}{-I_0 + I_0} = \infty$$

However, the apparent backward resistance is not infinite because for any value of reverse bias there does exist a small current I_0. For germanium and silicon junction diodes this resistance is quite large compared to the apparent forward resistance. In germanium diodes the ratio of apparent backward to forward resistance is 400,000 : 1 while for silicon diodes this ratio is 1,000,000 : 1.

Example 2.26. Find the dynamic resistance of a P-N junction diode at a forward current of 2 mA. Assume $\frac{k'T}{e}$ = 25 mV.

Solution : Forward current, I = 2 mA or 0.002 A

Volt equivalent of temperature, $V_T = \frac{k'T}{e} = 25$ mV or 0.025 V

From Eq. [2.111*a*], dynamic resistance is given approximately by $r = \frac{\eta V_T}{I}$

Substituting $\eta = 1$, $V_T = 0.025$ V and I = 0.002 A in above equation we have

$$\text{Dynamic resistance, } r = \frac{0.025}{0.002} = 12.5\ \Omega \quad \textbf{Ans.}$$

Example 2.27. Determine ac resistance for a germanium semiconductor diode having a forward bias of 200 mV and reverse saturation current of 1 μA at room temperature.

Solution : At room temperature (300 K),

$$V_T = 26 \text{ mV} = 0.026 \text{ V}$$

$$\text{Applied forward voltage, } V = 200 \text{ mV} = 0.2 \text{ V}$$

$$\text{Reverse saturation current, } I_0 = 1\ \mu A = 1 \times 10^{-6} \text{ A}$$

From Eq. (2.110), ac resistance for the diode,

$$r = \frac{\eta V_T}{I_0 e^{V/\eta V_T}} = \frac{0.026}{1 \times 10^{-6} \times e^{0.2/0.026}} = 11.86\ \Omega \quad \textbf{Ans.}$$

Taking $\eta = 1$ for Ge diode

Example 2.28. Explain the static and dynamic resistances in P-N junction diode. Determine dc resistance levels for the diode in following figure at (*i*) I_D = 2 mA, (*ii*) I_D = 20 mA, (*iii*) V_D = –10 V.

[*U.P. Technical Univ. 2006-07*]

Solution : (*i*) When $I_D = 2$ mA, $V_D = 0.5$ V

So resistance, $R_1 = \frac{V_D}{I_D} = \frac{0.5}{2 \times 10^{-3}} = 250\ \Omega$ **Ans.**

(*ii*) When $I_D = 20$ mA = 0.02 A, $V_D = 0.8$ V

So resistance, $R_2 = \frac{0.8}{0.02} = 40\ \Omega$ **Ans.**

(*iii*) When $V_D = -10$ V,

$I_D = -1\ \mu A = -1 \times 10^{-6}$ A

So resistance, $R_3 = \frac{-10}{-1 \times 10^{-6}} = 10\ M\Omega$ **Ans.**

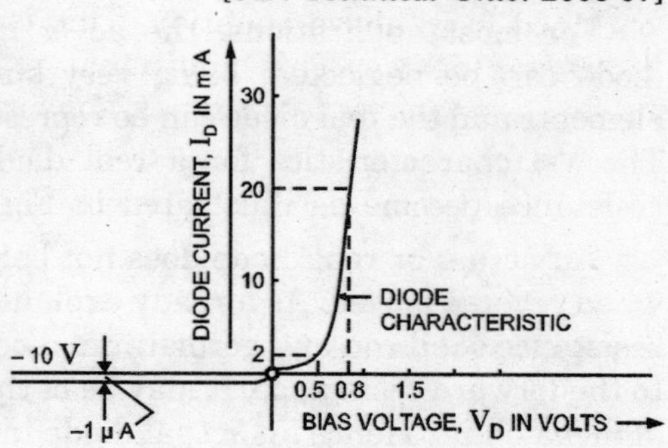

Fig. 2.47

2.23. Real Diode

As already mentioned in Art. 2.17, an actual diode neither behaves as a perfect conductor when forward biased nor as a perfect insulator when reverse biased.

When a diode is forward biased, it will not conduct till the potential barrier V_0 (0.7 V for Si and 0.3 V for Ge) at the junction is overcome. When the applied voltage exceeds the potential barrier V_0, the diode starts conducting. The forward current, I flowing through the diode causes a voltage drop in its internal resistance r'. Thus the voltage V applied across the diode is to overcome (*i*) the potential barrier V_0 and (*ii*) internal drop Ir'

So $V = V_0 + I r'$ ...(2.112)

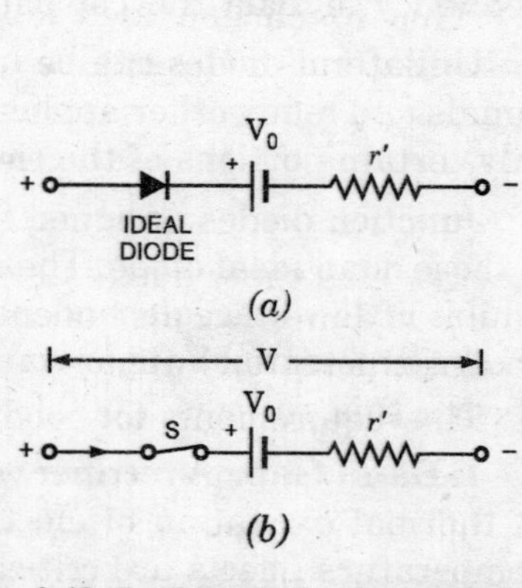

Fig. 2.48. Real Diode Representation Under Forward Bias Condition

Hence, a real diode may be considered as equivalent to an ideal diode in series with a barrier potential of V_0 volts and a forward bias ac resistance r', as illustrated in Fig. 2.48 (*a*). The equivalent circuit of a real diode under forward bias condition is shown in Fig. 2.48 (*b*). This circuit shows that a real diode still acts as a switch when forward biased but the voltage required to operate this switch is V given by Eq. (2.112).

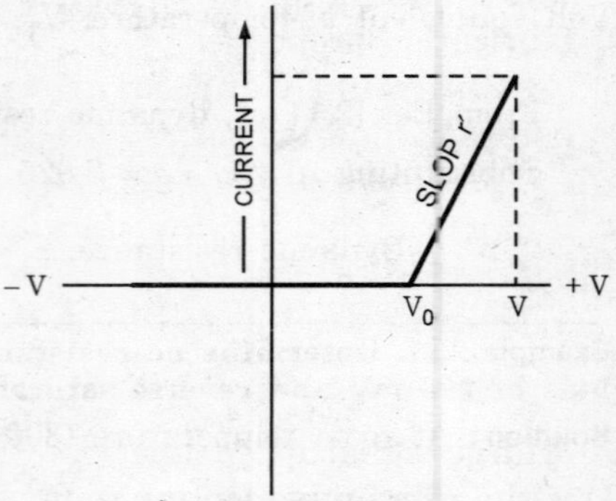

Fig. 2.49

Voltage-current characteristic of a real diode under forward bias condition is shown in Fig. 2.49 which follows the Eq. (2.112). Here, the reciprocal of the slope of this characteristic represents the ac or dynamic resistance r' of the diode.

For *large signal sources* whose voltage is much greater than the diode barrier potential V_0 (or knee voltage V_k), forward current I would be large and the ac or dynamic resistance r' of the diode will be equal to bulk resistance r_B of the diode, as already explained in Art. 2.22.

For *small signal sources* whose voltage is much smaller than V_0 or V_k, forward current I would be small and the ac or dynamic resistance of r' of the diode will be equal to apparent forward or junction resistance *r* given by Eq. (2.111*b*) because for smaller values of forward current r_B is negligible.

For most applications, the ac or dynamic resistance r' of a diode can be neglected, being very small as compared to other elements and the real diode can be represented as shown in Fig. 2.50. The V-I characteristics for a real diode, neglecting its dynamic resistance become as illustrated in Fig. 2.40.

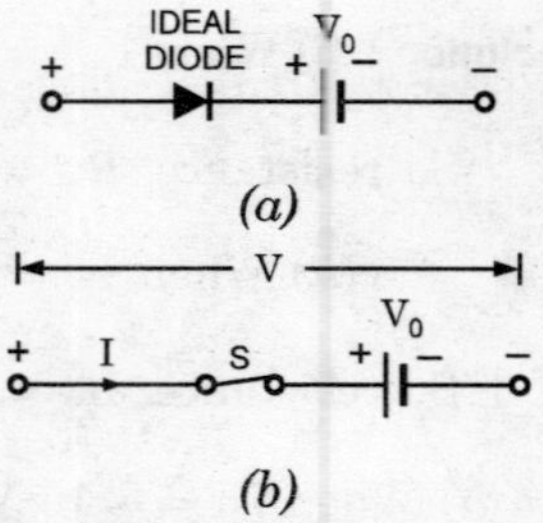

Fig. 2.50

An actual or real diode does not behave as a perfect insulator when reverse biased. As already explained in Art. 2.22, the reverse resistance of silicon and germanium diodes is quite large compared to the forward resistance. It may be of the order of many thousands of mega ohms. Hence an actual diode in the reverse bias condition can be thought of as equivalent to a resistor of high resistance.

The above discussion would be true only in case of small frequency signals. For large frequency signals the capacitive effects would have to be taken into account (Art. 2.24).

Unilateral diodes can be used for rectification of alternating currents, waveshaping of ac signals and many other applications. Waveshaping involves alteration of ac signals by passing only certain portions of the signal while blocking other portions.

Junction diodes designed for use as rectifiers must have volt-ampere characteristics close to those of an ideal diode. The reverse current should be negligible and forward current should exhibit little voltage dependence. The reverse breakdown voltage should be large and the offset or barrier potential V_0 in forward direction should be small. No diode can offer all these properties.

The requirements for good rectifier junctions are :

1. Band Gap. A rectifier with a wide band gap material can operate at higher temperatures as thermal excitation of electron-hole pairs (EHPs) is reduced by the increased band gap. Temperature effects are critical in rectifiers, which carry large currents in forward direction and are thereby subjected to appreciable heating. On the other hand, the contact potential and offset voltage generally increase with E_G. Silicon is preferred over germanium because of its wider band gap, lower leakage current, higher breakdown potential and also convenient fabrication properties.

2. Doping Concentration. This affects the avalanche breakdown voltage, the contact potential, and the series resistance of the diode.

3. Physical Geometry. Limitations on the diode area lead to difficulties in obtaining uniform starting material and junction processing over large areas. Premature reverse breakdown can sometimes occur. This effect can be controlled by beveling the edge or by diffusing a guard ring to isolate the junction from the edge of the sample.

2.23.1. Piecewise Linear Diode Characteristic

A large-signal approximation which usually leads to a sufficiently accurate engineering solution is the **piecewise linear** representation. For instance, the piecewise linear approximation for a semiconductor diode is given in Fig. 2.49. The break point is not at the origin, and, therefore, V_0 (or V_γ) is also known as the offset, or threshold, voltage. The diode behaves like an open circuit when the applied voltage is less than potential barrier V_0, and has a constant incremental resistance $r' = \frac{dV}{dI}$ when applied voltage exceeds potential barrier or threshold voltage. The equivalent circuit for this linear approximation is a constant voltage source in series with a resistor [Fig. 2.48 (*b*)]. This method models the diode with segments of straight lines : thus the name *piecewise linear model.*

Example 2.29. What is the current in the circuit shown in Fig. 2.51.

Solution : Since diode is silicon one, and the barrier potential for silicon diode is 0.7 V,

$$V_0 = 0.7 \text{ V}$$

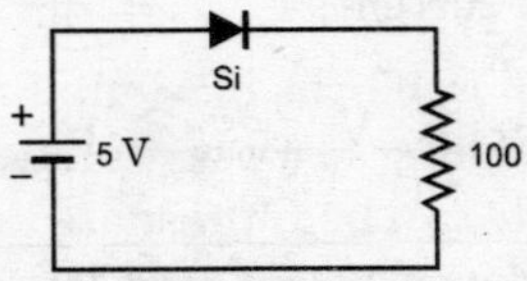

Fig. 2.51

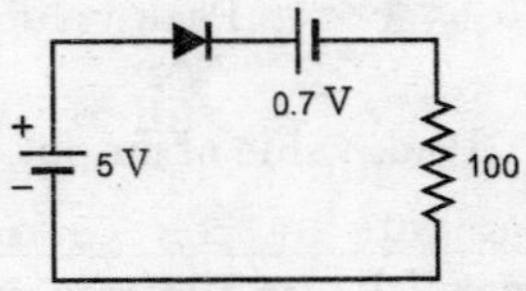

Fig. 2.52

This barrier potential acts in opposite direction to the applied voltage. A simplified circuit is shown in Fig. 2.52. The diode forward resistance is being assumed to be zero.

$$\text{Current flowing through the circuit, } I = \frac{V - V_0}{R} = \frac{5 - 0.7}{100} = 0.043 \text{ A or } 43 \text{ mA} \quad \textbf{Ans.}$$

Example 2.30. Find the voltage V_A of the circuit given in Fig. 2.53.

Solution : Assuming diodes of negligible forward resistance, the given circuit is redrawn, as given in Fig. 2.54.

Current through the circuit,

$$I = \frac{V - 2V_0}{R} = \frac{15 - 2 \times 0.7}{7 \text{ k}\Omega} = 1.943 \text{ mA}$$

$\because$ barrier potential of silicon diode, $V_0 = 0.7$ V

Potential V_A = Voltage drop across 7 kΩ

= 1.943 mA × 7 kΩ = 13.6 V **Ans.**

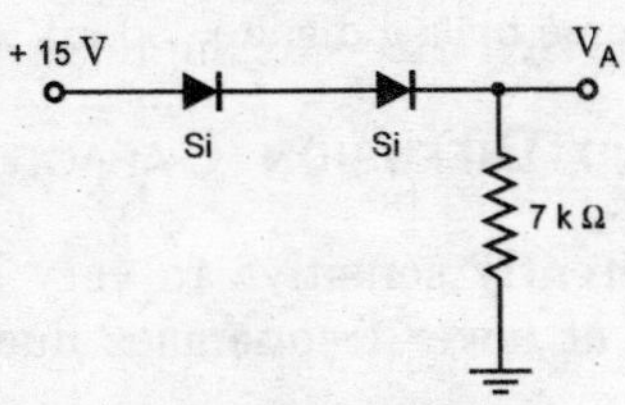

Fig. 2.53

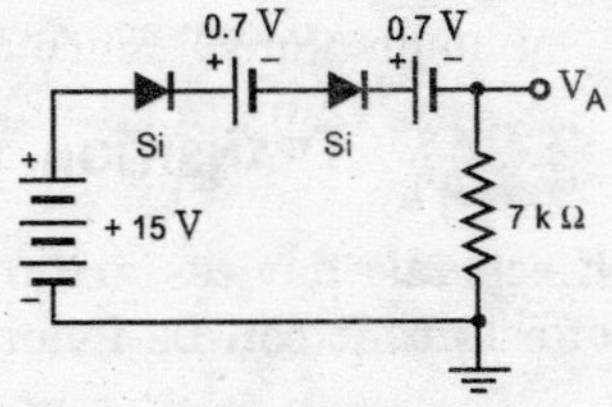

Fig. 2.54

Example 2.31. Determine the voltage V_A in the circuit shown in Fig. 2.55.

Solution : It appears that on switching on the supply both the diodes turn 'on', but this is not true. When the supply voltage is switched on, germanium diode whose barrier potential is 0.3 V will turn-on first and a level of 0.3 V will be maintained across the parallel branch. The silicon diode will

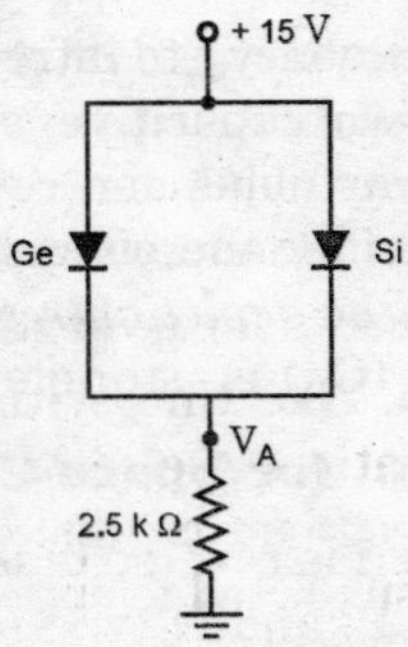

Fig. 2.55

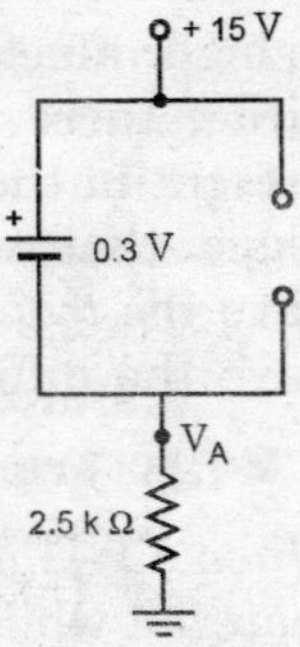

Fig. 2.56

never get opportunity to have 0.7 V across it and, therefore, remain in open-circuit state, as illustrated in Fig. 2.56.

So, $V_A = 15 - 0.3 = 14.7$ V **Ans.**

Example 2.32. An ac voltage of peak value 25 V is connected in series with a silicon diode and a load resistance of 600 Ω. If the forward resistance of the diode is 15 Ω determine *(i)* the peak value of current through the diode *(ii)* peak value of the output voltage.

What will be the peak values of current through the diode and output voltage if the diode is assumed to be ideal ?

Solution : Peak value of applied voltage, $V_{max} = 25$ V

Forward resistance of diode, $r_f = 15\ \Omega$

Load resistance, $R_L = 600\ \Omega$

Potential barrier voltage, $V_0 = 0.7$ V

The current through the diode will be maximum (peak) at the instant the input or applied voltage attains positive peak *i.e.* $V_{in} = 25$ V

(i) Peak value of current, $I_{max} = \dfrac{V_{max} - V_0}{R_L + r_f} = \dfrac{25 - 0.7}{600 + 15} = 0.0395$ A or 39.5 mA **Ans.**

(ii) Peak value of output voltage, $V_{out\ max} = I_{max} \times R_L = 0.0395 \times 600 = 23.7$ V **Ans.**

When the diode is ideal *i.e.* when $V_0 = 0$ and $r_f = 0$

Peak value of current, $I'_{max} = \dfrac{V_{max}}{R_L} = \dfrac{25}{600} = 0.04167$ A or 41.67 mA **Ans.**

Peak value of output voltage, $V_{out\ max} = I'_{max} \times R_L = 0.04167 \times 600 = 25$ V **Ans.**

Example 2.33. In the circuit shown below, find *(i)* current *i*, *(ii)* voltage across forward biased diode and reverse biased diode.

[M.D. Univ. Third Semester May 2005]

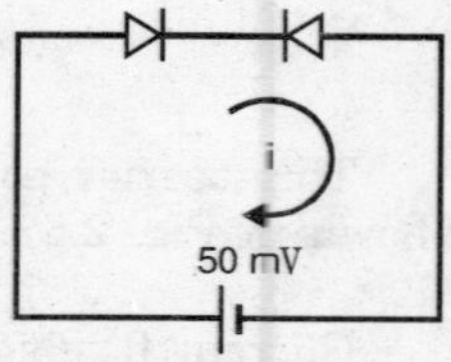

Fig. 2.57

Solution : Assume that the diodes are ideal. Then reverse biased diode will act as open-circuit and, therefore, current $i = 0$ **Ans.**

Voltage across forward biased diode = 0 V **Ans.**

Voltage across reverse biased diode = 50 mV **Ans.**

2.24. TRANSITION AND DIFFUSION CAPACITANCES

Electronic devices are inherently sensitive to very high frequencies. Most shunt capacitive effects that can be ignored at lower frequencies due to very high reactance (X_C being equal to $\frac{1}{2\pi f C}$) cannot be ignored at very high frequencies. X_C will become sufficiently small due to high value of frequency f to introduce a low reactance *shorting* path. In a P-N semiconductor diode, there are two capacitive effects to be considered. Both types of capacitances are present in the forward-bias and reverse-bias regions, but one so outweighs the other in each region that we consider the effects of only one in each region. In the reverse-bias region we have the *transition* or *depletion-region capacitance* (C_T), while in the forward bias region we have the diffusion (C_D) or storage capacitance.

2.24.1. Transition (or Space-Charge) Capacitance

When a P-N junction is reverse biased, the depletion region acts like an insulator or dielectric material while the P- and N-type regions on either side have a low resistance and act as the plates. Thus P-N junction may be considered a parallel plate capacitor. The junction capacitance is termed as space charge *capacitance* or *transition capacitance* or *depletion region capacitance* and is denoted by C_T.

As mentioned earlier, a reverse bias causes majority carriers to move away from the junction, thereby uncovering more immobile charges. So the thickness W of the depletion layer increases with the increase in reverse bias voltage. This increase in uncovered charge with applied voltage may be considered a *capacitive effect*. The incremental capacitance C_T may be defined as

$$C_T = \left|\frac{dQ}{dV}\right| \qquad ...(2.113)$$

where dQ is the increase in charge due to increase in voltage, dV.

$$\text{But } i = \frac{dQ}{dt} = C_T \frac{dV}{dt} \qquad ...(2.114)$$

Since the depletion region increases with the increase in reverse-bias potential, the resulting transition capacitance decreases as shown in Fig. 2.58.

The fact that the capacitance depends on the applied reverse bias, has application in a number of electronic systems.

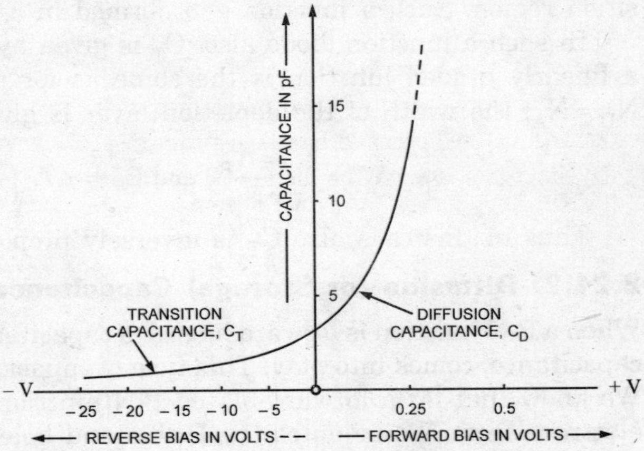

Fig. 2.58

1. *Step Graded Junction.* A junction is said to be step graded if there is an abrupt change from acceptor ion concentration on the P-side to donor ion concentration on the N-side. Such a junction gets formed in alloyed junction or fused junction diode. Usually the acceptor density N_A and the donor density N_D are kept unequal. The capacitance of a parallel plate capacitor, in general, is given as

$$C = \frac{\epsilon A}{d}$$

where ϵ is the absolute permittivity of the dielectric material between the plates of area A separated by a distance d.

So the transition capacitance C_T of a P-N diode is given as

$$C_T = \frac{\epsilon A}{W} \qquad ...(2.115)$$

where ϵ is the absolute permittivity of the semiconductor medium, A is the cross-sectional area of the junction and W is the width of the depletion layer and is given as

$$W^2 = \left[\frac{2 \epsilon V_B}{e}\right]\left[\frac{1}{N_A} + \frac{1}{N_D}\right] \qquad ...(2.116)$$

$$\text{or } W = \sqrt{\frac{2 \epsilon V_B}{e N_D}} \quad \text{when } N_A >> N_D \qquad ...(2.117)$$

$$\text{So } C_T = \epsilon A \sqrt{\frac{e N_D}{2 \epsilon V_B}} = A \sqrt{\frac{N_D}{V_B}} \cdot \sqrt{\frac{e \epsilon}{2}} \qquad ...(2.118)$$

Thus in a step graded junction, C_T varies inversely proportional to the square root of the junction voltage V_B where V_B is given as

$$V_B = V_0 - V_R \quad \text{...(2.119)}$$

where V_R is a negative number indicating the applied reverse bias and V_0 is the contact potential.

2. ***Linearly Graded Junction (or Growth Junction).*** A junction is said to be linearly graded if the charge density varies gradually (almost nearly) with the distance in the transition region. Such a junction gets formed in a growth junction diode.

In such a junction diode also, C_T is given by Eq. (2.115). Thus the expression for C_T for a linearly graded junction is the same as for a step graded junction. Since in this case $N_D = N_A$, the width of the depletion layer is given as

$$W = \sqrt{\frac{6\epsilon V_B}{eN_D}} \text{ and } C_T = \epsilon A\sqrt{\frac{eN_D}{6\epsilon V_B}} = A\sqrt{\frac{N_D}{V_B}} \cdot \sqrt{\frac{e\epsilon}{6}} \quad \text{...(2.120)}$$

Thus in this case also C_T is inversely proportional to the square root of V_B.

2.24.2. Diffusion (or Storage) Capacitance

When a P-N junction is forward biased, a capacitance, which is much larger than the transition capacitance, comes into play. This type of capacitance is called the *diffusion capacitance*, C_D. We know that for a forward-biased P-N junction, the potential barrier is reduced. Now the electrons from N-side enter the P-side and holes from P-side enter into the N-side. These charge carriers diffuse away from the junction and progressively recombine. The density of charge carriers is high near the junction and decays exponentially with distance. Thus a charge is stored on both sides of the junction when forward biased. It is observed that the amount of stored charge varies with the applied potential as for a true capacitor. It is convenient to introduce an incremental capacitance, called *diffusion* or *storage capacitance* C_D given by the equation $C_D = \frac{dQ}{dV}$ where dQ represents the change in number of minority carriers stored outside the depletion region when a change in voltage across diode, dV, is applied.

If τ is the mean lifetime of charge carriers, then a flow of charge Q yields a diode current I given as

$$I = \frac{Q}{\tau} \quad \text{or} \quad Q = \tau I$$

and from Eq. (2.99), $I = I_0(e^{V/\eta V_T} - 1)$

$$\text{So } Q = \tau I_0(e^{V/\eta V_T} - 1) = \tau I_0\, e^{V/\eta V_T} \qquad \because e^{V/\eta V_T} >> 1$$

So diffusion capacitance

$$C_D = \frac{dQ}{dV} = \frac{d\,\tau I_0 e^{V/\eta V_T}}{dV} = \frac{\tau I_0}{\eta V_T} e^{V/\eta V_T} = \frac{\tau(I+I_0)}{\eta V_T}$$

For a forward bias greater than a few tenths of a volt, $\left[\frac{V}{\eta V_T}\right] >> 1$ and according to Eq. (2.99) $I >> I_0$ and so the diffusion capacitance is given as

$$C_D = \frac{\tau I}{\eta V_T} \quad \text{...(2.121)}$$

Thus the diffusion capacitance is directly proportional to the forward current through the diode. This relationship occurs since the number of majority carriers injected into the depletion region is directly proportional to the forward current of the diode. The variation in diffusion capacitance with the variation in forward bias applied across a P-N junction is shown in Fig. 2.58.

This capacitance assumes great significance in the operation of devices which are required to switch rapidly from forward bias to reverse bias. Large value of C_D, delays both of the switching-on and the switching-off.

In fact in a forward biased diode, both the diffusion capacitance C_D and the transition capacitance C_T are present, but $C_D >> C_T$. Typically C_D is more than a million times greater than C_T. So in a forward biased diode C_T may be neglected and only C_D may be considered. Typical value of C_D is 0.02 μF.

Similarly in a reverse biased diode both C_D and C_T are present but $C_T >> C_D$. Hence in a reverse biased diode C_D is neglected and only C_T is considered.

The capacitive effects described above may be represented by a capacitor connected in parallel with an ideal diode, as illustrated in Fig. 2.59.

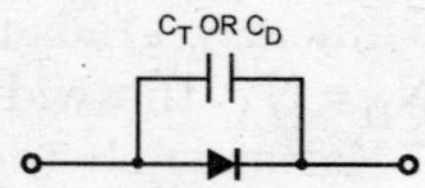

Fig. 2.59

Example 2.34. Consider an abrupt P-N junction with donor density N_D = 10^{17} atoms/cm³ and acceptor density N_A = 0.5 × 10^{16} atoms/cm³. Sketch the charge distribution about the junction and estimate the junction width when

(*i*) no external voltage is applied, (assume the junction barrier voltage to be 0.7 V),

(*ii*) with an external voltage of – 10 V applied.

Assume uniform charge distribution on both sides of the junction in the space-charge region. Asssume an ϵ_r of 10 for the material and ϵ_0 = 8.85 × 10^{-12} F/m. [*U.P.S.C. I.E.S. Electrical Engineering-I, 1999*]

Solution : Donor density, $N_D = 10^{17} \times 10^6$ atoms/m³ $= 10^{23}$ atoms/m³

Acceptor density, $N_A = 0.5 \times 10^{16} \times 10^6$ atoms/m³ $= 0.5 \times 10^{22}$ atoms/m³

Absolute permittivity for the material, $\epsilon = \epsilon_r \epsilon_0 = 10 \times 8.85 \times 10^{-12}$ F/m $= 8.85 \times 10^{-11}$ F/m

Electron charge, $e = 1.602 \times 10^{-19}$ C

Charge distribution is shown in Fig. 2.60.

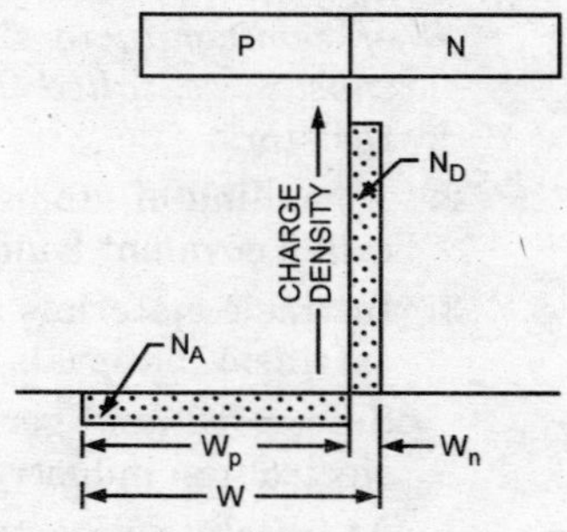

Fig. 2.60

(*i*) When no external voltage is applied *i.e.*, V = 0

Junction barrier voltage, $V_B = 0.7$ V

$$\text{Junction width, } W = \sqrt{\frac{2 \epsilon V_B}{e}\left[\frac{1}{N_A} + \frac{1}{N_D}\right]}$$

$$= \sqrt{\frac{2 \times 8.85 \times 10^{-11} \times 0.7}{1.602 \times 10^{-19}}\left[\frac{1}{0.5 \times 10^{22}} + \frac{1}{10^{23}}\right]}$$

$= 0.393 \times 10^{-6}$ m **Ans.**

(*ii*) When external voltage of – 10 V is applied

Junction voltage, $V_B = V_0 - V = 0.7 - (-10) = 10.7$ V

$$\text{Junction width, } W = \sqrt{\frac{2 \times 8.85 \times 10^{-11} \times 10.7}{1.602 \times 10^{-19}} \times \left[\frac{1}{0.5 \times 10^{22}} + \frac{1}{10^{23}}\right]} = 1.538 \times 10^{-6} \text{ m } \textbf{Ans.}$$

Example 2.35. Prove that the magnitude of maximum electric field E_m at a step graded junction with $N_A >> N_D$ is given by $E_m = \frac{2V_j}{W}$.

Solution : From Eq. (2.116)

$$W^2 = \left[\frac{2 \epsilon V_B}{e}\right]\left[\frac{1}{N_A} + \frac{1}{N_D}\right] = \frac{2 \epsilon V_B}{e} \times \frac{1}{N_D} \quad \because N_A >> N_D$$

$$\text{or Barrier potential, } V_B \text{ or } V_j = \frac{e \times N_D}{2 \epsilon} W^2 \qquad ...(i)$$

where e is electronic charge, N_D is donor concentration, $\in$ is the permittivity of the medium and W is the width of depletion or space charge region.

From Poisson's equation $\frac{d^2V}{dx^2} = \frac{-\rho}{\in} = \frac{-eN_D}{\in}$ where ρ is charge density and is equal to $e\ N_D$

Integrating above equation we have

$$\frac{dV}{dx} = \int_0^W \frac{-eN_D}{\in}\,dx = \frac{-eN_D}{\in}\int_0^W dx = \frac{-eN_D}{\in}\,.\,W$$

Since maximum electric field E_{max} can be defined as

$$E_{max} = -\left.\frac{dV}{dx}\right|_{x=0} = \frac{eN_D}{\in}\,.\,W$$

Maximum electric field can be given as

$$E_{max} = \frac{eN_D}{\in}\ W = \frac{2V_j}{W} \qquad \because \text{ from Eq. } (i)\ \frac{eN_D W}{\in} = \frac{2V_j}{W}$$

$$\text{or } E_{max} = \frac{2V_j}{W} \text{ as desired.}$$

Highlights

1. A semiconductor is a material that has a resistivity somewhere between that of a good conductor and that of an insulator. The resistivity of different semiconductor materials lies in a very wide range from 10^{-4} to about 0.5 Ω-m whereas the resistivities of conductors and insulators are of the order of 10^{-8} Ω-m and 10^{12} Ω-m respectively. Most semiconductor materials used in electronics industry have negative temperature coefficients.

   The semiconductors have another very important property—*the resistivity (or conductivity) of semiconductors changes considerably when even minute amounts of certain other substances, called the impurities, are added to them.* Semiconductor materials are nonlinear resistors.

2. A bonding of atoms, strengthened by sharing of electrons between neighbouring atoms, is called covalent bonding. The covalent bonding arrangement exist in silicon and germanium.

3. Intrinsic materials are those semiconductors that have very low level of impurities, whereas extrinsic materials are semiconductors that have been exposed to a doping process.

   Silicon (Si) and germanium (Ge) are the two most widely used semiconductor materials in electronics industry.

   At absolute zero temperature E_G for silicon and germanium are 1.21 eV and 0.785 eV respectively.

   Hole, strictly speaking, is not a particle; but for all practical purposes it is viewed as a positive charged particle capable of conducting current. The mobility of hole is about half that of an **electron.**

   The resistance of a semiconductor depends largely on (*a*) *temperature*—decreases with the increase in temperature, (*b*) *illumination*—decreases in brighter surroundings and (*c*) *magnitude of the electric field* — the current does not increase in proportion to the applied potential, as in case of conductors obeying Ohm's law but increases by far more than the voltage *i.e., the semiconductors are nonlinear*.

   In intrinsic semiconductors, the Fermi level lies between the conduction and valence bands.

   Conduction in a pure semiconductor consists of electron flow in the conduction band and hole flow in the valence band.

   A germanium device is seldom used at a temperature higher than 75°C, a silicon seldom higher than 175°C.

4. Depending on the type of impurity added, the extrinsic semiconductors can be divided into two classes viz. (*i*) N-type semiconductors and (*ii*) P-type semiconductors.

An N-type material is formed by adding donor atoms (such as arsenic, antimony, bismuth or phosphorus) that have five valence electrons to establish a high level of relatively free electrons. In an N-type material, the electron is the majority carrier and the hole is the minority carrier.

A P-type material is formed by adding acceptor atoms (such as boron, gallium, indium or aluminium) that have three valence electrons to establish a high level of holes in the material. In a P-type material, the hole is the majority charge carrier and the electron is the minority charge carrier.

5. Under equilibrium the product of concentration of free electrons and concentration of holes is constant and is independent of the amount of doping by donor and acceptor impurities. This is known as **mass action law.** Thus

$$np = n_i^2$$

6. Concentration of holes in N-type semiconductors is given by

$$p = \frac{n_i^2}{n} = \frac{n_i^2}{N_D} \quad \text{where } N_D \text{ is the concentration of donor atoms.}$$

Similarly in case of P-type semiconductor

$$n = \frac{n_i^2}{N_A} \quad \text{where } N_A \text{ is the concentration of acceptor atoms.}$$

7. Total current density due to drift of electrons and holes on application of electric field is given by

$$J = Ee(n\mu_e + p\mu_h)$$

The current due to diffusion of electrons and holes is given by

$$J_e = eD_e\frac{dn}{dx} \quad \text{and} \quad J_h = -eD_h\frac{dp}{dx}$$

The diffusion coefficient and mobility are related through the Einstein relationship, given by

$$\frac{D_h}{\mu_h} = \frac{D_e}{\mu_e} = V_T = \frac{T}{11{,}600}$$

8. Exact position of Fermi level in N-type extrinsic semiconductor is given by

$$N_D = n_C e^{-(E_C - E_F)/kT}$$

or

$$E_F = E_C - kT\log\frac{n_C}{N_D}$$

Similarly in case of P-type semiconductor

$$E_F = E_V + kT\log\frac{n_V}{N_A}$$

9. The Hall effect is a consequence of a charged carrier moving in the presence of perpendicular electric and magnetic fields. The charged carrier is deflected, inducing a Hall voltage. The polarity of the Hall voltage is a function of the semiconductor conductivity type. The majority carrier concentration and mobility can be determined from the Hall voltage.

Hall coefficient is given by

$$R_H = \frac{\text{Electric field due to Hall effect}}{\text{Current density} \times \text{magnetic field}} = \frac{E_H}{J \times B} = \frac{V_H/d}{J \times B} = \frac{BI}{\rho W} \times \frac{1}{d \times J \times B} = \frac{I}{\rho W d J} = \frac{1}{\rho} = \frac{1}{ne} = \frac{V_H W}{BI}$$

10. Most semiconductor devices employ one or more P-N junctions. The P-N junction is the control element for the performance of all semiconductor devices such as rectifiers, amplifiers, switching devices, linear and digital integrated circuits. The P-N junction is produced by placing a layer of P-type semiconductor next to the layer of N-type semiconductor. The interface separating the N and P regions is referred to as the metallurgical junctions.
11. On the formation of P-N junction some of the holes from P-type material tend to diffuse across the boundary into N-type material and some of the free electrons similarly diffuse into the P-type material, as illustrated in Fig. 2.25. This happens due to density gradient (as concentration of holes is higher on P-side than that on N-side and concentration of electrons is higher cn N-side than that on P-side). This process is known as *diffusion*.
12. A space charge region, or depletion region, is formed on either of the metallurgical junction separating the P and N junctions. This region is essentially depleted of any mobile electrons or holes. A net positive charge density, due to the positively charged donor impurity ions, exists in the N region and a negative charge density, due to the negatively charged acceptor impurity ions, exists in the P region.

    An electric field exists in the depletion region due to the net space charge density. The direction of the electric field is from the N region to the P region.

    A pd exists across the space-charge region. Under zero applied bias, the pd, known as built-in potential barrier, maintains thermal equilibrium and holds back majority carrier electrons in the N-region and majority carrier holes in the P region. The pd existing across the space charge region is known as the contact potential. The built-in potential barrier or the contact potential for the step junction is given by

$$V_{bi} \text{ or } V_0 = V_T \log_e \frac{N_A N_D}{n_i^2}$$

13. A semiconductor containing both donor and acceptor impurity atoms in the same region is called a compensated semiconductor. Thermal-equilibrium electron concentration is given by

$$n_0 = \frac{N_D - N_A}{2} + \sqrt{\frac{(N_D - N_A)^2}{4} + n_i^2}$$

14. A forward-biased PN junction has the positive terminal applied to the P side and the negative to the N side. The reverse is true for a reverse-biased junction.

    When a forward-bias voltage is applied across a PN junction, the potential barrier is lowered so that holes from the P-region and electrons from the N region can flow across the junction.
15. An applied reverse bias voltage (N region positive w.r.t. the P region) increases the potential barrier, increases the space-charge width, and increases the magnitude of the electric field.
16. Metal semiconductor junction can also form ohmic contacts, which are low-resistance junctions providing conduction in both directions with very little voltage drop across the junction.
17. A diode is a unidirectional device permitting the easy flow of current in one direction but restraining the flow in opposite direction. A semiconductor diode is simply a P-N junction with connecting leads or terminals on its two sides. P-N junction is also called the crystal diode because it is grown out of crystal.
18. An ideal diode is a two terminal polarity sensitive device that has zero resistance when it is forward biased and infinite resistance when reverse biased.
19. The current in a P-N diode is bipolar in character (because it is made up of both positive and negative charge carriers) and the total current is constant throughout the device, but the proportion of current due to holes and that due to electrons varies with the distance.
20. In the forward-bias condition the diode current increases exponentially with the increase in voltage across the diode. Diode current, when the diode is forward biased, is given by

$$I = I_0(e^{V/V_T} - 1) \text{ where } V_T = \frac{T}{11,600} \text{ and } I_0 = Ae\left[\frac{D_h}{L_h N_D} + \frac{D_e}{L_e N_A}\right] n_i^2$$

Reverse saturation current may also be given by

$$I_0 = A_0 T^3 e^{-V_{GO}/V_T}$$

21. The reverse saturation current I_0 will just double in magnitude for every 10-fold increase in temperature.

The temperature coefficient for the forward voltage of a P-N junction is approximately –1.8 mV/°C for silicon diode and –2.02 mV/°C for germanium diode.

22. The static and dynamic resistances of a diode may be determined from the characteristic curves and represent the dc and ac opposition to the flow of current offered by the diode.

The dc resistance of a diode is determined by the ratio of the diode voltage and current at the point of interest and is not sensitive to the shape of the curve. The dc resistance decreases with the increase in diode current or voltage.

The ac resistance of a diode is sensitive to the shape of V-I characteristic curve in the region of interest and decreases for higher level of diode current or voltage.

23. An equivalent circuit of a diode is a linearization of the diode's characteristic curve and typically consists of an ideal diode in series with a battery (representing the cut-in, offset or threshold voltage) and a resistance equal to the dynamic resistance of the diode over its linear portion.

The cut-in or threshold voltage is about 0.7 V for silicon diodes and 0.3 V for germanium diodes.

24. A P-N junction exhibits a transition capacitive effect of several picofarads in reverse bias, but have many microfarads of diffusion capacitance in the forward-biased state.

The capacitance of diode increases exponentially with the increase in the forward-bias voltage. Its lowest levels are in the reverse-bias region.

In fact in a forward biased diode, both the diffusion capacitance C_D and the transition capacitance C_T are present, but $C_D >> C_T$. Typically C_D is more than a million times greater than C_T. So in a forward biased diode C_T may be neglected and only C_D may be considered. Typical value of C_D is 0.02 μF.

Similarly in a reverse biased diode both C_D and C_T are present but $C_T >> C_D$. Hence in a reverse biased diode C_D is neglected and only C_T is considered.

The capacitive effects described above may be represented by a capacitor connected in parallel with an ideal diode, as illustrated in Fig. 2.59.

Exercises

1. Explain, with the help of a neat sketch why copper is a good conductor. Classify the following as conductor or semiconductor.
(*i*) Gold (*ii*) Germanium (*iii*) Silver (*iv*) Silicon. **[U.P. Technical Univ. June 2001]**
2. Define an intrinsic material, a negative temperature coefficient and covalent bonding.
3. Giving relevant defining equations, explain the phenomenon of drift and diffusion associated with carrier movement in semiconductors. **[I.E.S., 1999]**
4. Differentiate between the drift current and diffusion current.
[M.D. Univ. Third Semester May 2006; December 2006]
5. Explain the phenomenon of diffusion in semiconductor.
[M.D. Univ. Third Semester December 2007]
6. Distinguish between intrinsic and extrinsic semiconductors. Discuss why intrinsic semiconductor is not used in semiconductor devices.
[M.D. Univ. Third Semester December 2005]

7. Draw the atomic structure of a P type material and explain the effect of acceptor impurities on the energy band structure.
8. Distinguish between intrinsic and extrinsic semiconductors. What is the effect of temperature on the conductivity of a semiconductor ?
9. Sketch the energy-band picture for (*i*) an intrinsic (*ii*) an *n*-type and (*iii*) a *p*-type semiconductor. Indicate the positions of the Fermi, the donor and the acceptor levels.
10. How does conduction take place in an intrinsic semiconductor ? Explain with examples. What are limitations of intrinsic semiconductors and how are these taken care of by extrinsic semiconductors ? Show that the minimum value of conductivity of a semiconductor with impurity is $\sigma_{min} = 2n_e e \mu_e$ where n_e-concentration of conduction electrons (m^{-3}), e-electron charge (C), μ_e-mobility of electrons ($m^2 V^{-1} s^{-1}$).

[U.P.S.C. I.E.S. Elec. Engineering-I, 1997]

11. Show that a hole can take part in transport of current. Define mobility of a charge carrier. Show that $\mu/D = e/kT$.
12. Define the terms conductivity, intrinsic concentration and energy gap of semiconductors.
13. An N-type semiconductor has excess of electrons and P-type excess of holes. Does that mean N-type materials have a net negative charge and P-type materials have a net positive charge. Comment. **[Pb. Technical Univ. Electronic Devices and Circuits 2005-06]**
14. What do you mean by mobility of a carrier ? How does it depend on temperature and doping concentrations.
15. Explain Hall effect in semiconductors. What is Hall angle ?

[U.P.S.C. I.E.S. Electrical Engineering-I, 1996]

16. Explain the Hall effect in semiconductors. Derive the expression for Hall coefficient of P-type semiconductor.

[M.D.U. Rohtak Electrical Engineering Materials and Semiconductor Devices 2006-07]

17. Explain the concept of continuation equation in a semiconductor.

[M.D. Univ. Third Semester December 2007]

18. Explain working of P-N junction diode ? **[M.D. Univ. Third Semester May 2007]**
19. What is p-n junction diode ? How potential barrier is formed in a p-n junction diode ?

[M.D. Univ. Third Semester May 2006; December 2006]

20. Explain P-N junction theory.

[M.D.U. Rohtak Electrical Engineering Materials and Semiconductor Devices 2006-07]

21. What do you understand by barrier potential across a junction ? What is its significance ?

[Pb. Technical Univ. Electronic Devices and Circuits, 2003]

22. What is barrier potential in a P-N junction ? What happens when the N-type material is made more positive than the P-type material by an external voltage source ? What ultimately happens if the applied voltage is gradually increased ?
23. What is a P-N junction ? What happen when P-N junction is biased in forward direction ?

[U.P. Technical Univ. Electrical and Electronics Engineering Materials 2006-07]

24. What are junction diodes ? Explain the flow of current across forward and reverse biased junctions.

[M.D.U. Rohtak Electrical Engineering Materials and Semiconductors Devices 2006-07]

25. Draw the energy band diagram for a P-N junction under open-circuited condition. Clearly indicate various energy levels in P-region, space region and N-region. How will it be modified if P-N Junction is forward biased ?
26. Discuss the formation and volt-amp characteristics of P-N junction. What happen when a P-N junction is biased in (*i*) forward direction, (*ii*) backward direction.

27. Explain the behaviour of P-N junction at no bias, reverse bias and forward bias. Sketch V-I characteristics of P-N junction diode. **[U.P. Technical Univ. 2006-07]**

28. Draw sketches to show the method of fabricating alloy diodes and diffused diodes. Briefly explain.

29. Draw the *v-i* characteristics of an ideal diode and explain.

30. What is an ideal diode ? How it can be represented as a switch ? Draw the equivalent circuit and its characteristics.

31. Plot the hole current, the electron current, and the total current as a function of distance on the both sides of a P-N junction. Indicate the transition region.

32. Give quantitative theory of P-N diode current. Show that reverse saturation current, $I_0 = K_1 T^2 e^{-V_{GO}/V_T}$.

33. Draw the characteristics of a p-n junction diode with proper voltage and current labels and regions of forward and reverse bias. Write the relation between forward current and forward voltage.

34. Discuss V-I characteristics of P-N junction diode.

[M.D. Univ. Third Semester December 2005]

35. Draw the characteristics of a junction diode at four different temperatures *e.g.*, –100ºC, 25ºC, 100ºC, 200ºC and comment on the curves.

36. Derive an expression for the current-voltage relation in an ideal P-N junction diode.

[U.P. Technical Univ. Semiconductor Materials and Devices 2006-07]

37. Explain the ac and dc resistance of a semiconductor diode and derive expression of diode resistance.

38. Draw the piecewise linear equivalent circuit of diode and explain it briefly.

[Pb. Technical Univ. Electronic Devices and Circuits 2005]

39. What is diffusion and transition capacitance of P-N junction ?

[M.D. Univ. Third Semester December 2005; May 2007]

40. Differentiate between diffusion capacitance and transition capacitance.

[M.D. Univ. Third Semester May 2006; December 2006]

41. Discuss the various capacitances used in a diode.

[M.D. Univ. Third Semester December 2007]

42. How does the transition capacitance CT vary with the depletion layer width with the applied reverse voltage ? What is the order of magnitude of C_T ?

43. What is the difference between diffusion and transition capacitance ? How will you represent the capacitive effect of a practical diode on an ideal diode ?

[U.P. Technical Univ. 2006-07]

44. Write short notes on the following :

(*i*) Intrinsic and extrinsic semiconductors.

(*ii*) P-type semiconductor. **[M.D. Univ. May 2008]**

(*iii*) Formation of P-N junction.

(*iv*) Continuation equation.

(*v*) V-I characteristics of P-N junction.

(*vi*) Diffusion and transition capacitances.

Short Answer Type Questions With Answers

Q. 1. Is hole a fundamental particle in an atom ? **[U.P. Technical Univ. 2003-04]**

Ans. Hole is not a fundamental particle in an atom. Holes may be thought of as positive particles, and as such, they move through an electric field in a direction opposite to that of electrons.

Q. 2. Define a hole in a semiconductor. **[U.P. Technical Univ. 2001-02]**

Ans. When an energy is supplied to a semiconductor, a valence electron is lifted to a higher energy level, the departing electron leaves a vacancy in the valence band. This vacancy is called a hole. Thus, a vacancy left in the valence band because of lifting of an electron from valence band to conduction band, is known as hole.

Q. 3. What is hole current ?

Ans. The movement of the hole (positively charged vacancy in the valence band) from positive terminal of the supply to the negative terminal through semiconductor constitutes hole current.

Q. 4. What is intrinsic semiconductor ?

Ans. An intrinsic semiconductor is one which is made of the semiconductor material in the extremely pure form (impurity content not exceeding one part in 100 million parts of semiconductors).

Q. 5. Why silicon and germanium are the two widely used semiconductor materials ?

Ans. Because the energy required to release an electron from their valence band (*i.e.*, to break their covalent bonds) is very small (1.12 eV for Si and 0.72 eV for Ge).

Q. 6. Which of the two semiconductor materials Si or Ge has larger conductivity at room temperature ? Why ?

Ans. Since energy required in transferring electrons from valence band to conduction band is more in case of silicon than that in case of germanium, the conductivity of Ge will be more than that of Si at room temperature.

Q. 7. Why does a pure semiconductor behave like an insulator at absolute zero temperature ? **[U.P. Technical Univ. 2000-01]**

Ans. For a pure semiconductor at a temperature of absolute zero (– 273.15°C) the valence band is usually full and there may be no electron in the conduction band and it is difficult to provide additional energy required for lifting electron from valence band to conduction band by applying electric field. Hence the conductivity for a pure semiconductor at absolute zero temperature is zero and it behaves like an insulator.

Q. 8. What is the main factor for controlling the thermal generation and recombination ?

Ans. Temperature, because with the increase in temperature, concentrations of free electrons and holes increase and the rate of recombination is proportional to the product of concentrations of free electrons and holes and also the rate of production of electron-hole pairs (thermal generation) increases with the rise in temperature.

Q. 9. Define mean lifetime of a carrier.

Ans. The amount of time between the creation and disappearance of a free electron is called the lifetime. It varies from a few nanoseconds to several microseconds depending how perfect the crystal is and other factors.

Q. 10. In which bands do the movement of electrons and holes take place ?

Ans. Free electrons move in conduction band while holes in valence band.

Q. 11. What is the mechanism by which conduction takes place inside the semiconductor ?

Ans. Conduction occurs in any given material when an applied electric field causes electrons to move in a desired direction within the material. This may be due to one or both of two processes, electron motion and hole transfer. In case of former process, free electrons in the conduction band move under the influence of the applied electric field. Hole transfer involves electrons which are still attached to the atoms *i.e.* those in valence band. For details refer to Art. 1.9.

Q. 12. What do you mean by drift velocity and mobility of a free electron ?

Ans. The average velocity of a free electron is known as drift velocity whereas mobility of an electron is defined as the drift velocity per unit electric field.

Q. 13. Define mobility of a carrier. Show that the mobility constant of electron is larger than that of a hole.

Ans. Mobility is defined as the average particle drift velocity per unit electric field.

The mobility of electrons is more than that of holes because the probability of an electron having the energy required to move to an empty state in the conduction band is much greater than the probability of an electron having the energy required to move to the empty state in valence band. The mobility of electron is about double that of an hole.

Q. 14. Define diffusion current in a semiconductor.

Ans. The diffusion of charge carriers is as a result of a gradient of carrier concentration (*i.e.*, the difference of carrier concentration from one region to another). In this case concentrations of charge carriers (either electrons or holes) tend to distribute themselves uniformly throughout the semiconductor crystal. This movement continues until all the carriers are evenly distributed throughout the material. This type of movement of charge carriers is called the diffusion current.

Q. 15. Define drift current in a semiconductor.

Ans. The steady flow of electrons in one direction caused by applied electric field constitutes an electric current, called the drift current. For details refer to Art. 1.10.1.

Q. 16. What happens to the conductivity of the semiconductor with the rise in temperature? Compare with the conductivity of metals.

Ans. With the increase in temperature, the concentration of charge carriers increases resulting in increase of conductivity of semiconductors. The conductivity of metal decreases with the increase in temperature.

Q. 17. Why the temperature coefficient of resistance of a semiconductor is negative?

[U.P. Technical Univ. 2004-05]

Ans. With the increase in temperature, the concentration of charge carriers (electrons and holes) increases. As more charge carriers are made available, the conductivity of a pure semiconductor increases *i.e.* resistivity of a pure semiconductor decreases with the rise in temperature *i.e.* semiconductors have negative temperature coefficient of resistance.

Q. 18. What is meant by Fermi level in semiconductor ? Where does the Fermi level lie in an intrinsic semiconductor ?

Ans. Fermi level in a semiconductor can be defined as the maximum energy that an electron in a semiconductor have at absolute zero temperature.

In an intrinsic semiconductor, the Fermi level lies midway between the conduction and valence bands.

Q. 19. Differentiate between intrinsic semiconductors and extrinsic semiconductors ? Give examples.

Ans. An intrinsic semiconductor is one which is made of the semiconductor material in its extremely pure form.

When a small amount of impurity is added to a pure semiconductor crystal during the crystal growth in order to increase its conductivity, the resulting crystal is called the extrinsic semiconductor.

Q. 20. Give reasons — Doping is done in semiconductors.

Ans. Intrinsic (or pure) semiconductor by itself is of little significance as it has little current conduction capability at ordinary room temperature but if very small amount of impurity (of the order of one atom per million atoms of pure semiconductor) is added to it in the process of crystallization, the electrical conductivity is increased many times.

Q. 21. Describe the difference between P-type and N-type semiconductor materials.

Ans. When a small amount of trivalent impurity (such as boron, gallium, indium or aluminium) is added to a pure semiconductor crystal during crystal growth, the resulting crystal is called the P-type semiconductor.

When a small amount of pentavalent impurity (such as arsenic, antimony, bismuth or phosphorus) is added to a pure semiconductor crystal during crystal growth, the resulting crystal is called the N-type semiconductor.

Q. 22. What do you mean by donor and acceptor impurities?

Ans. Donor impurities (such as arsenic, antimony, bismuth or phosphorus) when added to a pure semiconductor lattice, form N-type extrinsic semiconductor. The pentavalent impurities are called the donor type impurities as such impurities donate electrons to the lattice.

Acceptor impurities (such as boron, gallium, indium or aluminium) when added to a pure semiconductor lattice form P-type extrinsic semiconductor. The trivalent impurities are called the acceptor type impurities because such impurities accept electrons from the lattice.

Q. 23. Explain the term doping and its need.

Ans. The electrical conductivity of intrinsic semiconductor, which has little current conduction capability at ordinary room temperature and so is of little use, can be increased many times by adding very small amount of impurity (of the order of one atom per million atoms of pure semiconductor) to it in the process of crystallization. This process is called the doping.

Q. 24. What are the majority and minority carriers in P-type semiconductor materials?

Ans. Holes are majority carriers and free electrons are minority carriers in P-type semiconductor materials.

Q. 25. What is the effect of temperature on extrinsic semiconductors?

Ans. With the increase in temperature of an extrinsic semiconductor, the number of thermally generated carriers is increased resulting in increase of concentration of minority carriers. At temperature exceeding critical temperature the extrinsic semiconductor behaves like an intrinsic semiconductor but with higher conductivity.

Q. 26. What are the charge carriers in N-type and P-type semiconductors?

Ans. Free electrons in N-type semiconductors and holes in P-type semiconductors are the charge carriers.

Q. 27. For the same order of doping, why does N-type semiconductor exhibit larger conductivity than P-type semiconductor?

Ans. Since the mobility of electrons is greater than that of holes, for same level of doping, N-type semiconductor exhibits larger conductivity.

Q. 28. What is the ratio of majority and minority carriers in intrinsic and extrinsic semiconductors?

Ans.

Types of Semiconductor Material	*Ratio of Majority and Minority Carriers*
Intrinsic	Unity
Extrinsic	Very large

Q. 29. State law of mass action.

Ans. Under thermal equilibrium conditions, the product of concentration n of free electrons and concentration p of holes is constant and is independent of the amount of doping by donor and acceptor impurities. *i.e.* $n\,p = n_i^2$. This is known as mass action law.

Q. 30. Define the term intrinsic concentration.

Ans. Intrinsic concentration of holes or free electrons is called the intrinsic concentration, n_i, which varies with temperature according to the relation $n_i = A_0 T^3 e^{-E_{Go}/kT}$.

where A_o is a constant and independent of temperature, k is a Boltzmann constant in eV per K and E_{Go} is the energy gap in eV at absolute zero temperature, T is the temperature in degrees absolute and n_i is the concentration in number/m^3. A_o is taken as 2.33×10^{43}.

Q. 31. Define diffusion constant of an electron and hole.

Ans. Diffusion constant of any charge carrier (electron or hole) may be defind from the Einstein relationship

$$\frac{D_h}{\mu_h} = \frac{D_e}{\mu_e} = V_T$$

where D_h and D_e are diffusion constant, μ_h and μ_e are mobilities and V_T is the volt-equivalent of temperature.

Q. 32. Give the continuation equation for electrons.

Ans. Continuation equation indicates that the rate of thermal generation of electrons equals the rate of electrons lost due to recombination under equilibrium condition.

Q. 33. Describe Hall effect.

Ans. When a specimen (metal or semiconductor) is placed in a transverse magnetic field and a direct current is passed through it, then an electric field is induced across its edges in the perpendicular direction of current as well as magnetic field. This phenomenon is called the Hall effect.

Q. 34. "Hall effect has played a decisive role in revealing the mechanism of conduction in semiconductors." Explain the statement. **[U.P.S.C. I.E.S. 2003]**

Ans. The conductivity of a semiconductor does determine the mobility and density of charge carriers separately. But in case there is only one type of charge carriers (either electrons or holes), the density of charge carriers can be found from a measurement of the Hall coefficient of the material. Hence, the given statement is justified.

Q. 35. What properties of a semiconductor are determined from a Hall effect?

Ans. The Hall effect may be used to determine (*i*) whether a semiconductor is of N-type or P-type (*ii*) the carrier concentration and (*iii*) measure the conductivity of a material and then compute mobility.

Q. 36. What is Hall coefficient? Where is it used?

Ans. Hall coefficient $R_H = \frac{1}{\rho} = \frac{\mu}{\sigma}$

where ρ is the charge density, μ is the mobility and σ is the conductivity of specimen.

It is used to determine mobility of specimen.

Q. 37. What is a P-N junction?

Ans. The contact surface between the layers of P-type and N-type semiconductor pieces placed together so as to form a P-N junction is called the P-N junction.

Q. 38. How do the transition region width and contact potential across a P-N junction vary with the applied bias voltage?

Ans. When the P-N junction is forward biased, the transition region width is reduced and the contact potential is also reduced with the increase in applied bias voltage.

When the P-N junction is reverse biased, the transition is widened, and the contact potential is increased with the increase in applied bias voltage.

Q. 39. Which type of charges are present on the two opposite faces of the junction?

Ans. Positive charge on N-side and negative charge on P-side of the junction.

Q. 40. What types of carriers are present in space charge region ?

Ans. No mobile carrier is present in the space charge region.

Q. 41. Why is space region called the depletion region ?

Ans. The region around the junction is completely ionised on formation of P-N junction. As a result, there are no free electrons on the N-side nor the holes on the P-side. Since the region around the junction is depleted of mobile charges, it is called the depletion region.

Q. 42. Why an electric field is produced in a depletion region of a P-N junction ?

Ans. The separation of positive and negative space charge densities in a P-N junction results in an electric field.

Q. 43. What is space charge width ?

Ans. The space charge region extends into the N and P-regions from the metallurgical junction. The distance is known as the space charge width.

Q. 44. The electric field in the space charge region decreases with forward bias and increases with reverse bias. Why ?

Ans. Because applied electric field opposes built-in field.

Q. 45. Define cut-in voltage of a p-n junction diode ?

Ans. The forward voltage, at which the current through the P-N junction starts increasing rapidly, is called the cut-in voltage.

Q. 46. What do you understand by reverse saturation current of a diode ?

Ans. Reverse saturation current of a diode is due to minority carriers and is caused when the diode is reverse biased. Only a very small voltage is required to direct all minority carriers across the junction, and when all minority carriers are flowing across, further increase in bias voltage will not cause increase in current. This current is referred to as a reverse saturation current.

Q. 47. What is the effect of temperature on the reverse current of a P-N junction ?

Ans. Reverse current of a P-N junction increases with the increase in junction temperature.

Q. 48. Define peak inverse voltage ?

Ans. Peak inverse voltage is the maximum voltage that can be applied to the P-N junction without damage to the junction. If the reverse voltage across the junction exceeds its peak-inverse voltage (PIV), the junction may get destroyed owing to excessive heat.

Q. 49. Define the limitations in the operating conditions of P-N junction.

Ans. Every P-N junction has limiting values of (*i*) maximum forward current (*ii*) peak inverse voltage (PIV) and (*iii*) maximum power rating. The P-N junction provides satisfactory performance when operated within these limiting values. The P-N junction diode may get destroyed due to excessive heat if any of these values is exceeded.

Q. 50. What is a P-N junction ?

Ans. The contact surface between the layers of P-type and N-type semiconductor pieces placed together so as to form a P-N junction is called the P-N junction.

Q. 51. Why is silicon preferred over germanium in the manufacture of semiconductor devices ?

Ans. The silicon semiconductor devices have, in general, higher PIV and current ratings and wider temperature range than germanium semiconductor devices and, therefore, silicon is preferred over germanium in the manufacture of semiconductor devices.

Q. 52. What does the arrowhead represent in a schematic symbol of a P-N junction ?

Ans. The arrowhead in the schematic symbol of a P-N junction indicates the direction of conventional current flow when the diode is forward biased.

Q. 53. What is an ideal diode ?

Ans. An ideal diode is a two terminal polarity sensitive device that has zero resistance when it is forward biased and infinite resistance when reverse biased.

Q. 54. Is reverse saturation current of a junction diode independent of reverse bias voltage ?

Ans. Yes.

Q. 55. Germanium is more temperature dependent than silicon. Why ?

Ans. Because the reverse saturation current in case of a germanium diode is approximately 1,000 times larger.

Q. 56. What is the effect of temperature on the reverse saturation current of a diode ?

Ans. Reverse saturation current, theoretically, increases by 8 percent per °C for silicon and 11 percent per °C for germanium. But from experimental data it is found that the reverse saturation current increases 7 percent per °C for both silicon and germanium. This is because a surface leakage current component of reverse saturation current is independent of temperature. Since $(1.07)^{10} \simeq 2.0$, the reverse saturation current approximately doubles for every 10°C rise in temperature.

Q. 57. Explain the effect of temperature on the V-I characteristics of the diode.

Ans. Reverse saturation current I_0 and volt-equivalent of temperature V_T increase with the increase in temperature. I_0 doubles for every 10°C rise in temperature. This causes lowering of threshold or cut-in voltages. With the increase in temperature, forward characteristic becomes more ideal. In reverse-bias, the breakdown voltage increases with the increase in temperature. Refer to Fig. 2.45.

Q. 58. What is static resistance of a diode ?

Ans. The static or dc resistance of a diode is the resistance offered by it to the direct current. It is defined as the ratio of the diode voltage and current at the point of interest and is not sensitive to the shape of V-I characteristic curve. It decreases with the increase in diode current or voltage.

Q. 59. Define dynamic resistance of a P-N junction diode in forward biased condition.

Ans. The resistance offered by a P-N junction diode to the changing forward current is defined as the dynamic resistance.

It may be given as

$$\text{AC or dynamic resistance, } r = \frac{\text{Small change in forward voltage}}{\text{Small change is forward current}}$$

Thus ac resistance of a diode is sensitive to the shape of *v-i* characteristic curve in the region of interest. It decreases for higher level of diode current or voltage.

Q. 60. Explain what do you understand by the expression "linearization of a nonlinear device".

Ans. A non-linear device, such as a diode, has a non-linear voltage-current characteristic. In a diode V and I are related by equation

$$I = I_0\,(e^{V/\eta V_T} - 1) \qquad \text{...Refer to Eq. (2.99)}$$

Diode shows nonlinearity until the threshold or cut-in-voltage V_T is reached. Voltage-current characteristics of the diode follow a linear relationship after it has attained the cut-in-voltage, just as for a resistance of very low magnitude. It means that diode exhibits non-linearity for lower voltages ($V < V_T$) but linearity for voltages exceeding V_T.

The equivalent circuit for the linear approximation is a constant-voltage source in series with a resistor [Fig. 2.50 (*b*)].

Q. 61. What is step graded junction ?

Ans. A junction is said to be step graded if there is an abrupt change from acceptor ion concentration on the P-side to donor ion concentration on the N-side such as alloyed or fused junctions.

Q. 62. What is linear graded junction ?

Ans. A junction is said to be linearly graded if the charge concentration varies gradually with the distance in its transition region such as a growth junction.

Q. 63. What are the two types of capacitances across P-N junction ? Which of these is more important in case of forward bias ? **[Pb. Technical Univ. 2003]**

Ans. In a P-N junction, there are two capacitive effects which are to be considered. Thus, there are two types of capacitances namely transition or depletion region capacitance and diffusion or storage capacitance. In the case of a forward biased diffusion capacitance is much larger than transition capacitance (more than a million times).

Q. 64. Define transition and diffusion capacitances.

Ans. When a P-N junction is reverse biased, the depletion region acts like an insulator or dielectric material while the P- and N-type regions on either side have a low resistance and act as the plates. Thus P-N junction may be considered a parallel plate capacitor. The junction capacitance is termed as space charge *capacitance* or *transition capacitance* or *depletion region capacitance* and is denoted by C_T.

When a P-N junction is forward biased, a capacitance which is much larger than the transition capacitance, comes into play. This type of capacitance is called the diffusion capacitance C_D.

Q. 65. How will you represent capacitive effect of a practical diode on an ideal diode ?

Ans. The capacitive effect of a practical diode on an ideal diode may be represented by a capacitor connected in parallel with an ideal diode.

Q. 66. How does diffusion capacitance C_D vary with dc diode current ?

Ans. In a P-N junction diode, the diffusion capacitance is directly proportional to the diode forward current.

Q. 67. How does transition capacitance vary with junction voltage in a P-N junction diode ?

Ans. Transition capacitance is inversely proportional to the square root of the junction voltage.

Problems

1. For silicon the intrinsic concentration is approximately 10^{16} carriers/m^3. If an impurity concentration of 10^{22} donor atoms/m^3 is doped, determine the electron and hole concentration. **[Ans. 10^{22}; 10^{10}]**

2. A silicon wafer (intrinsic carrier concentration 1×10^{16} m^{-3}) is doped with 2×10^{22} aluminium atoms/m^3 and 1×10^{22} arsenic atoms/m^3. Determine the minority carrier concentration. Assume complete dopant ionization. **[Ans. 1×10^{10}/m^3]**

3. The electron and hole concentration in a P-type semiconductor is 10^{12}/m^3 and 10^{22}/m^3 respectively. If the mobility of electrons and holes are 0.04 and 0.06 m^2/V-s respectively, determine the conductivity of the material. **[Ans. 96/Ω-m]**

4. Find the concentration of holes and electrons in an N-type silicon at 300 K if the conductivity is 0.1 (Ω-cm)$^{-1}$. Given that n_i at 300 K for silicon is 1.5×10^{10}/cm^3. μ_e at 300 K for silicon is 1,300 cm^2/V-s. **[Ans. $n = 4.8 \times 10^{14}$ electrons/cm^3, $p = 4.688 \times 10^5$ holes/cm^3]**

5. A sample of pure germanium at room temperature has a density of charge carriers of 2.5×10^{19}/m^3. It is doped with donor impurity atoms at the rate of one impurity atom for every 10^6 atoms of germanium. The density of germanium atoms is 4.2×10^{28} atoms/m^3. Determine hole density. **[Ans. 1.42×10^{16}/m^3]**

6. What is the donor concentration in a sample of N-type germanium having a resistivity of 0.1 Ω-m at 300 K.

   [Hint : $N_D = \dfrac{\sigma_n}{e\,\mu_n}$ and $\mu_n = 3{,}800$**]** **[Ans. 1.645×10^{16}/m^3]**

7. The intrinsic resistivity of silicon at 27°C is 2.8×10^3 Ω-m. The electron and hole mobilities are 0.38 and 0.18 m^2/V-s respectively. Calculate intrinsic carrier density at the given temperature. **[Ans. 3.986×10^{15}/m^3]**

8. A germanium diode has each of the accepter and donor impurities of concentration 9.5 × 10^{14}/cm^3. Assume n_i = 2.5 × 10^{13}/cm^3. At room temperature of 300 K, calculate the height of the potential barrier under open-circuit conditions. Boltzmann constant, k = 8.61 × 10^{-5} eV/°C. **[Ans.** 0.188V]

9. A germanium diode draws 50 mA with a forward bias of 0.27 V. The junction is at room temperature of 27°C. Determine the reverse saturation current of the diode. **[Ans.** 1.545 μ A]

10. The Hall coefficient of a specimen of a doped silicon is found to be 3.66 × 10^{-4} m^3 C^{-1}; the resistivity of the specimen is 8.93 × 10^{-3} Ω-m. Find the mobility and the density of the charge carriers, assuming single carrier conduction. **[I.E.S. 2002]** **[Ans.** 0.041 m^2/V-s; 0.17 × 10^{23}/m^3]

11. Determine ac resistance for a semiconductor diode with a forward bias of 0.25 V. Reverse saturation current at room temperature is of 1.2 μA. **[Ans.** 1.445 Ω]

12. Determine the current flowing in the circuit shown in Fig. 1. **[Ans.** 1.94 m A]

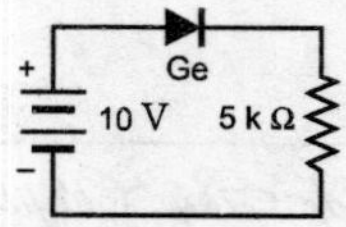

Fig. 1

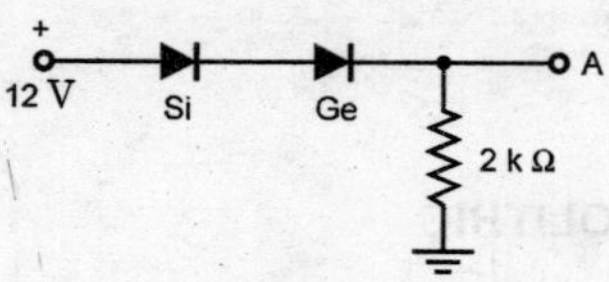

Fig. 2

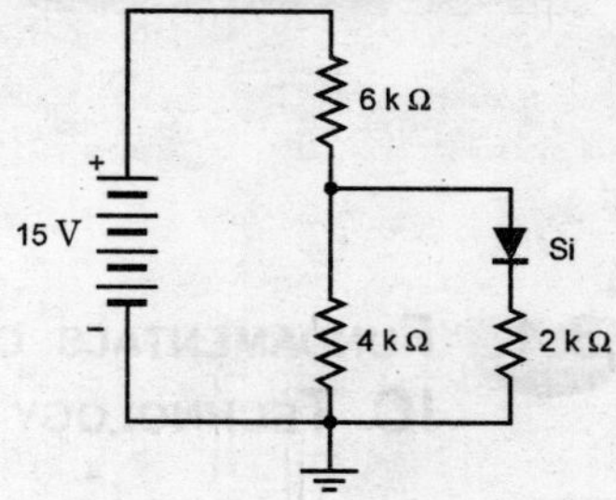

Fig. 3

13. Determine the current flowing in the circuit shown in Fig. 2. Also determine the potential of point A. **[Ans.** 5.5 mA, 11 V]

14. Determine the current through 2 kΩ resistor in the circuit shown in Fig. 3. **[Ans.** 1.205 mA]

[Hint. $R_{Th} = \dfrac{6 \times 4}{6+4} = 2.4\ k\Omega$; $V_{Th} = \dfrac{15}{6+4} \times 4 = 6$ V

$$I = \frac{V_{Th} - V_0}{R_{Th} + R} = \frac{6 - 0.7}{2.4 + 2} = 1.205 \text{ mA}]$$

❑❑❑

CHAPTER 3

Construction and Characteristics of Devices

3.1. FUNDAMENTALS OF MONOLITHIC IC TECHNOLOGY

Inside This Chapter

The word 'monolithic' is derived from the Greek monos, meaning 'single' and lithos, meaning 'stone'. Thus monolithic circuit is built into a single stone or single crystal *i.e.* in monolithic ICs, all circuit components, (both active and passive) and their interconnections are formed into or on the top of a single chip of silicon as shown in Figs. 3.1 (*a*) and (*b*). This type of technology is ideal for manufacturing identical ICs in large quantities and, therefore, provides lowest per unit cost and highest order of reliability. Monolithic ICs are by far the most common type of ICs used in practice, because of mass production, lower cost and higher reliability.

Since their invention, manufacturers have been manufacturing monolithic ICs to carry out all types of functions. Commercially available ICs of this type can be used as amplifiers, voltage regulators, crowbars, AM receivers, TV circuits, and computer circuits.

However, the monolithic circuits have the following limitations or drawbacks :

(*i*) Low power rating. Since monolithic ICs are of about the size of a discrete small-signal transistor, they typically have a maximum power rating of less than 1 watt. This limits their use to low-power applications.

(*ii*) Poorer isolation between components.

(*iii*) No possibility of fabrication of inductors.

(*iv*) Small range of values of passive components used in the ICs.

(*v*) Lack of flexibility in circuit design as for making any variation in the circuit, a new set of masks is required.

The fabrication of discrete devices such as resistor, diode, transistor or an integrated circuit in general can be done by the same technology. The various processes usually take

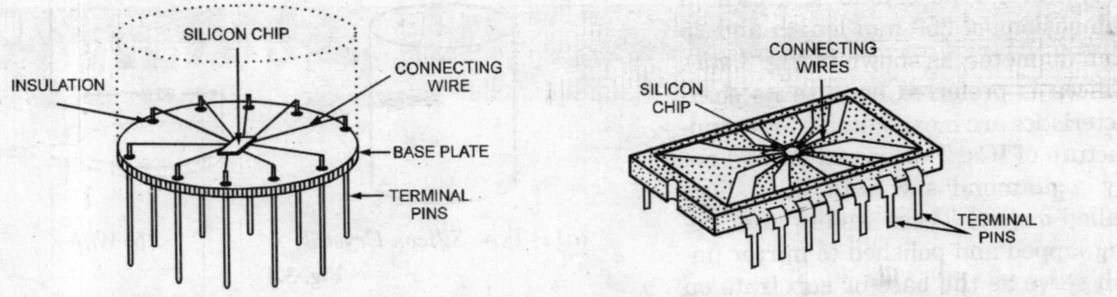

(a) *Monolithic IC in Can-Type Enclosure* (b) *Monolithic IC in Plastic Package*

Fig. 3.1. ***Basic Fabrication Techniques For ICs***

place through a single plane and, therefore, the technology is referred to as *planar technology*. A simple circuit shown in Fig. 3.2 when fabricated by silicon planar technology will have the cross-section view shown in Fig. 3.3.

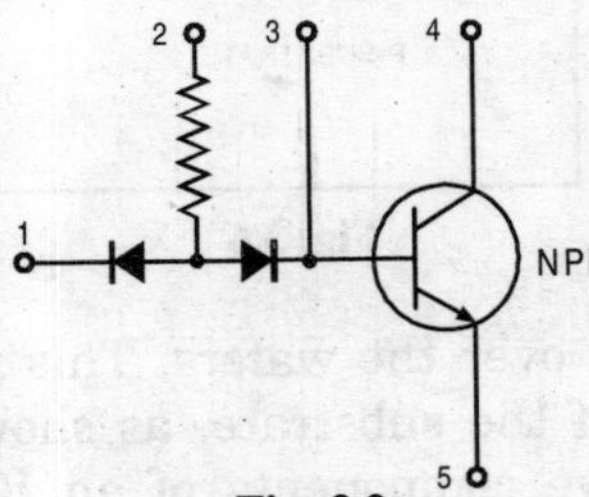

Fig. 3.2.

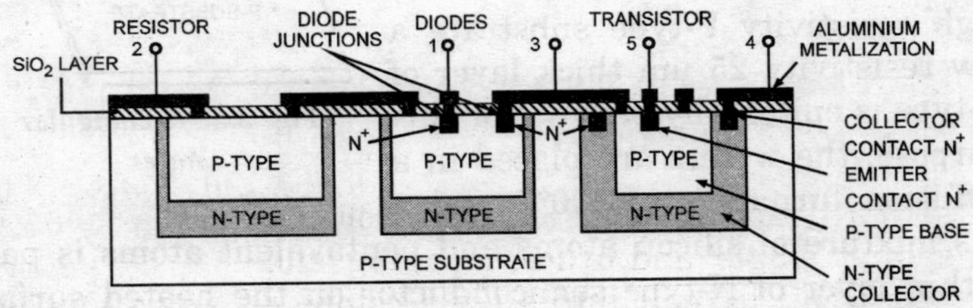

Fig. 3.3. ***Cross-Sectional View of the Circuit Shown in Fig. 3.2 When Transformed into a Monolithic Form***

The basic structure of an IC consists of the four layers of different materials.

The bottom most layer, typically 200 μm thick, is a P-type silicon and serves as the body or *substrate* upon which the complete IC is built.

The second layer, typically 25 μm thick, is of N-type silicon material which is grown as a single crystal extension of the substrate. All active and passive elements like diodes, transistors, resistors and capacitors are fabricated or built in this layer employing series of P-type and N-type impurity diffusions.

The third layer is of silicon dioxide (SiO_2) material and acts as a barrier in the selective diffusion in the second layer. This SiO_2 layer is etched away from the region where diffusion is desired to be permitted, the rest of the wafer remaining protected against diffusion. The etching is done by subjecting the SiO_2 layer to a photolithographic process. This SiO_2 layer serves the additional function of protecting the silicon wafer against contamination.

The fourth layer is a metallic layer of aluminium. This is added at the top in selected regions to permit interconnections between the different fabricated components.

3.2. Monolithic or Planar Processes of IC Production

A *monolithic IC* is one in which all circuit components and their interconnections are formed on a single thin wafer, called the *substrate*. The basic production processes for the monolithic ICs are given below :

1. *Crystal Growth of Substrate.* As already mentioned in the previous article, it is the bottom most layer that serves as the body or substrate upon which the complete IC is built.

A typical P-type crystal is grown in dimensions of 250 mm length and 25 mm diameter, as shown in Fig. 3.4(*a*). Silicon is preferred because its characteristics are more suitable for manufacture of ICs. The crystal is then cut by a diamond saw into thin slices called *wafers*. These wafers after being lapped and polished to mirror finish serve as the base or substrate on which hundreds of ICs are produced. The enlarged view of circular wafer is shown in Fig. 3.4(*b*). The wafer may also be rectangular in shape as shown in Fig. 3.5.

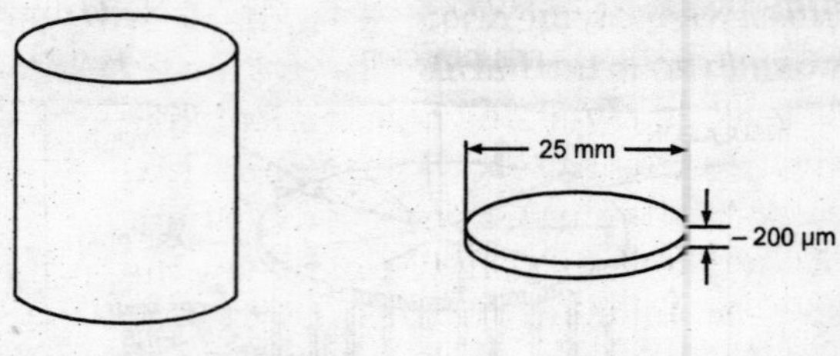

(*a*) *P-Type Silicon Crystal* (*b*) *Wafer*

Fig. 3.4

Fig. 3.5. *Rectangular Wafer*

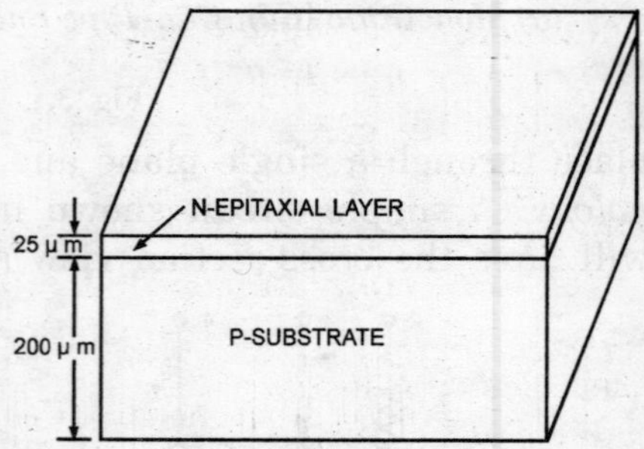

Fig. 3.6

2. *Epitaxial Growth.* On the high resistivity P-type substrate a low resistivity 25 µm thick layer of N-type is epitaxially grown. For this purpose, the wafers are placed in a diffusion furnace at 1,200°C and a gas mixture of silicon atoms and pentavalent atoms is passed over the wafers. This forms a thin layer of N-type semiconductor on the heated surface of the substrate, as shown in Fig. 3.6. It is this expitaxial layer that all active and passive components of an IC are formed. This layer ultimately becomes the collector for a transistor or an element for a diode or a capacitor. The resistivity of P-type substrate for $N_A = 1.4 \times 10^{21}$ atoms/m^3 is typically 10 Ω-cm. The resistivity of N-type epitaxial layer is suitably chosen in the range of 0.1 – 0.5 Ω-cm. This layer is finally polished and cleaned.

3. *Oxidation.* In order to prevent the contamination of the epitaxial layer, a thin layer of SiO_2 is formed over the entire surface, as illustrated in Fig. 3.7. The SiO_2 layer is grown by exposing the epitaxial layer to an oxygen atmosphere to about 1,000°C. This surface layer of SiO_2 will prevent any impurities from entering the N-type epitaxial layer. However, selective etching of this layer will permit the diffusion of the proper impurity into designed areas of the N-type epitaxial region of the silicon wafer.

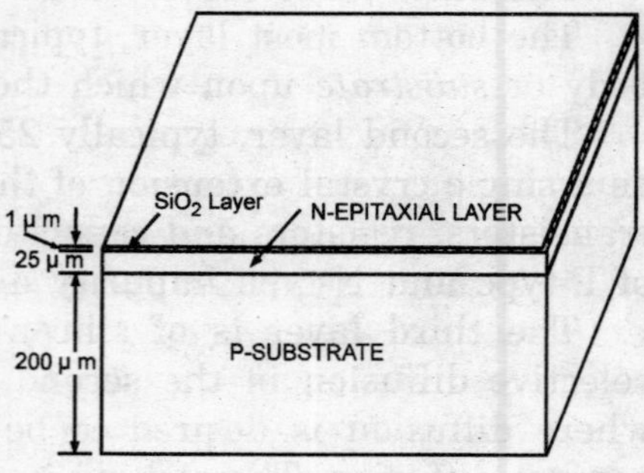

Fig. 3.7

4. *Photolithographic Process.* The monolithic technique requires the selective removal of the silicon–dioxide (SiO_2) to form openings through which impurities may be diffused, if required. The photolithographic process shown in Fig. 3.8(*a*) is used for this purpose. During the process the wafer is coated with a thin layer of photosensitive material, commonly known as photoresist (such as Kodak photoresist KPR). A large black and white layout of the desired pattern of openings or windows is made and then reduced photographically. This negative, or stencil, of the required dimensions is placed as a mask over the photoresist, as illustrated in Fig. 3.8(*a*).This wafer surface with mask is then exposed to the ultraviolet light. Due to ultraviolet light, the photoresist below the transparent portions of the mask becomes polymerised. The mask is now removed, and the wafer is developed by using a chemical like trichlorolethylene. The chemical dissolves the unpolymerised portions of the photoresist film and leaves the surface as shown in Fig. 3.8(*b*). The oxide not covered by polymerised photoresist is then removed by immersing the chip in an etching solution of HCl. Those portions of the SiO_2 which are protected by the photoresist remain unaffected by the acid. After etching and diffusion of impurities, the resist mask is stripped off with a chemical solvent like hot sulphuric acid (H_2SO_4) and by means of a mechanical abrasion process. The appropriate impurities are then diffused through oxide free windows.

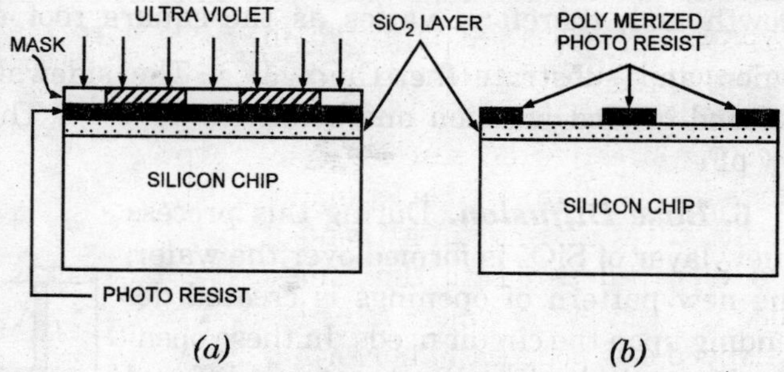

Fig. 3.8. ***Photolithographic Process***

5. ***Isolation Diffusion.*** SiO_2 layer is removed from the desired areas (four selected portions from the wafer, as illustrated in Fig. 3.9) using photolithographic etching process explained above. The remaining SiO_2 layer serves as mask for the diffusion of acceptor impurities. The wafer is now subjected to *isolation diffusion* at a suitably high temperature and for appropriate time period allowing P-type impurity (boron in this case) to penetrate into the N-type epitaxial layer through the openings in SiO_2 layer and ultimately reach the P-type substrate. The temperature and time period of diffusion are required to be carefully controlled. The process results in formation of N-type regions, called the *isolation islands*. The name is given as they are separated by back-to-back P-N junctions. Their purpose is to permit electrical isolation between various components of IC. Each electrical element is later on formed in a separate isolation island. The bottom of the N-type isolation island ultimately forms the collector of an N-P-N transistor. The P-type substrate is always kept negative w.r.t. the isolation islands and provided with reverse bias at P-N junctions. If P-N junctions are forward biased, the isolation will get lost.

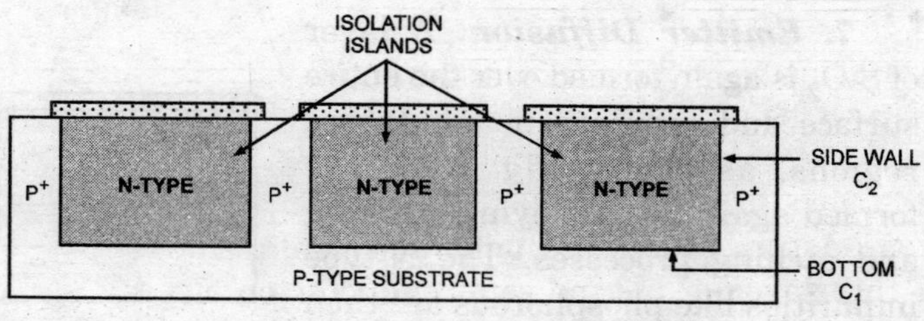

Fig. 3.9. ***Isolation Diffusion***

Isolation diffusion is controlled so as to cause high acceptor concentration P^+ (typically $N_A = 5 \times 10^{26}$ atoms/m^3) in the region between the isolation islands. This concentration is much higher than that of P-type substrate. This is for preventing the depletion region of the reverse-biased isolation island-to-substrate junction from extending into P^+ region and from possibly connecting two adjacent isolation islands.

Two adjoining isolation islands are connected to the P-type substrate by a barrier capacitance or transition capacitance. This is undesirable and is called the *parasitic capacitance*. It adversely affects the performance of the IC and puts a limitation on its use. The parasitic capacitance has two components; the capacitance C_1 from the bottom of the N-type region to the substrate and capacitance C_2 from the sidewalls of the isolation islands to the P-region. The bottom component C_1 is essentially due to step junction formed by epitaxial growth and, therefore, varies as the square root of the voltage V between the isolation region and substrate (*i.e.* $C_1 \propto \sqrt{V}$). The sidewall capacitance C_2 is associated with a diffused graded junction and so varies as $V^{-1/2}$. The total capacitance is of the order of a few pF.

6. *Base Diffusion.* During this process a new layer of SiO_2 is formed over the wafer. The new pattern of openings is created depending upon the circuit needs. In these openings P-type impurities like boron are diffused under regulated environments to form P-regions. This forms the base region of an N-P-N transistor or as well as resistors, the anode of diode, and junction capacitor. In this case, the diffusion time is so controlled that the P-type impurities do not reach the substrate. The resistivity of the base layer is usually much higher than that of the isolation regions.

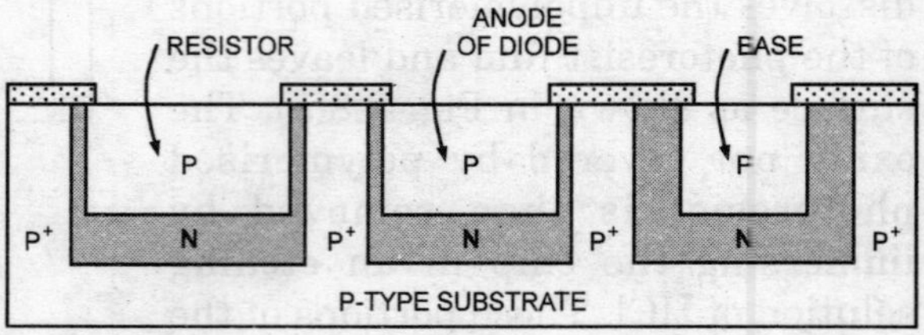

Fig. 3.10. ***Base Diffusion***

7. *Emitter Diffusion.* A layer of SiO_2 is again formed over the entire surface and openings in the P-type regions, as shown Fig. 3.11, are formed again by employing masking and etching processes. The N-type impurities like phosphorous are then diffused through these windows under controlled environments to form the transistor emitters, the cathode regions for diodes, and junction capacitors. Additional windows (such as W_1 and W_2 in Fig. 3.11) are usually made into the N-regions to permit aluminium metallic connections.

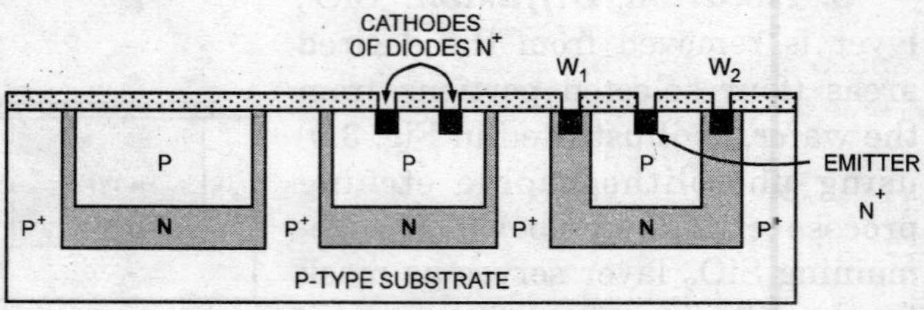

Fig. 3.11, ***Emitter Diffusion***

8. *Aluminium Metalization.* For making electrical connection between various components of the IC, several windows are opened on a newly created SiO_2 layer. Now a thin layer of aluminium is deposited on the entire top surface. Further, photoresist technique is used to etch away all the unwanted aluminium areas. The structure then provides the connected strips to which leads are attached, as illustrated in Fig. 3.12. Figure 3.12 represents the complete IC layout of the circuit shown in Fig 3.2.

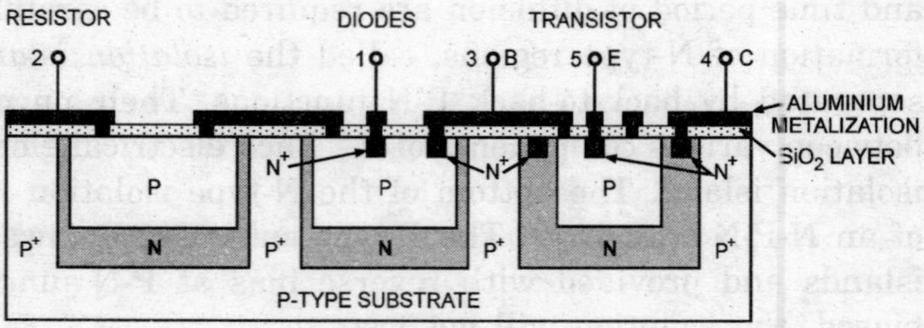

Fig. 3.12. ***Aluminium Metalization***

9. *Scribing and Mounting.* In practice, a large number (several hundreds) of identical ICs are manufactured simultaneously on a silicon wafer. After the completion of the metalization process, the wafer is scribed with a diamond-tipped tool and separated into individual chips. Each chip is then mounted on a ceramic wafer and is attached to a suitable header. Next the package leads are connected to the IC chip by stitch bonding of 25 micron aluminium or gold wire from the terminal pad on the IC chip to the package lead.

3.3. FABRICATION OF COMPONENTS OF MONOLITHIC

The outstanding feature of an IC is that it comprises a number of circuit elements inseparably associated in a single small package to perform a complete electronic circuit. We shall now see how various circuit elements (such as transistors, diodes, resistors etc.) can be fabricated in an IC form.

1. *Transistors and Diodes.* Transistors and diodes are usually formed using the epitaxial planar diffusion process described in previous article (Art. 3.2). Collector, base, and emitter regions are diffused into a P-type silicon substrate, as shown in Fig. 3.13, and surface terminals are provided for connections. In discrete transistors the substrate is normally employed as collector. If this were done with transistors in monolithic IC, all transistors fabricated on one substrate would have their collectors connected together. This is the reason that separate collector regions are diffused into the substrate.

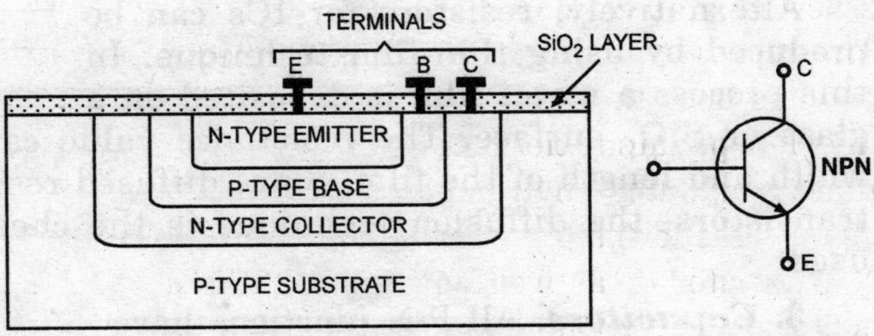

Fig. 3.13. ***Transistor***

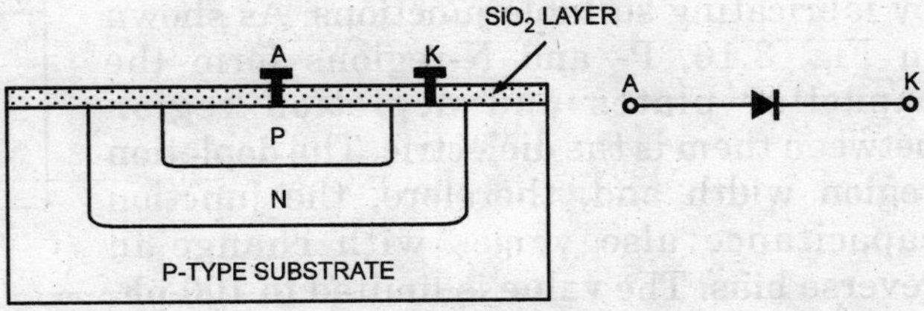

Fig. 3.14. ***Diode***

Even though separate collector regions are formed, they are not completely isolated from the substrate. For proper functioning of the circuit it is necessary that the P-type substrate is always kept negative w.r.t. the transistor collector. This is achieved by connecting the substrate to the most negative terminal of the circuit supply.

The unwanted or *parasitic junctions*, even when reverse-biased, can still affect the circuit performance adversely. The junction reverse leakage current can cause a serious problem in circuits operating at very low current levels. The capacitance of the reverse-biased junction may affect the circuit high-frequency performance, and the junction breakdown voltage imposes limits on the usable level of supply voltage. All these adverse effects can be reduced to the minimum if highly resistive material is employed for the substrate. If the substrate is very lightly doped, it will behave almost as an insulator.

Integrated circuit diodes are usually fabricated by diffusion exactly as transistors. Only two of the regions are used to form one P-N junction. In Fig. 3.14, collector-base junction of the transistor is used as a diode. Anode of the diode is formed during the base diffusion of the transistor and the collector region of the transistor becomes the cathode of the diode. For high speed switching emitter-base junction is used as a diode.

2. *Diffused Resistors.* In IC resistors, the resistance value can be controlled by varying the concentration of doping impurity and depth of diffusion. The range of resistor values that may be produced by the diffusion process varies from ohms to hundreds of kilo ohms. The typical tolerance, however, may be no better than ±5%, and may even be as high as ±20%. On the other hand, if all the resistors are diffused at the same time, then the tolerance ratio may be good.

Most resistors are formed during the base diffusion of the integrated transistor, as shown in Fig. 3.15. This is because it is the highest resistivity region. For low resistance values, emitter region is used as it has much lower resistivity.

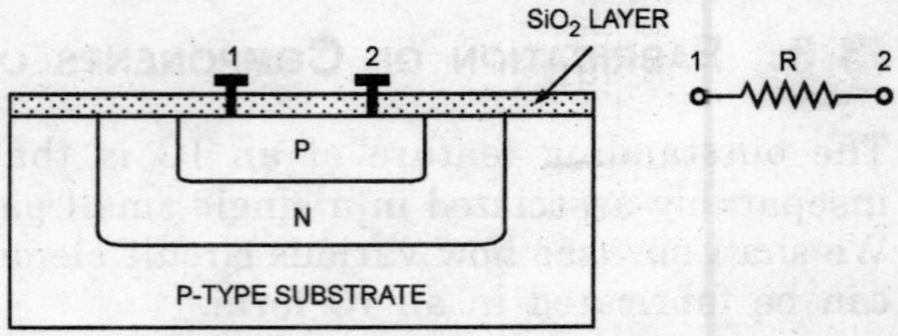

Fig. 3.15. *IC Diffused Resistor*

Alternatively, resistors for ICs can be produced by using thin-film technique. In this process a metal film is deposited on a glass or SiO_2 surface. The resistance value can be controlled by varying thickness, width and length of the film. Since diffused resistors can be processed while diffusing transistors, the diffusion technique is the cheapest and, therefore, the most widely used.

3. *Capacitors.* All P-N junctions have capacitance so capacitors may be produced by fabricating suitable junctions. As shown in Fig. 3.16, P- and N-regions form the capacitor plates and depletion region between them is the dielectric. The depletion region width and, therefore, the junction capacitance also varies with change in reverse bias. The value is limited to 100 pF.

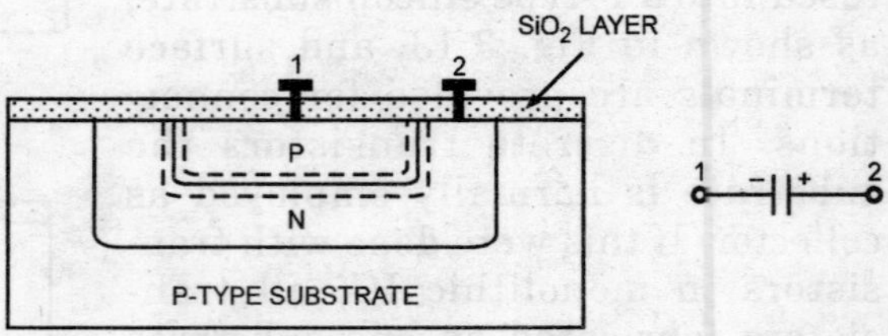

Fig. 3.16. *IC Diffused Capacitor*

IC capacitors may also be fabricated by utilizing the SiO_2 surface layer as a dielectric as shown in Fig. 3.17. A heavily-doped N-region is diffused to form one plate of the capacitor. The other plate is formed by depositing a film of aluminium on the SiO_2 formed on the wafer surface. The breakdown voltage of such a capacitor is much larger than for diffused capacitor and voltage of any polarity can be used.

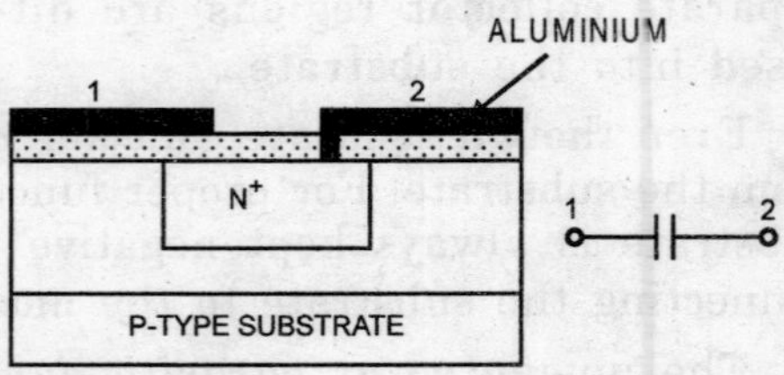

Fig. 3.17. *IC Thin Plate Capacitor*

4. *Inductors.* There is no feasible process for fabrication of inductors as part of monolithic structure. Normally, an equivalent circuit of inductor is used but if it cannot be avoided then terminals are provided to connect it externally.

3.4. METAL-SEMICONDUCTOR JUNCTIONS

Metal-semiconductor junctions are those junctions which are fabricated by forming a compatible contact between a metal and a semiconductor. These junctions may be *rectifying (non-ohmic) contact type* or a *non-rectifying (ohmic) contact type.* The rectifying (non-ohmic) contact type metal-semiconductor is formed by depositing a metal film on a semiconductor surface followed by making of contact photolithographically. On the other hand, the ohmic contacts are made by heavy doping of the semiconductors in contact region. An ohmic contact is a low-resistance junction providing conduction in both directions between the metal and the semiconductor. Ideally, the current through the ohmic contact is a linear function of applied voltage, and the applied voltage should be very small. The importance lies due to the fact that they are easy to fabricate, useful for high speed rectification and for formation of non-rectifying (ohmic) contacts to semiconductors.

3.5. Breakdown Mechanisms in P-N Junction

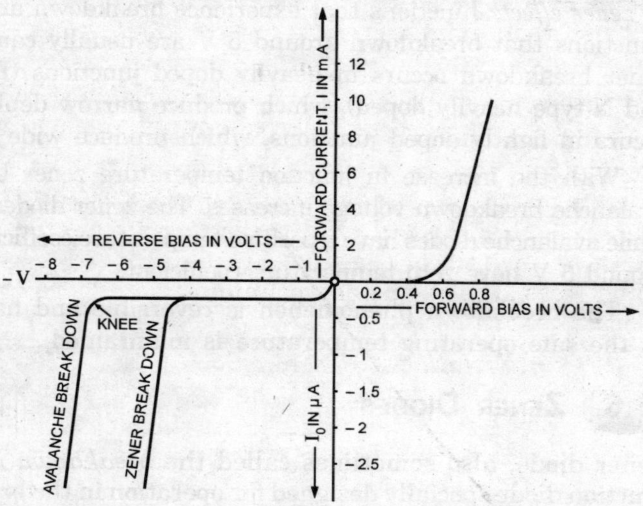

Fig. 3.18

When an ordinary P-N junction diode is reverse biased, normally only very small reverse saturation current flows. This current is due to movement of minority carriers. It is almost independent of the voltage applied. However, if the reverse bias is increased, a point is reached when the junction breaks down and the reverse current increases abruptly, as shown in Fig. 3.18. This current could be large enough to destroy the junction. If the reverse current is limited by means of a suitable series resistor, the power dissipation at the junction will not be excessive, and the device may be operated continuously in its breakdown region to its normal (reverse saturation) level. It is found that for a suitably designed diode, the breakdown voltage is very stable over a wide range of reverse currents. This quality gives the *breakdown diode* many useful applications as a *voltage reference source.*

The critical value of the voltage, at which the breakdown of a P-N junction diode occurs is called the *breakdown voltage.* The breakdown voltage depends on the width of the depletion region, which, in turn, depends on the doping level. The junction offers almost zero resistance at the breakdown point.

There are two mechanisms by which breakdown can occur at a reverse biased P-N junction : *avalanche breakdown* and *zener breakdown.*

The minority carriers, under reverse biased conditions, flowing through the junction acquire a kinetic energy which increases with the increase in reverse voltage. At a sufficiently high reverse voltage (say 5 V or more), the kinetic energy of minority carriers becomes so large that they knock out electrons from the covalent bonds of the semiconductor material. As a result of collision, the liberated electrons in turn liberate more electrons and the current becomes very large leading to the breakdown of the crystal structure itself. This phenomenon is called the *avalanche breakdown.* The breakdown region is the '*knee*' of the characteristic curve. Now the current is not controlled by the junction voltage but rather by the external circuit.

Under a very high reverse voltage, the depletion region expands and the potential barrier increases leading to a very high electric field across the junction. The electric field will break some of the covalent bonds of the semiconductor atoms leading to a large number of free minority carriers, which suddenly increase the reverse current. This is called the *zener effect.* The breakdown occurs at a particular and constant value of reverse voltage called the *breakdown voltage*. It is found that zener breakdown occurs at electric field intensity of about 3×10^7 V/m.*

* From basic circuit theory electric field intensity E is given as $\frac{V}{d}$ where V is the applied reverse voltage in volts and d is the width of depletion region in metres. The narrower the depletion region, the smaller the required bias to cause zener breakdown. A small reverse bias can produce a sufficiently strong electric field in a narrow depletion region.

Either of the two (zener breakdown or avalanche breakdown) may occur independently, or both of these may occur simultaneously. Diode junctions that breakdown below 5 V are caused by *zener effect.* Junctions that experience breakdown above 5 V are caused by *avalanche effect.* Junctions that breakdown around 5 V are usually caused by combination of two effects. The zener breakdown occurs in heavily doped junctions (P-type semiconductor moderately doped and N-type heavily doped), which produce narrow depletion layers. The avalanche breakdown occurs in lightly doped junctions, which produce wide depletion layers.

With the increase in junction temperature zener breakdown voltage is reduced while the avalanche breakdown voltage increases. The zener diodes have a negative temperature coefficient while avalanche diodes have a positive temperature coefficient. Diodes that have breakdown voltages around 5 V have zero temperature coefficient.

The breakdown phenomenon is reversible and harmless so long as the safe operating temperature is maintained.

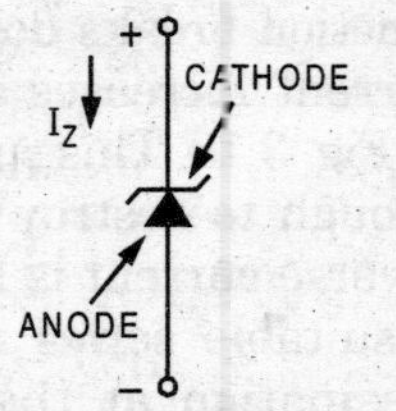

Fig. 3.19. ***Zener Diode Symbol***

3.6. ZENER DIODES

Zener diode, also sometimes called the *breakdown diode*, is a P-N junction diode specially designed for operation in the breakdown region in reverse bias condition. The diode may use either zener breakdown mechanism or avalanche breakdown mechanism, already discussed in detail in Art. 3.5. Despite the fact that avalanche and zener constitute two types of breakdown diodes, the name zener is commonly applied to both types of diodes. The breakdown diode may be silicon or germanium one but silicon is preferred over germanium because of higher operating temperature and current capability. The knee point is also more sharp in case of a silicon diode. These diodes are analogous to gas diodes in which large current appears on reaching the breakdown potential.

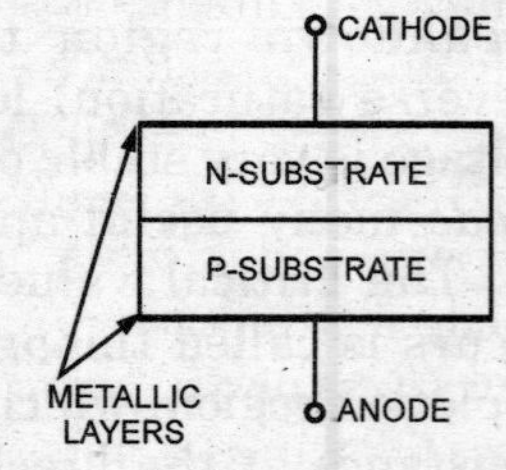

Fig. 3.20. ***Diffused Structure of a Zener Diode***

The symbol of a zener diode is given in Fig. 3.19. This is similar to that of a normal diode except that the line representing the cathode is bent at both ends *i.e.* the bar is turned into Z-shape; stands for zener.

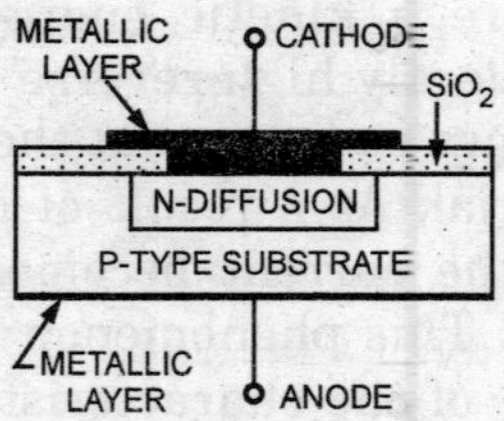

Fig. 3.21. ***Diffused and Passivated Structure of a Zener Diode***

Construction. Zener diode is like an ordinary P-N junction diode except that it is properly doped so as to have a sharp breakdown voltage. There are several methods used in the manufacture of zener diodes, such as diffused structure, diffused and passivated structure and alloy diffused structure.

Diffused structure zener diode consists of two N and P substrates diffused together and has metallic layers deposited on both sides to connect anode and cathode terminals, as illustrated in Fig. 3.20. The drawback of such a construction is that the edges of the junction are exposed to contamination.

In passivated structure of a zener diode, the edges of the junction are covered by layer of SiO_2, as illustrated in Fig. 3.21.

Similarly, all junctions are protected in alloy diffused structures, as shown in Fig. 3.22.

In general, the alloy diffused structure gives better zener characteristics at lower zener voltages. On the other hand, diffused and passivated zener diode is used at higher zener voltages.

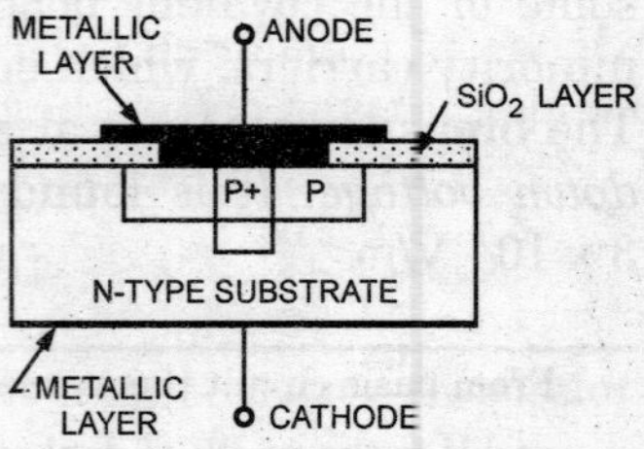

Fig. 3.22. ***Alloy Diffused Structure of a Zener Diode***

Volt-Ampere Characteristics. It has already been discussed in Art. 3.5 that when the reverse bias on a crystal diode is gradually increased, a point is reached when the junction breaks down and a reverse current increases abruptly, as illustrated in Fig. 3.23. The breakdown region is the knee of the reverse characteristic, as illustrated in Fig. 3.23. This breakdown of junction was first explained by the American scientist, C. Zener. So, the breakdown voltage is sometimes called, *zener voltage* V_Z and the sharply increased current is called the *zener current* I_Z.

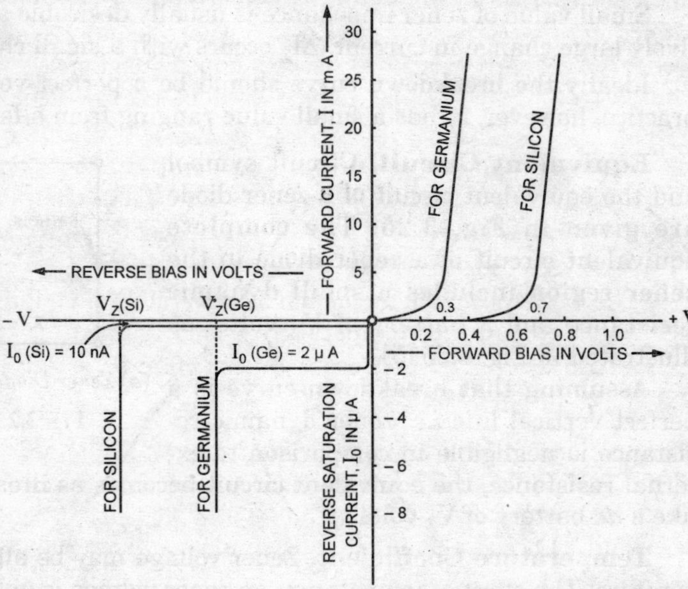

Fig. 3.23. *Volt-Ampere Characteristics of Typical Silicon and Germanium Zener Diodes*

Volt-ampere characteristics of typical silicon and germanium diodes are illustrated in Fig. 3.23. Breakdown diode is operated over the breakdown region and the current is limited by an external resistance. The breakdown or zener voltage V_Z depends upon the amount of doping. A heavily doped diode will have very thin depletion layer, so zener voltage will be very low. By controlling the thickness of depletion layer and doping concentration it is possible to make the zener diode to breakdown at a specified voltage, called the zener voltage (ranging from 2 V to 200 V). The forward characteristics of a zener diode are similar to those of an ordinary P-N junction diode, as illustrated.

During the operation in the breakdown region, it does not burn out immediately. As long as the current through the diode is limited by the external circuit within permissible values, it does not burn out.

Zener Impedance. This is very important parameter of a zener diode. Zener impedance Z_Z is essentially the dynamic resistance of a zener diode (similar to dynamic resistance of an ordinary P-N diode). It is defined as the reciprocal of the slope of the zener curve

$$i.e.\ r_Z = \frac{\Delta V_Z}{\Delta I_Z} \qquad ...(3.1)$$

where ΔV_Z and ΔI_Z are the small variations in voltage and current respectively.

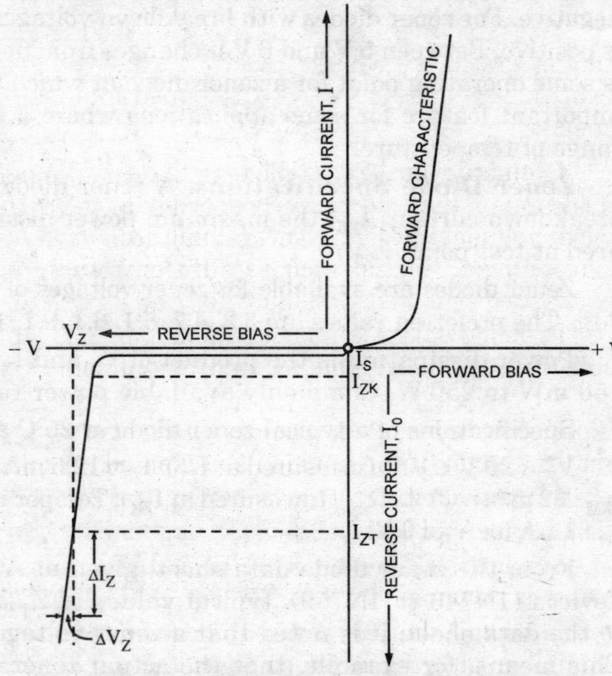

Fig. 3.24. *Zener Diode Characteristic With Enlarged View of Breakdown Region*

It is usually measured just above the knee in zener region and is named as Z_{ZK}. When measured at test point, it is designated Z_{ZT}. It decreases with increase in zener current.

Small value of zener impedance is usually desirable as it indicates steep curve where relatively large change in current, ΔI_Z occurs with a small change in voltage ΔV_Z.

Ideally the breakdown curve should be a perfect vertical line and r_Z should be zero. In practice, however, r_Z has a small value ranging from a few ohms to several hundred ohms.

Equivalent Circuit. Circuit symbol and the equivalent circuit of a zener diode are given in Fig. 3.25. The complete equivalent circuit of a zener diode in the zener region includes a small dynamic resistance and a battery of V_Z volts, as illustrated in Fig. 3.25 (*b*).

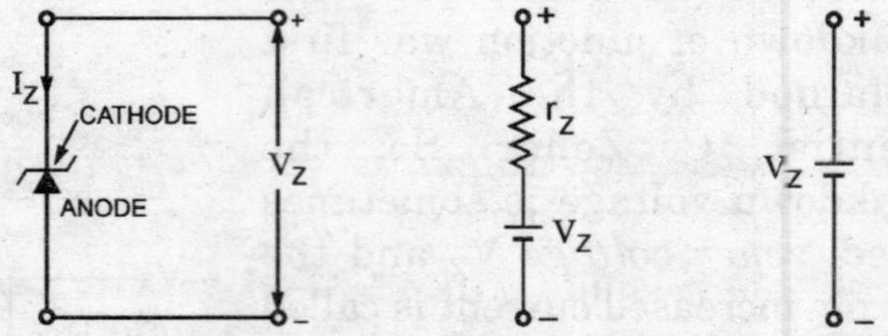

(a) Zener Diode Symbol *(b) Complete* *(c) Approximate*

Fig. 3.25. *Equivalent Circuits of Zener Diode*

Assuming that breakdown curve is a perfect vertical line *i.e.* zener dynamic resistance is negligible in comparison to external resistance, the equivalent circuit becomes as illustrated in Fig. 3.25 (*c*) *i.e.* it appears like a *dc* battery of V_Z volts.

Temperature Coefficient. Zener voltage may be affected by change in the ambient temperature. The effect of temperature on zener voltage is given in terms of *temperature coefficient*, which is defined as the percentage change in nominal zener voltage for each degree centigrade of change in junction temperature. Mathematically it is given as $\frac{\Delta V_Z}{V_Z (t - t_0)} \times 100$ where ΔV_Z is the change in zener voltage due to temperature variation of $(t - t_0)$.

For zener diodes with breakdown voltages of less than 5 V, the temperature coefficient is negative. For zener diodes with breakdown voltages exceeding 6 V, the temperature coefficient is positive, Between 5 V and 6 V it changes from negative to positive value: it means that there is some operating point for a zener diode at which temperature coefficient is zero. This is very important feature for some applications where a solid zener voltage is required over a wide range of temperature.

Zener Diode Specifications. A zener diode is specified by its breakdown voltage V_Z, breakdown current, I_{ZK}, the maximum power dissipation $P_{Z\,(max)}$ and zener impedance measured at test point, Z_{ZT}.

Zener diodes are available for zener voltages of 2.4 to 200 V with accuracies between 5 and 20%. The preferred values are 3.3, 4.7, 5.1, 6.1, 9.1, 10, 11, 12, 13, 15, to 200 V.

Power dissipation is the product of V_Z and I_Z. The maximum power rating varies from 150 mW to 250 W. Commonly available power ratings are 0.4, 1, 3, 10, and 50 watts.

Specifications of a typical zener diode at 25°C ambient are given below.

V_{ZT} : 20 V ± 10% (measured at I_{ZT}); I_{ZT} : 12.5 mA for V_{ZT} = 20 V; I_{ZK} = 0.25 mA for V_{ZK} = 12 V; I_{ZM} : 32 mA; r_{ZK} : 22 Ω_{max} (measured at I_{ZK}); Temperature coefficient : 0.075% per °C; $P_{Z\,(max)}$: 1 W; I_R : 1 μA for V_R : 6 V.

From the zener diode data sheet, given at Appendix 1.2, it is seen that for low-power devices (1N746 to 1N759), typical values of Z_{TT} range from 5 Ω to 30 Ω. Referring again to the data sheet, it is noted that a zener voltage tolerance of ± 5% or ± 10 % is specified. This means, for example, that the actual zener voltage of 1N759 is 12 V ± 10% i.e. 10.8 to 13.2 V. The value of V_Z will remain stable at whatever voltage it happens to be within this range.

Example 3.1. Determine the temperature coefficient of a 5 V zener diode at 25°C, if the nominal voltage drops to 4.8 V at a temperature of 100°C.

Solution : Change in zener voltage, $\Delta V_Z = 4.8 - 5 = -0.2$ V

Change in temperature, $(t - t_0) = 100 - 25 = 75°C$.

$$\text{Temperature coefficient of zener diode} = \frac{\Delta V_Z}{V_Z(t-t_0)} \times 100 = \frac{-0.2}{5 \times 75} \times 100 = -0.0533\% \textbf{ Ans.}$$

3.7. Zener Diode Applications

Zener diodes have numerous applications in transistor circuitry. Some of their common applications are explained below.

1. Zener Diode Shunt Regulator (or Voltage Regulation). Voltage regulation is a measure of a circuit's ability to maintain a constant output voltage even when either input voltage or load current varies.

Figure 3.26 shows how a zener diode can be used as a voltage regulator to provide a constant voltage from a source whose voltage may vary appreciably. A resistor R_S is necessary to limit the reverse current through the diode to safer value. The voltage source V_S and the resistor R_S are so selected that the diode operates in the breakdown region. The diode voltage in this region, which is also the voltage across the load R_L, is called *zener voltage* V_Z and the diode current is called the *zener current* I_Z. The series resistor R_S absorbs the output voltage fluctuations so as to maintain voltage across the load constant. Zener diode is reverse-connected across the input voltage whose variations are to be regulated.

As long as voltage across the load resistor R_L is less than the breakdown voltage V_Z, the zener diode does not conduct, the resistors R_S and R_L constitute a potential divider across V_S. At an increased supply voltage V_S, the voltage drop across load resistor (or zener diode) becomes greater than the zener breakdown voltage. It then operates in its breakdown region. The series resistor R_S limits the zener current I_Z from exceeding its rated maximum $I_{Z\,max}$. Current through resistor R_S is given as

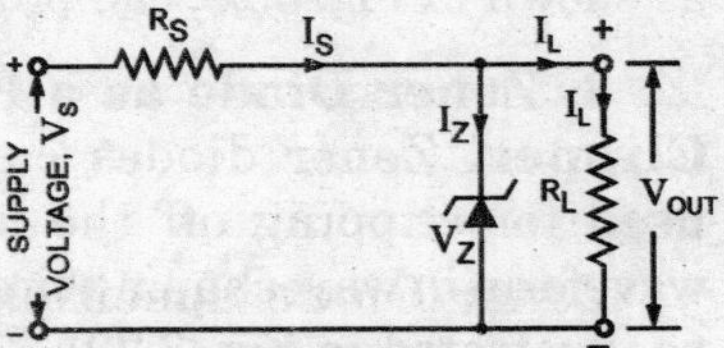

Fig. 3.26

$$I_S = \frac{V_S - V_Z}{R_S} \quad \text{...(3.2)}$$

The current from the power supply splits at the junction of zener diode and the load resistor R_L.

$$\text{So } I_S = I_Z + I_L \quad \text{...(3.3)}$$

When the zener diode operates in its breakdown region, the voltage across it, V_Z remains fairly constant even though the current I_Z flowing through it may vary considerably.

When the supply voltage V_S increases, the current through both the zener diode and load resistance R_L increases. At the same time, however, the zener diode resistance decreases and the current through the diode increases more than proportionately. As a result, a greater voltage drop will occur across the series resistor R_S and the output voltage V_{OUT} (voltage across the diode or load resistance R_L) will become very close to the original value. The reverse is also true. Thus a zener diode can maintain the output voltage V_{OUT} within a fraction of a volt when the supply or input voltage V_S may vary over a range of several volts. The zener diode will maintain a constant voltage across the load as long as the supply voltage is more than the zener voltage.

Let us examine the other cause of the output voltage variation. When the load resistance R_L decreases for constant input voltage V_S, load current I_L increases. This additional current is not supplied from the source of supply but the demand of additional load current is met by decrease in zener current I_Z. This keeps the voltage drop across series resistance R_S constant and so the output voltage V_{OUT}. It may be noted that the diode current I_Z attains its maximum $I_{Z\,max}$ when load current I_L becomes zero. Thus the value of series resistor R_S can be determined from the following equation

$$R_S = \frac{V_S - V_{OUT}}{I_{Z\,max}} \qquad \text{...(3.4)}$$

The advantages of zener diode voltage regulators over other voltage regulators are that they are smaller, lighter, more rugged and have a longer life. They are also simpler and inherently cheaper.

Such regulators have the following drawbacks :

1. Their efficiency is low for heavy load currents because there is a considerable power loss in series resistor R_S and the diode.
2. The output voltage varies slightly due to zener impedance r_Z.
3. The output voltage cannot be choosen independently but depends upon the breakdown voltage of the zener diode.

This regulator is employed when there are small variations in load current and supply voltage.

2. Meter Protection. Figure 3.27 shows how does a zener diode protect a meter (AVO meter or multimeter) movement or circuit against damage from an accidental application of excessive voltage. In this arrangement zener diode is connected in parallel with the meter, as shown in Fig. 3.27. In case of accidental overload most of the current will pass through the zener diode and so the meter movement will be saved from damage. Two zener diodes connected, as shown in Fig. 3.28, can provide overload protection regardless of the applied polarity.

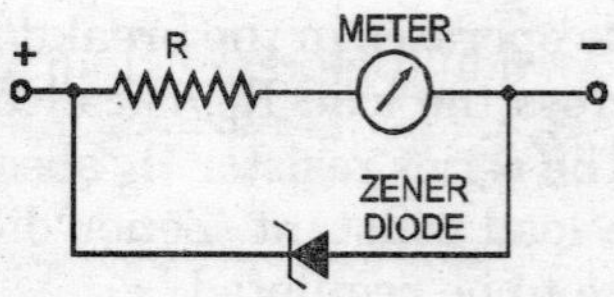

Fig. 3.27

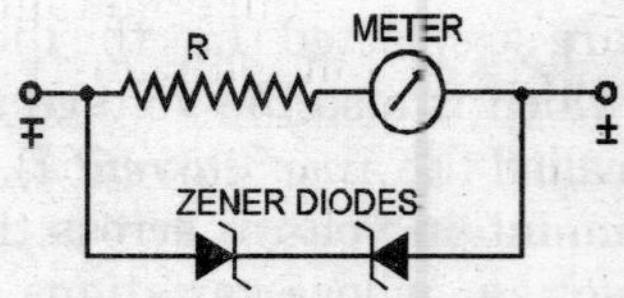

Fig. 3.28

3. Zener Diode as a Peak Clipper. Zener diodes can be used for clipping off the input waveform in waveshaping circuits, as illustrated in Fig. 3.29.

Two similar zener diodes D_1 and D_2 are shunted across the input sinusoidal voltage of peak value of V_{max} volts. Let the zener voltage of each of zener diodes be V_Z volts (V_Z being lesser than V_{max}).

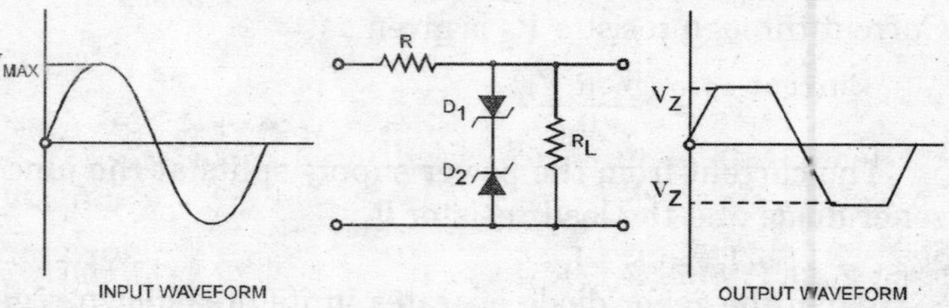

Fig. 3.29

We have seen that a zener diode acts like a 'short' (or very low resistance) when forward biased and 'open' (or very high resistance) when reverse biased till it goes into breakdown at V_Z. The action of the circuit is explained below.

For positive input half cycle diode D_1 is forward biased and acts like a 'short' while diode D_2 acts like an open (being reverse biased) up to V_Z. Thereafter diode D_2 goes into breakdown and holds the output voltage constant at V_Z till input voltage falls below V_Z in the latter part of the positive half cycle. This is because during this duration of time both of the diodes act as 'short'. When the supply voltage falls below V_Z, diode D_2 comes out of the breakdown and again acts as an open across the load circuit. During the negative input half cycle, the actions of diodes D_1 and D_2 are reversed. The result is that the output voltage waveform is "clipped" off on both peaks, as illustrated in Fig. 3.29.

4. Switching Operation. Zener diode can produce a sudden change from low current to high current, so it is useful for switching operations. The use of zener diode in switching operations has made possible an extremely fast performance in computer applications.

5. Zener Diode as a Reference Element. There are many applications in which it is desirable that a constant voltage is maintained between two points in a circuit, and use this voltage as reference for comparing other voltages against it. The difference between the two voltages (the voltage under comparison and reference voltage) is usually amplified and then used to perform some control function. Such an arrangement is employed in power supply voltage regulator circuits, measurement circuits and servomechanism circuits. The constant voltage characteristic of a zener diode in its breakdown region makes it desirable for such an application.

A circuit employing zener diode as a reference element is given in Fig. 3.30. The value of R is so chosen that the diode operates well within its breakdown region. V_{REF} equals the zener breakdown voltage V_Z. The difference of input voltage V_{IN} and reference voltage V_{REF} (*i.e.* $V_{IN} - V_{REF}$) gives the control output. Obviously V_{IN} should be more than V_{REF}.

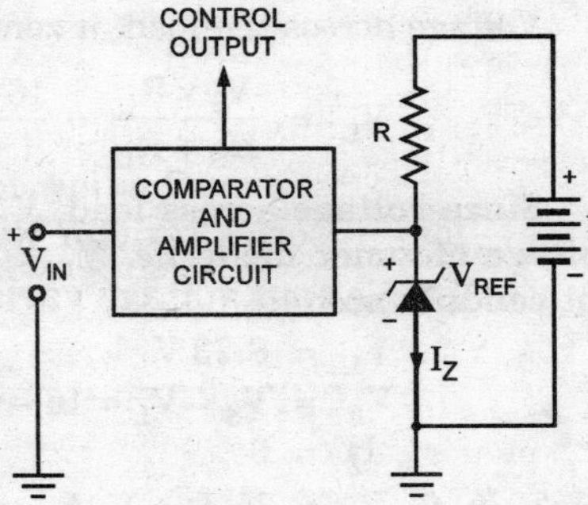

Fig. 3.30. ***Zener Diode as Reference Element***

Example 3.2. For the circuit shown in Fig. 3.31, find

(*a*) the output voltage,

(*b*) the voltage drop across R_s,

(*c*) the current through zener.

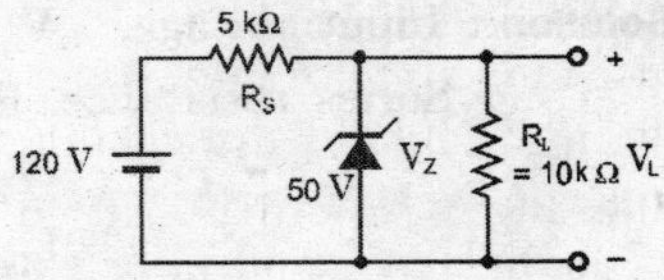

Fig. 3.31

Solution : From the Fig. 3.31 we have

(*a*) Output voltage, $V_L = V_Z = 50$ V **Ans.**

(*b*) Voltage drop across $R_S = I_S \times R_S = V_S - V_Z = 120 - 50 = 70$ V **Ans.**

(*c*) Load current, $I_L = \dfrac{V_L}{R_L} = \dfrac{50\text{ V}}{10\text{ k}\Omega} = 5$ mA

Current through R_s, $I_S = \dfrac{70\text{ V}}{5\text{ k}\Omega} = 14$ mA

Hence, current through zener diode is given by

$$I_Z = I_S - I_L = 14\text{ mA} - 5\text{ mA} = 9\text{ mA} \quad \textbf{Ans.}$$

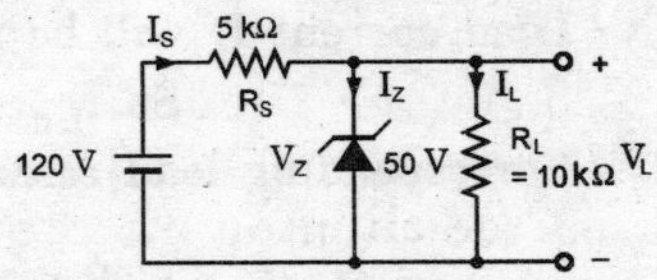

Fig. 3.32

Example 3.3. For the circuit shown in Fig. 3.33, find the maximum and minimum values of zener diode current.

Solution : Output voltage, $V_{OUT} = V_Z = 50$ V

Load current, $I_L = \dfrac{V_{OUT}}{R_L} = \dfrac{50}{10\text{ k}\Omega} = 5$ mA

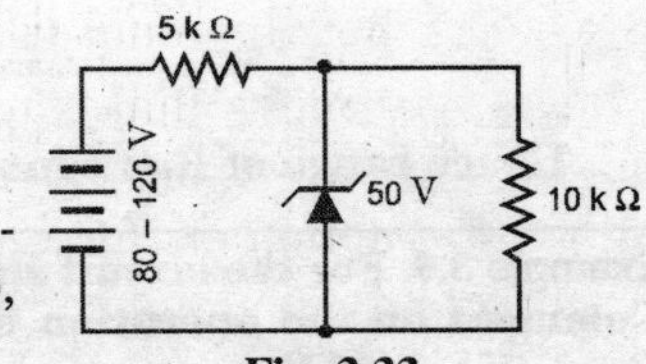

Fig. 3.33

The zener current will be maximum when input voltage is maximum *i.e.* 120 V. Corresponding current through series resistance (5 kΩ),

$$I_{S\,max} = \frac{V_{S\,max} - V_{OUT}}{R_S} = \frac{120 - 50}{5\text{ k}\Omega} = 14\text{ mA}$$

Corresponding zener current,

$$I_{Z\,max} = I_{S\,max} - I_L = 14 - 5 = 9\text{ mA} \quad \textbf{Ans.}$$

The zener current will be minimum when input voltage is minimum *i.e.* 80 V. Corresponding current through series resistance (5 kΩ),

$$I_{S\,min} = \frac{V_{S\,min} - V_{OUT}}{R_S} = \frac{80-50}{5\text{ k}\Omega} = 6\text{ mA}$$

Corresponding zener current, $I_{Z\,min} = I_{S\,min} - I_L = 6 - 5 = 1.0$ mA **Ans.**

Example 3.4. For a zener diode network shown below in Fig. 3.34. Determine V_L, V_R, I_Z and P_Z.

[*Pb. Technical Univ. Electronics Devices and Circuits 2005-06*]

Solution : Assume breakdown voltage of zener diode 10 V.

Voltage across the load, if zener diode is open-circuited, by voltage divide rule,

$$V_L = \frac{V_S \times R_L}{R_S + R_L} = \frac{16 \times 1.2}{1 + 1.2} = 8.73\text{ V}$$

Since voltage across load, V_L is less than breakdown voltage of zener diode *i.e.* $V_L < V_Z$, the zener diode will not conduct and

$$V_L = 8.73\text{ V}$$

$$V_R = V_S - V_L = 16 - 8.73 = 7.27\text{ V} \quad \textbf{Ans.}$$

$$I_Z = 0$$

$$P_Z = V_Z I_Z = V_L I_Z = 8.73 \times 0 = 0\text{ Watt} \quad \textbf{Ans.}$$

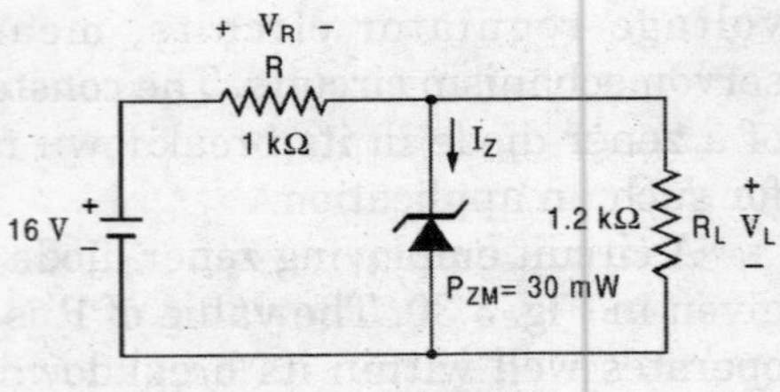

Fig. 3.34

Example 3.5. Explain the working of diode as a shunt regulator. For the circuit of Fig. 3.35, determine the range of R_L and I_L that will result in V_{RL} being maintained at 10 V.

Solution : Input voltage, $V_{in} = 50$ V

Series resistance, $R = 1$ kΩ

$$V_Z = 10\text{ V}$$

$$I_{Zm} = 32\text{ mA}$$

$$\text{Supply current, } I_R = \frac{V_{in} - V_L}{R} = \frac{50-10}{1{,}000} = 40\text{ mA}$$

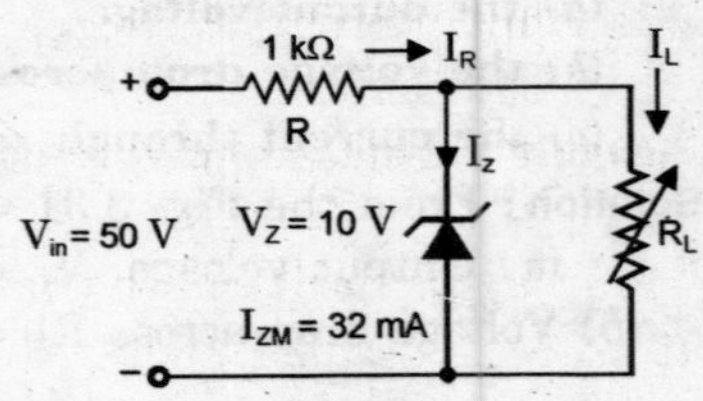

Fig. 3.35

Load current I_L will be maximum when zener current $I_Z = 0$

$$\text{So } I_{L\,max} = I_R - I_{Z\,min} = 40 - 0 = 40\text{ mA}$$

Corresponding load resistance will be minimum and

$$R_{L\,min} = \frac{V_L}{I_{L\,max}} = \frac{V_Z}{I_{L\,max}} = \frac{10}{40 \times 10^{-3}} = 250\ \Omega$$

Load current I_L will be minimum when I_Z is maximum *i.e.* 32 mA.

$$\text{So } I_{L\,min} = I_R - I_{Z\,max} = 40 - 32 = 8\text{ mA}$$

Corresponding load resistance will be maximum and

$$R_{L\,max} = \frac{V_L}{I_{L\,min}} = \frac{10}{8 \times 10^{-3}} = 1.25\text{ k}\Omega$$

Hence range of R_L : From 250 Ω to 1.25 kΩ **Ans.** and range of I_L : From 8 mA to 40 mA **Ans.**

Example 3.6. For the circuit shown is Fig. 3.36, determine, V_L, I_L, I_Z and I_R with R_L = 200 Ω and R_L = 50 Ω. Comment on the operation of the circuit.

V_{in} = 20 V

R_S = 220 Ω

V_Z = 10 V

$P_{Z\,max}$ = 400 mW.

(*i*) R_L = 50 Ω

(*ii*) R_L = 200 Ω

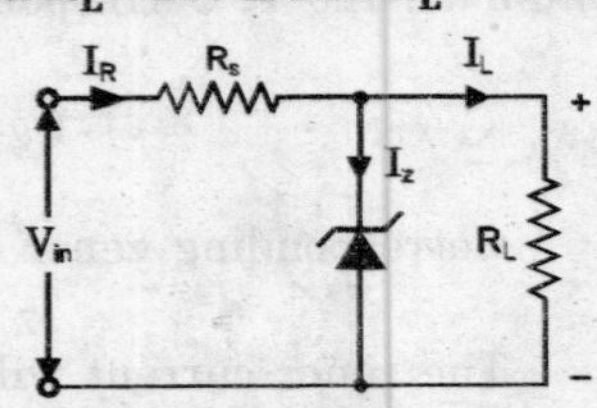

Fig. 3.36

Solution : (*i*) For $R_L = 200\ \Omega$

Voltage across the load, if zener diode is open-circuited,

$$V_L = \frac{V_{in} \times R_L}{R_S + R_L} = \frac{20 \times 200}{220 + 200} = 9.52\ V$$

Since voltage across load, V_L is less than breakdown voltage of zener diode *i.e.* $V_L < V_Z$, the zener diode will not conduct and $V_L = 9.52$ V **Ans.**

$$I_L = I_R = \frac{V_{in}}{R_S + R_L} = \frac{20}{220 + 200} = \frac{1}{21}\ \text{A or 47.62 mA} \quad \textbf{Ans.}$$

$$I_Z = 0 \quad \textbf{Ans.}$$

(*ii*) For $R_L = 50\ \Omega$

Voltage across load, if zener diode is open-circuited

$$V_L = \frac{V_{in} \times R_L}{R_S + R_L} = \frac{20 \times 50}{220 + 50} = 3.7\ V$$

Since voltage across load, V_L is less than breakdown voltage of zener diode *i.e.* $V_L < V_Z$, the zener diode will not conduct and $V_L = 3.7$ V **Ans.**

$$I_L = I_R = \frac{V_{in}}{R_S + R_L} = \frac{20}{220 + 50} = \frac{2}{27}\ \text{A or 74.07 mA} \quad \textbf{Ans.}$$

$$I_Z = 0$$

Example 3.7. Describe the physical mechanism for zener breakdown. For the circuit shown in Fig. 3.37, find the voltage drop across the resistance of 5 kΩ.

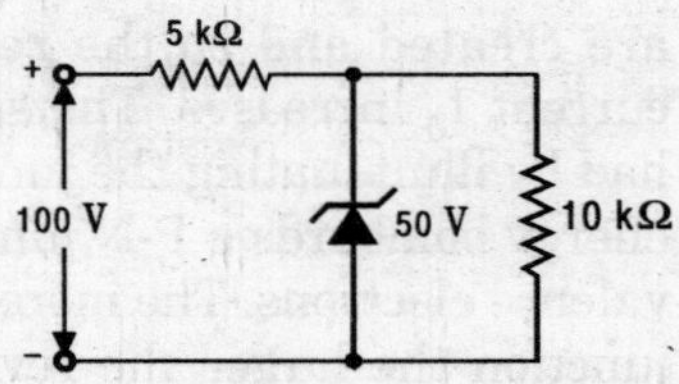

Fig. 3.37

Solution : Removing the zener diode from the circuit, we have

$$V = \frac{10}{10 + 5} \times 100 = 66.67\ V \qquad \therefore\ V > V_Z\ (i.e.\ 50\ V)$$

So zener diode is in ON state.

Thus the voltage drop across the 10 kΩ load resistor

$$V_L = V_Z = 50\ V$$

Applying kVL in the circuit we have

$$100 - V_R - 50 = 0$$

or $\quad V_R = 50$ V

Hence the voltage drop across the 5 kΩ resistor $= 50$ V **Ans.**

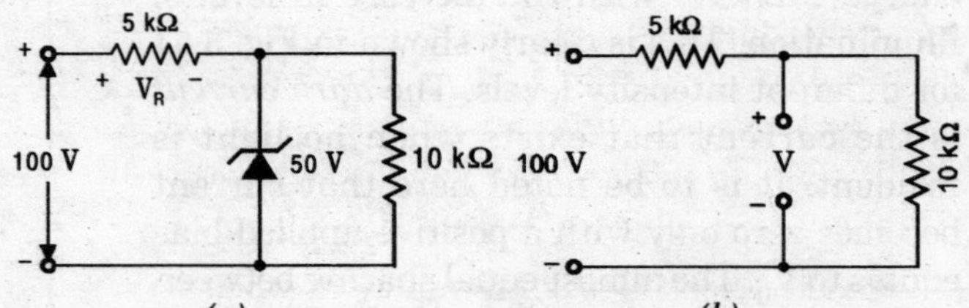

(*a*) (*b*)

Fig. 3.38

Example 3.8. The zener diode regulator circuit, shown in Fig. 3.39, has a fixed voltage drop of 12 V across it as long as I_Z is maintained between 20 mA and 200 mA. Find R_s so that V_S remains 12 V while V_{DC} varies from 15 V to 19.5 V.

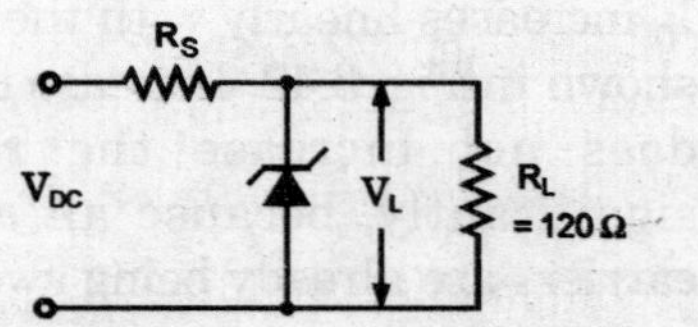

Fig. 3.39

Solution : From Fig. 3.40, we have

$$\text{Load current, } I_L = \frac{12\ V}{120\ \Omega} = 100\ \text{mA} \qquad ...(i)$$

For $V_{DC(min)} = 15$ V

Voltage drop across R_S, $V_{S(min)} = V_{DC(min)} - V_Z$

$$= (15 - 12)\ V = 3\ V \qquad ...(ii)$$

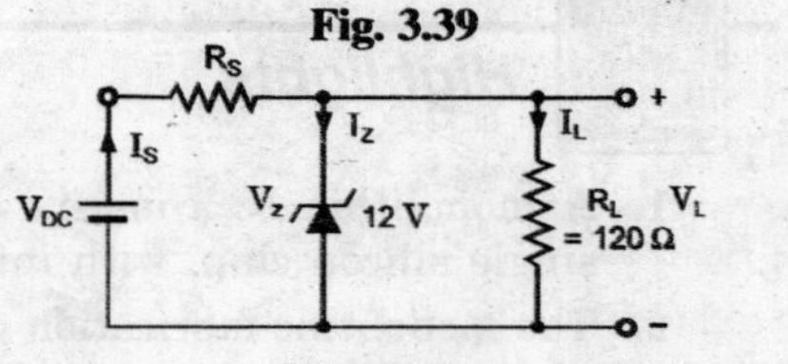

Fig. 3.40

$$\text{For } V_{DC_{(max)}} = 19.5 \text{ V}$$

$$\text{Voltage drop across } R_S, V_{S_{(max)}} = V_{DC_{(max)}} - V_Z = (19.5 - 12 \text{ V}) = 7.5 \text{ V} \quad ...(iii)$$

$$\text{Minimum current through } R_S = I_{Z(min)} + I_L$$

$$\text{or } I_{S_{(min)}} = 20 \text{ mA} + 100 \text{ mA} = 120 \text{ mA}$$

$$\text{Source resistance, } R_S = \frac{V_{S(min)}}{I_{S(min)}} = \frac{3 \text{ V}}{120 \text{ mA}} = 25 \ \Omega \textbf{ Ans.}$$

$$\text{Similarly for } V_{DC_{(max)}}; R_S = \frac{V_{S(max)}}{I_{Z(max)} + I_L} = \frac{7.5 \text{ V}}{(200 + 100) \text{ mA}} = 25 \ \Omega \textbf{ Ans.}$$

3.8. ELECTRICAL AND OPTICAL EXCITATION IN DIODES

When the P-N junction is reverse-biased, a reverse saturation current flows due to thermally generated holes and electrons being swept across the junction as the minority carriers. With the increase in temperature of the junction more and more hole-electron pairs are created and so the reverse saturation current I_0 increases. The same effect can be had by illuminating the junction. When light energy bombards a P-N junction, it dislodges valence electrons. The more light striking the junction the larger the reverse current in a diode. It is due to generation of more and more charge carriers with the increase in level of illumination. This is clearly shown in Fig. 3.41 for different intensity levels. The *dark current* is the current that exists when no light is incident. It is to be noted here that current becomes zero only with a positive applied bias equals to V_0. The almost equal spacing between the curves for the same increment in luminous flux reveals that the reverse saturation current I_0 increases linearly with the luminous flux as shown in Fig. 3.42. Increase in reverse voltage does not increase the reverse current significantly, because all available charge carriers are already being swept across the junction.

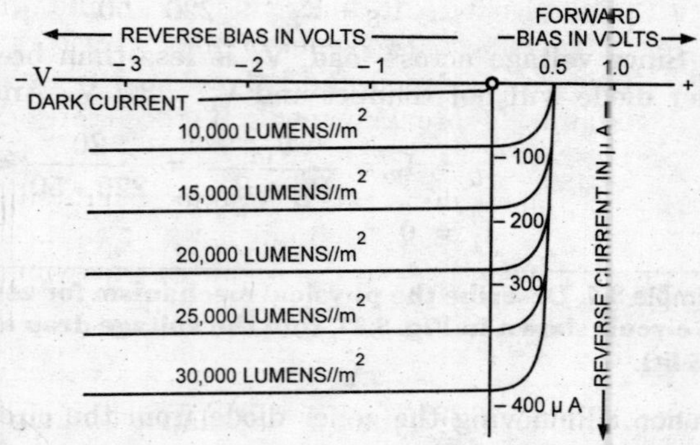

Fig. 3.41. ***Photodiode Characteristics***

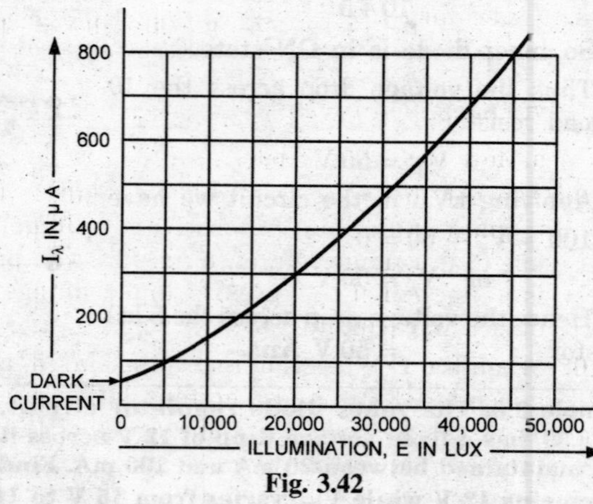

Fig. 3.42

Highlights

1. A monolithic IC consists of active and passive components formed by diffusion into a single silicon chip, with interconnections provided by an aluminium metallization process.
2. The monolithic fabrication process consists of wafer preparation, epitaxial growth, formation of insulation layer over the entire surface, photolithographic process, isolation diffusion, base and emitter diffusion, aluminium metallization, scribing and mounting.

3. Diodes are obtained by using the P-N junctions available in a diffused transistor structures.
4. Metal-semiconductor junctions are those junctions which are fabricated by forming a compatible contact between a metal and a semiconductor. These junctions may be *rectifying (non-ohmic) contact type* or a *non-rectifying (ohmic) contact type.* The rectifying (non-ohmic) contact type metal-semiconductor is formed by depositing a metal film on a semiconductor surface followed by making of contact photolithographically. On the other hand, the ohmic contacts are made by heavy doping of the semiconductors in contact region.
5. There are two mechanisms by which breakdown can occur at a reversed biased P-N junction : avalanche breakdown and zener breakdown. The breakdown phenomenon is reversible and harmless so long as the safe operating temperature is maintained.
Either of the two (zener breakdown or avalanche breakdown) may occur independently, or both of these may occur simultaneously. Diode junctions that breakdown below 5 V are caused by *zener effect.* Junctions that experience breakdown above 5 V are caused by *avalanche effect.* Junctions that breakdown around 5 V are usually caused by combination of two effects. The zener breakdown occurs in heavily doped junctions (P-type semiconductor moderately doped and N-type heavily doped), which produce narrow depletion layers. The avalanche breakdown occurs in lightly doped junctions, which produce wide depletion layers.
6. A zener diode is a P-N junction operated in the reverse-biased mode to take advantage of its sharply defined breakdown voltage. The direction of conduction for a zener diode is opposite to that of the arrow in the symbol, and the zener voltage has a polarity opposite to that of a forward-biased diode.
7. The zener voltage V_Z is specified at some test value of current I_{ZT}, at which the diode will exhibit some dynamic impedance Z_Z known as zener impedance and given by $Z_Z = \frac{\Delta V_Z}{\Delta I_Z}$ where ΔV_Z and ΔI_Z are the small variations in voltage and current respectively.
8. Zener diodes have a temperature coefficient, which generally is negative for V_Z below 5 V but positive above 6 V, and is expressed in percent of V_Z per °C with the change in zener voltage given by E_g. $\alpha_Z = \frac{\Delta V_Z}{V_Z(t - t_0)} \times 100$. Reference zener diodes are available with α_Z as low as 0.0005% per °C.
9. A zener diode may be used to regulate the load voltage at the value V_Z by acting as a "bypass value" to counteract line voltage or load current variations. Zener diodes are also useful in voltage reference circuits. The percent regulation, a figure of merit for the circuit, is a function of the range of input voltage and load resistance values, and of the individual device parameters.
10. When the P-N junction is reverse-biased, a reverse saturation current flows due to thermally generated holes and electrons being swept across the junction as the minority carriers. With the increase in temperature of the junction more and more hole-electron pairs are created and so the reverse saturation current I_0 increases. The same effect can be had by illuminating the junction. When light energy bombards a P-N junction, it dislodges valence electrons. The more light striking the junction the larger the reverse current in a diode. It is due to generation of more and more charge carriers with the increase in level of illumination.

Exercises

1. Explain the fabrication process of monolithic ICs. Discuss the advantages and disadvantages of monolithic ICs.

2. Explain photolithographic process used in monolithic IC production.
3. Explain in detail planar technology for device fabrication. **[M.D. Univ. Third Semester December 2005]**
4. Explain in detail the various steps involved in planar technology for device fabrication. **[M.D. Univ. Third Semester May 2005; May 2006; December 2006]**
5. Explain briefly planar technology for device fabrication. **[M.D. Univ. Third Semester December 2007]**
6. Discuss in brief planar technology of device fabrication. **[M.D. Univ. Third Semester May 2007]**
7. Using planar technology write down various steps required for fabrication of NPN transistor. **[M.D. Univ. Third Semester December 2007]**
8. What is metal-semiconductor junction ? How rectifying and non-rectifying contact type metal semiconductor junctions are formed ? Explain.
9. Explain various types of breakdown mechanism in P-N junction. **[M.D. Univ. Third Semester May 2006]**
10. Differentiate between Zener breakdown and Avalanche breakdown. **[M.D. Univ. Third Semester December 2006]**
11. Explain the breakdown mechanism in a diode. **[M.D. Univ. Third Semester December 2007]**
12. Explain Zener and Avalanche breakdown mechanisms in a P-N junction diode. **[M.D. Univ. Third Semester May 2008]**
13. Explain with reference to the zener diode characteristic, zener voltage, breakdown current and dynamic impedance. How does the dynamic impedance vary with zener current ? Draw the equivalent circuit of an ideal zener diode in the breakdown region.
14. Breakdown diodes with a breakdown voltage greater than 5 V generally have a positive temp. coefficient but those with lower breakdown voltage have a negative temp. coefficient. Do you agree with this statement ? Justify.
15. Explain the construction and working of zener diode. **[M.D. Univ. Third Semester May 2006]**
16. What is a zener diode ? How does a zener diode work as a regulator ? **[M.D. Univ. Third Semester May 2007]**
17. Discuss optical and electrical excitation in diodes. **[M.D. Univ. Third Semester December 2006]**
18. Write short notes on the following :
    (*i*) Planar technology of device fabrication. **[M.D. Univ. Third Semester December 2006]**
    (*ii*) Metal semiconductor junction.
    (*iii*) Breakdown mechanism in P-N junctions.
    (*iv*) Zener diode. **[M.D. Univ. Third Semester December 2005; May 2008]**
    (*v*) Optical and electrical excitation in diodes.

Short Answer Type Questions With Answers

Q. 1. What is monolithic IC ?

Ans. A monolithic IC is one in which all circuit components and their interconnections are formed on a single thin wafer, called the substrate.

Q. 2. Why is SiO_2 layer formed over the entire surface in a monolithic IC ?

Ans. SiO_2 layer is formed over the entire surface so as to prevent the contamination of the epitaxial layer.

Q. 3. How is SiO_2 layer formed in a monolithic IC ?

Ans. The SiO_2 layer is grown by exposing the epitaxial layer to an oxygen atmosphere to about 1,000°C.

Q. 4. In ICs the substrate is not employed as collector. Why ?

Ans. If in ICs the substrate is employed as collector, all transistors fabricated on one substrate would have their collectors connected together.

Q. 5. Why is the diffusion technique of formation of resistors most widely used ?

Ans. The diffused resistors can be processed while diffusing transistors, so the diffusion technique is the cheapest and, therefore, is most widely used.

Q. 6. What is metal-semiconductor junction ?

Ans. Metal-semiconductor junction is that which is fabricated by forming a compatible contact between metal and a semiconductor.

Q. 7. What are the types of metal semiconductor junctions ?

Ans. Metal-semiconductor junctions are of two types viz. rectifying (non-ohmic) contact type and non-rectifying (ohmic) contact type.

Q. 8. How are the ohmic and non-ohmic contact type metal-semiconductor junctions formed ?

Ans. The ohmic contacts are made by heavy doping of the semiconductors in contact region. On the other hand non-ohmic contacts are formed by depositing a metal film on a semiconductor surface followed by making of contact photolithographically.

Q. 9. What are the two mechanisms of breakdown in a P-N junction ?

Ans. Avalanche and zener breakdowns.

Q. 10. Name the breakdown mechanism in a lightly doped P-N junction under reverse biased condition.

Ans. Avalanche breakdown.

Q. 11. Name the breakdown mechanism in a heavily doped P-N junction under reverse biased condition.

Ans. Zener breakdown.

Q. 12. What is a zener diode ?

Ans. Zener diode is a P-N junction diode specially designed for operation in its breakdown region.

Q. 13. What is zener voltage ?

Ans. The voltage at which the zener diode breaks down is called the zener voltage.

Q. 14. What is meant by the temperature coefficient ?

Ans. The effect of temperature on zener voltage is given in terms of temperature coefficient which is defined as the percentage change in nominal zener voltage for each degree centigrade of change in junction temperature.

Mathematically it is given as $\frac{\Delta V_Z}{V_Z(t - t_0)} \times 100$ where ΔV_Z is the change in zener voltage due to temperature variation of $(t - t_0)$.

Q. 15. What happens to the series current, load current and zener current when the dc input voltage of a zener regulator increases ?

Ans. Zener current and series current increase while the load current remains unchanged.

Q. 16. Why is zener diode used as voltage regulator ?

Ans. Zener diode has the property of behaving like a dc battery in 'on' state (*i.e.* when the voltage across zener diode exceeds its zener voltage rating V_Z). In 'on state', the voltage across zener diode remains constant until the voltage across it drops less than V_Z. This property of zener diode makes its use as voltage regulator.

Q. 17. Explain how zener diode maintains constant voltage across the load ?

Ans. Zener diode has the property of behaving like a dc battery in 'on' state. If the zener diode is shunted across the load R_L and the voltage across zener diode is more than zener voltage V_Z then zener diode is in 'on' state, and any variation in voltage across the zener diode due to variations either in supply voltage or in load resistance is not able to change the output voltage. Thus zener diode maintains voltage constant across the load.

Problems

1. A zener diode shunt regulator is shown in Fig. 1.

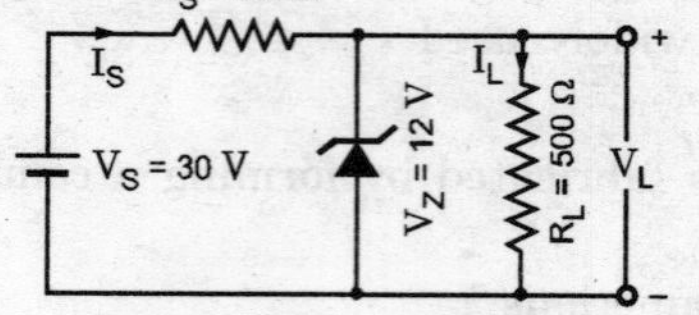

Fig. 1

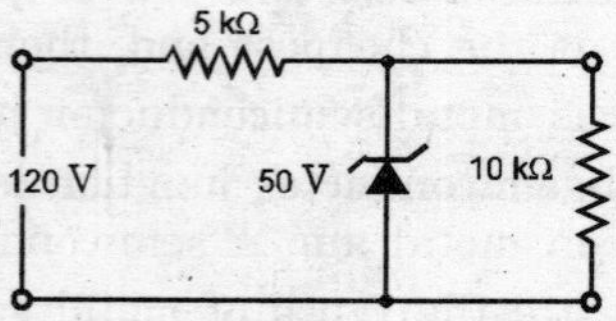

Fig. 2

Determine

(*i*) load voltage,

(*ii*) voltage drop across series resistance,

(*iii*) current through the diode. **[Ans.** (*i*) 12 V (*ii*) 18 V (*iii*) 51 mA]

2. For the circuit shown in Fig. 2 find

(*a*) the output voltage,

(*b*) the voltage drop across series resistance,

(*c*) the current through zener diode. **[Ans.** (*a*) 50 V (*b*) 70 V (*c*) 9 mA]

3. Determine the maximum and minimum values of zener current if in the circuit shown in Fig. 3.26 the load resistance R_L = 5 kΩ, series resistance is 2 kΩ, input voltage varies between 80 V and 100 V and the output voltage is 50 V. **[Ans.** 15 mA and 5 mA]

4. In the circuit shown in Fig. 3.26 the series resistance is 2 kΩ, load resistance is 5 kΩ, input voltage is 60 V and output voltage is 35 V. Determine (*i*) voltage drop across series resistance (*ii*) load current (*iii*) current through series resistance and (*iv*) zener diode current.

[Ans. (*i*) 25 V (*ii*) 7.0 mA (*iii*) 12.5 mA (*iv*) 5.5 mA]

5. Determine the temperature coefficient of a 7.5 V zener diode at 25°C if the nominal voltage increases to 7.8 V at a temperature of 105°C. **[Ans.** + 0.05%]

6. For the circuit of Fig. 3, find (*a*) the output voltage (*b*) voltage across R_S and (*c*) the current through zener diode. **[Ans.** (*a*) 8 V (*b*) 4 V (*c*) 0 V]

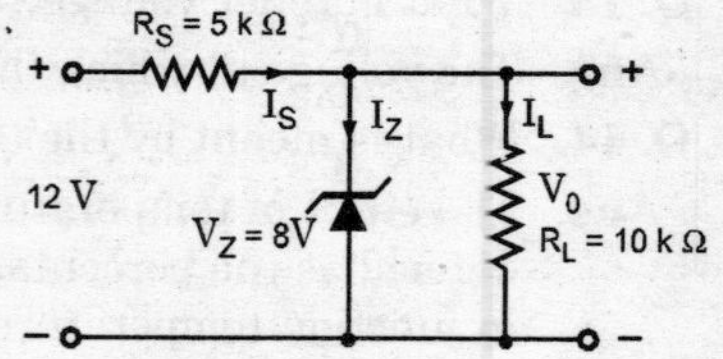

Fig. 3

❑❑❑

SECTION-C

Transistors

Contents of This Section

BIPOLAR JUNCTION TRANSISTORS : Fundamentals of BJT Operation; Amplification with BJTs.

FIELD-EFFECT TRANSISTORS : Metal-Semiconductor Field-Effect Transistors (MESFETs); Metal-Insulator Semiconductor Field-Effect Transistors (MISFETs); Metal-Oxide Semiconductor Field-Effect Transistors (MOSFETs) : Construction, Operation and Characteristics of Above Devices.

CHAPTER 4

Bipolar Junction Transistors (BJTs)

Inside This Chapter

4.1. INTRODUCTION

The transistor is a solid state device, whose operation depends upon the flow of electric charge carriers within the solid. Transistor is capable of amplification and in most respect it is analogous to a vacuum triode. The main difference between the two is that the transistor is a current controlled device whereas vacuum triode is a voltage controlled device. The transistor is only about 6 decade old, yet it is replacing vacuum tubes in almost all applications. The reasons are obviously its advantages over vacuum tubes such as compact size, light weight, rugged construction, more resistive to shocks and vibrations, instantaneous operation (no heating required), low operating voltage, high operating efficiency (no heat loss) and long life with essentially no ageing effect if operated within permissible limits of temperature and frequency. However, transistors, in comparison to vacuum triodes, have some drawbacks also such as loud hum noise, restricted operating temperature (up to 75°C) and operating frequency (up to a few MHz only).

It consists of a silicon or germanium (preferably silicon because of its smaller cut-off current I_{CBO}, smaller variations in I_{CBO} due to variations in temperature and higher operating temperature) crystal in which a layer of N-type material is sandwiched between two layers of P-type material, as shown in Fig. 4.1 (*a*). Alternatively a transistor may consist of a layer of P-type material sandwiched between two layers of N-type material, as shown in Fig. 4.1 (*b*). In the former case the transistor is referred to as a P-N-P *transistor* and in latter case, as an N-P-N *transistor*. Each type of transistor has two P-N junctions—

one junction between the emitter and base, called the *emitter-base junction* or simply the *emitter junction* and the other junction between the base and collector, called the *collector-base junction* or simply the *collector junction.* Thus a transistor is like two P-N junction diodes connected back to back. Each of these P-N junction diodes have all the characteristics discussed in earlier chapters. The two junctions give rise to three regions provided with three terminals called the *emitter*, *base* and *collector*, as shown in Fig. 4.1. The emitter, base and collector correspond in a general way to the cathode, grid and plate or anode of a vacuum triode.

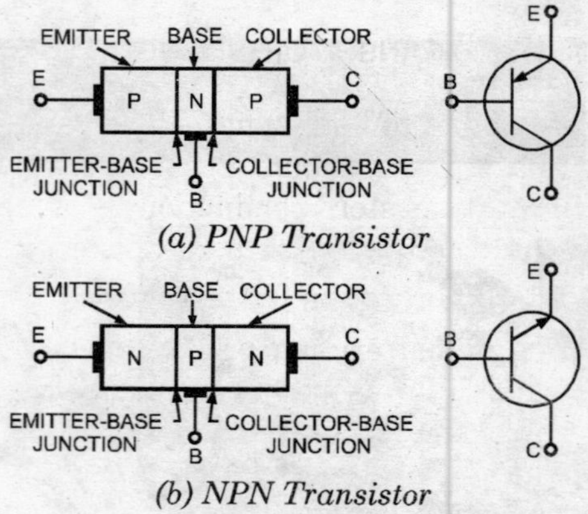

(a) *PNP Transistor*

(b) *NPN Transistor*

Fig. 4.1

4.2. Transistor Terminals

As mentioned above, transistor is a single crystal in which there are two P-N junctions, as shown in Fig. 4.1. The idea behind is to have first section to supply the charges (either holes or electrons) to be collected by the third section through the middle section. One side section supplying free charges is called the *emitter*, other side section collecting these charges is called the *collector* and the middle section which is formed between the emitter and collector is called the *base*.

1. *Emitter*. It is left hand section (or region) of the transistor and its main function is to supply majority charge carriers (electrons in case of NPN transistors and holes in case of P-N-P transistors) to the base. The emitter is always forward biased w.r.t. base so that it is able to supply majority charge carriers to the base. The emitter is heavily doped so that it may be able to inject a large number of charge carriers.

2. *Collector*. It is the right hand section of the transistor and its main function is to collect majority charge carriers. Collector is always reverse biased so as to remove the charge carriers away from its junction with the base. It is moderately doped.

3. *Base*. It is the middle section of the transistor and is very lightly doped and is very thin (of the order of μm) in comparison to either emitter or collector so that it may pass most of the injected charge carriers to the collector.

Base forms two junctions *i.e.* emitter junction and collector junction, each having its own barrier voltage. The forward biased emitter-base junction offers low resistance to the emitter current whereas the collector-base junction, which is reverse biased, offers high resistance to the collector current. As the resistance of emitter-base junction is very small as compared to that of collector-base junction, therefore, the forward bias applied to the emitter-base junction is usually very small whereas the reverse bias on the collector-base junction is much large.

In most of the transistors, the collector region is made physically larger than emitter region because collector is to dissipate much power.* Due to this difference, collector and emitter are not interchangeable. However, for sake of convenience, it is customary to show emitter and collector to be of equal size.

As regards the symbols, arrowhead is always at the emitter. The direction indicates the conventional direction of current flow (from emitter to base in case of P-N-P transistor and from base to emitter in case of NPN transistor).** Generally, no arrow head is marked for collector since its reverse leakage current is always opposite to the direction of emitter current.

* During operation of a transistor, much heat is generated at the collector junction.

** In a P-N-P transistor emitter is positive w.r.t. base while in case of an N-P-N transistor base is positive w.r.t. emitter.

4.3. TRANSISTOR ACTION

Transistors of both types (P-N-P and N-P-N) behave exactly in the same way except change in biasing and majority carriers. In P-N-P transistors the conduction is by holes whereas in N-P-N transistors conduction is by electrons. However, N-P-N transistors are preferred due to their better high frequency response.

4.3.1. Unbiased Transistor

When no battery is connected between the different terminals of a transistor, the transistor is said to be in an unbiased state or in an open-circuit state.

We know that when a semiconductor is doped with donor or pentavalent impurities, it becomes an N-type semiconductor. The additional free electrons tend to raise the Fermi (average electron energy) level within the crystal. Similarly, when a semiconductor is doped with acceptor or trivalent impurities, it becomes a P-type semiconductor. The additional holes tend to lower the Fermi level. This is illustrated in Fig. 4.2 (*b*).

(a) Depletion Regions

(b) Isolated Semiconductor Crystals

(c) Energy Diagram of An N-P-N Transistor

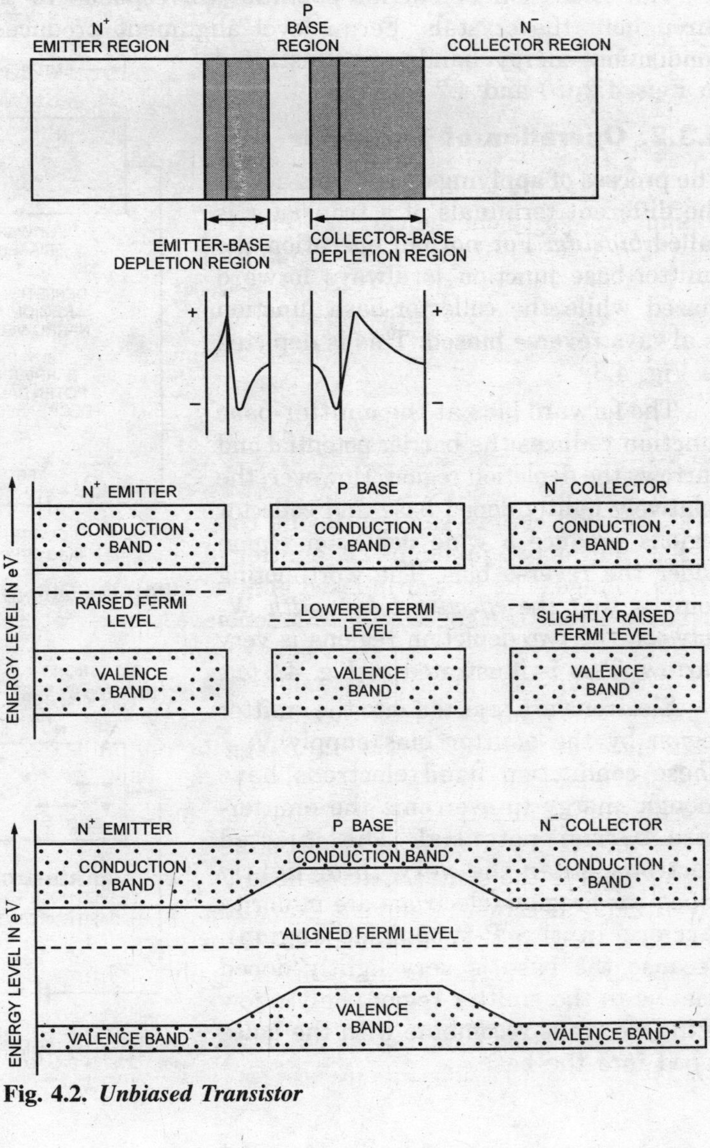

Fig. 4.2. *Unbiased Transistor*

We also know that when an N-type region is formed next to a P-type region, a barrier potential (0.7 V in case of Si and 0.3 V in case of Ge at room temperature) is produced at the P-N junction. The free electrons in the N-region diffuse into the adjacent P-region to annihilate holes. Consequently, a layer of positive ions is formed on the N side, and a layer of negative ions is formed on the P-side. This serves to create the barrier potential.

The same phenomenon occurs at both the P-N junctions found in the N-P-N transistor structure [Fig. 4.2 (*a*)]. The depletion regions (ionization layers), which are formed, extend into the semiconductor crystal as a function of the doping levels. Specifically, it extends deeply into the lightly doped regions, and slightly into the heavily doped regions. Since emitter is heavily doped than base and collector, the emitter-base depletion region width is smaller than that of collector-base depletion region width. Note that the barrier potentials are negative on the base and positive on the emitter and collector for the N-P-N device. In case of P-N-P device it would be just reverse *i.e.*, positive on the base and negative on emitter and collector. This is illustrated in Fig. 4.2 (*a*).

The formation of barrier potential corresponds to the alignment of the Fermi levels throughout the crystals. Fermi level alignment produces a distortion in the valence and conduction energy bands, as illustrated in Figs. 4.2 (*b*) and 4.2 (*c*).

4.3.2. Operation of Transistor

The process of applying dc voltages across the different terminals of a transistor is called *biasing*. For normal operation the emitter-base junction is always forward biased while the collector-base junction is always reverse biased. This is depicted in Fig. 4.3.

The forward bias at the emitter-base junction reduces the barrier potential and narrows the depletion region. However, the relatively lightly doped base and collector regions produce a wide depletion region under the reverse bias. The worthnoting point is that the *effective base width* W_b between the two depletion regions is very narrow. This is illustrated in Fig. 4.3 (*a*).

Electrons are injected into the emitter region by the emitter bias supply V_{EB}. These conduction band electrons have enough energy to overcome the emitter-base barrier potential. The injected electrons enter the very thin, lightly doped base region (electrons are minority carriers in the P-type base region). Because the base is very lightly doped relative to the emitter region, only a few of the electrons recombine with the holes doped into the base.

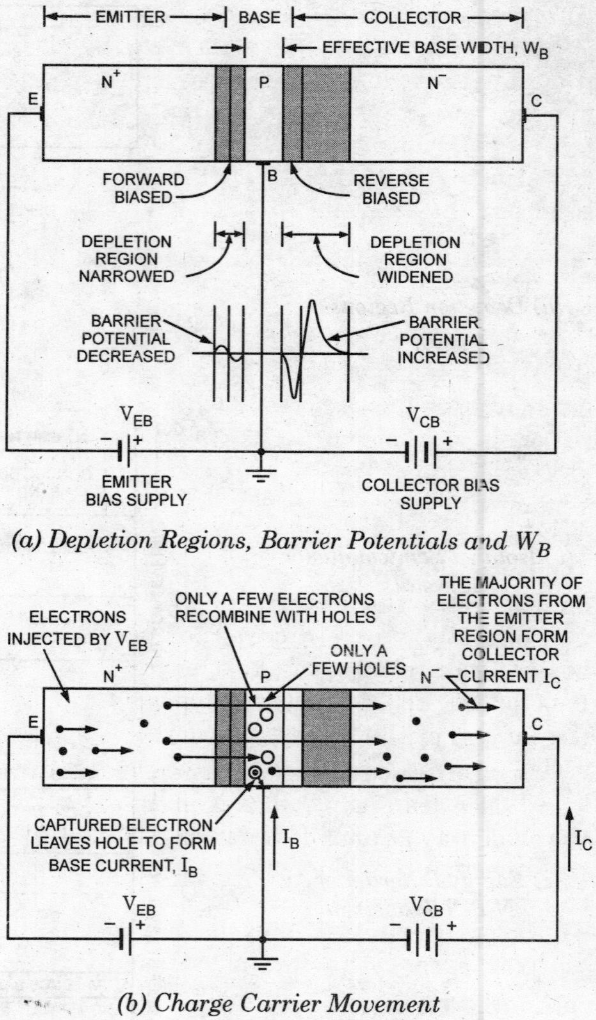

(*a*) *Depletion Regions, Barrier Potentials and* W_B

(*b*) *Charge Carrier Movement*

Fig. 4.3

Injection of electrons makes the electrons concentration on the emitter junction very large and on the collector junction the concentration of the electrons is extremely small and as such the gradient of electron concentration is very large in base. Injected electrons diffuse into collector region due to extremely small thickness of base which is much less than the diffusion length. Most of the electrons cross into the collector region. Collector is reverse biased and creates a strong electrostatic field between base and collector. The field immediately collects the diffused electrons which enter the collector junction. Flow of electrons into the base region when confronted with the holes, a few electrons (say 1 to 5%) combine and neutralize and rest of the electrons say 95 to 99% of the injected electrons diffuse into the collector region and collected by the collector electrode. To maintain the base neutrality base electrode provides equal number of electrons which have combined with holes and results in a base current. Thus the emitter current I_E is equal to the sum of collector current I_C and base current I_B. The ratio of collector current I_C to emitter current I_E is called the α and it is measure of the possible current amplification. In a transistor α cannot be higher than unity, but practical values of 0.95 to 0.99 are attained in commercial transistors.

The operation of the transistor may also be examined by means of its energy band diagram. Since the forward bias tends to align the valence and conduction energy bands, while reverse bias tends to increase their misalignment, the diagram shown in Fig. 4.4 is arrived at. As can be seen from Fig. 4.4, electrons must lose energy to travel from the emitter region to the base region. Further, the electrons must also lose energy in moving from base region into collector region. Only a few of the electrons in the base region lose energy to recombine with the (valence band) hole. Typically, at least 95% of the electrons have a long enough lifetime to diffuse into the collector region.

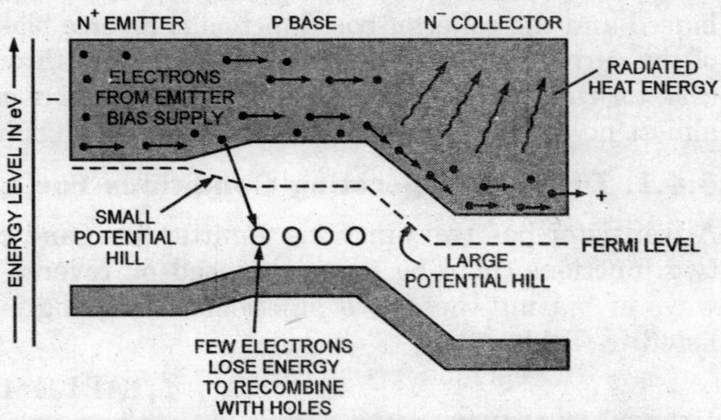

Fig. 4.4. ***Energy Band Diagram For an N-P-N Transistor in its Active Region***

It is observed from Fig. 4.4 that the potential hill (bend in the Fermi level) between the base and the collector region is very steep. Specifically, the electrons must lose a considerable amount of energy as they move from the base region to the collector region. Typically, the electrons give up their energy in the form of heat. Consequently, the collector region must be able to dissipate this heat energy. It is for this reason that it is typically the largest of the three regions.

Operation of P-N-P Transistor. The P-N-P transistor behaves exactly the same way as an N-P-N device, with the exception that the majority charge carriers are holes. As illustrated in Fig. 4.5, holes are emitted from the P-type emitter across the forward-biased

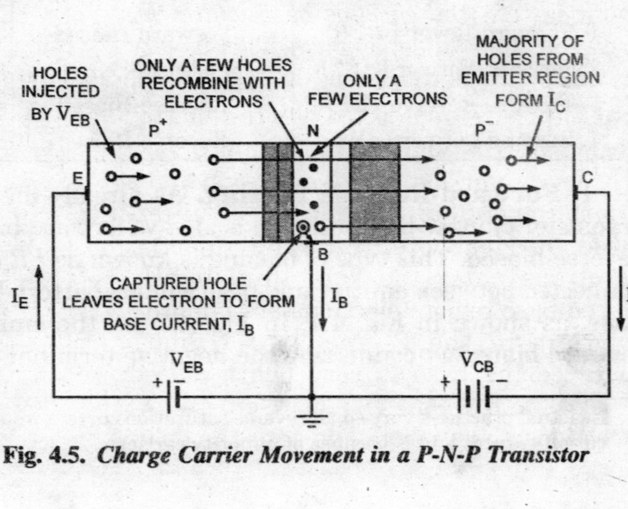

Fig. 4.5. ***Charge Carrier Movement in a P-N-P Transistor***

emitter-base junction into the base. In the lightly doped N-type base, the holes find few electrons to recombine with. Some of the holes flow out via the base terminal, but most are drawn across the collector by the positive-negative electric field at the reverse-biased collector-base junction. As in the case of NPN device, the forward bias at the emitter base junction controls the collector and emitter currents.

4.4. Transistor Biasing

As already mentiond above, for normal operation of a transistor emitter-base junction is always forward biased and collector-base junction is always reverse-biased.

For a moment if the biasing of emitter-base junction is ignored and only collector-base junction, which is reverse biased [Fig. 4.6 (*a*)] is considered, then due to high resistance of reverse biased collector diode, practically* no current would flow in the collector circuit.

Now if the biasing of collector-base junction is ignored and only emitter-base junction, which is forward biased [Fig. 4.6 (*b*)] is considered then practically whole of the emitter current flows through the base *i.e.* $I_B = I_E$.

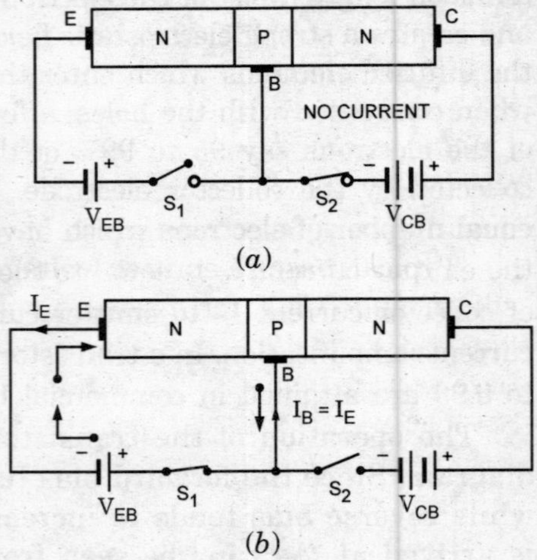

Fig. 4.6

However, if the emitter-base junction is forward biased and the collector-base junction is reverse biased it is seen that almost whole (about 95%) of the current constituted by the emitter flows through collector circuit. Thus the magnitude of collector current I_C depends upon the magnitude of emitter current I_E. No emitter current means almost no collector current. This is precisely what happens in a transistor.

4.4.1. Different Operating Conditions For a Transistor

A transistor has two junctions (emitter-base and collector-base junctions), and each of these two junctions, may be forward-biased or reverse-biased, therefore, there are four possible ways of biasing these two junctions. Accordingly it may operate in different conditions as listed in Table 4.1.

TABLE 4.1

Condition	*Emitter-Base (EB) Junction*	*Collector-Base (CB) Junction*	*Region of Operation*
1. Forward-Reverse (FR)	Forward-Biased	Reverse-Biased	Active
2. Forward-Forward (FF)	Forward-Biased	Forward-Biased	Saturation
3. Reverse-Reverse (RR)	Reverse-Biased	Reverse-Biased	Cut-off
4. Reverse-Forward (RF)	Reverse-Biased	Forward-Biased	Inverted

1. Forward-Reverse Biasing. As already mentioned above, for normal operation of a transistor emitter-base junction is always forward biased and collector-base junction is always reverse-biased. This type of biasing is known as *FR biasing*. For this purpose a battery V_{EB} is connected between emitter and base while a battery V_{CB} is connected between the collector and base, as shown in Fig. 4.7. In Fig. 4.7 (*a*), the emitter-base junction of N-P-N transistor is forward biased by connecting the negative terminal of battery V_{EB} to the emitter and positive

* In actual practice a very small reverse saturation current would flow in the collector circuit. This is called collector cut-off current and is because of minority carriers.

terminal to the base while collector-base junction is reverse-biased by connecting the positive terminal of the battery V_{CB} to the collector and negative terminal to the base. Similarly in Fig. 4.7 (*b*) the emitter-base junction of P-N-P transistor is forward biased by connecting the positive terminal of battery V_{EB} to the emitter and negative terminal to the base while collector-base junction is reverse-biased by connecting the negative terminal of battery V_{CB} to the collector and positive terminal to the base.

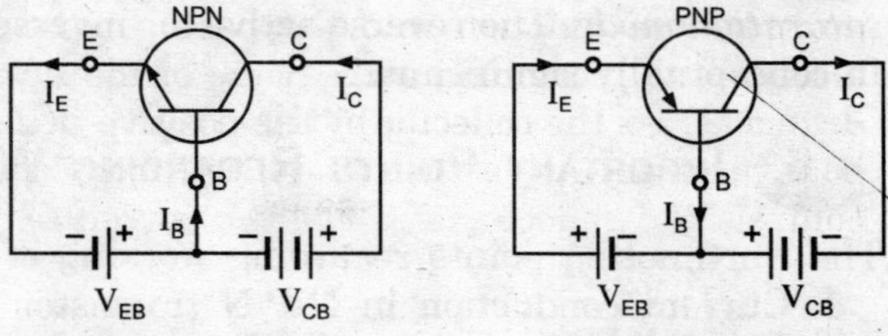

(*a*) *NPN Transistor* (*b*) *PNP Transistor*

Fig. 4.7. ***Active Region***

Typical base-to-emitter voltages for both N-P-N and P-N-P transistors are 0.7 V for silicon and 0.3 V for germanium. Typical collector-to-base voltages might be anything between 3 V and 20 V for most types of transistors, although in many cases the collector-to-base voltage may exceed 20 volts even.

The voltages across the terminals of a transistor follow the usual double subscript voltage convention. For example V_{BE} represents the voltage of base with respect to the emitter. If V_{BE} is equal to 0.7 V it means the base is 0.7 V more positive.

In FR biasing the transistor is in active region and the collector current I_C, depends upon the emitter current I_E. Generally transistor is operated in this region for amplification.

For connecting batteries properly in the transistor circuits the following scheme may be remembered.

The battery polarities to the emitter and base are the same as the letter of the emitter and base impurity designations. Thus, in the P-N-P type, the emitter receives the P (positive) terminal, and the base receives the N (negative) terminal. Similarly, in N-P-N type, the emitter receives the N (negative) terminal, the base receives the P (positive) terminal. The collector receives a battery polarity opposite to its impurity type designation. Thus P-type collector receives the N (negative) terminal of the battery and the N-type collector receives the P (positive) terminal.

2. Forward-Forward Biasing. In this biasing, both collector and emitter are forward biased, as illustrated in Fig. 4.8. The transistor operates in saturation region. In this region, the collector current becomes independent of the base current. Thus the transistor acts like a close switch.

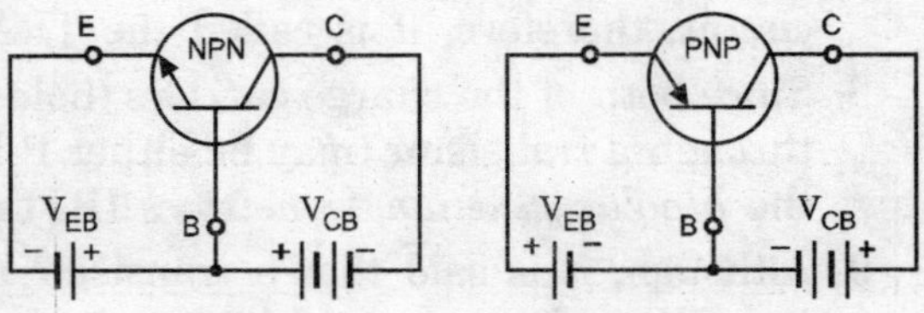

Fig. 4.8. ***Saturation Region***

3. Reverse-Reverse Biasing. In this biasing, both the junctions are reverse biased. In this biasing, the transistor has practically zero current because the emitter does not emit charge carriers into the base and no charge carriers are collected by the collector except a few thermally generated minority carriers. Thus, the transistor acts like an open switch.

4. Reverse-Forward Biasing. In this biasing, the emitter junction is reverse biased while the collector junction is forward biased. As the collector is not doped to the extent as the emitter is doped, therefore, it cannot inject the majority charge carriers to the base. So, in this region, the action of transistor is poor. This region is of little importance.

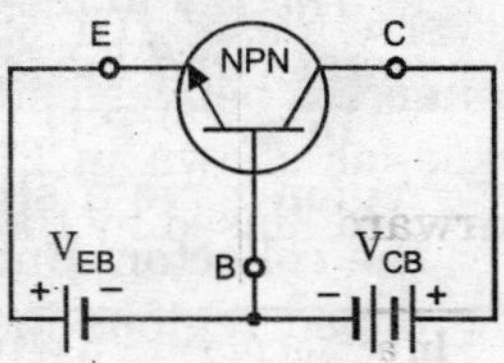

Fig. 4.9. ***Cut-off Region***

The *active mode* which is also called forward active mode, is the one used if the transistor is to operate as an amplifier. Switching operations (*e.g.*, logic circuits) utilize both the *cutoff mode* and the *saturation mode*. The reverse active (or inverse active) mode has very limited application but in conceptually significant.

4.5. Important Points Regarding Working Of Transistors

The worthnoting points regarding working of transistors are summarized below.

1. Current conduction in N-P-N transistor is by electrons and the conventional current flow will be in the opposite direction. Current conduction in P-N-P transistor is by holes but in external leads, the current will be by flow of electrons.
2. The collector current I_C is always less than emitter current I_E, being the difference of emitter current I_E and base current I_B.
3. The base current is only a small fraction (usually 5%) of emitter current.
4. As a standard convention, all the currents entering into the transistor are considered to be positive. Conversely currents that flow out of the transistor are taken to be negative. It means that if the actual conventional current flows in the outward direction, a negative sign is included along with its magnitude. Thus in an N-P-N transistor, the emitter current I_E is to be taken negative because it flows out of the transistor [Figs. 4.3 (*b*) and 4.7 (*a*)] while both the base current I_B and collector current I_C are to be taken positive because they flow into the transistor. Similarly in a P-N-P transistor I_E is to be taken positive and both base current I_B and collector current I_C are to be taken negative [Figs. 4.5 and 4.7(*b*)]. However, to avoid confusion, the actual direction of flow of current is indicated in the diagrams.
5. Emitter-base junction is always forward-biased and collector-base junction is always reverse-biased for active region of operation.
6. The input circuit (*i.e.* emitter-base junction), because of forward bias, offers low resistance and so needs usually very small bias (approximately 0.7 V for Si and 0.3 V for Ge). The output circuit (*i.e.* collector-base junction), because of reverse bias, offers high resistance and, therefore, needs much higher bias (3 to 20 V).
7. Transistor transfers the input signal from a low resistance circuit to a high resistance circuit, therefore, it is called the TRANsfer reSISTOR (TRANSISTOR).
8. Since both of the charge carriers (holes as well as electrons) are involved in current flow through a transistor (may be either P-N-P or N-P-N), these devices are sometimes called the *bipolar junction transistors* (BJTs).
9. Although, it is said that a transistor is a device with two PN diodes connected back-to-back, but it does not mean that two discrete diodes connected back-to-back, as shown in Fig. 4.10, can work as a transistor. In such a case, each diode has two equally doped regions, so that the overall circuit has four equally doped regions. This would not work as base region is not the same as in a transistor.

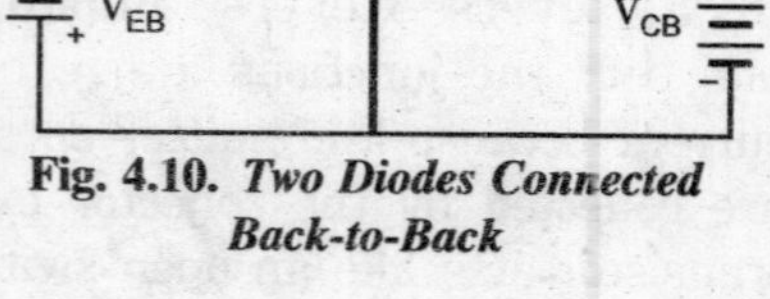

Fig. 4.10. ***Two Diodes Connected Back-to-Back***

The key to the transistor action is the lightly doped thin base between the heavily doped emitter and moderately doped collector.

In an N-P-N transistor, the free electrons passing through the base to the collector region have a short lifetime. As long as the base is thin, the free electrons can reach the collector. But in case of two discrete back-to-back connected diodes there are four doped regions instead of three and there is nothing that resembles a thin base region between an emitter and a collector.

Hence two discrete diodes connected back-to-back can never work as a transistor.

10. The choice of N-P-N transistor is made more often because majority charge carriers are electrons whose mobility is much more (almost double) than that of holes.

4.6. Transistor Current Components

The various current components which flow across the forward-biased emitter junction and, reverse-biased collector junction are indicated in Fig. 4.11. The current flowing into the emitter is referred to as the *emitter current* and designated as I_E. The emitter current I_E consists of two constituents (holes current I_{hE} due to crossing of holes from emitter into the base and electron current I_{eE} due to crossing of electrons from base into the emitter).

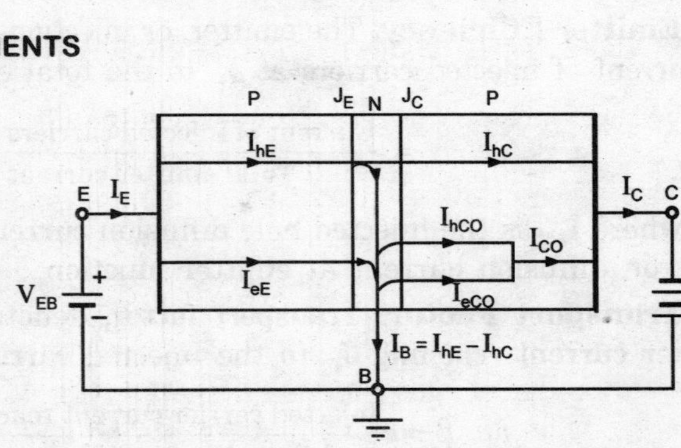

Fig. 4.11. ***Current Components in a PNP Transistor For a Forward-Biased Emitter Junction and a Reverse Biased Collector Junction***

$$i.e.\ I_E = I_{hE} + I_{eE} \quad ...(4.1)$$

In a commercial transistor the emitter is heavily doped and the base is very lightly doped and, therefore, electron current I_{eE} is negligibly small in comparison to hole current I_{hE}. Thus the entire emitter current I_E in a P-N-P transistor is due to flow of holes from emitter to the base. It is to be noted that the indicated direction of I_E external to the transistor is the conventional current direction.

All the holes crossing the emitter junction J_E do not reach the collector junction J_C because some of them combine with the electrons in the N-type base. If I_{hC} is the hole current at collector junction J_C, there must be a bulk recombination current I_{hE} less I_{hC} leaving the base, as shown in Fig. 4.11. In fact electrons enter the base region through the base lead to replenish those electrons which have been lost by recombination with the holes injected into the base across the emitter junction J_E. The holes on reaching the collector junction J_C cross this junction and enter the collector P-region.

If the emitter were open-circuited, I_E would have been zero and then I_{hC} also would have been zero. Under these conditions, the base and collector would act as a reverse biased diode and the collector current I_C would be equal to that reverse saturation current I_{CBO} (or simply I_{CO}).

When the emitter circuit is closed *i.e.* when $I_E \neq 0$, then from Fig. 4.11.

$$\text{Collector current, } I_C = I_{hC} + I_{CO} \quad ...(4.2)$$

For a P-N-P transistor, I_{CO} consists of holes moving across junction J_C from base region to collector region and electrons crossing junction J_C in the opposite direction.

$$\text{Thus } I_{CO} = I_{hCO} + I_{eCO} \quad ...(4.3)$$

In the active region operation with emitter junction J_E forward biased, the collector current is given as

$$I_C = \alpha I_E + I_{CO} \quad ...(4.4)$$

where α is the fraction of the total current I_E which constitutes I_{hC}.

It is to be noted that the indicated direction of emitter current I_E external to the transistor is the conventional current direction. Base current I_B and collector current I_C are also shown external to the transistor conventional current direction. According to the Kirchhoff's first law total current flowing into the transistor must be equal to the total current flowing out of it. So emitter current I_E is equal to the sum of collector and base currents I_C and I_B.

$$i.e.\ I_E = I_C + I_B \quad ...(4.5)*$$

* Some authors taking standard transistor convention, write this equation as $I_E = -(I_C + I_B)$

Let us now define different parameters which relate the current components discussed above.

Emitter Efficiency. The emitter, or injection, efficiency denoted by γ is defined as the ratio of current of injected carriers at J_E to the total emitter current

$$\text{i.e.}\quad \gamma = \frac{\text{Current of injected carriers at } J_E}{\text{Total emitter current}} = \frac{I_{hE}}{I_{hE} + I_{eE}} = \frac{I_{hE}}{I_E} \qquad \text{...(4.6)}$$

where I_{hE} is the injected hole diffusion current at emitter junction and I_{eE} is the injected electron diffusion current at emitter junction.

Transport Factor. Transport factor, denoted by β, is defined as the ratio of injected carrier current reaching J_C to the injected carrier current at J_E

$$\text{i.e.}\quad \beta = \frac{\text{Injected carrier current reaching } J_C}{\text{Injected carrier current at } J_E} = \frac{I_{hC}}{I_{hE}} \qquad \text{...(4.7)}$$

Large Signal Current Gain α. The term α has already been defined above. However, α can alternatively be defined from Eq. (4.4). From Eq. (4.4) α may be defined as the ratio of collector current increment from cut-off value ($I_C = I_{CO}$) to the emitter current increment from cut-off value ($I_E = 0$) to I_E. Thus, we may write

$$\alpha = \frac{I_C - I_{CO}}{I_E - 0} = \frac{I_C - I_{CO}}{I_E} \qquad \text{...(4.8)}$$

Term α is called the *large signal current gain* of a common base transistor. The parameter α is extremely important in transistor theory. It is to be pointed here that α is not constant, but varies with emitter current I_E, collector voltage V_{CB} and the temperature. Typically α lies in the range 0.95 to 0.99.

From Eqs. (4.2) and (4.8) we have

$$\alpha = \frac{I_{hC}}{I_E} = \frac{I_{hC}}{I_{hE}} \times \frac{I_{hE}}{I_E} = \beta\gamma \qquad \text{...(4.9)}$$

$$\because \text{ From Eq. (4.7) } \frac{I_{hC}}{I_{hE}} = \beta \text{ and from Eq. (4.6) } \frac{I_{hE}}{I_E} = \gamma$$

For an efficient transistor we would like β and γ to be very near unity *i.e.* the emitter current should be due mostly to holes (γ = 1) and most of the injected holes should eventually participate in the collector current (β 1).

The product βγ is defined as the factor α, called the *current gain ratio,* which represents the emitter-to-collector current amplification. There is no real amplification between these currents, since α is lesser than unity. On the other hand, the relation between collector current and base current is more promising for amplification.

DC Current Gain. If I_{CO} is negligibly small in comparison to I_C, α approximately equals I_C/I_E. This is referred to as the *dc current gain* of the common base (CB) transistor and is denoted by α_{dc}

$$\text{Thus}\qquad \alpha_{dc} = \frac{I_C}{I_E} \qquad \text{...(4.10)}$$

α_{dc} is always positive and less than unity.

Small Signal Current Gain α_{ac}. Small signal current gain α_{ac} is defined as the ratio of change in collector current to change in emitter current

$$i.e.\ \alpha_{ac} = \frac{\Delta I_C}{\Delta I_E} \text{ for constant collector-base voltage } V_{CB} \quad ...(4.11)$$

α_{ac} is always positive but less than unity (very close to unity). It also varies with I_E, V_{CB} and temperature.

Generalized Expression For Collector Current. In the active region the collector current I_C is given by Eq. (4.4) and is essentially independent of collector-base voltage and depends only on emitter current I_E. For having a generalized expression for I_C (which will be valid not only for reverse-biased collector junction J_C but for any voltage across J_C), I_{CO} will have to be replaced by the current in a P-N diode (that consisting of the base and collector regions). This current is given by the volt-ampere relationship of Eq. (2.99), with I_0 replaced by $-I_{CO}$ and V by V_C, where V_C represents the voltage drop across junction J_C from the P-side to the N-side. The complete expression for I_C may be written as

$$I_C = \alpha I_E + I_{CO}\left[1 - e^{V_C/V_T}\right] \quad ...(4.12)$$

If V_C is negative and is of large magnitude in comparison with that of V_T, Eq. (4.12) reduces to Eq. (4.4).

The physical interpretation of Eq. (4.12) is that the P-N junction diode current crossing the collector junction J_C is supplemented by a fraction α of the current I_E flowing in the emitter.

Example 4.1. A transistor has an emitter current of 10 mA and a collector current of 9.95 mA. Calculate its base current.

Solution : Emitter current, $I_E = 10$ mA

Collector current, $I_C = 9.95$ mA

Base current, $I_B = I_E - I_C = 10 - 9.95 = 0.05$ mA **Ans.**

4.7. CURRENT AMPLIFICATION FACTORS

There are two current amplification factors, the alpha factor (α) and the beta factor (β) defined below :

The alpha factor (α) is also called the *current amplification factor* and is the ratio of collector current (I_C) to emitter current (I_E).

$$\text{In general, } \alpha = \frac{\Delta I_C}{\Delta I_E} \; V_{CB} \text{ remaining constant} \quad ...(4.13)$$

The beta factor (β) is the current gain factor (*also called the transport factor*) of a common emitter circuit and is defined as the ratio of collector current (I_C) and base current (I_B).

$$\text{In general, } \beta = \frac{\Delta I_C}{\Delta I_B} \quad ...(4.14)$$

4.8. RELATIONSHIP BETWEEN α AND β

$$\text{As defined above } \beta = \frac{\Delta I_C}{\Delta I_B}$$

$$\text{and } \alpha = \frac{\Delta I_C}{\Delta I_E} = \frac{\Delta I_C}{\Delta I_C + \Delta I_B} \quad \because \Delta I_E = \Delta I_B + \Delta I_C$$

$$\text{or } \frac{1}{\alpha} = \frac{\Delta I_C + \Delta I_B}{\Delta I_C} = 1 + \frac{\Delta I_B}{\Delta I_C} = 1 + \frac{1}{\beta} = \frac{1+\beta}{\beta}$$

$$\text{Hence } \alpha = \frac{\beta}{1+\beta} \quad \text{and} \quad \beta = \frac{\alpha}{1-\alpha} \qquad ...(4.15)$$

$$\text{Thus, if } \alpha = 0.98 \qquad \beta = \frac{0.98}{0.02} = 49$$

$$\text{and if } \alpha = 0.99 \qquad \beta = \frac{0.99}{0.01} = 99$$

i.e. a small variation in α corresponds to a large variation in β. It is, therefore, better to determine β experimentally and calculate therefrom the corresponding value of α by means of expression

$$\alpha = \frac{\beta}{1+\beta} \qquad ...(4.16)$$

Example 4.2. In a certain transistor, collector current is 0.98 mA and base current is 20 µA. Determine the values of (*i*) emitter current, (*ii*) current amplification factor and (*iii*) current gain factor.

Solution : Collector current $I_C = 0.98$ mA

Base current $I_B = 20\ \mu A = 0.02$ mA

(*i*) Emitter current, $I_E = I_B + I_C = 0.98 + 0.02 = 1.0$ mA **Ans.**

(*ii*) Current amplification factor, $\alpha = \frac{I_C}{I_E} = \frac{0.98}{1.0} = 0.98$ **Ans.**

(*iii*) Current gain factor, $\beta = \frac{I_C}{I_B} = \frac{0.98}{0.02} = 49$ **Ans.**

Example 4.3. Determine the values of emitter current and collector current of a transistor having $\alpha_{dc} = 0.98$ and collector-to-base leakage current $I_{CBO} = 4\ \mu A$. The base current is 50 µA.

Solution : Base current, $I_B = 50\ \mu A = 0.05$ mA

Collector-to-base leakage current, $I_{CBO} = 4\ \mu A = 0.004$ mA

We know that $I_C = \alpha_{dc}\, I_E + I_{CBO} = \alpha_{dc}\,(I_C + I_B) + I_{CBO}$ $\qquad \because I_E = I_C + I_B$

or $I_C\,(1 - \alpha_{dc}) = \alpha_{dc}\, I_B + I_{CBO}$

$$\text{or} \quad I_C = \frac{\alpha_{dc}\, I_B}{1-\alpha_{dc}} + \frac{I_{CBO}}{1-\alpha_{dc}} = \frac{0.98 \times 0.05}{1-0.98} + \frac{0.004}{1-0.98} = 2.45 \text{ mA} + 0.2 \text{ mA} = 2.65 \text{ mA } \textbf{Ans.}$$

Emitter current, $I_E = I_C + I_B = 2.65 + 0.05 = 2.7$ mA **Ans.**

Example 4.4. A BJT has $I_B = 10\ \mu A$, $\beta = 99$ and $I_{CO} = 1\ \mu A$. What is its collector current I_C?

Solution : Base current, $I_B = 10\ \mu A$

Current gain factor, $\beta = 99$

$I_{CO} = 1\ \mu A$

We know that $I_C = \alpha I_E + I_{CO} = \alpha\,(I_C + I_B) + I_{CO}$ $\qquad \because I_E = I_C + I_B$

or $I_C\,(1 - \alpha) = \alpha I_B + I_{CO}$

$$\text{or} \quad I_C = \frac{\alpha\, I_B}{1-\alpha} + \frac{I_{CO}}{1-\alpha}$$

$$= \beta\, I_B + (1+\beta)\, I_{CO} \qquad \because \text{ from Eq. (4.15) } \frac{\alpha}{1-\alpha} = \beta \text{ and } \frac{1}{1-\alpha} = (1+\beta)$$

$$= 99 \times 10 + (1 + 99) \times 1 = 1{,}090\ \mu A \text{ or } 1.09 \text{ mA } \textbf{Ans.}$$

Example 4.5. The collector and base current of an N-P-N transistor are measured as $I_C = 5$ mA, $I_B = 50$ µA and $I_{CBO} = 1\ \mu A$.

(*i*) Determine α, β and I_E

(*ii*) Determine the new level of I_B required to produce $I_C = 10$ mA.

Solution : (*i*) Given that $I_C = 5$ mA

$I_B = 50$ μA

$I_{CBO} = 1$ μA

Since $I_C = \beta I_B + (\beta + 1) I_{CBO}$

$$\text{or } \beta = \frac{I_C - I_{CBO}}{I_B + I_{CBO}} = \frac{5\times10^{-3} - 1\times10^{-6}}{50\times10^{-6} + 1\times10^{-6}} = 98 \text{ Ans.}$$

$$\alpha = \frac{\beta}{1+\beta} = \frac{98}{99} = 0.99 \text{ Ans.}$$

$$I_E = I_B + I_C = 50\times10^{-6} + 5\times10^{-3} = 5.05 \text{ mA Ans.}$$

$$(ii)\ I_B = \frac{I_C - (\beta+1) I_{CBO}}{\beta} = \frac{10\times10^{-3} - (98+1)\times1\times10^{-6}}{98} = 101\ \mu\text{A Ans.}$$

4.9. EBERS-MOLL MODEL

Device modelling aims at relating physical device parameters to device terminal characteristics. Device modelling is especially important for integrated circuits, since simple and accurate device models are required to perdict the performance of a circuit. Generally, by making a model more accurate, we make it more complex. Thus, there is a trade-off between accuracy and complexity.

Ebers-Moll model is a transistor model which describes the operating states of a transistor. This model is useful to obtain information about the dc characteristics of a transistor. The transistor operates in active region when emitter junction is forward biased and collector junction is reverse biased. This model generalizes the behaviour of a transistor by taking into account the inverted mode of operation of transistor.

The dependence of the currents in a transistor upon the junction voltages, or vice-versa may be obtained by starting with Eq. (4.12), reproduced here

$$\text{i.e. } I_C = -\alpha_n I_E + I_{CO}\left[1 - e^{V_C/V_T}\right]$$

$$\text{or } I_C = -\alpha_n I_E - I_{CO}\left[e^{V_C/V_T} - 1\right] \quad \text{...(4.17)}$$

Subscript n to α has been added in order to indicate the normal operation of the transistor. Minus sign has been introduced with $\alpha_n I_E$ to take into account its direction.

With inverted mode of operation in mind, equation corresponding to Eq. (4.17) may be written as

$$I_E = -\alpha_i I_C - I_{EO}\left[e^{V_E/V_T} - 1\right] \quad \text{...(4.18)}$$

Here α_i is the inverted common-base current gain, just as α_n in Eq. (4.17) is the current gain in normal operation. I_{EO} is the emitter-junction reverse saturation current, and V_E is the voltage drop from P-side to N-side at the emitter junction and is positive for a forward-biased emitter.

***Ebers-Moll model* interpreting Eqs. (4.17) and (4.18) in terms of a circuit is given in Fig. 4.12 for a P-N-P transistor. It involves two ideal diodes placed back-to-back with reverse saturation currents $-I_{EO}$ and $-I_{CO}$ and two dependent current-controlled**

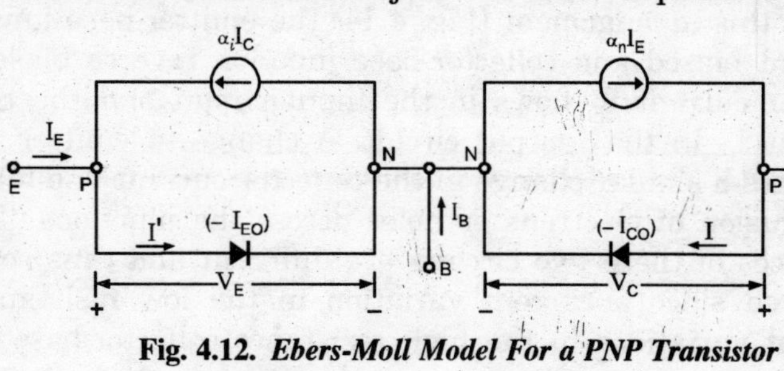

Fig. 4.12. *Ebers-Moll Model For a PNP Transistor*

current sources shunting the ideal diodes. This model is valid for both forward- and reverse-biased static voltages applied across the transistor junctions.

Here, base spreading resistance has been omitted and the difference between I_{CBO} and I_{CO} have been neglected.

From Fig. 4.12 it is evident that dependent current sources can be eliminated from the figure provided $\alpha_n = \alpha_i = 0$. For example, by making the width of the base much larger than the diffusion length of the minority carriers in the base, all the minority carriers will recombine in the base and none will survive to reach the collector. In such a condition transport factor β and hence also current gain factor α will be zero. Thus under these conditions, the transistor action ceases, and we simply have two diodes placed back-to-back.

From the above discussion it is obvious that a transistor cannot be constructed by simply connecting two separate (isolated) diodes back-to-back.

4.10. Transistor Circuit Configurations

As already mentioned, a transistor is a three-terminal device (having three terminals namely emitter, base and collector) but we require four terminals—two for the input and two for the output for connecting it in a circuit. Hence one of the terminals of the transistor is made common to the input and output circuits. Thus there are three types of configurations for operation of a transistor. These configurations are *common base* (CB), *common emitter* (CE) and *common collector* (CC) configurations. The term 'common' is used to denote the electrode that is common to input and output circuits. Since the common terminal is usually grounded, these configurations are also referred to as *grounded base*, *grounded emitter* and *grounded collector* configurations, as shown in Figs. 4.13 (*a*), 4.13 (*b*) and 4.13 (*c*) respectively. Each configuration has its own merits and demerits. It is to be remembered that regardless of the circuit configurations, the emitter is always forward biased and the collector is always reverse biased.

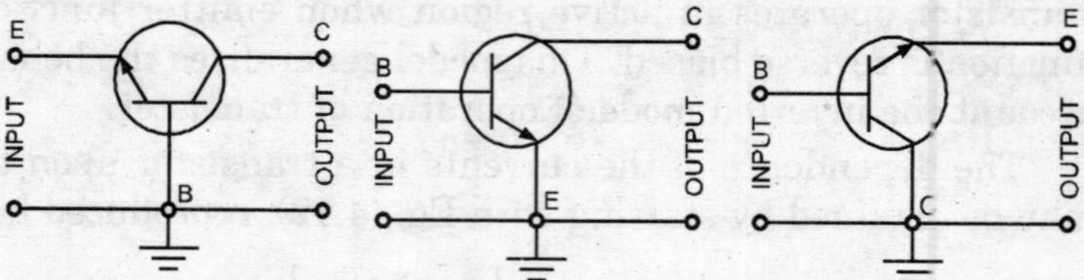

(*a*) *Common Base Configuration* (*b*) *Common Emitter Configuration* (*c*) *Common Collector Configuration*

Fig. 4.13. ***Different Circuit Configurations For NPN Transistor***

4.11. Common Base (CB) Configuration

In common base (CB) configuration, input is connected between emitter and base and output is taken across collector and base, as illustrated in Fig. 4.14 for an N-P-N transistor. Thus base is common between input and output circuits.

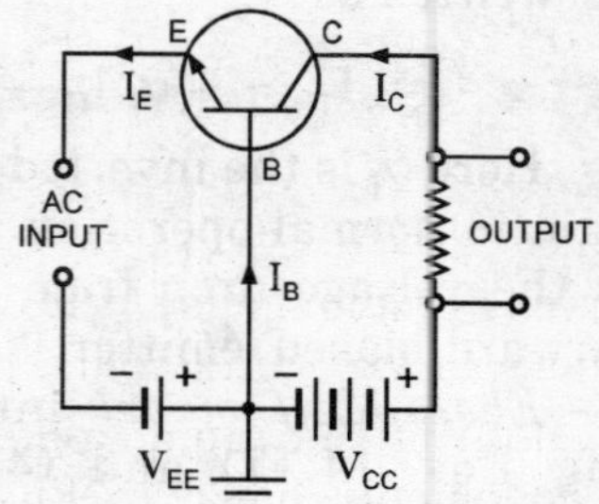

Fig. 4.14. ***Common Base NPN Transistor***

In this arrangement (Fig. 4.14) the emitter-base junction is forward biased and collector-base junction reverse biased. The emitter current I_E flows in the input circuit and the collector current I_C in the output circuit. A change in emitter current produces a similar change in the collector current due to change in diffusion of electrons or holes across the junctions. The impedances of these two circuits are different and thus some voltage and power gains can be achieved since a current variation in the low resistance emitter-base circuit has similar current variations in the high resistance collector-base circuit.

The ratio of collector current I_C (output current) to the emitter current I_E (input current) is called the *current amplification* factor α. If there is no input ac signal then the ratio of I_C to I_E is called the dc alpha (α_{dc}).

i.e. $\alpha_{dc} = \frac{-I_C}{I_E}$ (negative sign signifies that I_E flows out of transistor while I_C into transistor).*

If we write α_{dc} simply by α and ignore –ve sign,

$$\alpha = \frac{I_C}{I_E} \quad \text{...(4.19)}$$

Incidently, there is also an ac alpha (α_{ac}) for a transistor and it refers to the ratio of change in collector current to change in emitter current

i.e. $\alpha_{ac} = \frac{-\Delta I_C}{\Delta I_E}$ for constant collector base voltage V_{CB}. ...(4.20)

α** of a transistor is a measure of its quality–higher the value of α better is the transistor in the sense that I_C approaches I_E, its value ranges from 0.95 to 0.99. The value of α can be increased by making base thin and lightly doped.

The collector current consists of two parts (*i*) the current produced by normal transistor action *i.e.* component depending upon emitter current (αI_E) which is produced by majority carriers and (*ii*) the leakage current due to movement of minority carriers across base-collector junction on account of it being reverse biased. This leakage current flows in the same direction as that of I_C and is abbreviated as I_{CBO}*** (collector-base current with emitter open).

∴ Total collector current, $I_C = \alpha I_E + I_{CBO}$ ...(4.21)

The current I_{CBO}, sometimes referred to as I_{CO}, is usually very small and may be neglected in transistor calculations.

From Eq. (4.21)

$$\alpha = \frac{I_C - I_{CO}}{I_E} \simeq \frac{I_C}{I_E} \quad \text{...(4.22)}$$

neglecting I_{CO} in comparison with I_C

$$I_B = I_E - I_C = I_E - (\alpha I_E + I_{CO})$$

$$= (I - \alpha) I_E - I_{CO} \quad \text{...(4.23)}$$

∵ from Eq. (4.21)

$$I_C = \alpha I_E + I_{CO}$$

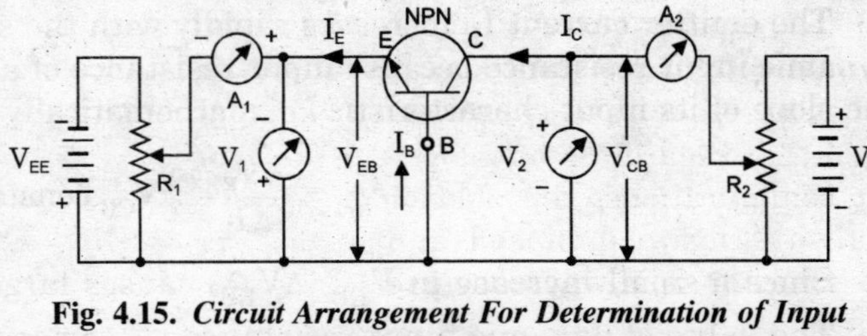

Fig. 4.15. ***Circuit Arrangement For Determination of Input and Output Characteristic Curves of a Transistor in CB Configuration***

4.11.1. Characteristics of Common Base Configuration

The performance of transistors, when connected in a circuit, may be determined from their characteristic curves that relate different dc currents and voltages of a transistor. Such curves are known as *static characteristic curves*. There are two important characteristics of a transistor viz. input characteristics and output characteristics.

Both of the input and output characteristics can be determined with the same circuit arrangement shown in Fig. 4.15.

---

* In case of a P-N-P transistor emitter current I_E flows into the transistor and collector current I_C flows out of it.

** For all practical purposes α_{dc}, α_{ac} and α have the same values.

*** I_{CBO} is exactly like the reverse saturation current I_S or I_0 of a reverse biased diode discussed earlier and is extremely temperature dependent as it is made up of thermal generated minority carriers.

Input Characteristics. The curve drawn between emitter current I_E and emitter-base voltage V_{EB} for a given value of collector-base voltage V_{CB} is known as input characteristic. For determination of the input characteristics, the output voltage V_{CB} is maintained constant, and the input voltage V_{EB} is set at several convenient levels. For each level of input voltage, the input current I_E is noted. Then the curve is plotted between emitter current I_E and emitter base voltage V_{EB}, as shown in Fig. 4.16.

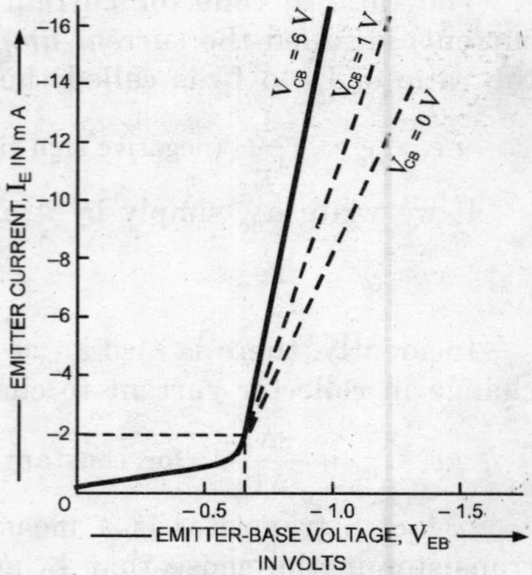

Fig. 4.16. ***Input Characteristics For Common Base NPN Transistor***

For a given value of V_{CB} the curve is just like that of a forward biased P-N junction because emitter-base junction in forward bias mode is similar to a P-N junction. This characteristic may be used to determine input resistance of the transistor.

From Fig. 4.16 it is seen that with the increase in V_{CB}, it conducts better, although the effect is not very significant. This is because large collector-base (reverse-bias) voltages cause the depletion layer at the collector-base junction to penetrate deeper into the base of the transistor, thus reducing the distance and the resistance between the emitter-base and collector-base regions.

There exists a *cut-in*, *offset* or *threshold* voltage V_{EB} below which the emitter current is very small. Substantial emitter current flows only when emitter-base junction is forward biased beyond the knee of the characteristic (approximately 0.7 V for Si and 0.3 V for Ge).

The emitter current I_E increases rapidly with the small increase in V_{EB} which means low dynamic input resistance because input resistance of a transistor is given as the reciprocal of the slope of its input characteristic *i.e.* mathematically

$$r_{in} = \frac{\Delta V_{EB}}{\Delta I_E} \text{ ...} V_{CB} \text{ constant} \qquad \text{...(4.24)}$$

Since a small increase in V_{EB} (ΔV_{EB}) causes large change in I_E (ΔI_E), r_{in} is quite low.

The value of dynamic input resistance r_{in} varies from point-to-point in the initial part of the characteristic because it is nonlinear. Its value over the linear part of the characteristic is about 100 Ω but for lower values of V_{EB}, it is considerably larger. This dependence of r_{in} on V_{EB} causes distortion of the signals handled by the transistors.

Output Characteristics. The curve drawn between collector current I_C and collector-base voltage V_{CB} for a given value of emitter current I_E is known as output characteristic.

For determination of output characteristics, the emitter current is held constant at each of several levels. For each fixed level of emitter current I_E, the output voltage V_{CB} is adjusted in convenient steps, and the corresponding levels of collector current I_C are noted. From these data output characteristics can be drawn. The curves so drawn are shown in Fig. 4.17.

The worthnoting points regarding output characteristics of common base (CB) configuration are given below :

(*i*) The collector current I_C varies with V_{CB} only for very low voltage (below 1 V) but transistor is never operated in this region.

(*ii*) In active region (emitter forward biased and collector reverse biased) collector current I_C is almost equal to I_E and appears to remain constant when V_{CB} is increased. In fact, there is very small increase in I_C with increase in V_{CB}. This is because the increase in V_{CB} expands the collector-base depletion region and thus shortens the distance between the two depletion regions. With emitter current I_E held constant, however, the increase in I_C is so small that it is usually noticeable only for large variations in V_{CB}. I_C is slightly lesser than I_E in magnitude.

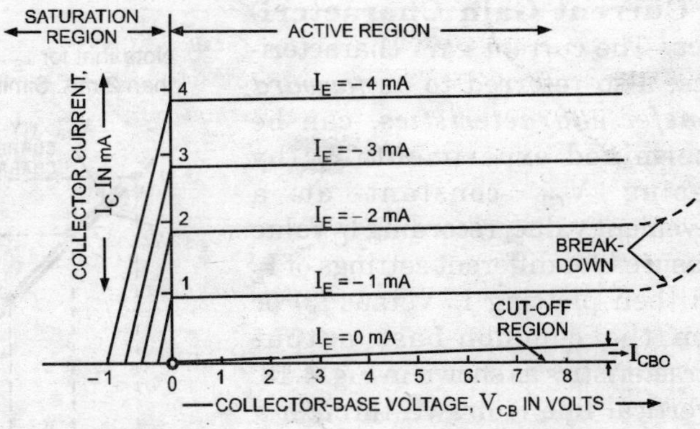

Fig. 4.17. ***Output Characteristics For Common Base NPN Transistor***

Transistor is normally operated in active region.

(*iii*) Although the collector current I_C is practically independent of V_{CB} over the transistor operating range, however, if V_{CB} is increased beyond a certain value, I_C eventually increases rapidly because of avalanche or zener (or both) effects. This condition is known as *punch-through* or *reach-through*. When it occurs large currents can flow, possibly destroying the device. The extension of the depletion region is, of course, the direct consequence of the increase in V_{CB}. Thus it is very essential to maintain V_{CB} below the maximum safe limit specified by the manufacturer of the device.

(*iv*) A very large change in collector voltage causes a very small change in collector current *i.e.* output resistance of CB configuration is very high (of the order of few hundred kΩ)—the dynamic output resistance r_{out} being given as the ratio of ΔV_{CB} and ΔI_C for a given value of I_E.

(*v*) In *cut-off* region (emitter and collector junctions both reverse-biased) small collector current I_C flows even when emitter current $I_E = 0$. This is the collector leakage current I_{CBO} or I_{CO}.

(*vi*) In saturation region (both emitter and collector junctions forward-biased) collector current I_C flows even when $V_{CB} \simeq 0$. Even when the externally applied bias voltage is reduced to zero, there is still a barrier potential existing at the collector-base junction, and this assists in the flow of I_C. To stop it the collector-base junction has to be forward biased. Consequently, as illustrated in Fig. 4.17, collector current I_C is reduced to zero when V_{CB} is increased negatively.

We have already determined that a small signal voltage impressed in the low-resistance input (emitter) circuit of a transistor causes a relatively large emitter current. Almost the same amount of current will flow in the high-resistance output (collector) circuit of transistor, where the voltage may be very high. Evidently, then, both the output voltage and power can be quite large, as compared to the tiny input voltage and power present at the emitter.

CB configuration is rarely used in audio-frequency (AF) circuits because its current gain is less than unity and its input and output resistances are quite different.

Current Gain Characteristics. The current gain characteristics, also referred to as *forward transfer characteristics,* can be determined experimentally (by keeping V_{CB} constant at a convenient value, recording I_C value measured for different settings of I_E and then plotting I_C versus I_E) or from the common-base output characteristics as shown in Fig. 4.18. A vertical line is drawn through a selected value of V_{CB} and corresponding values of I_E and I_C are read off along this line. Then a graph is plotted between I_E versus I_E, and the characteristic is labelled with the V_{CB} used.

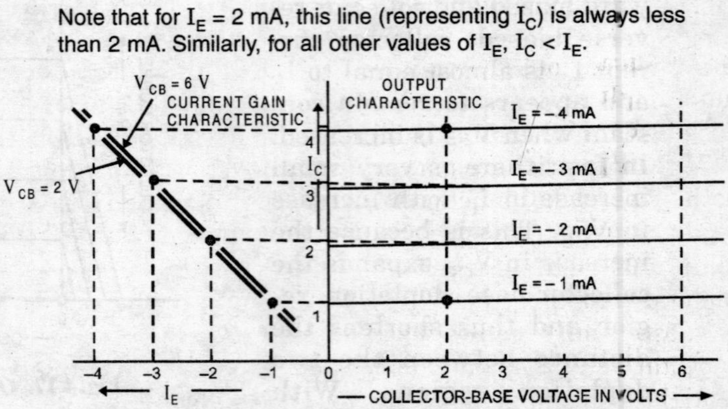

Fig. 4.18. *Current Gain Characteristics For Common Base NPN Transistor*

Example 4.6. A transistor is connnected in CB configuration. When the emitter voltage is changed by 200 mV, the emitter current changes by 5 mA. During this variation, the collector-to-base voltage is kept fixed. Calculate the dynamic input resistance of transistor.

Solution : The dynamic input resistance for CB configuration is given as

$$r_{in} = \left.\frac{\Delta V_{EB}}{\Delta I_E}\right|_{V_{CB} = \text{constant}} = \frac{200 \text{ mV}}{5 \text{ mA}} = 40 \ \Omega \textbf{ Ans.}$$

Example 4.7. The reverse saturation current of an N-P-N transistor in common base circuit is 12.5 μA. For an emitter current of 2 mA, collector current is 1.97 mA. Determine the current gain and base current.

Solution : Emitter current $I_E = 2$ mA

Collector current, $I_C = 1.97$ mA

Collector-to-base leakage current,

$$I_{CBO} = 12.5 \ \mu\text{A} = 0.0125 \text{ mA}$$

$$\text{Since } I_C = \alpha I_E + I_{CBO}$$

$$\therefore \text{ Current gain, } \alpha = \frac{I_C - I_{CBO}}{I_E} = \frac{1.97 - 0.0125}{2} = 0.979 \textbf{ Ans.}$$

$$\text{Base current, } I_B = I_E - I_C = 2 - 1.97 = 0.03 \text{ mA } \textbf{Ans.}$$

Example 4.8. In a grounded base configuration (Fig. 4.14) the voltage drop across load resistance of 4 kΩ is 3 V. Determine base current. α = 0.96.

Solution : Load resistance $R_L = 4$ kΩ

Voltage drop across load resistance, $I_C R_L = 3$ V

$$\text{Collector current, } I_C = \frac{I_C R_L}{R_L} = \frac{3 \text{ V}}{4 \text{ k}\Omega} = 0.75 \text{ mA}$$

$$\text{Emitter current, } I_E = \frac{I_C}{\alpha} = \frac{0.75}{0.96} = 0.78 \text{ mA}$$

$$\text{Base current, } I_B = I_E - I_C = 0.78 - 0.75 = 0.03 \text{ mA } \textbf{Ans.}$$

4.12. EARLY EFFECT AND BASE-WIDTH MODULATION

In active region, the emitter-base (EB) junction is forward biased and collector-base (CB) junction is reverse-biased, so the barrier width at EB junction is negligible in comparison with the space charge width at CB junction, as illustrated in Fig. 4.19.

The transition or space-charge region at a junction is the region of uncovered charges on both sides of the junction. As the voltage applied across the junction increases,the transition region penetrates deeper into the collector and base. Because neutrality of the charge is to be maintained, the number of uncovered charges on each side remains equal. Since the doping in the base is substantially smaller than that of collector, the penetration of the transition region into the base is much larger than that into the collector region. Hence the collector*depletion region is neglected and all the immobile charge is indicated in the base region.

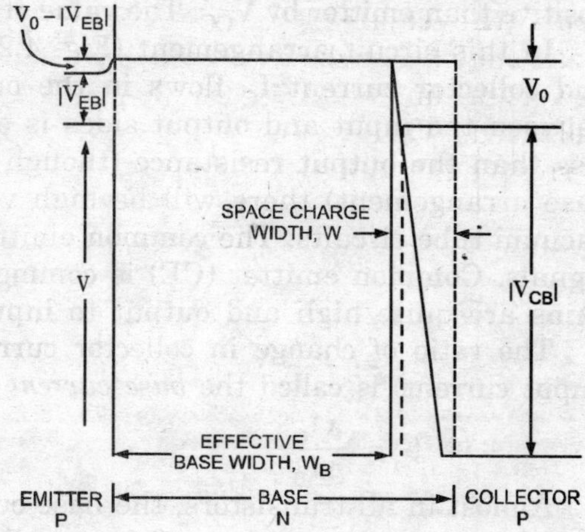

Fig. 4.19. ***The Potential Variation Through P-N-P Transistor Under Biased Conditions***

As obvious from Fig. 4.19 the effective base width is metallurgical base width less space charge width W. This *modulation of the effective base width* by the collector voltage V_{CB} is known as the *Early effect*.

An increase in magnitude of collector voltage increases space-charge width W [$\because$ from Eq. (2.117) $V_B = \dfrac{eN_D W^2}{2\in}$] at the output junction diode and this, in turn, causes the effective base width W_B to decrease.

The decrease in effective base width W_B with increasing reverse bias has three consequences.

(*i*) There is less chance for recombination within the base region. Hence the transport factor β* and also α, increase with an increase in the magnitude of CB junction voltage.

(*ii*) The concentration gradient of minority carriers is increased within the base, and consequently, the current of minority carriers injected across the emitter junction increases, with increasing reverse collector voltage.

(*iii*) For extremely large voltages, W_B may be reduced to zero causing voltage breakdown. When effective base width W_B is reduced to zero, the emitter barrier becomes V', which is smaller than normal value $V_0 - |V_{EB}|$ because the collector voltage has reached through the base region. This lowering of the emitter-junction voltage causes an excessively large emitter current, thus placing an upper limit on the magnitude of the collector voltage. This phenomenon is known as *punch through*.

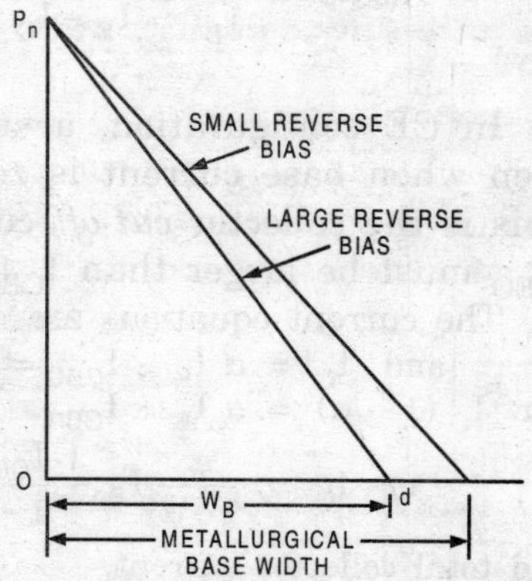

Fig. 4.20. ***Minority Carrier Charge Density Within The Base***

The punch-through (or reach-through) phenomenon may be defined as the mechanism by which a transistor usefullness may be terminated by increasing the collector voltage.

---

* When no signal is applied, then the ratio of I_C to I_B is called the dc beta (β_{dc}) of a transistor. With signal applied, the ratio of ΔI_C to ΔI_B is called the ac beta (β_{ac}).

4.13. Common Emitter (CE) Configuration

In common emitter configuration input is connected between base and emitter while the output is taken between collector and emitter, as shown in Fig. 4.21. Thus emitter is common to input and output circuits.

In this configuration, the bias voltages are applied between collector and emitter and base and emitter. Emitter base junction is forward biased and the base is made more positive than the emitter by V_{BB}, collector-emitter junction is reverse-biased and the collector is made more positive than emitter by V_{CC}. The value of V_{CC} must be greater than that of V_{BB}.

In this circuit arrangement (Fig. 4.21) the base current I_B flows in the input circuit and collector current I_C flows in the output circuit. In this arrangement current gain between the input and output sides is obtained; and since the input resistance is again less than the output resistance (though difference is not as much as in case of common base arrangement) there will be high voltage and power gains like those in equivalent vacuum tube circuits. The common emitter produces a reversal between input and output signals. Common emitter (CE) is commonly used because its current, voltage and power gains are quite high and output to input impedance ratio is moderate.

The ratio of change in collector current (output current) and change in base current (input current) is called the *base current amplification factor* β

$$i.e.\ \beta = \frac{\Delta I_C}{\Delta I_B} \quad \ldots(4.25)$$

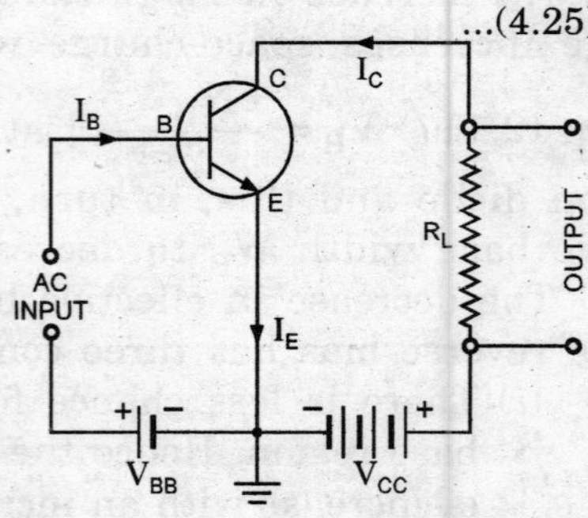

Fig. 4.21. *Common Emitter NPN Transistor*

Almost in all transistors, the base current is less than 5% of emitter current I_E so β is usually higher than 20. Usually β ranges from 20 to 100.

$$\text{Also } \beta = \frac{I_C}{I_B} = \frac{I_C}{I_E - I_C} = \frac{I_C/I_E}{1 - \frac{I_C}{I_E}} = \frac{\alpha}{1-\alpha} \quad \ldots(4.26)$$

In CE configuration, a small collector current flows even when base current is zero (*i.e.* base lead is open). This is the collector *cut-off* current and denoted by I_{CEO}. I_{CEO} must be larger than I_{CBO}.

The current equations are $I_E = I_B + I_C$ ...(4.27)

$$\text{and } I_C = \alpha I_E + I_{CBO} = \alpha (I_B + I_C) + I_{CBO}$$

$$\text{or } I_C (1 - \alpha) = \alpha I_B + I_{CBO}$$

$$\text{or } I_C = \frac{\alpha}{1-\alpha} I_B + \frac{I_{CBO}}{1-\alpha} \quad \ldots(4.28)$$

and total collector current,

$$I_C = \beta I_B + I_{CEO} \quad \ldots(4.29)$$

Comparing Eqs. (4.28) and (4.29) we have

$$\beta = \frac{\alpha}{1-\alpha} \quad \text{as already derived in Eq. (4.26)}$$

$$\text{and } I_{CEO} = \frac{1}{1-\alpha} I_{CBO} \quad \ldots(4.30)$$

Substituting the value of $I_{CEO} = \frac{1}{1-\alpha} I_{CBO}$ in Eq. (4.29) we have

$$I_C = \beta I_B + \frac{1}{1-\alpha} I_{CBO} = \beta I_B + (\beta + 1) I_{CBO} \quad \ldots(4.31) \quad \because \text{ from Eq. (4.26) } \frac{1}{1-\alpha} = \beta + 1$$

4.13.1. Characteristics of Common Emitter Configuration

The circuit used for determination of transistor common emitter characteristics is shown in Fig. 4.22. The input voltage is applied between base and emitter terminals, and the output is taken from collector and emitter terminals. Input and output voltages and currents are measured by voltmeters and ammeters as shown.

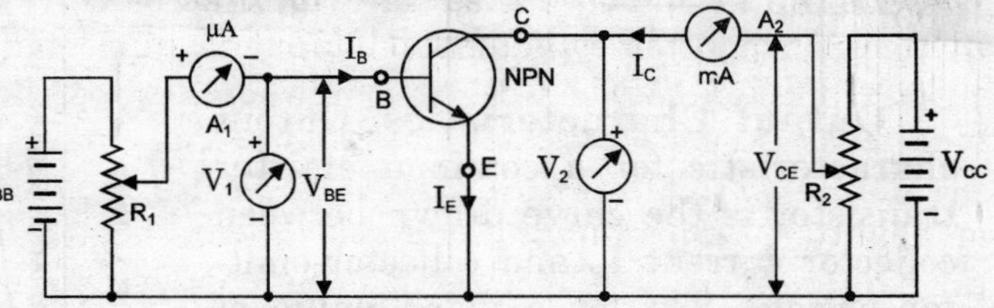

Fig. 4.22. *Circuit Arrangement For Determination of Input and Output Characteristics of a Transistor in CE Configuration*

Input Characteristics. The curve drawn between base current I_B and base-emitter voltage V_{BE} for a given value of collector-emitter voltage V_{CE} is known as the input characteristic. For determination of input characteristics, collector-emitter voltage V_{CE} is held constant and base current I_B is recorded for different values of base-emitter voltage V_{BE}. Now the curves are drawn between base current I_B and base-emitter voltage V_{BE} for different levels of V_{CE}, as shown in Fig. 4.23. The worthnoting points regarding input characteristics are given below :

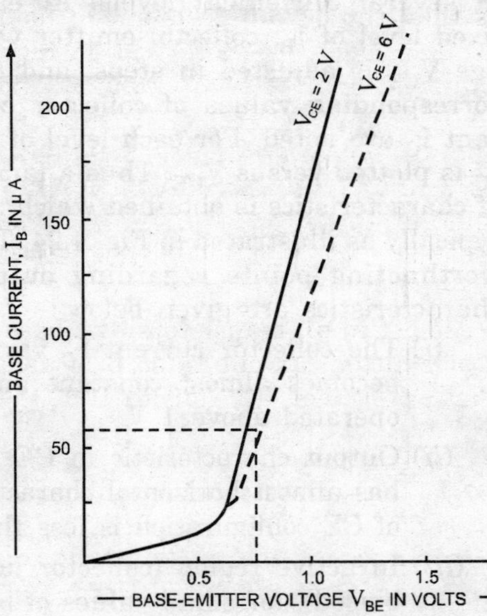

Fig. 4.23. *Input Characteristics For Common Emitter NPN Transistor*

(*i*) The input characteristics of CE transistors are quite similar to those of a forward biased diode because the base-emitter region of the transistor is a diode and it is forward biased.

(*ii*) In comparison to common base (CB) arrangement base (or input) current increases less rapidly with the increase in base-emitter voltage, V_{BE}. This indicates that input resistance is larger in common emitter configuration than that in common base configuration. Due to initial nonlinearity of the curve, input resistance varies from point to point in the initial part of the characteristic. Its value over the linear part of the curve is of the order of few hundred ohms.

(*iii*) An increment in value of V_{CE} causes the input current I_B to be lower for a given level of V_{BE}. This is because the higher levels of V_{CE} provide greater collector-base junction reverse bias, causing greater depletion region penetration into the base and thus reducing the distance between the collector-base and emitter-base regions. As a result more of the charge carriers from the emitter flows across the collector-base junction, and few flow out through the base lead. Usually effect of change in V_{CE} on the input characteristic is ignored.

The ratio of change in base-emitter voltage (ΔV_{BE}) to the resulting change in base current (ΔI_B) at constant collector-emitter voltage (V_{CE}) is known as *dynamic input resistance* (r_{in}). It is calculated as the reciprocal of the slope of the input characteristic at a given V_{BE}.

$$i.e., \; r_{in} = \left. \frac{\Delta V_{BE}}{\Delta I_B} \right|_{V_{CE} = \text{constant}} \qquad ...(4.32)$$

Output Characteristics. Output characteristic for a common emitter transistor is the curve drawn between collector current I_C and collector-emitter voltage V_{CE} for a given value of base current I_B. For determination of common emitter output characteristics, base current I_B is maintained constant at several convenient levels. At each fixed level of I_B, collector-emitter voltage V_{CE} is adjusted in steps, and the corresponding values of collector current I_C are noted. For each level of I_B, I_C is plotted versus V_{CE}. Thus a family of characteristics is obtained which are typically as illustrated in Fig. 4.24. The worthnoting points regarding output characteristics are given below :

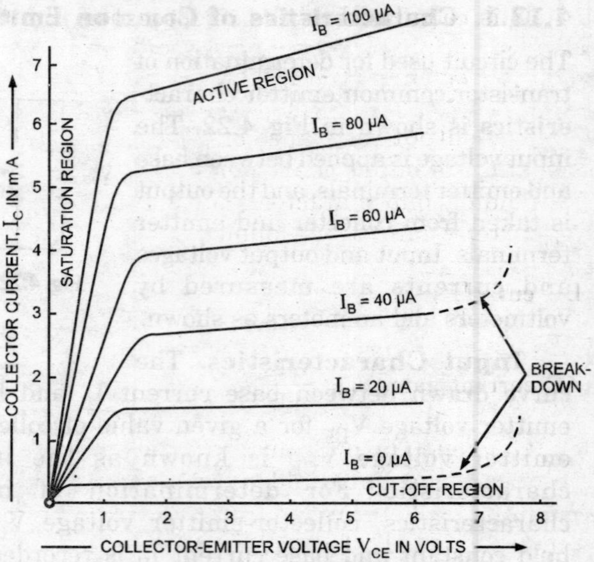

Fig. 4.24. ***Output Characteristics For Common Emitter NPN Transistor***

(*i*) The collector current I_C varies with V_{CE} for V_{CE} between 0 and 1 V and then becomes almost constant and independent of V_{CE}. The transistors are always operated above 1 V.

(*ii*) Output characteristic in CE configuration has some slope while CB configuration has almost horizontal characteristics. This indicates that output resistance in case of CE configuration is less than that in CB configuration.

(*iii*) In active region (collector junction reverse biased and emitter junction forward biased), for small values of base current I_B the effect of collector voltage V_C over I_C is small but for large values of I_B this effect increases.

The shape of the characteristic is same as in CB configuration but with the difference that collector current I_C is larger than input current (base current I_B). Thus the current gain for this configuration is larger than unity. The transistor must always be operated in the active region, when employed as an amplifying device.

(*iv*) With low values (ideally zero) of V_{CE} the transistor is said to be operated in saturation region and in this region base current I_B does not cause a corresponding change in collector current I_C.

(*v*) With much higher V_{CE}, the collector-base junction completely breaks down and because of this avalanche breakdown collector current I_C increases rapidly and the transistor gets damaged.

(*vi*) In cutoff region, small amount of collector current I_C flows even when base current $I_B = 0$. This is called I_{CEO}. Since main current I_C is zero, the transistor is said to be cut-off.

(*vii*) Moderate output to input impedance ratio makes this configuration an ideal one for coupling between various transistor stages.

The ratio of change in collector-emitter voltage (ΔV_{CE}) to the change in collector current (ΔI_C) at constant base current is known as *dynamic* ouput resistance.

It is calculated as the reciprocal of the slope of output characteristic at a given V_{CE}.

$$i.e., \; r_{out} = \left.\frac{\Delta V_{CE}}{\Delta I_C}\right|_{I_B = \text{constant}} \qquad ...(4.33)$$

The output resistance of CE configuration is less than that of CB configuration as the slope of the ouput characteristic is more in this case. Its value is of the order of 50 kΩ.

The output characteristics may also be used to determine the dc current gain β and ac current gain β_0 as follows

$$\text{DC current gain, } \beta = \frac{I_C}{I_B} \qquad ...(4.34)$$

$$\text{AC current gain, } \beta_0 = \left.\frac{\Delta I_C}{\Delta I_B}\right|_{V_{CE} = \text{constant}} \qquad ...(4.35)$$

Current Gain Characteristics. Like the common-base current gain characteristics, they may be obtained experimentally, or determined from the output characteristics. For experimental determination of current gain characteristics, V_{CE} is kept constant at a convenient value and value of collector current I_C is observed and recorded for different values of base current I_B and then a graph is plotted between I_C and I_B.

The current gain characteristics can be obtained from the output characteristics as shown in Fig. 4.25, and detailed in Art. 4.11 except that I_C is plotted against I_B rather than I_C against I_E.

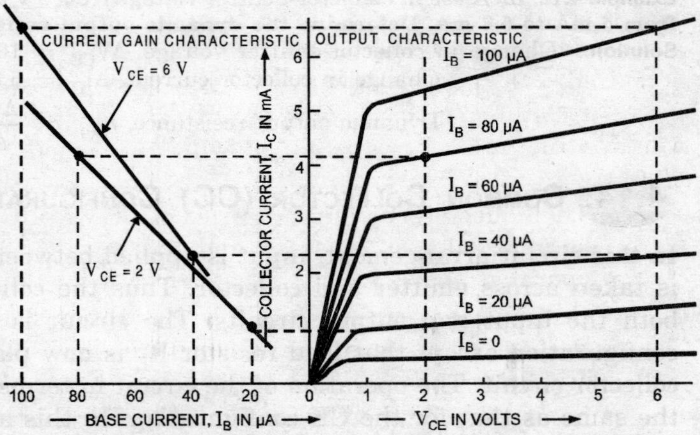

Fig. 4.25. ***Current Gain Characteristics For Common Emitter NPN Transistor***

Example 4.9. A transistor is connected in common emitter configuration (Fig. 4.21). Collector supply voltage V_{CC} is 10 volts, load resistance R_L is 800 Ω, voltage drop across load resistance is 0.8 V and current gain α = 0.96. Determine collector-emitter voltage and base current.

Solution : Collector supply voltage, V_{CC} = 10 V

Voltage drop across load resistance, $I_C R_L$ = 0.8 V

Collector-emitter voltage, $V_{CE} = V_{CC} - I_C R_L = 10 - 0.8 = 9.2$ **Ans.**

$$\text{Collector current, } I_C = \frac{I_C R_L}{R_L} = \frac{0.8}{800} = 1.0 \text{ mA}$$

$$\text{Current gain factor, } \beta = \frac{\alpha}{1-\alpha} = \frac{0.96}{1-0.96} = 24$$

$$\therefore \text{ Base current, } I_B = \frac{I_C}{\beta} = \frac{1.0 \text{ mA}}{24} = 41.67 \; \mu\text{A} \textbf{ Ans.}$$

Example 4.10. A germanium transistor with α_0 = 0.98 gives a reverse saturation current I_{CO} = 10 μA is CB configuration. When transistor is used in CE configuration with a base current of 0.22 mA, calculate the collector current.

Solution : Collector-to-base leakage current,

$$I_{CO} = 10\ \mu A = 0.01\ mA$$

AC current gain, $\alpha_o = 0.98$

Base current, $I_B = 0.22$ mA

From Eq. (4.28) collector current,

$$I_C = \frac{\alpha}{1-\alpha} \cdot I_B + \frac{I_{CO}}{1-\alpha} = \frac{0.98}{1-0.98} \times 0.22 + \frac{0.01}{1-0.98} = 10.78 + 0.5 = 11.28 \text{ mA } \textbf{Ans.}$$

Example 4.11. A change of 250 mV in base-emitter voltage causes a change of 1 mA in the base current. Determine the dynamic input resistance.

Solution : Change in base-emitter voltage, $\Delta V_{BE} = 250$ mV

Change in base current, $\Delta I_B = 1$ mA

$$\text{Dynamic input resistance, } r_{in} = \frac{\Delta V_{BE}}{\Delta I_B} = \frac{250 \times 10^{-3}}{1 \times 10^{-3}} = 250\ \Omega \textbf{ Ans.}$$

Example 4.12. Increase in collector-emitter voltage from 5 V to 10 V causes increase in collector current from 5 mA to 5.8 mA. Determine the dynamic output resistance.

Solution : Change in collector-emitter voltage, $\Delta V_{CE} = 10 - 5 = 5$ V

Change in collector current, $\Delta I_C = 5.8 - 5.0 = 0.8$ mA

$$\text{Dynamic output resistance, } r_{out} = \frac{\Delta V_{CE}}{\Delta I_C} = \frac{5}{0.8 \times 10^{-3}} = 6.25\ k\Omega \textbf{ Ans.}$$

4.14. Common Collector (CC) Configuration

In this circuit arrangement, input is applied between base and collector while the output is taken across emitter and collector. Thus the collector forms the terminal common to both the input and output circuits. The circuit is essentially the same as that of CE configuration except that load resistor R_L is now placed in the emitter circuit instead of collector circuit. The operation of the circuit in terms of the currents that flow is basically the same as that for the CE configuration. In this arrangement base current I_B flows in the input circuit and emitter current I_E flows in the output circuit. So the change in emitter current ΔI_E to the change in base current ΔI_B gives the current amplification factor γ. The circuit provides nearly the same current gain as in case of CE arrangement because $\Delta I_E \simeq \Delta I_C$.

With base current I_B equal to I_{CO}, the emitter current I_E is zero so no current flows in the load resistor R_L. With the increase in input current I_B, the transistor passes through the active region and finally reaches saturation. Almost the complete supply voltage, except for a small voltage drop across the transistor appears across the load resistor R_L.

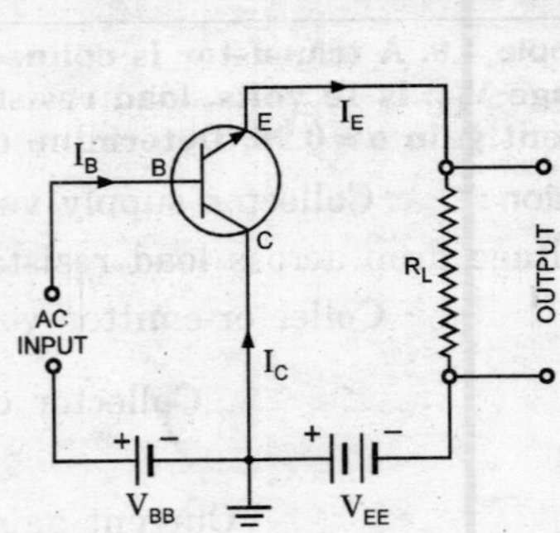

Fig. 4.26. *Common Collector NPN Transistor*

The current equations are

$$I_C = \alpha\ I_E + I_{CBO}$$

and $I_E = I_B + I_C = I_B + \alpha\ I_E + I_{CBO}$

or $I_E\ (1 - \alpha) = I_B + I_{CBO}$

$$\text{or } I_E = \frac{I_B}{1-\alpha} + \frac{I_{CBO}}{1-\alpha} = (\beta + 1)\ I_B + (\beta + 1)\ I_{CBO} \quad \ldots(4.36) \quad \because \frac{1}{1-\alpha} = \beta + 1$$

$$\text{Current gain, } \gamma = \frac{\Delta I_E}{\Delta I_B} = \frac{\Delta I_E}{\Delta I_E - \Delta I_C} = \frac{\Delta I_E / \Delta I_C}{\frac{\Delta I_E}{\Delta I_C} - 1} = \frac{1/\alpha}{1/\alpha - 1} = \frac{1}{1-\alpha} = \beta + 1 \quad \ldots(4.37)$$

The input and output characteristics are shown in Figs. 4.27 and 4.28 respectively.

The common collector (CC) arrangement gives very high input impedance and very low output impedance and, therefore, its voltage gain is always less than unity. Hence this configuration is seldom used for amplification. However, owing to relatively high input impedance and low output impedance, this configuration is primarily used for impedance matching *i.e.* for driving low impedance from a high impedance source. This configuration is also called *emitter follower*.

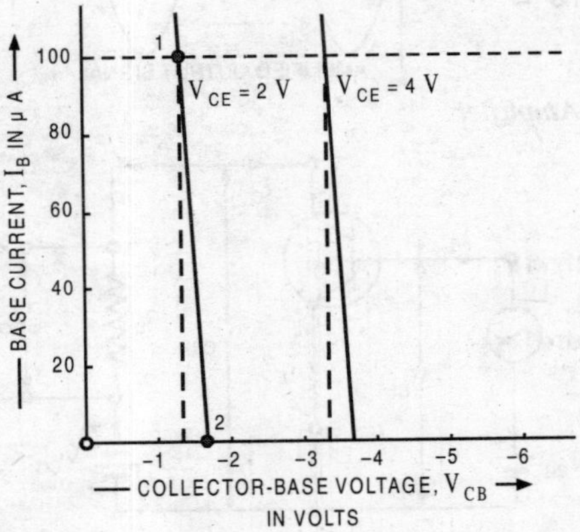

Fig. 4.27. ***Input Characteristics For Common Collector NPN Transistor***

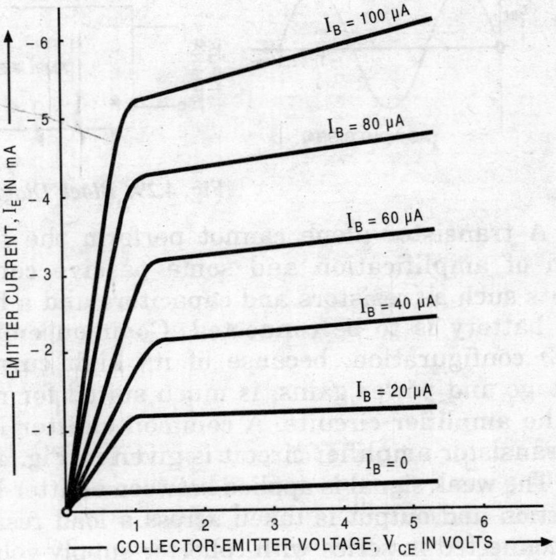

Fig. 4.28. ***Output Characteristics For Common Collector NPN Transistor***

4.15. Comparison of Characteristics of Transistors in Different Configurations

The essential characteristics of each of basic configurations are tabulated below :

Characteristics \ *Configuration*	*Common Base*	*Common Emitter*	*Common Collector*
Input impedance	Low (about 100 Ω)	Medium (about 800 Ω)	Very high (about 750 kΩ)
Output impedance	Very high (about 500 kΩ)	High (about 50 kΩ)	Low (about 50 Ω)
Current gain	Less than unity but usually more than 0.9 (about 0.98)	High (about 80)	High (about 100)
Voltage gain	About 150	About 500	Less than unity
Leakage current	Very small (5 μA for Ge and 1 μA for Si)	Very large (500 μA for Ge and 20 μA for Si)	Very large
Output Signal phase	In phase with input	Reverse	In phase with input
Applications	For high frequency applications	For AF applications	For impedance matching

4.16. Transistor as an Amplifier

The main utility of a transistor lies in its ability of amplifying weak signals. The weak signal is applied at the input terminals and the amplified output is obtained across the output terminals, as illustrated in Fig. 4.29.

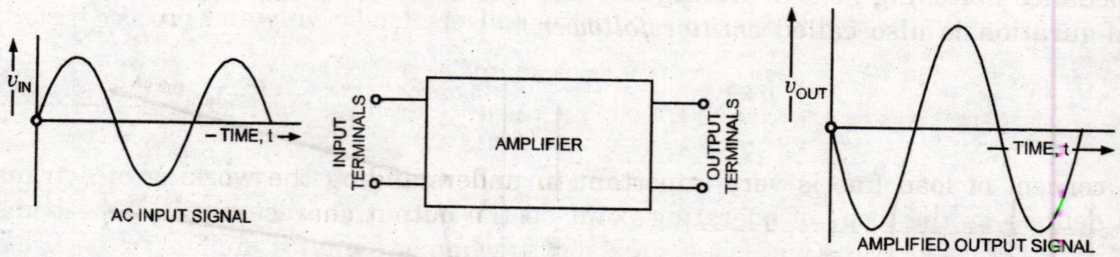

Fig. 4.29. ***Block Diagram of An Amplifier***

A transistor alone cannot perform the function of amplification and some passive components such as resistors and capacitors and a biasing battery is to be connected. Common emitter (CE) configuration, because of its high current, voltage and power gains, is much suited for most of the amplifier circuits. A common emitter N-P-N transistor amplifier circuit is given in Fig. 4.30.

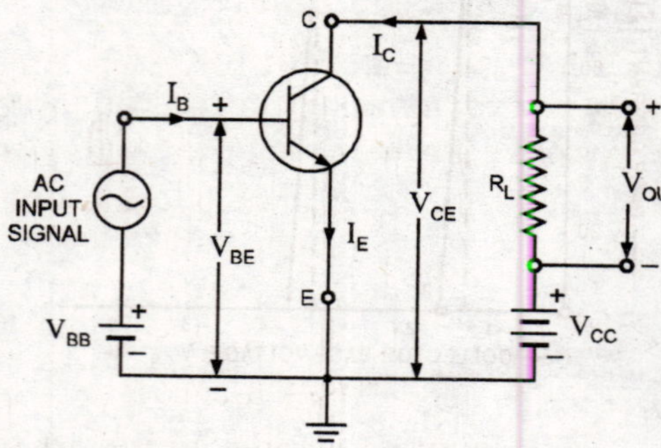

Fig. 4.30. ***Common Emitter NPN Transistor Amplifier Circuit***

The weak signal is applied between emitter-base junction and output is taken across a load resistor R_L connected in series with collector supply voltage V_{CC} (Fig. 4.30). In order to obtain faithful amplification, it is necessary that the input circuit remains always forward biased regardless the polarity of the ac input signal. So a battery V_{BB} is inserted in the circuit with the polarity indicated in addition to the signal voltage. This dc voltage is called the *bias voltage* and its magnitude is such as to keep the input circuit always forward biased regardless the polarity of the input signal.

The input circuit being forward biased has low resistance and a small change ΔV_{in} in input signal voltage causes a relatively large change ΔI_E in emitter current. This causes almost the same change in collector current because of transistor action. The collector current flowing through a high load resistance R_L develops a large voltage across it. The change in output voltage across load resistance R_L may be many times the change in input signal voltage. Thus voltage amplification $A = \frac{\Delta V_{out}}{\Delta V_{in}}$ will be greater than unity, and the transistor acts as an amplifier. This is further illustrated below by considering typical circuit values.

Let the load resistance R_L be of 10 kΩ and a change of 0.1 V in input signal voltage cause a change 0.5 mA in emitter current. This change of 0.5 mA in emitter current will also change collector current I_C by approximately 0.5 m A. This change of 0.5 mA in collector current will produce a change of $0.5 \times 10^{-3} \times 10 \times 10^3$ *i.e.* 5 V in output voltage appearing across the load resistor R_L of 10 kΩ. Thus the change of 0.1 V in input signal voltage causes a change of 5 V in the output voltage giving a voltage amplification of $\frac{5}{0.1} = 50$.

Necessity of Biasing. For most of the applications, transistors are required to operate as *linear amplifiers* (*i.e.* to amplify output voltage as a linear function of the input voltage).

To achieve this, it is necessary to operate the transistor over region of its characteristic curves which are linear, parallel and equi-spaced for equal increments of the parameters. Such an operation can be ensured by proper selection of zero signal operating point and limiting the operation of the transistor over the linear portion of the characteristics. For proper selection of zero signal operating point, proper biasing *i.e.* application of dc voltages at emitter-to-base junction and collector-to-base junction is required.

If the transistor is not biased properly, it would work inefficiently and produce distortion in the output signal.

4.17. TRANSISTOR LOAD LINES

The concept of load line is very important in understanding the working of a transistor. It is defined as the locus of operating point on the output characteristic of the transistor. It is the line on which the operating point moves when ac signal is applied to the transistor.

4.17.1. DC Load Line

In the circuit shown in Fig. 4.30, V_{CC} is the supply voltage to collector, R_C (or R_L) is the collector resistance (or load resistance) and V_{CE} is the collector-to-emitter voltage. Applying Kirchhoff's second law to the output or collector circuit we have,

$$V_{CC} = V_{CE} + I_C R_C \quad ...(4.38)$$

$$\text{or} \quad I_C = \frac{V_{CC} - V_{CE}}{R_C}$$

$$\text{or} \quad I_C = \frac{-V_{CE}}{R_C} + \frac{V_{CC}}{R_C} \quad ...(4.39)$$

This equation is to be plotted on the output characteristic of the transistor, V_{CE} and I_C are variables.

Identify this equation as $y = mx + c$ where $m = \frac{-1}{R_C}$ as the slope of the line and $I_C = \frac{V_{CC}}{R_C}$ as intercept of the line on vertical current axis (Fig. 4.31).

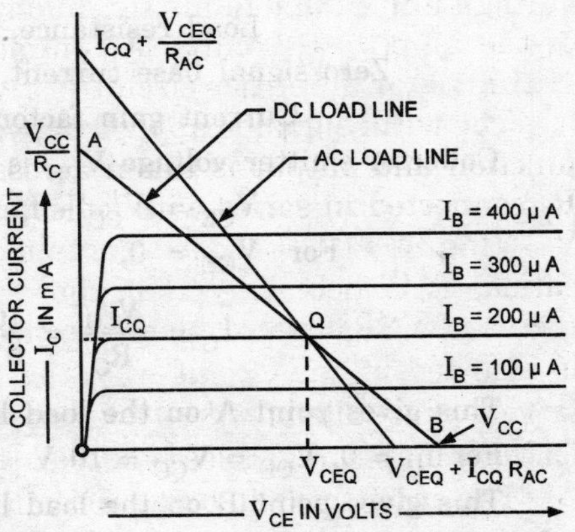

Fig. 4.31. *Common Emitter Transistor Output Characteristics With DC and AC Load Lines*

Consider the following two particular situations :

(*i*) When $V_{CE} = 0$, $I_C = \frac{V_{CC}}{R_C}$ ...saturation point A

(*ii*) When $I_C = 0$, $V_{CE} = V_{CC}$ ...cut-off point B

By joining these two points A and B, *dc load line* is obtained. The dc load line represents the dynamic characteristic of the device. The dc load line gives the values of collector current I_C and collector-emitter voltage V_{CE} corresponding to zero signal conditions.

4.17.2. Quiescent Point

It is a point on the dc load line which represents dc collector-emitter voltage V_{CE} and collector current I_C in the absence of ac signal. It is also called the *operating point* because the variations in V_{CE} and I_C take place, at about this point when signal is applied. The best position for this point is midway between cut-off and saturation points where $V_{CE} = \frac{1}{2} V_{CC}$. Point Q is quiescent point marked on the output characteristics shown in Fig. 4.31. Selection of the operating point is done as per application for which the device is to be used. For example in case of a small signal amplifier, in which power is conserved, operating point is selected so as to give lowest quiescent value of I_C, while for an amplifier operated to deliver small amount of power, operating point is selected so that available quiescent current is about one-half of the maximum permissible collector current I_C.

4.17.3. AC Load Line

When an ac signal is applied, the transistor voltage V_{CE} and collector current I_C vary above and below the quiescent point Q. So point Q is common to both dc and ac load lines. The *ac load line* gives the values of V_{CE} and I_C when an ac signal is applied. For drawing ac load line, take a convenient collector current change ΔI_C and compute the corresponding collector-emitter voltage change $\Delta V_{CE} = -\Delta I_C . R_C$ to obtain another point lying on the ac load line. Now the ac load can be drawn by joining this point and point Q. AC load is steeper than dc load line but the two lines intersect at the quiescent point Q determined by the biasing dc voltages and currents. AC load line takes into account the ac load resistance while the dc load line considers only the dc load resistance.

Example 4.13. In CE configuration (Fig. 4.30) collector-supply voltage V_{CC} = 10 V, load resistance R_C is 8 kΩ. Draw dc load line. Determine the operating point Q for zero signal if base current is 15 μA and β is 40.

Solution : Collector supply voltage, $V_{CC} = 10$ V

Load resistance, $R_C = 8$ kΩ

Zero signal base current, $I_B = 15$ μA

Current gain factor, $\beta = 40$

Collector-emitter voltage V_{CE} is given as

$$V_{CE} = V_{CC} - I_C R_C$$

For $V_{CE} = 0$,

$$I_C = \frac{V_{CC}}{R_C} = \frac{10\text{ V}}{8\text{ k}\Omega} = 1.25\text{ mA}$$

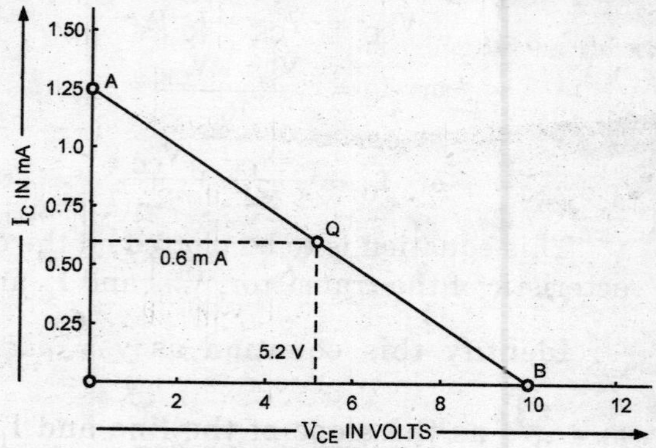

Fig. 4.32

This gives point A on the load line.

For $I_C = 0$, $V_{CE} = V_{CC} = 10$ V

This gives point B on the load line. AB is the required dc load line.

Zero signal collector current,

$$I_C = \beta I_B = 40 \times 15 \times 10^{-6} = 0.6\text{ mA}$$

Zero signal collector-emitter voltage,

$$V_{CE} = V_{CC} - I_C R_C = 10 - 0.6\text{ mA} \times 8\text{ k}\Omega = 5.2\text{ V}$$

∴ Operating point Q is (5.2 V, 0.6 mA) **Ans.**

Example 4.14. In a transistor circuit load resistance is 5 kΩ and quiescent current is 1.2 mA. Determine the operating point when the battery voltage V_{CC} = 12 V. How will the Q-point change when the load resistance is changed from 5 kΩ to 7.5 kΩ?

Solution : Zero signal collector current, $I_C = 1.2$ mA

Load resistance in collector circuit, $R_L = 5$ kΩ

Collector supply voltage, $V_{CC} = 12$ V

Zero signal collector-emitter voltage, $V_{CE} = 12 - 1.2\text{ mA} \times 5\text{ k}\Omega = 6$ V

Hence operating point is (6 V, 1.2 mA) **Ans.**

When load resistance is changed from 5 kΩ to 7.5 kΩ

Zero signal collector-emitter voltage, $V_{CE} = 12 - 1.2\text{ mA} \times 7.5\text{ k}\Omega = 3$ V

So operating point is (3 V, 1.2 mA) **Ans.**

Example 4.15. Draw the load line for the following figure (Fig. 4.33). What is I_C at saturation point? Find V_{CE} at cut-off point.

Solution : (*i*) Applying Kirchhoff's voltage law to the output loop we have

$$V_{CC} = I_C R_C + V_{CE}$$

For cut-off point, $I_C = 0$

and, therefore, $V_{CE} = V_{CC} = 20\text{ V}$ **Ans.**

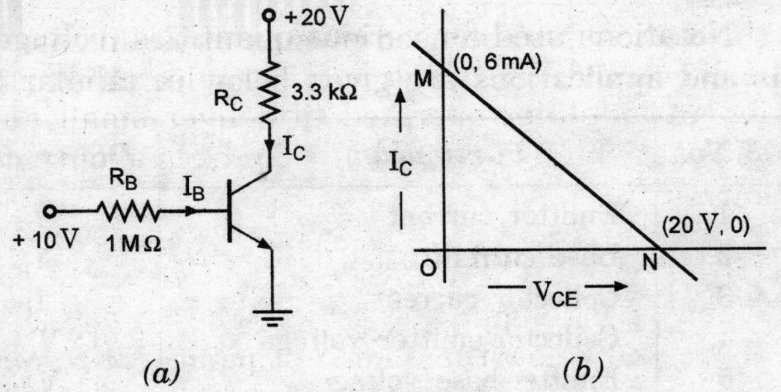

Fig. 4.33

∴ For saturation point, $V_{CE} = 0$

$$\text{So } I_C = \frac{V_{CC}}{R_C} = \frac{20}{3.3 \times 10^3} = 6\text{ mA } \textbf{Ans.}$$

Therefore, load line coordinates are given as (20 V, 0 mA) and (0 V, 6 mA) as shown in Fig. 4.33 (*b*).

Example 4.16. Determine V_C and V_B for the network given below with $\beta = 45$ and $V_{BE} = 0.7$ volt.

Solution : $\beta = 45$ and $V_{BE} = 0.7$ volt

Applying Kirchhoff's voltage law in input loop we have

$$I_B R_B + V_{BE} + V_{EE} = 0$$

Putting values in above equation we have

$$I_B \times 100 \times 10^3 + 0.7 - 9 = 0$$

$$\text{or } I_B = \frac{9 - 0.7}{10^5} = 83\ \mu\text{A}$$

$$I_C = \beta \times I_B = 45 \times 83 \times 10^{-6}\text{ mA} = 3.7\text{ mA}$$

Collector voltage, $V_C = V_{CC} - I_C R_C$

$= 0 - 3.7 \times 10^{-3} \times 1.2 \times 10^3$

$= -4.44\text{ V}$ **Ans.**

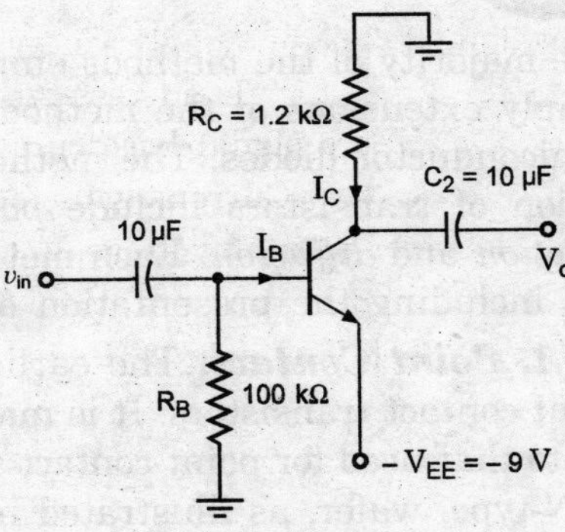

Fig. 4.34

$$V_B = V_{BE} + V_{EE} = 0.7 - 9 = -8.3\text{ volts } \textbf{Ans.}$$

4.18. STANDARD NOTATIONS FOR VOLTAGES AND CURRENTS

Standard notations or symbols are usually used for voltages and currents in transistor circuit analysis and applications.

1. Subscripts E (or *e*), B (or *b*) and C (or *c*) usually refer to emitter, base and collector quantities respectively.
2. Upper case (capital) letters with capital subscripts of the proper electrode symbol are used for dc or average values of voltages and currents.
3. Upper case (capital) letters with lower subscripts of the proper electrode symbol are used for rms values of ac voltages and currents.
4. Lower case (small) letters with lower subscripts of proper electrode symbol are used for instantaneous values of ac voltages and currents.

Notations used for various quantities (voltages and currents) in transistor circuit analysis and applications are given below in tabular form.

S.No.	Particular	DC or average	RMS ac	Instantaneous ac
1.	Emitter current	I_E	I_e	i_e
2.	Base current	I_B	I_b	i_b
3.	Collector current	I_C	I_c	i_c
4.	Collector-emitter voltage	V_{CE}	V_{ce}	v_{ce}
5.	Emitter-base voltage	V_{EB}	V_{eb}	v_{eb}
6.	Collector-base voltage	V_{CB}	V_{cb}	v_{cb}
7.	Emitter supply voltage	V_{EE}	—	—
8.	Base supply voltage	V_{BB}	—	—
9.	Collector supply voltage	V_{CC}	—	—
10.	Emitter voltage to ground	V_E	V_e	v_e
11.	Base voltage to ground	V_B	V_b	v_b
12.	Collector voltage to ground	V_C	V_c	v_c

4.19. Transistor Fabrication

The majority of the methods employed in transistor fabrication are simply extensions of the methods employed in the manufacture of semiconductor diodes. The methods most commonly used for fabrication of transistors include *point-contact*, *alloy junction*, *grown junction* and *diffusion*. Each method will be discussed here in brief, but including the presentation of the fundamental steps involved.

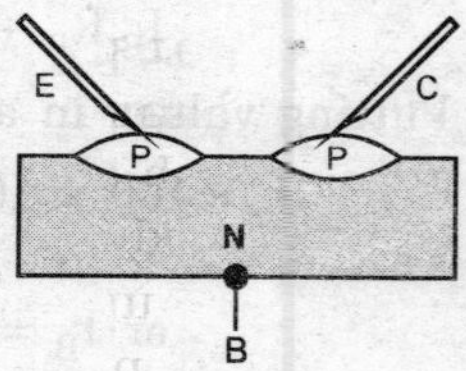

Fig. 4.35. ***PNP Point-Contact Transistor***

1. *Point Contact.* The earliest transistors developed were the point contact transistors. It is manufactured in a manner very similar to that used for point-contact semiconductor diodes. In this case two wires are placed next to N-type, wafer, as illustrated in Fig. 4.35. Electrical pulses are then applied to each wire resulting in a P-N junction at the boundary of each wire and the semiconductor wafer. Thus a P-N-P transistor shown in Fig. 4.35 is formed. Such transistors suffer from poor reliability and reproducibility and are, therefore, no longer commercially used.

2. *Alloy Junction.* This is also an extension of the alloy method of manufacturing semiconductor diodes. In this method a very thin layer of about 250 μm of N-type germanium crystal wafer is taken. It is lightly doped to serve as the base of the transistor. The small dots of indium are attached to the opposite sides of the semiconductor wafer. The entire structure is heated to a high temperature (above the melting point of indium but below that of germanium). The indium melts and dissolves in germanium beneath it and forms a saturated solution. On cooling, the germanium in contact with the base material recrystallizes with enough indium concentration to change it from N-type to P-type. The emitter, base and collector leads are then connected to this alloy and the whole assembly is then properly enclosed, as shown in Fig. 4.36.

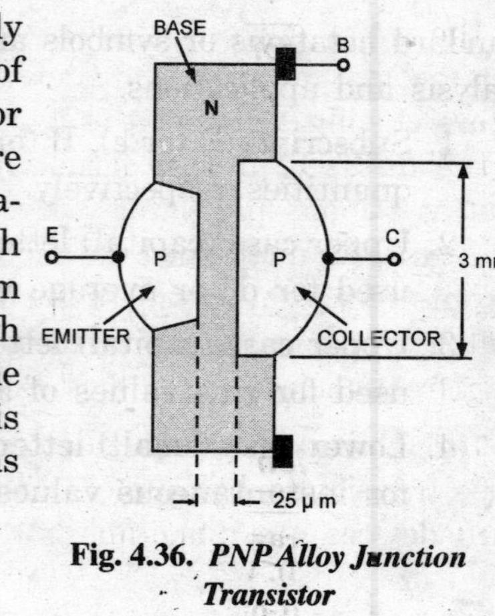

Fig. 4.36. ***PNP Alloy Junction Transistor***

The collector dot and resulting junction are made larger than the emitter so as to withstand the heavy current and power dissipation at the collector-base junction.

The alloy junction transistor is one of the earliest types of transistors that is still in use. It is relatively cheaper and gives high current gain. It can be fabricated for operation at high current and power levels.

3. *Grown Junction.* Such a transistor is formed by first growing germanium (or silicon) crystal of P-type from a melt of Ge (or Si) and afterwards P-type impurity is changed to N-type by adding N-type impurity in sufficiently large quantity to the melt and after a very small time interval, again the impurity is changed to P-type by adding P-type impurity in sufficient quantity. Thus a very thin layer of N-type semiconductor is formed sandwiched between P-type material on either side, as shown in Fig. 4.37. The process requires that the impurity control be such as to ensure the proper base width and doping levels of the N- and P-type materials. Transistors of this type are, in general, limited to less than $\frac{1}{4}$ W rating.

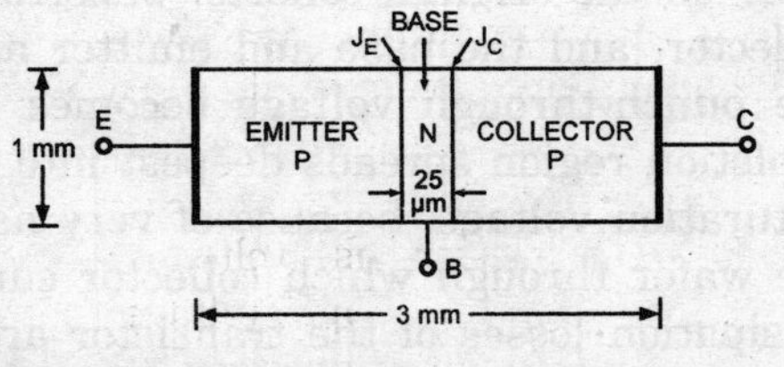

Fig. 4.37. *PNP Grown Junction Transistor*

4. *Diffused Junction.* The most frequently employed method of manufacturing transistors is the diffusion technique. The diffusion technique is employed in the production of mesa and *planar transistors*, each of which can be of the *diffused* or *epitaxial* type.

(*a*) Diffused Mesa. In the production of diffused mesa transistors, the thin wafer is kept as a whole disc (*i.e.* it is not sliced first). Several thousand transistors are simultaneously formed on the wafer by *diffusion process.* As illustrated in Fig. 4.38, the main body of the wafer becomes the N-type collectors, the diffused P-regions become the bases, and the final N-regions are the emitters. Metal strips are deposited on the base and emitter surfaces to form contacts.

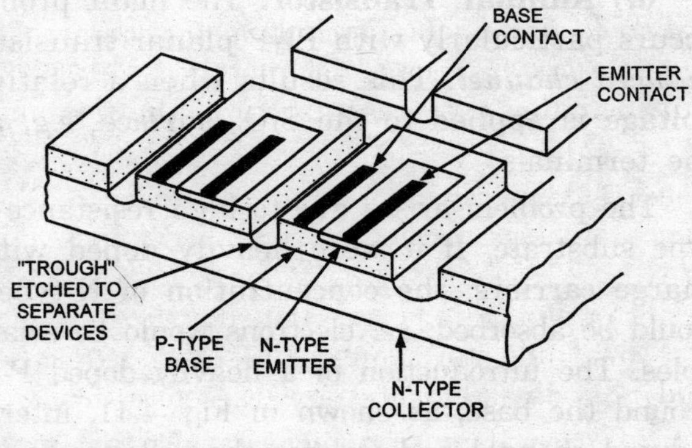

Fig. 4.38. *NPN Diffused Mesa Transistor*

The individual transistors could be separated by the usual process of scribing lines on the wafer surface and breaking the wafer into individual units. This would give a very rough edge, however, and there would likely be high leakage between collector and base. So, before cutting the disc, the transistors are isolated by etching away the unwanted portions of the diffused area to form separating troughs between devices. This leaves the base and emitter regions projecting above the main wafer forming the collector region. This is *mesa* structure. As mentioned earlier in the discussion of diode fabrication, the diffusion technique permits very tight control of the doping levels and thickness of various regions. The narrow base widths, which can be achieved by the diffusion process, make the mesa transistor useful at very high frequencies.

(*b*) Epitaxial Mesa. One of the drawbacks of diffused mesa transistors is that diffused mesa transistors, because of the highly resistive collector region, have a high *saturation voltage.* Such devices are unsuitable for saturated switching applications. The same characteristic of

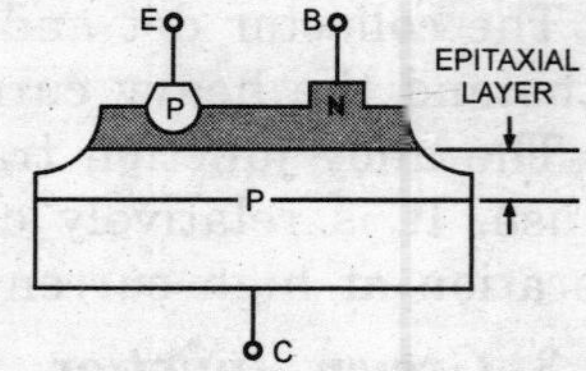

Fig. 4.39. *PNP Epitaxial Mesa Transistor*

high collector resistance is desirable to give high punch-through voltages. Epitaxial process provides both high *punch-through voltage* and low saturation levels, as shown in Fig. 4.39.

The major difference between the epitaxial mesa transistor and the diffused mesa transistor is the addition of an epitaxial layer on the original collector substrate. This layer becomes the collector, and the base and emitter are diffused as before. Now the punch-through voltage becomes high as the collector-base depletion region spreads deepest into the lightly doped collector. Saturation voltage, because of very narrow collector region, becomes low. The main body of the wafer through which collector current must flow has also a very low resistance. Thus dissipation losses of the transistor are reduced.

(c) Diffused Planar Transistor. The planar and epitaxial planar transistors are fabricated using two diffusion processes to form the base and emitter regions. The planar transistor, as shown in Fig. 4.40, has a flat surface, which accounts for the term "planar". A silicon dioxide layer is added, as illustrated in the figure, to eliminate exposed junctions. This construction gives a very low collector-base leakage current I_{CBO} of the order of 0.1 nA.

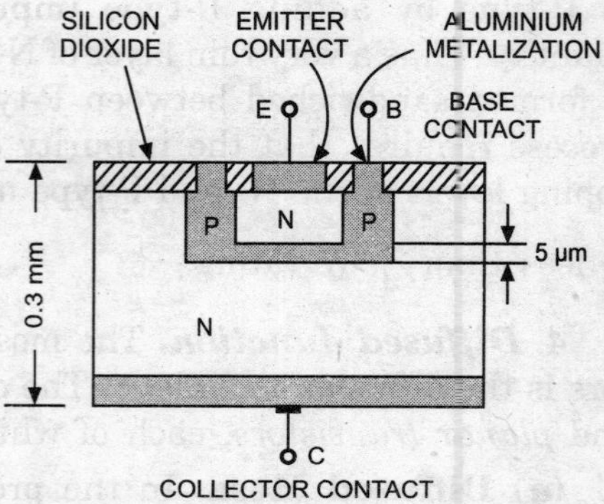

Fig. 4.40. *PNP Diffused Planar Transistor*

(d) Annular Transistor. The main problem that occurs particularly with PNP planar transistor is the *induced channel*. This results when a relatively high voltage is applied to the SiO_2 surface, *e.g,* at one of the terminals.

The problem arises due to high resistance of the P-type substrate. If it were heavily doped with P-type charge carriers, the concentration of N-type carriers would be absorbed; *i.e.* electrons would be swallowed by holes. The introduction of a heavily doped P-type ring around the base, as shown in Fig. 4.41, interrupts the induced channel and isolates the collector-base junction from the surface of the device. The *annular transistor* is, therefore, a high-voltage device with the low collector-base leakage of the planar transistor.

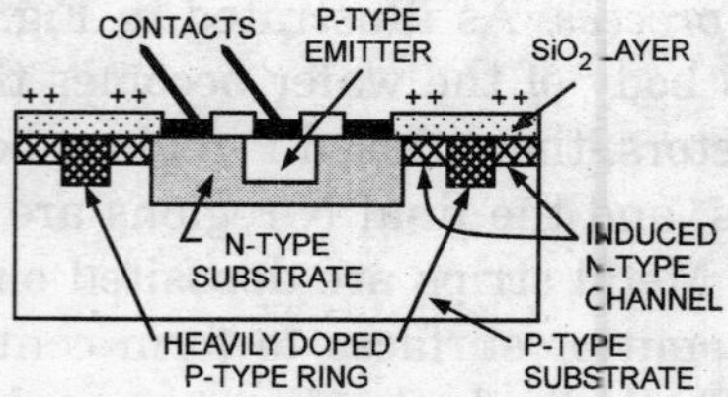

Fig. 4.41. *PNP Annular Transistor*

4.20. Transistor Maximum Ratings

The standard transistor data sheet will include at least three maximum ratings—*collector dissipation, collector voltage* and *collector current*.

Following are maximum ratings, for the 2N 3903 and 2N 3904 transistors, at 25°C free air temperature.

$$P_{C\,max} = 310 \text{ mW}$$
$$I_{C\,max} = 200 \text{ mA}$$
$$V_{CE\,max} = 40 \text{ V}$$

The *power* or *dissipation rating* of a transistor is defined as the power that a transistor can handle without destruction. It is given as the product of collector voltage and collector current. For CE configuration

$$P_{C\,max} = V_{CE}\,I_C \qquad ...(4.40)$$

Small-signal transistors can dissipate half a watt or less whereas *power transistors* can dissipate more than half a watt.

Specifying of total device dissipation as 310 mW at a maximum free air temperature of 25°C means that ($V_{CE} \times I_C$) must not exceed 310 mW. For example, if the maximum collector current ($I_{C\,max}$) of 200 mA is to be used, then $V_{CE\,max}$ should be $\frac{310\text{ mW}}{200\text{ mA}}$ *i.e.*, 1.55 V.

It is important to note that if the free air temperature exceeds 25°C, the device power or dissipation rating must be derated. The derating factor for 2N 3903 – 2N 3904 transistors is 2.81 mW/°C.

4.21. Importance of Collector-Emitter Voltage, V_{CE}

The collector-emitter voltage, V_{CE} plays a significant role in indication of region of operation of transistor *i.e.* whether the transistor is operating in cut-off, saturation or well into saturation regions.

Equality of V_{CE} and V_{CC} (*i.e.* $V_{CE} = V_{CC}$) indicates that the transistor is operating in *cut-off mode* (*i.e.* the transistor is turned-*off*). Zero value of V_{CE} indicates that the transistor is operating in saturation mode (*i.e.* the transistor is turned-*on* fully). Negative value (less than zero value) of V_{CE} indicates that the transistor is *well in saturation.* In practice, both of these conditions are avoided, except in switching circuits.

In practice, the transistor is operated with $V_{CE} = \frac{1}{2} V_{CC}$ (*i.e.* at approximately $\frac{1}{2}$ on).

Thus, variations in base current I_B in either direction controls collector current I_C well in both directions. It means that collector current I_C increases or decreases with the increase or decrease of base current I_B. However, with I_B *off* ($I_B = 0$), I_C is also *off* ($I_C = 0$). On the other hand, if the collector has been turned fully *on* (*i.e.* transistor is operating in saturation), the collector current is maximum, and the further increase in I_B will not reflect in I_C.

4.22. Transistor Amplifiers

One of the most important characteristics of a signal is its amplitude. Sometimes, a signal may have the desired waveform and frequency, and yet cannot be used because of its small amplitude. This is because the circuits or devices, that are to be operated with such signals as their input, need definite levels of amplitude of signal for proper operation. Thus the amplification or strengthening of the input signal becomes imperative. The device that amplifies the input signal is called the *amplifier*.

Almost no electronic system can work without an amplifier. For instance radio receivers amplify very weak signals until they are strong enough to make the sound audible. The transducers employed in medical and scientific investigations generate signals in microvolt or millivolt range which are incapable to operate indicating instruments. So these signals need amplification thousands and million times so as to be strong enough to operate indicating instruments.

We have studied that a properly biased transistor increases the amplitude of a weak signal and so acts as an amplifier. So an amplifier may be defined as a device that increases the current, voltage or power of an input signal with the help of a transistor by furnishing the additional power from a separate source of supply.

When only one transistor with associated circuit is used for increasing the strength of a weak signal, the circuit is known as *single stage transistor amplifier*. It may be emphasised here that a practical amplifier is always a multistage amplifier *i.e.* an amplifier using a number of stages of amplification.

We know that a transistor alone cannot perform the function of amplification and some passive components, such as resistors and capacitors, and a biasing supply is to be connected. When an ac signal is applied between the base and emitter terminals of a properly biased transistor, a small base current starts flowing. Because of transistor action, a much larger ac current (β times the base current) flows through the collector resistor R_C. Since the value of collector resistance is quite high (usually 5–10 kΩ), a large voltage appears across R_C. Thus, a weak signal applied between base and emitter circuit appears in the amplified form between collector and emitter. This explains the *amplifying action of a transistor*. When the input signal is quite weak and produces *small fluctuations* in the collector current in comparison to its quiescent value, the amplifier is called the *small signal amplifier* (or *voltage amplifier*). On the other hand, when the fluctuations produced in collector current are quite large (beyond the linear portion of the output characteristics), the amplifier is called the *large signal* amplifier (or *power amplifier*). Small signal or voltage amplifier is used as the first stage of the amplifier used in radio and TV receivers, tape-recorders, stereos and measuring instruments.

A dc voltage source connected to an amplifier is shown in Fig. 4.42. The amplifier consists of transistors that must be biased in the forward-active region so that the transistors can act as amplifying devices. It is desirable that the output signal is linearly proportional to the input signal so that the output of the speakers is an exact (as much as possible) reproduction of the signal generated from the CD player. Thus, it is desired that the amplifier be linear one.

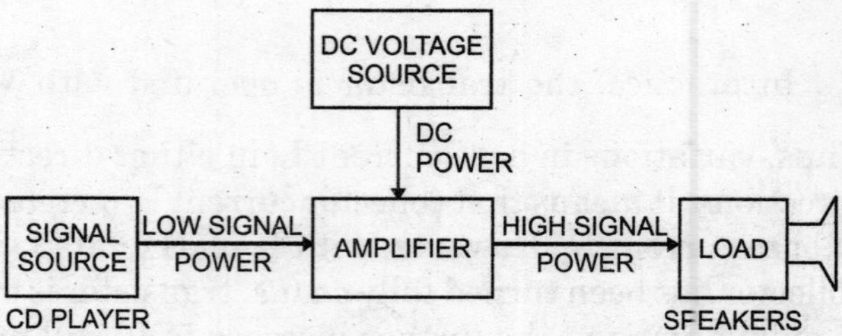

Fig. 4.42 *Block Diagram of a Compact Disc Player System*

4.23. Transistor Amplifier Practical Circuit

A practical circuit of a single stage transistor amplifier in CE configuration is shown in Fig. 4.43.

The resistors R_1, R_2 and R_E form the *biasing* and stabilization circuits. The biasing circuit must establish a proper operating point otherwise a part of the –ve half cycle of the signal may be cut-off in the output. The resistor R_L connected across the output terminals is called the load. When a number of stages are employed then R_L represents the input resistance for the next stage.

An electrolytic capacitor, called the *input capacitor*, C_{in} of capacity of about 10 μF is used to couple the signal to the transistor base. In the absence of this capacitor, the signal source resistance will come across R_2 and thus change the bias. This capacitor allows only ac signal to pass but isolates the signal source from R_2. Another capacitor, called the emitter bypass capacitor C_E, of capacity of about 100 μF is used in parallel with emitter resistance R_E in order to provide a low reactance path to the amplified ac signal. In the absence of this capacitor, amplified ac signal flowing through R_E will cause a voltage drop across it which in turn will feedback the input side and reduce the output voltage.

For coupling of one stage of amplifier to the next stage another capacitor C_C, called the coupling or blocking capacitor, of capacity of about 10 μF is used. Because of its presence, the output across the load resistance R_L is free from the collector dc voltage. In its absence R_C will come in parallel with the resistor R_1 of the biasing network of next stage and thereby change the biasing conditions of the next stage.

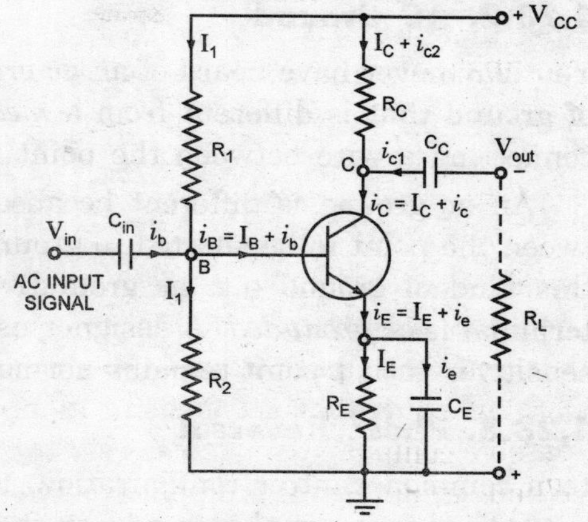

Fig. 4.43. ***Practical Circuit of a Transistor Amplifier in CE Configuration With Self Biasing***

Various currents flowing in different branches of the amplifier circuit are indicated in Fig. 4.43.

With the application of input signal

Base current, i_B = DC base current due to biasing circuit + ac base current due to ac input signal

$$= I_B + i_b \quad \text{...(4.41)}$$

Collector current, i_C = DC collector current due to biasing circuit + ac collector current due to ac input signal

$$= I_C + i_c = \beta\,(I_B + i_b) \quad \text{...(4.42)}$$

Emitter current i_E = DC emitter current + ac emitter current due to ac input signal

$$= I_E + i_e \quad \text{...(4.43)}$$

4.23.1. Emitter Bypass Capacitor

As mentioned above, it is included in the circuit to provide a low reactance path to the amplified ac signal. As shown in Fig. 4.43, it is not connected in series like a coupling capacitor, instead, it is connected across the emitter resistor R_E. The reason for doing so is to bypass or shunt ac current away from the emitter resistor R_E. When the frequency is high enough it looks like a short. Thus emitter bypass capacitor C_E is used to prevent *negative feedback* in the emitter circuit.

Very simply, negative feedback involves feeding back a portion of the amplifier *output* signal to its *input* circuit. Further, the signal is fedback in such a manner that it *opposes* the input signal. The **removal of bypass capacitor** results in an increase in the amplifier's *input resistance,* a reduction in its *voltage gain,* and an *increase* in the *output resistance* "looking into the collector". Obviously, an increase in input resistance is a step in the right direction for voltage amplifier. However, a reduction in the voltage gain and increase in the output resistance are generally undesirable effects. Yet there are some advantages offered by these sacrifices. First, even though the voltage gain is reduced, but *stability* is increased. Specifically, it tends to be less dependent on the transistor's parameters (*e.g.*, its g_m). Second, the increase in the collector-to-ground output resistance makes our approximation of R_{out} equal to R_C even *more realistic.*

4.23.2. AC Ground

You may never have heard of an *ac ground* before, but it really does exist as a separate kind of ground that is different from a *mechanical ground.* A mechanical ground is achieved by connecting a wire between the point to be grounded and ground.

An ac ground is different because it is produced by connecting a *bypass capacitor* between the point to be shorted to ground, as far as ac signals are concerned, and the ground, this kind of ground (*i.e.* ac ground) exists only at high frequencies. In Fig. 4.43 emitter terminal is *ac grounded.* A designer uses an ac ground when any point needs to be frequency sensitive. Such a point remains normal at low frequencies but grounded at high frequencies.

4.23.3. Phase Reversal

In a common emitter configuration, the output voltage increases in the negative direction when the input signal increases in the positive direction and vice-versa. This is called phase reversal* and causes a phase difference of 180° between the input signal voltage V_{in} and output voltage V_{out}.

Consider a CE amplifier circuit shown in Fig. 4.44, the signal being fed at the input terminals *i.e.* between base and emitter and the output is taken from the collector and emitter ends.

The instantaneous output voltage v_{CE} is given as

$$v_{CE} = V_{CC} - i_C R_C \quad \text{...(4.44)}$$

Reactance of coupling capacitor C_C is negligible at ordinary signal frequencies and so it may be considered a short for the output signal.

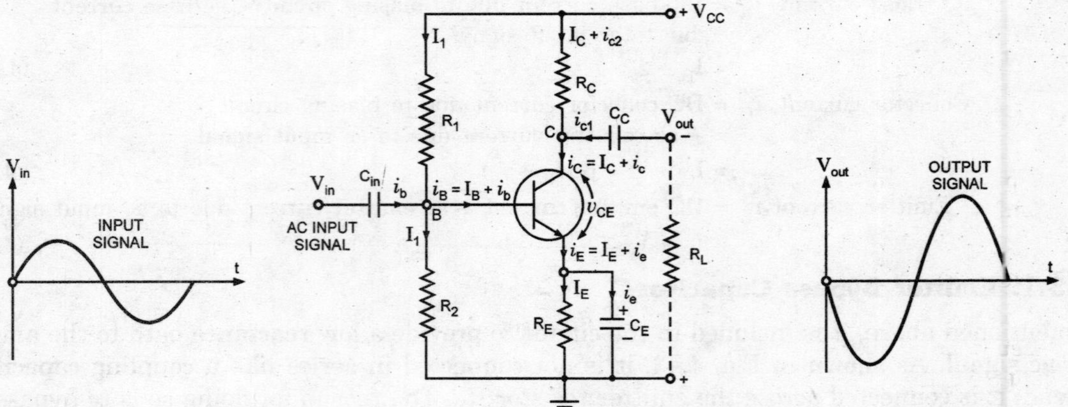

Fig. 4.44. *Practical Circuit of a Transistor Amplifier in CE Configuration With Self Biasing*

With the increase in the signal voltage in the positive half cycle, the base current increases causing increase in collector current and so the voltage drop $i_C R_C$. As V_{CC} is constant, therefore, output voltage v_{CE} decreases. Thus we can say that as the signal voltage increases in positive direction, the output voltage increases in the negative direction *i.e.* there is a phase difference of 180° between output and input. Conversely on negative half cycle of the input signal voltage, lesser current flows through collector resistance R_C and so voltage drop across it decreases. For this reason output voltage v_{CE} increases.

The above phenomenon can be proved mathematically as below :

* This will be so when output is taken from collector and emitter ends, as is always done. However, if the output is taken across collector resistance R_C, it will be in phase with the input.

Differentiating Eq. (4.44), we have

$$dv_{CE} = 0 - di_C R_C$$

$$\text{or} \quad dv_{CE} = - di_C R_C$$

The negative sign indicates that the output voltage is 180º out of phase with the input voltage.

The above fact *i.e.* phase reversal in CE configuration can be illustrated graphically with the aid of output characteristics and ac load line, as shown in Fig. 4.45. The base current varies between I_{B_1} and I_{B_2} with I_B as the zero signal base current. From figure we see that when the base current is maximum in the positive direction, v_{CE} becomes maximum in the negative direction and vice-versa. Thus, the input and output voltages are in phase opposition *i.e.* the transistor has produced a phase reversal of output voltage w.r.t. the input signal voltage.

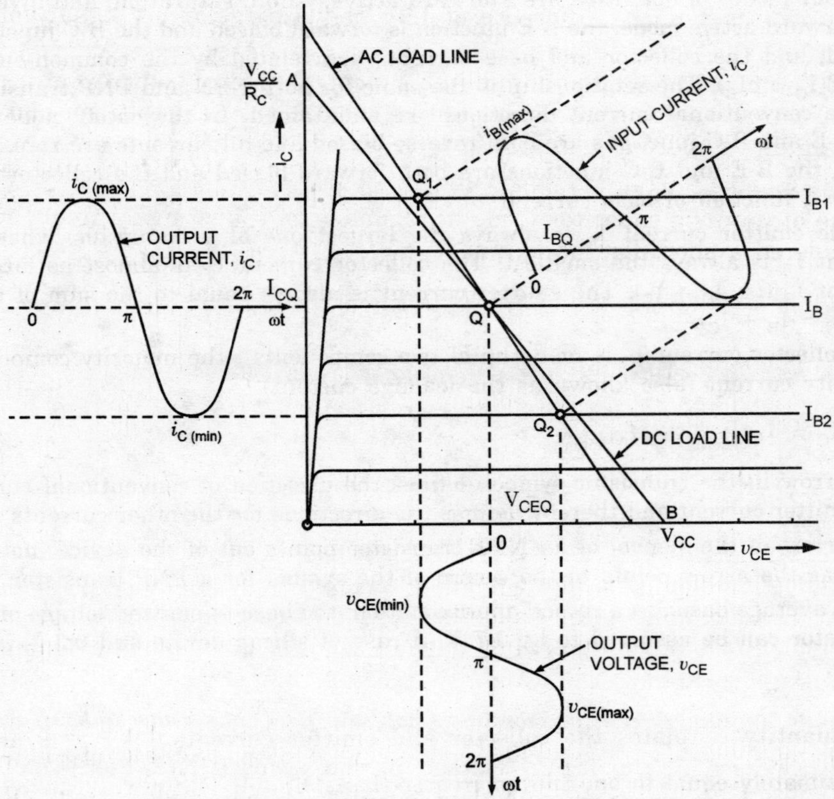

Fig. 4.45. ***Graphical Illustration of Phase Reversal***

The worthnoting points are that common base and common collector amplifiers do not produce reversal, and input and output currents are in phase in all the three configurations.

Highlights

1. Semiconductor devices have the following advantages over vacuum tubes : They are (*i*) smaller in size (*ii*) lighter in weight, (*iii*) more rugged and (*iv*) more efficient. In addition, they have (*i*) no warm up period, (*ii*) no heater requirement and (*iii*) lower operating voltages.

2. There are two complementary bipolar transistors—NPN and PNP. Each transistor has three separately doped regions and two PN junctions. The center region (base) is very narrow, so the two PN junctions are said to be interacting junctions.
3. Since a transistor consists of either PNP or NPN slices, a very simplified model for biasing purposes comprises two diodes "pointing inward or outward" respectively.
4. For normal operation, base-emitter junction is forward-biased and the base-collector junction is reverse biased. Under these conditions, majority carriers from the emitter are injected into the base where they become minority carriers. These minority carriers diffuse across the base into the B-C space charge region where they are swept into the collector.

   In this mode of operation, the current at one terminal of the transistor (collector current) is controlled by the voltage applied across the other two terminals of the transistor (base-emitter voltage). This is the basic transistor action.
5. The modes of operation of a BJT are determined by the biases applied to the two junctions. The four modes of operation are : forward active, cutoff, saturation, and inverse active. In the forward-active mode, the B-E junction is forward biased and the B-C junction is reverse biased, and the collector and base currents are related by the common-emitter current gain $\beta(I_C = \beta I_B)$. The relationship is the same for both NPN and PNP transistors, as long as the conventional current directions are maintained. In the cutoff mode of operation, the B-E and B-C junctions are both reverse-biased and all currents are zero. In saturation mode, the B-E and B-C junctions are both forward-biased and the collector current is no longer a function of base current.
6. The dc emitter current I_E is always the largest one of a transistor, whereas the base current I_B is always the smallest. The collector current I_C is almost as large as emitter current I_E (*i.e.* I_C I_E). The emitter current is always equal to the sum of the other two (*i.e.* $I_E = I_B + I_C$).
7. The collector current I_C is made up of two components : the majority component and the minority current (also known as the leakage current).

   i.e., $I_C = I_{C_{majority}} + I_{CO_{minority}}$
8. The arrow in the transistor symbol defines the direction of conventional current flow for the emitter current and thereby defines the directions for the other currents in the device. The arrow in the symbol of an NPN transistor points out of the device (not pointing in), whereas the arrow points to the centre of the symbol for a PNP transistor (pointing in).
9. On an average basis, as a second approximation, the base-to-emitter voltage of an operating transistor can be assumed to be 0.7 V in case of silicon device and 0.3 V in germanium device.
10. The quantity α relates the collector and emitter currents $\left(\alpha = \frac{I_C}{I_E}\right)$ and is always approximately equal to one (never greater than 1).
11. The impedance between terminals of a forward-biased junction is always relatively small, whereas the impedance between terminals of a reverse-biased junction is usually quite large.
12. The quantity beta β provides an important relationship between the base and collector currents and is usually between 50 and 400.
13. The dc beta is defined by a simple ratio of dc currents at an operating point $\left(i.e.\ \beta_{dc} = \frac{I_C}{I_B}\right)$, whereas the ac beta β_{ac} is sensitive to the characteristics in the region of interest. For most applications, however, the two are considered equivalent as a first approximation.

14. A three-terminal device needs two sets of characteristics to completely define its characteristics.
The transistor input characteristic is essentially that of a forward-biased junction, with some dependence upon the transistor's output voltage.
15. For a transistor used in CB configuration, the collector or output characteristics consist of practically horizontal lines of I_C versus V_{CB} with I_E as the controlling parameter.
16. The CB ac current gain, $\alpha_{ac} = \left.\frac{\Delta I_C}{\Delta I_E}\right|_{V_{CB} = \text{Constant}}$ while $\alpha_{dc} = \frac{I_C}{I_E}$ and is less than unity.
17. The reverse saturation current in a CB configuration $= I_{CBO} = I_{CO}$, while in a CE configuration it is $I_{CEO} = (\beta + 1)I_{CBO}$.
18. The collector or output characteristics for a transistor used in the CE configuration have the collector current I_C versus collector voltage curves sloping upward slightly with base current as the controlling parameter.
19. The CE ac current gain, $\beta_{ac} = \left.\frac{\Delta I_C}{\Delta I_B}\right|_{V_{CE} = \text{Constant}}$, while $\beta_{dc} = \frac{I_C}{I_B}$.
20. Transistors have maximum ratings of voltages, currents and powers. Small-signal transistors can dissipate 1W or less. Power transistors can dissipate more than 1W. Temperature can change the value of the transistor characteristics. Maximum power decreases with the increase in temperature. Also current gain varies greatly with temperature.
21. The device that amplifies the amplitude of the input signal is called the amplifier. An amplifier may be defined as a device that increases the current, voltage or power of an input signal with the help of a transistor by furnishing the additional power from a separate source of supply.
Amplification in the ac domain cannot be had without applying the dc biasing level.
22. In a CE configuration, the output voltage increases in the negative direction when the input signal voltage increases in the positive direction and vice-versa. This is called the phase reversal and causes a phase difference of 180° between the input signal voltage and output voltage.

Exercises

1. Draw NPN and PNP transistors. Label all the currents and show the direction of flow. How are all the currents of a transistor related?
2. What is a transistor? Show the transistor action for the N-P-N and P-N-P transistors.
[Pb. Technical Univ. Electronic Devices and Circuits 2004-05]
3. Explain the working of a PNP transistor with the help of its energy band diagram. Explain 'emitter injection efficiency' and base 'transport factor'. How do they influence the operation of a transistor?
4. If a transistor has its collector-base junction forward biased and the other junction reverse biased, will it work? Explain.
5. If two P-N junction diodes are connected back to back as shown in Fig. 1, will it work as a transistor with the terminals designated in the figure? Explain.

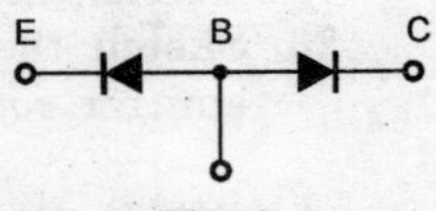

Fig. 1

6. Define with respect to BJT the following : (*i*) I_{CO}, (*ii*) α, (*iii*) β, (*iv*) I_{CEO}, (*v*) I_{CBO}

7. Differentiate between I_{CO} and I_{CBO}. What is the effect of temperature on I_{CBO} ?

8. Explain different components of currents flowing through the structure of an N-P-N transistor. How the emitter injection efficiency and base transport factor influences the amplification factor ? **[U.P. Technical Univ. 2005-06]**

9. In reference to a bipolar junction transistor, define the following quantities :

(*i*) The emitter injection efficiency.

(*ii*) Current transfer ratio.

(*iii*) Base-to-collector current amplification factor.

(*iv*) Early effect. **[U.P. Technical Univ. 2006-07]**

10. What are the factors that affect the amplification factor of the BJT. **[U.P. Technical Univ. 2007-08]**

11. Draw a relation between α and β in case of BJT. **[M.D. Univ. Third Semester May 2005]**

12. In how many modes the BJT works ? Also explain the biasing pattern for each of them.

13. Explain the following regions in a transistor : (*i*) Active, (*ii*) Saturation, (*iii*) Cut-off.

14. Define Eber's Moll model. Justify the following statement giving Eber's Moll model :

There is no essential reason theoretically which constraints us from using a BJT in an inverted mode.

15. Draw the Ebers-Moll model for a transistor and give the equation for emitter current and collector current.

16. Explain why a transistor action cannot be achieved by connecting two back-to-back diodes. In a transistor explain, why emitter region is heavily doped, base width is small and collector area is large. **[U.P. Technical Univ. 2006-07]**

17. Explain construction, working and characteristics of BJT. **[M.D. Univ. Third Semester December 2006]**

18. Give the output characteristics of NPN transistor. **[M.D. Univ. Third Semester May 2005]**

19. Differentiate between CE, CC and CB amplifiers. **[Pb. Technical Univ. Electronic Devices and Circuits 2004-05]**

20. Sketch a family of CB output characteristics for a transistor. Indicate the active, cut-off and saturation regions. Explain the shapes of the curves qualitatively. **[Pb. Univ. Applied Electronics-I, May 1994]**

21. Sketch typical CB input characteristic curves for an NPN transistor. Label all variables. How would you calculate the input dynamic resistance of the transistor ?

22. What are the characteristics and uses of a common base transistor amplifier ?

23. What is Early effect ? Explain how it affects the BJT characteristics in CB configuration.

24. Explain base-width modulation (Early effect) with the aid of plots of potential and minority concentration throughout the base region.

25. Explain the operation of NPN transistor in CB configuration. Show that $\alpha = \dfrac{\beta}{1+\beta}$. Also give the reason why the input impedance of CB is less than input impedance of CE configuration.

26. Sketch the typical input-output characteristics of a BJT when connected in (*i*) common emitter configuration and (*ii*) common base configuration. **[Pb. Technical Univ. Electronic Devices and Circuits 2004-05]**

27. Compare the CB, CE and CC configurations of a transistor. For CE configuration, prove that $I_C = \beta\, I_B + (\beta + 1)\, I_{CO}$.

28. Draw the circuits of various transistor configurations. List their important features. Why CE configuration is mainly used ? What are uses of other configurations ?

29. Draw a simple circuit of a common-collector amplifier and explain its operation. Obtain an expression for its voltage gain.

30. Compare CE, CB and CC configurations of BJT. Also discuss the input and output characteristics of a transistor in CE configuration.

[M.D. Univ. Third Semester December 2007; May 2008]

31. Draw the load line on the output characteristics of a common-emitter amplifier. What is its significance? Show how the transistor acts as an amplifier.

[Pb. Univ. Applied Electronics I, Dec. 1995]

32. Explain the difference between ac and dc load lines. Derive the load line equation of a BJT in CE configuration.

33. Explain with neat sketches the diffused mesa and epitaxial mesa transistors. Discuss the main difference between their construction and give advantages and disadvantages of the mesa transistors.

34. Explain the manufacturing process for diffused planner and annular transistors. Give their merits and demerits.

35. Explain working of a transistor as an amplifier. **[M.D. Univ. Third Semester May 2007]**

36. Draw the circuit diagram of a single stage transistor amplifier, stating the function of each component used in the circuit.

37. Explain the effects of coupling and bypass capacitors in a circuit.

38. Write short notes on the following :

(*i*) Use of CE configuration of transistors for amplifiers.

(*ii*) Fabrication of transistor by planar technology.

(*iii*) Transistor as an amplifier.

(*iv*) Phase reversal.

Short Answer Type Questions With Answers

Q. 1. Explain why an ordinary junction transistor is called bipolar?

Ans. Because the transistor operation is carried out by two types of charge carriers (majority and minority carriers), an ordinary transistor is called bipolar.

Q. 2. What is done to the base region of a transistor to improve its operation?

Ans. Base is made thin and very lightly doped in comparison to either emitter or collector so that it may pass most of the injected charge carriers to the collector.

Q. 3. Why transistor is called current controlled device?

Ans. The output voltage, current or power is controlled by the input current in a transistor so it is called the current controlled device.

Q. 4. What is the significance of arrowhead in the transistor symbol?

Ans. Arrowhead is always marked on the emitter. The direction indicates the conventional direction of current flow (from emitter-to-base in case of P-N-P transistor and from base-to-emitter in case of N-P-N transistor). Generally, no arrowhead is marked for collector since its reverse leakage current is always opposite to the direction of emitter current.

Q. 5. Discuss the need for biasing the transistor.

Ans. For normal operation, base-emitter junction should be forward-biased and the collector-base junction reverse-biased. The amount of bias required is significant for establishment of the operating or Q-point which is dictated by the mode of operation desired.

In case the transistor is not biased properly, it would (*i*) work inefficiently, (*ii*) produce distortion in the output signal and (*iii*) with the change in transistor parameters or temperature rise, the operating point may shift and the amplifier output will be unstable.

Q. 6. What are 'emitter injection efficiency' and 'base transport factor' and how do they influence the transistor operation ?

Ans. The ratio of current of injected carriers at emitter junction to the total emitter current is called the emitter injection efficiency. The ratio of collector current to base current is known as transport factor

$$i.e., \beta = \frac{I_C}{I_B}$$

The larger the value of emitter injection efficiency, the larger the injected carriers at emitter junction and this increases the collector current. The larger the β value larger the injected carriers across collector junction and hence collector current increases.

Q. 7. Which of the transistor currents is always the largest ? Which is always the smallest ? Which two currents are relatively close in magnitude ? **[U.P. Technical Univ. 2006-07]**

Ans. The emitter current I_E is always the largest one. The base current I_B is always the smallest. The collector current I_C and emitter current I_E are relatively close in magnitude *i.e.* $I_C \simeq I_E$.

Q. 8. Why silicon type transistors are more often used than germanium type ?

Ans. Because silicon transistor has smaller cut-off current I_{CBO}, small variations in I_{CBO} due to variations in temperature and high operating temperature as compared to those in case of germanium type.

Q. 9. Why collector is made larger than emitter and base ?

Ans. Collector is made physically larger than emitter and base because collector is to dissipate much power.

Q. 10. Why the width of the base region of a transistor is kept very small compared to other regions ?

Ans. Base region of a transistor is kept very small and very lightly doped so as to pass most of the injected charge carriers to the collector.

Q. 11. Why emitter is always forward biased ?

Ans. Emitter is always forward biased w.r.t. base so as to supply majority charge carriers to the base.

Q. 12. Why collector is always reverse-biased w.r.t. base ?

Ans. Collector is always reverse-biased w.r.t. base so as to remove the charge carriers away from the base-collector junction.

Q. 13. Can a transistor be obtained by connecting two semiconductor diodes back-to-back ?

Ans. No. Because in case of two discrete back-to-back connected diodes there are four doped regions instead of three and there is nothing that resembles a thin base region between an emitter and a collector.

Q. 14. What is α_{dc} ?

Ans. α_{dc} is the *dc* current gain and is equal to the ratio of collector current to the emitter current.

Q. 15. What is β ?

Ans. The beta factor (β) is the current gain factor of a common-emitter circuit and is defined as the ratio of collector current and base current.

Q. 16. How α and β are related to each other ?

Ans. α and β are related as below :

$$\alpha = \frac{\beta}{1+\beta} \quad \text{or} \quad \beta = \frac{\alpha}{1-\alpha}$$

Q. 17. Define beta of a transistor.

Ans. The β factor of a transistor is the common emitter current gain of that transistor and is defined as the ratio of collector current to base current.

$$i.e., \beta = \frac{I_C}{I_B}.$$

Q. 18. What are the three configurations of a transistor amplifier ?

Ans. The three configurations of a transistor amplifier are common-base, common-emitter and common-collector configurations.

Q. 19. Why is there a maximum limit of collector supply voltage for a transistor ?

Ans. Although collector current is practically independent of collector supply voltage over the transistor operating range, but if V_{CB} is increased beyond a certain value collector current I_C eventually increases rapidly and possibly destroys the device.

Q. 20. For a PNP transistor in the active region what is the sign (positive or negative) of I_E, I_C, I_B, V_{CB} and V_{EB} ? **[Pb. Univ. May 1994]**

Ans. Signs of I_E, I_C, I_B, V_{CB}, and V_{EB} for a PNP transistor are as follows :

Quantity	Sign
I_E	Positive
I_C	Negative
I_B	Negative
V_{CB}	Negative
V_{EB}	Positive

Q. 21. Explain why $I_{CEO} >> I_{CBO}$?

Ans. The collector cut-off current denoted by I_{CEO} is much larger than I_{CBO}. I_{CEO} is given as

$$I_{CEO} = \frac{1}{1-\alpha} I_{CBO}$$

Because α is nearly equal to unity (slightly less than unity), $I_{CEO} >> I_{CBO}$.

Q. 22. Why CE configuration is most popular in amplifier circuits ?

Ans. CE configuration is mainly used because its current, voltage and power gains are quite high and the ratio of output impedance and input impedance are quite moderate.

Q. 23. Why is CC configuration seldom used ?

Ans. CC configuration is seldom used because its voltage gain is always less than unity.

Q. 24. What are the main purposes for which a common collector amplifier may be used ? **[I.E.S. 1997]**

Ans. For common collector configuration current gain is high (about 100) but voltage gain is less than unity, input impedance is the highest and output impedance is the lowest. This circuit finds wide application as a buffer amplifier between a high impedance source and a low impedance load.

Q. 25. Which configuration among CE, CB, CC gives highest input impedance and no voltage gain ?

Ans. Common collector configuration has the highest input impedance and voltage gain less than unity.

Q. 26. Which of the configurations (CB, CE, CC) has the (*i*) highest R_i (*ii*) lowest R_i (*iii*) highest R_0 (*iv*) lowest R_0 (*v*) lowest A_i ? **[Pb. Univ. May 1994]**

Ans.

	Characteristic	*Configuration*
(*i*)	Highest input resistance, R_{in}	CC
(*ii*)	Lowest input resistance, R_{in}	CB
(*iii*)	Highest output resistance, R_{out}	CB
(*iv*)	Lowest output resistance, R_{out}	CC
(*v*)	Lowest current gain, A_i	CB

Q. 27. Explain base-width modulation (Early effect).

Ans. The modulation of the effective base-width by the collector voltage is known as Early effect. An increase in collector voltage increases the space charge width at the ouput junction diode and thus the effective base width W_b is reduced.

Q. 28. What is punch through effect ?

Ans. When on increasing collector voltage effective base-width is reduced to zero the emitter barrier becomes smaller than normal value $V_0 - |V_{EB}|$ because the collector voltage has reached through the base region. This lowering of emitter-junction voltage causes an excessively large emitter current, thus placing an upper limit on the magnitude of collector voltage. This phenomenon is known as punch through.

Q. 29. What do you understand by collector reverse saturation ? In which configuration does it have a greater value ?

Ans. When input current (I_E in case of CB configuration and I_B in case of CE configuration) is zero, collector current I_C is not zero although it is very small. In fact this is the reverse leakage current or collector reverse saturation current (I_{CBO} or simply I_{CO} in CB configuration and I_{CEO} in CE configuration). In case of CE configuration it is much more than that in case of CB configuration.

Q. 30. What is quiescent point ?

Ans. Quiescent point is a point on the *dc* load line which represents V_{CE} and I_C in the absence of *ac* signal and variations in V_{CE} and I_C take place around this point when ac signal is applied.

Q. 31. Why point contact transistors are not commercially used ?

Ans. Because of poor reliability and reproducibility.

Q. 32. What is the major difference between the epitaxial mesa transistor and the diffused mesa transistor ?

Ans. The major difference between the epitaxial mesa transistor and diffused mesa transistor is that an epitaxial layer is added on the original collector substrate in epitaxial mesa transistor.

Q. 33. Explain how a BJT can be used as an amplifier.

Ans. A transistor operates as an amplifier by transfer of the current from low impedance loop to high impedance loop. For detail refer to Art. 4.16.

Q. 34. What limits the maximum current a transistor is capable of handling ?

Ans. The junction area of the collector region limits the maximum current a transistor can handle. Large power ratings are obtained from larger junction areas. More is the surface area, more is the effective heat dissipation and more is the magnitude of current, a transistor can handle.

Q. 35. What is an amplifier ?

Ans. The device that amplifies the amplitude of the input signal is called the amplifier. An amplifier may be defined as a device that increases the current, voltage or power of an input signal with the help of a transistor by furnishing the additional power from a separate source of supply.

Q. 36. What is meant by small signal amplifier ?

Ans. When the input signal is quite weak and produces small fluctuations in the output current in comparison to its quiescent value, the amplifier is called the small signal or voltage amplifier.

Q. 37. What is meant by phase reversal ?

Ans. In a CE configuration, the output voltage increases in the negative direction when the input signal voltage increases in the positive direction and vice-versa. This is called the phase reversal and causes a phase difference of 180° between the input signal voltage and output voltage.

Problems

1. Determine emitter current I_E, collector current I_C for a transistor with $\alpha_{dc} = 0.97$ and collector to base leakage current 10 μA. Base current is 50 μA **[Ans.** 2.00 mA, 1.95 mA]
2. For a certain transistor collector current is 6.25 mA, base current is 100 μA and collector-to-base leakage current 5 μA. Determine emitter current and α_{dc}. **[Ans.** 6.35 m A, 0.983]
3. In a particular transistor the collector current is 5.6 mA and the emitter current is 5.75 mA. Determine α_{dc}. **[Ans.** 0.974]
4. A transistor has $\alpha = 0.98$. If emitter current of the transistor is 1 mA, determine the base current and gain factor β. **[Ans.** 20 μ A; 49]
5. Determine β if $\alpha = 0.99$ **[Ans.** 99]
6. A bipolar junction transistor has a collector current of 4 mA and base current of 20 μA. Determine its β. **[Ans.** 200]
7. A BJT has a base current of 200 μA and emiter current of 20 mA. Determine collector current and β. **[Ans.** 19.8 mA; 99]
8. In a transistor $I_C = 0.95$ mA, $I_E = 1$ mA. Determine base current, α and β in CB configuration. **[Ans.** 50 μA; 0.95; 19]
9. In a transistor CE configuration the voltage across 5 kΩ load resistance is 5 V. Determine collector current, base current and emitter current if $\beta = 50$. **[Ans.** 1.0 mA; 20 μA; 1.02 mA]
10. In a grounded base configuration the voltage drop across load resistance of 4 kΩ is 3.2 V. Determine collector current, emitter current and base current if $\alpha = 0.96$. **[Ans.** 0.8 mA; 0.8333 mA; 0.0333 mA]
11. A CE transistor amplifier has input resistance of 100 Ω and output resistance of 6 kΩ. Determine the voltage gain and power gain if current gain factor is 90. **[Ans.** 5,400; 486,000]
12. Increase in collector-emitter voltage from 4 V to 8 V causes increase in collector current from 5 mA to 6 mA. Determine the output resistance. **[Ans.** 4 kΩ]
13. A change of 200 mV in base-emitter voltage causes a change of 60 μA in the base current. Determine its input resistance. **[Ans.** 3.33 kΩ]
14. A transistor is connected in CB configuration. When the emitter voltage is changed by 200 mV, the emitter current changes by 5 mA. During this variation, the collector to base voltage is kept fixed. Calculate the dynamic input resistance of transistor. **[Ans.** 40 Ω]

❑❑❑

CHAPTER 5

Field-Effect Transistors

5.1. INTRODUCTION

Inside This Chapter

We know that P-N junctions have two dominant features *viz.* the injection of minority carriers with forward bias and a variation of the depletion width W with reverse bias. These two P-N junction properties are used in two important types of transistors *viz. bipolar junction* transistor abbreviated as BJT and the junction *field-effect transistor abbreviated as* JFET. The BJT discussed in chapter 4 makes use of the injection of minority carriers across a forward-biased junction, while JFET discussed in this chapter depends on the control of a junction depletion width under reverse bias. FET is a majority carrier device, and is, therefore, often called a *unipolar* transistor while BJT operates by the injection and collection of minority carriers. Since in this device, both holes and electrons take part in conduction, it is called a *bipolar transistor*. Like its bipolar counterpart, the FET is a three-terminal (namely drain, source and gate) semiconductor device in which current conduction is by one type of majority carriers (electrons in case of an N-channel FET or holes in a P-channel FET). Unlike the BJT, a FET needs virtually no input (bias signal) current. Field-effect devices are controlled by an electric field (or voltage) and give an extremely *high input impedance*–most important advantage over BJT devices.

Field-effect transistors (FETs) are of several types such as junction field-effect transistors (JFETs), metal-semiconductor field-effect transistors (MESFETs), and metal-insulator field-effect transistors (MISFETs).

In a junction field-effect transistor (JFET) the gate voltage controls the depletion width of a reverse biased P-N junction. A metal-semiconductor field-effect transistor (MESFET) is the same basic device except that the P-N junction is replaced by a Schottky barrier rectifying junction. Metal-insulator semiconductor field-effect transistor (MISFET) is one in which the metal gate electrode may be separated from the semiconductor by an insulator. The metal-oxide-semiconductor field-effect transistor (MOSFET) is a special type of MISFET which makes use of an oxide layer as insulator.

Although the bipolar junction transistors (BJTs) reigned supreme in the early days of semiconductor integrated electronics, it has been gradually supplanted in most applications by the silicon MOSFETs. The main reason is, unlike BJTs, the different types of FETs are characterised by high input impedance, as the control voltage is applied to a reverse-biased junction or Schottky barrier, or across the insulator. Such devices are particularly suitable for controlled switching between a conducting state and a non-conducting state and that is why they are quite useful in digital circuits. They are also well suited for integration of many devices on a single chip. The MOSFET is, without doubt, the core of the integrated circuit (IC) design at present time.

5.2. Junction Field-Effect Transistors (JFETs)

The junction field-effect transistor (JFET), in conjunction with other circuit elements, is capable of voltage gain and signal-power gain. As mentioned in Art. 5.1, there are two general categories of junction field-effect transistors. The first is the PN junction FET, or PN JFET, and the second is the metal-semiconductor field-effect transistor (MESFET). The PN JFET is fabricated with PN junction whereas the MESFET is fabricated with a Schottky barrier rectifying junction.

The current in a JFET is through a semiconductor region called the *channel,* with ohmic contacts at each end. The basic transistor action is the modulation of the channel conductance by the electric field that is perpendicular to the channel. The modulating electric field is induced in the space charge region of a reverse-biased P-N junction or a reverse-biased Schottky barrier junction and is, therefore, a function of voltage applied to the gate.

5.3. PN Junction Field-Effect Transistors

PN JFETs are of two types *viz.* N-channel, JFETs and P-channel JFETs, Generally N-channel JFETs are preferred. The P-channel JFET is usually a lower frequency device than the N-channel JFET due to lower hole mobility. N-channel and P-channel JFETs are shown in Figs. 5.1(*a*) and 5.1(*b*) respectively.

5.3.1. Basic Construction

Its structure is quite simple. In an N-channel JFET an N-type silicon bar, referred to as the *channel*, has two smaller pieces of P-type silicon material diffused on the opposite sides of its middle part, forming P-N junctions, as illustrated in Fig. 5.1 (*a*). The two P-N junctions forming diodes or gates are connected internally and a common terminal,* called the *gate terminal*, is brought out. Ohmic contacts (direct electrical connections) are made at the two ends of the channel—one lead is called the *source terminal* S and the other *drain terminal* D.

The silicon bar behaves like a resistor between its two terminals D and S. The gate terminal is analogous to the base of an ordinary transistor (BJT). It is used to control the flow of current from source to drain. Thus, source and drain terminals are analogous to emitter and collector terminals respectively of a BJT.

In Fig. 5.1 (*a*), the gate is P-region, while the source and the drain are N-regions. Because of this, a PN JFET is similar to two diodes. The gate and the source form one of the diodes, and the gate and the drain form the other diode. These two diodes are usually referred as the *gate-*

* When a manufacturer connects a separate lead to each gate, the device is called a dual-gate JFET. The main use of a dual-gate JFET is with a mixer, a special circuit used in communication equipment.

source diode and the *gate-drain diode*. Since JFET is a silicon device, it takes only 0.7 V for forward bias to get significant current in either diode.

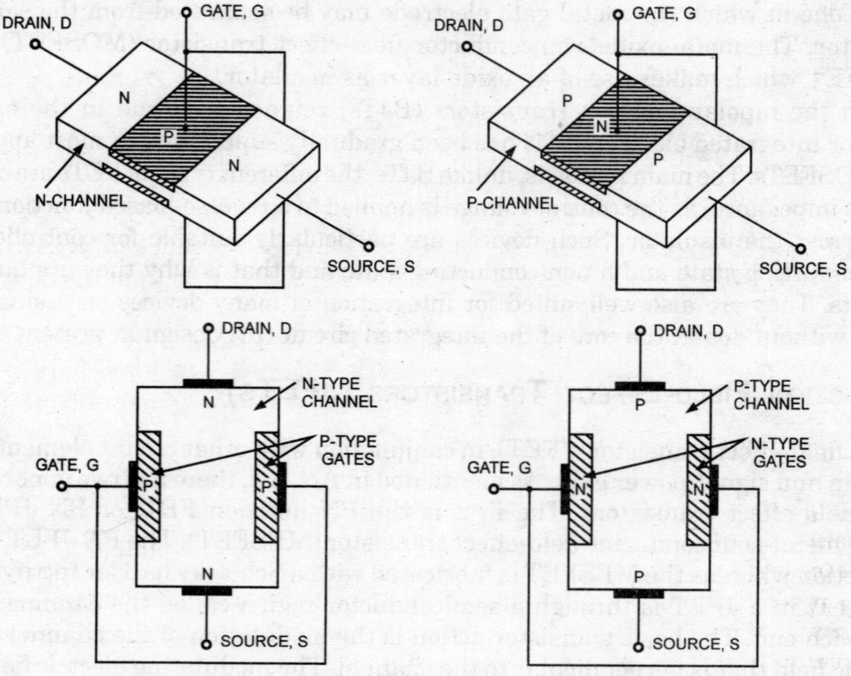

(a) N-Channel JFET *(b) P-Channel JFET*

Fig. 5.1. ***PN JFETs***

With the gate terminal not connected, and a potential applied (+ ve at the drain and – ve at the source), a current called the *drain current*, I_D flows through the channel located between the two P-regions. This current consists of only majority carriers–electrons in this case.

P-channel PN JFET is similar in construction to N-channel PN JFET except that P-type semiconductor material is sandwiched between two N-type junctions, as shown in Fig. 5.1 (*b*). In this case majority carriers are holes.

5.3.2. Standard Notations

Worthnoting FET notations are as follows :

1. ***Source***. The terminal through which the majority carriers enter the channel, is called the *source* terminal S and the conventional current entering the channel at S is designated as I_S.
2. ***Drain***. The terminal, through which the majority carriers leave the channel, is called the *drain* terminal D and the conventional current leaving the channel at D is designated as I_D.
   The drain-to-source voltage is called V_{DS}, and is positive if D is more positive than S.
3. ***Gate***. There are two internally connected heavily doped impurity regions formed by alloying, by diffusion, or by any other method available to create two P-N junctions. These impurity regions are called the gate G. A voltage V_{GS} is applied between the gate and source in the direction to reverse-bias the P-N junction. Conventional current entering the channel at G is designated as I_G.
4. ***Channel***. The region between the source and drain, sandwiched between the two gates is called the *channel* and the majority carriers move from source to drain through this channel.

5.3.3. Schematic Symbols

The schematic symbols for N-type and P-type JFETs are shown in Figs. 5.2 (*a*) and 5.2 (*b*) respectively. The vertical line in the symbol may be thought as channel and source S and drain D connected to the line.

Note that the direction of the arrow at the gate indicates the direction in which the gate current flows when the gate junction is forward biased. Thus for the N-channel JFET, the arrow at the gate junction points into the device and in P-channel JFET, it is away from the device.

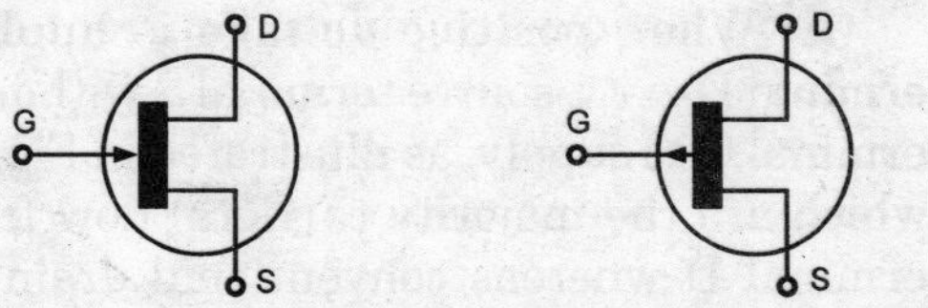

(a) N-Channel JFET (b) P-Channel JFET

Fig. 5.2. ***Schematic Symbols For JFETs***

5.3.4. Polarity Conventions

The polarities for N-channel and P-channel JFETs are shown in Figs. 5.3 (*a*) and 5.3 (*b*) respectively. In both of the cases the voltage between gate and source is such that the gate is reverse biased. This is the normal method of connection of JFETs. The drain and source terminals are interchangeable *i.e.* either end can be used as a source and the other end as a drain. The source terminal is always connected to that end of the drain voltage supply which provides the necessary charge carriers *i.e.* in an N-channel JFET source terminal, S is connected to the – ve end of the drain voltage supply for obtaining electrons whereas in P-channel JFET source terminal, S is connected to the + ve end of the drain voltage supply for getting holes.

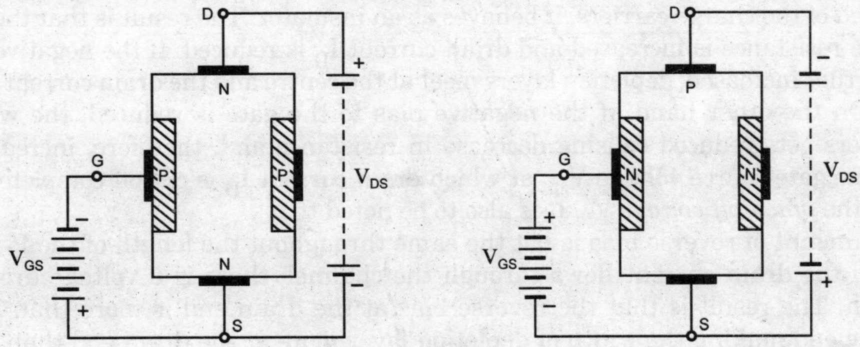

(a) N-Channel JFET (b) P-Channel JFET

Fig. 5.3. ***Polarity Conventions For JFETs***

Other worthnoting points are as follows :

(*i*) For N-channel JFET, I_D and V_{DS} are positive while V_{GS} is negative.

(*ii*) For P-channel JFET, I_D and V_{DS} are negative while V_{GS} is positive.

(*iii*) Assumed positive directions of I_S, I_D and I_G are into the JFET.

5.3.5. Operation

Let us consider an N-channel JFET for discussing its operation.

(*i*) When neither any bias is applied to the gate (*i.e.* when $V_{GS} = 0$) nor any voltage to the drain w.r.t. source (*i.e.* when $V_{DS} = 0$), the depletion regions around the P-N junctions are of equal thickness and symmetrical.

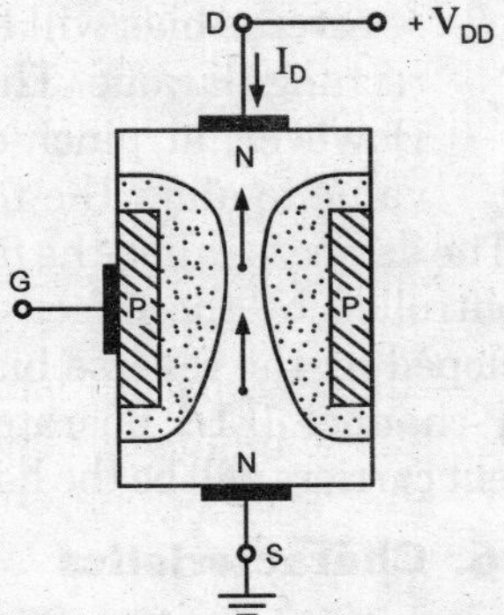

Fig. 5.4. ***JFET With No Bias Voltage***

(*ii*) When positive voltage is applied to the drain terminal D w.r.t. source terminal S without connecting gate terminal G to supply, as illustrated in Fig. 5.4, the electrons (which are the majority carriers) flow from terminal S to terminal D whereas conventional drain current I_D flows through the channel from D to S. Due to flow of this current, there is a uniform voltage drop across the channel resistance as we move from terminal D to terminal S. This voltage drop reverse biases the diode. The gate is more negative with respect to those points in the channel which are nearer to D than to S. Hence, depletion layers penetrate more deeply into the channel at points lying closer to D than to S. Thus wedge-shaped depletion regions are formed, as shown in Fig. 5.4, when V_{DS} is applied. The size of the depletion layer formed determines the width of the channel and hence the magnitude of current I_D flowing through the channel.

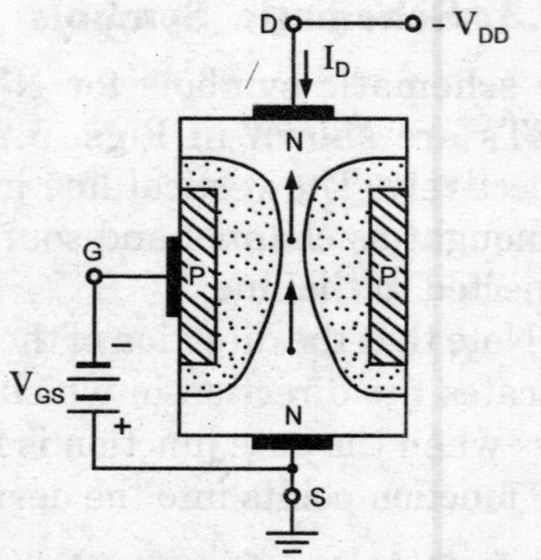

Fig. 5.5. ***JFET With Small Negative Gate Source Bias***

To see how the width of the channel varies with the variation in gate voltage, let us consider the situation when the gate is biased negative with respect to the source while the drain is applied with positive bias with respect to the source, as illustrated in Fig. 5.5. Now the P-N junctions are reverse biased and depletion regions are formed. P-regions are heavily doped compared to the N-channel, so the depletion regions penetrate deeply into the channel. Since a depletion region is a region depleted of the charge carriers, it behaves as an insulator. The result is that the channel is narrowed, the resistance is increased and drain current I_D is reduced. If the negative voltage at the gate is further increased, depletion layers meet at the centre and the drain current I_D is cut-off completely. On the other hand, if the negative bias to the gate is reduced, the width of the depletion layers gets reduced causing decrease in resistance and, therefore, increase in drain current I_D. The gate-source voltage V_{GS} at which drain current I_D is cut-off completely (pinched off), is called the *pinch-off voltage* V_P. It is also to be noted that

1. The amount of reverse bias is not the same throughout the length of the P-N junction. When the drain current flows through the channel, there is a voltage drop along its length. The result is that the reverse bias at the drain end is more than that at the source end making the width of depletion layer more at the drain end than that at the source end. Thus the channel becomes narrower at the drain end in comparison to that at source end, as shown in Fig. 5.5.
2. The channel is not completely closed at the drain end. Because in that case there will be no drain current, so there will be no voltage drop along the channel length and amount of reverse bias will become uniform and the wedge shaped depletion region will become rectangular one. The channel will open and the drain current will flow.
   However, at pinch-off voltage, the channel width is reduced to a constant minimum value to allow the flow of drain current.

The device is called the *field-effect transistor* (FET) because the drain current (output current) is controlled by the effect of the extension of the field associated with the depletion region developed by the reverse bias at the gate.

P-channel JFET operates in the same manner as an N-channel JFET except that channel current carriers will be the holes in place of electrons and the polarities of V_{GS} and V_{DS} are reversed.

5.3.6. Characteristics

There are two types of static characteristics viz. (1) output or drain characteristic and (2) transfer characteristic.

1. *Output or Drain Characteristic.* The curve drawn between drain current I_D and drain-source voltage V_{DS} with gate-to-source voltage V_{GS} as the parameter is called the *drain* or *output characteristic*. This characteristic is analogous to collector characteristic of a BJT.

(*a*) Drain Characteristic With Shorted Gate. The circuit diagram for determining the drain characteristic with shorted-gate for an N-channel JFET is given in Fig. 5.6 (*a*) and the drain characteristic with shorted-gate is shown in Fig. 5.6 (*b*).

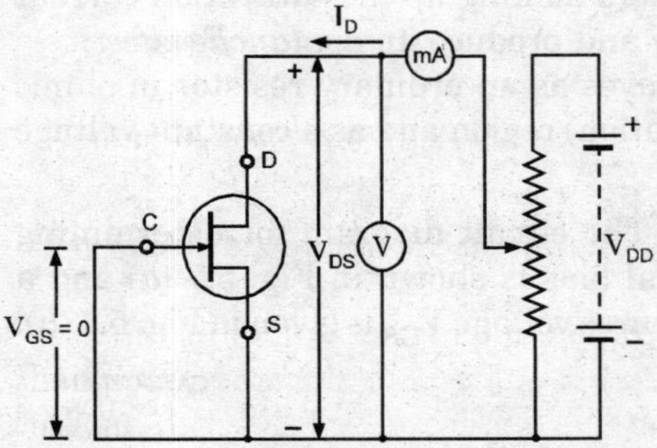

(a) Circuit Diagram For Determining Drain Characteristic With Shorted Gate For An N-Channel JFET

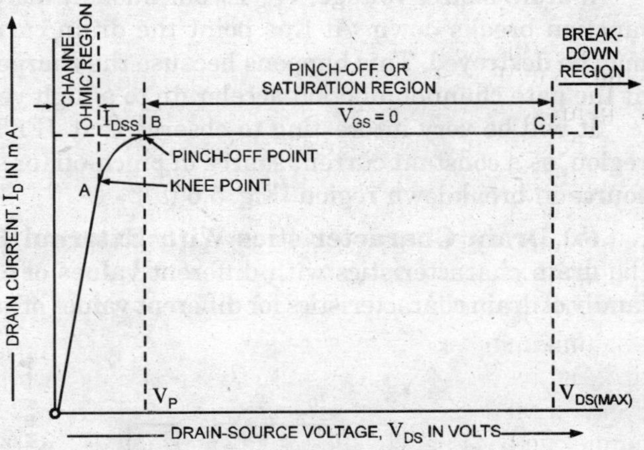

(b) JFET Drain Characteristic With Shorted Gate

Fig. 5.6

Initially when drain-source voltage V_{DS} is zero, there is no attracting potential at the drain, so no current flows inspite of the fact that the channel is fully open. This gives drain current $I_D = 0$. For small applied voltage V_{DS}, the N-type bar acts as a simple semiconductor resistor, and the drain current I_D increases linearly with the increase in V_{DS}, up to the knee point. This region (to the left of the knee point) of the curve is called the *channel ohmic region*, because in this region the FET behaves like an ordinary resistor.

With the increase in drain current I_D, the ohmic voltage drop between the source and channel region reverse-biases the gate junction. The reverse-biasing of the gate junction is not uniform throughout. The reverse-bias is more at the drain end than that at the source end of the channel, so with the increase in V_{DS}, the conducting portion of the channel begins to constric more at the drain end. Eventually a voltage V_{DS} is reached at which the channel is *pinched-off*. The drain current I_D no longer increases with the increase in V_{DS}. It approaches a constant saturation value. The value of voltage V_{DS} at which the channel is pinched off (*i.e.* all the free charges from the channel get removed), is called the *pinch-off voltage* V_P. The pinch-off voltage V_P, not too sharply defined on the curve, where the drain current I_D begins to level off and attains a constant value. From point A (knee point) to the point B (pinch-off point) the drain current I_D increases with the increase in voltage V_{DS} following a reverse square law. The region of the characteristic in which drain current I_D remains fairly constant is called the *pinch-off region*. It is also sometimes called the *saturation region* or *amplifier region*. In this region the JFET operates as a *constant current device* since drain current (or output current) remains almost constant. It is the normal operating region of the JFET when used as an amplifier. The drain current in the pinch-off region with $V_{GS} = 0$ is referred to the *drain-source saturation current*, I_{DSS}.

It is to be noted that in the pinch-off (or saturation) region the channel resistance increases in proportion to increase in V_{DS} and so keeps the drain current almost constant and the reverse bias required by the gate-channel junction is supplied entirely by the voltage drop across the channel resistance due to flow of I_{DSS} and not by the external bias because $V_{GS} = 0$.

Drain current in the pinch-off region is given by Shockley's equation

$$I_D = I_{DSS}\left(1 - \frac{V_{GS}}{V_P}\right)^2 = I_{DSS}\left(1 - \frac{V_{GS}}{V_{GS\,(OFF)}}\right)^2 \quad ...(5.1)$$

where I_D is the drain current at a given gate-source voltage V_{GS}, I_{DSS} is the drain current with gate shorted to source and $V_{GS\,(OFF)}$ is the gate-source cut-off voltage.

If drain-source voltage, V_{DS} is continuously increased, a stage comes when the gate-channel junction breaks down. At this point the drain current increases very rapidly, and the JFET may be destroyed. This happens because the charge carriers making up the saturation current at the gate channel junction accelerate to a high velocity and produce an *avalanche effect*.

It will be very interesting to observe that JFET behaves as an ordinary resistor in ohmic region, as a constant current source in pinch-off (or saturation) region and as a constant voltage source in breakdown region [Fig. 5.6 (*b*)].

***(b)* Drain Characteristics With External Bias.** The circuit diagram for determining the drain characteristics with different values of external bias is shown in Fig. 5.7 (*a*) and a family of drain characteristics for different values of gate-source voltage V_{GS} is given in Fig. 5.7 (*b*).

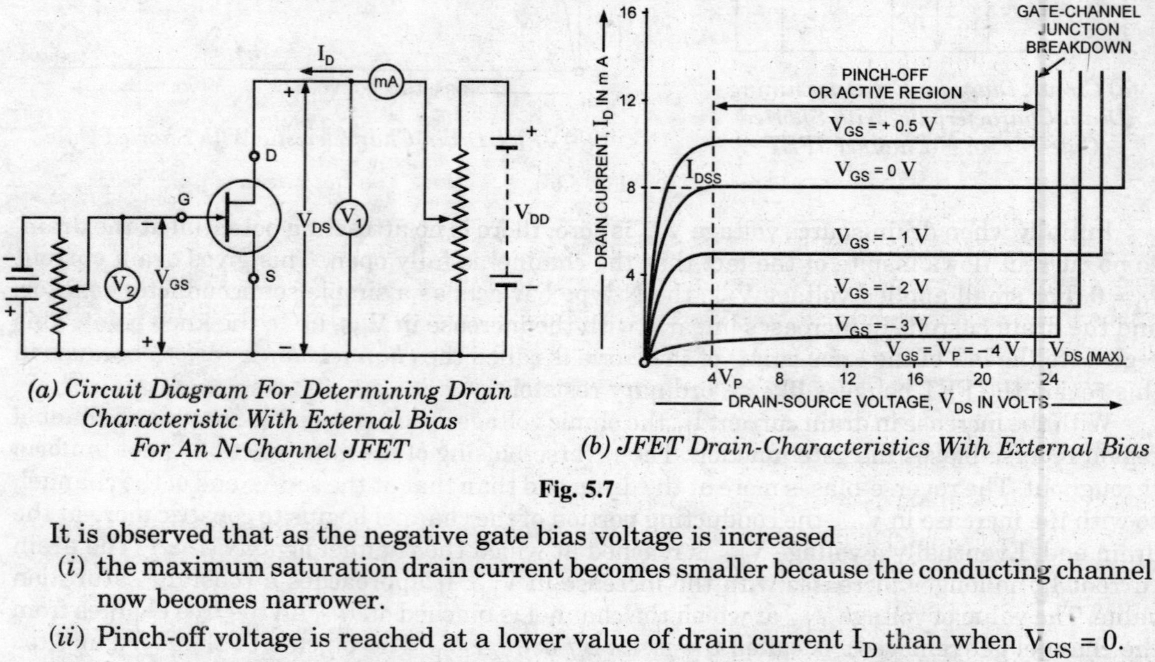

(*a*) *Circuit Diagram For Determining Drain Characteristic With External Bias For An N-Channel JFET*

(*b*) *JFET Drain-Characteristics With External Bias*

Fig. 5.7

It is observed that as the negative gate bias voltage is increased

(*i*) the maximum saturation drain current becomes smaller because the conducting channel now becomes narrower.

(*ii*) Pinch-off voltage is reached at a lower value of drain current I_D than when $V_{GS} = 0$.

When an external bias of say − 1 V is applied between the gate and the source, the gate-channel junctions are reverse biased even when drain current, I_D is zero. Hence the depletion regions are already penetrating the channel to a certain extent when drain-source voltage, V_{DS} is zero. Due to this reason, a smaller voltage drop along the channel (*i.e.* smaller than that for $V_{GS} = 0$) will increase the depletion regions to the point where they pinch-off the current. Consequently, the pinch-off voltage V_P is reached at a lower drain current, I_D when $V_{GS} = 0$.

(*iii*) The ohmic region portion decreases.

(*iv*) Value of drain-source voltage V_{DS} for the avalanche breakdown of the gate junction is reduced.

Value of drain-source voltage, V_{DS} for breakdown with the increase in negative bias voltage is reduced simply due to the fact that gate-source voltage, V_{GS} keeps adding to the reverse bias at the junction produced by current flow. Thus the maximum value of V_{DS} that can be applied to a FET is the lowest voltage which causes avalanche breakdown.

It is also observed that with $V_{GS} = 0$, I_D saturates at I_{DSS} and the characteristic shows V_P = 4 V. When an external bias of – 1 V is applied, the gate-channel junctions still require –4 V to achieve pinch-off. It means that a 3 V drop is now required along the channel instead of the previous 4.0 V. Obviously, this drop of 3 V can be achieved with a lower value of drain current. Similarly, when V_{GS} = – 2 V and – 3 V, pinch-off is achieved with 2 V and 1 V respectively, along the channel. These drops of 2 V and 1 V are, of course, achieved with further reduced values of drain current, I_D. It is further observed that when the gate-source bias is numerically equal to pinch-off voltage, V_P (–4 V in this case), no channel drop is required and, therefore, drain current, I_D is zero. The gate-source bias voltage required to reduce drain current, I_D to zero is designated the gate-source cut-off voltage, $V_{GS\,(OFF)}$ and, as explained,

$$V_{GS\,(OFF)} = V_P$$

Hence for working of JFET in the pinch-off or active region it is necessary that the following conditions be fulfilled.

$$V_P < V_{DS} < V_{DS\,(MAX)} \quad \text{...(5.2)}$$

$$V_{GS\,(OFF)} < V_{GS} < 0 \quad \text{...(5.3)}$$

$$0 < I_D < I_{DSS} \quad \text{...(5.4)}$$

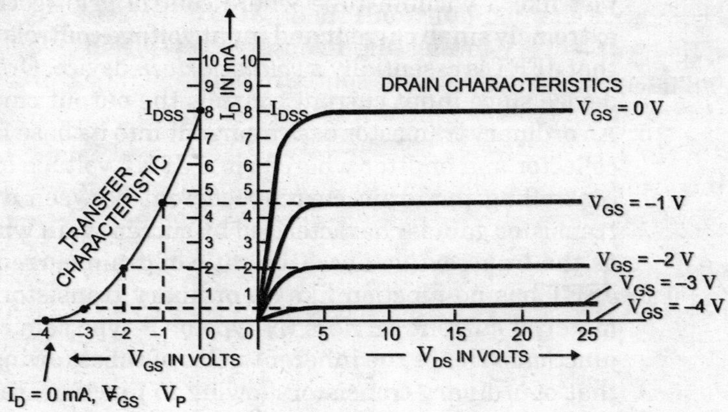

Fig. 5.8. *JFET Transfer Characteristic*

2. *Transfer Characteristic*. The transfer characteristic for a JFET can be determined experimentally, keeping drain-source voltage, V_{DS} constant and determining drain current, I_D for various values of gate-source voltage, V_{GS}. The circuit diagram is shown in Fig. 5.7 (*a*). The curve is plotted between gate-source voltage, V_{GS} and drain current, I_D, as illustrated in Fig. 5.8. It is observed that

(*i*) drain current decreases with the increase in –ve gate-source bias,

(*ii*) drain current, $I_D = I_{DSS}$ when $V_{GS} = 0$

(*iii*) drain current, $I_D = 0$ when $V_{GS} = V_P$

The transfer characteristic follows Eq. (5.1).

The transfer characteristic can also be derived from the drain characteristic by noting values of drain current, I_D corresponding to various values of gate-source voltage, V_{GS} for a constant drain-source voltage and plotting them.

It may be noted that a P-channel JFET operates in the same way and have the similar characteristics as an N-channel JFET except that channel carriers are holes instead of electrons and the polarities of V_{GS} and V_{DS} are reversed.

5.4. Merits and Demerits of JFETs

Junction field effect transistors combine several merits of both conventional (or bipolar) transistors and vacuum tubes. Some of these are enumerated below :

1. Its operation depends upon the flow of majority carriers only. It is, therefore, a *unipolar* (one type of carrier) device. On the other hand in an ordinary transistor both majority and minority carriers take part in conduction and, therefore, ordinary transistor is

sometimes called the *bipolar* transistor. The vacuum tube is another example of a unipolar device.

2. It is simpler to fabricate, smaller in size, rugged in construction and has longer life and higher efficiency. Simpler to fabricate in IC form and space requirement is also lesser.
3. It has high input impedance (of the order of 100 MΩ), because its input circuit (gate to source) is reverse biased, and so permits high degree of isolation between the input and output circuits. However, the input circuit of an ordinary transistor is forward biased and, therefore, ordinary transistor has low input impedance.
4. It carries very small current because of reverse biased gate and, therefore, it operates just like a vacuum tube where control grid (corresponding to gate in JFET) carries extremely small current and input voltage controls the output current. This is the reason that JFET is essentially a *voltage driven* device. Ordinary transistor is a current operated device since input current controls the output current.
5. An ordinary transistor uses a current into its base for controlling a large current between collector and emitter whereas in a JFET voltage on the gate (base) terminal is used for controlling the drain current (current between drain and source). Thus, an ordinary transistor gain is characterised by current gain whereas the JFET gain is characterised as the *transconductance* (the ratio of drain current and gate-source voltage).
6. JFET has no junction like an ordinary transistor and the conduction is through bulk material current carriers (N-type or P-type semiconductor material) that do not cross junctions. Hence the inherent noise of tubes (owing to high temperature operation) and that of ordinary transistors (owing to junction transitions) is not present in JFET.
7. It is relatively immune to radiation.
8. It has –ve temperature coefficient of resistance and, therefore, has better thermal stability.
9. It has high power gain and, therefore, the necessity of employing driver stages is eliminated.
10. It exhibits no offset voltage at zero drain current and, therefore, makes an excellent signal chopper.
11. It has square-law characteristics and, therefore, it is very useful in the tuners of radio and TV receivers.
12. It has got high frequency response.

The main drawback of JFET is its relative small gain-bandwidth product in comparison with that of a conventional transistor and greater susceptibility to damage in its handling. Other disadvantages of JFET are low voltage gains because of small transconductance and higher cost than that of BJTs.

5.5. PRACTICAL FET STRUCTURE

The FET structure given in Fig. 5.1 is not practical because of the difficulties involved in diffusing impurities into both sides of a semiconductor wafer. Fig. 5.9 gives a single-ended geometry junction FET where diffusion is from one side only. It uses a P-type substrate on which N-type channel is epitaxially grown. A P-type gate is then diffused into the N-type channel. The substrate serves as the second gate. Both the

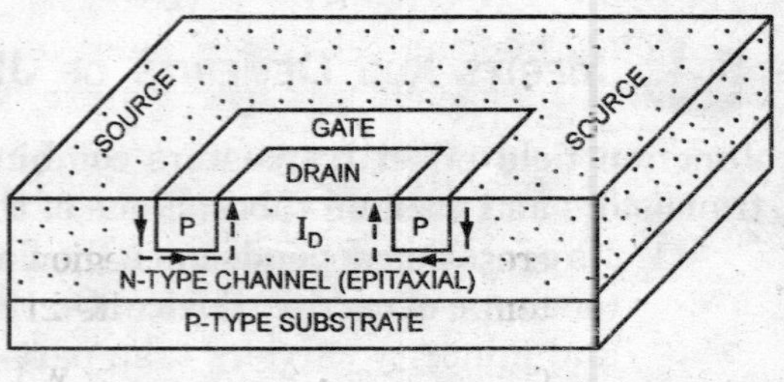

Fig. 5.9. ***A Metallic Crystal Structure***

P-type diffused gate and the P-type substrate serving as the gate are of low resistivity material. Thus penetration of depletion region into the N-type channel takes place. Drain current I_D follows a path shown in the figure.

5.6. FET CONFIGURATIONS

FETs (JFETs and MOSFETs) have three configurations just as BJTs have. The three configurations of FETs—the common-gate, common source, and common drain configurations have been given in Fig. 5.10. Input and output terminals of different configurations have been given in Table 5.1.

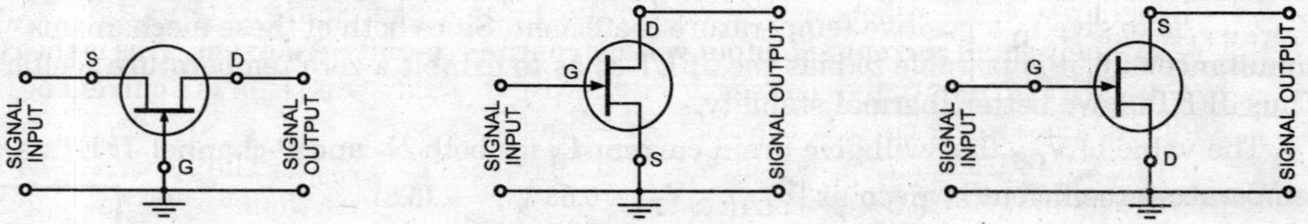

(a) Common Gate Configuration *(b) Common Source Configuration* *(c) Common Drain Configuration*

Fig. 5.10. ***N-Channel FET Configurations***

The three FET configurations have characteristics that are similar to the three corresponding BJT configurations, as given in Table 5.1.

TABLE 5.1. FET Configurations

Configuration	Input Terminal	Output Terminal
Common gate	Source	Drain
Common source	Gate	Drain
Common drain	Gate	Source

TABLE 5.2. FET-BJT Correspondence

FET Configuration	BJT Configuration
Common gate	Common base
Common source	Common emitter
Common drain	Common collector

5.7. JFET TEMPERATURE EFFECTS

The JFET, like other solid-state devices, is subject to the effect of temperature. However, it is possible to bias the JFET such that it will exhibit a *zero temperature coefficient i.e.* drain current I_D will be independent of temperature. This is explained below :

There are two primary mechanisms underlying the temperature sensitivity of the conductivity of a JFET channel.

1. Decrease in the depletion region width at the channel-gate P-N junction with the increase in temperature. This results in increase of channel thickness.
2. Decrease in carrier mobility with the increase in temperature.

Increase in channel thickness with the increase in temperature makes drain current I_D to increase. Another way of looking into the situation is to note that $V_{GS\,(OFF)}$ increases in magnitude with the increase in temperature. So far a fixed value of V_{GS}, I_D will increase with the temperature. $V_{GS\,(OFF)}$ $[V_P]$ has a positive temperature coefficient of about 2.2 mV/°C.

The second factor [decrease in carrier mobility with the increase in temperature] makes the channel conductivity to decrease with the increase in temperature. The result is that drain current I_D decreases (assuming V_{DS} constant) with the increase in temperature. Thus we see that I_{DSS} tends to fall as the temperature rises.

Thus we have two distinct mechanisms affecting drain current I_D as a function of temperature. Carrier mobility gives I_D a negative temperature coefficient while the effect of $V_{GS\,(OFF)}$ is to give I_D a positive temperature coefficient. Since both of these mechanisms occur simultaneously, it is possible to bias the JFET so as to exhibit a zero temperature coefficient. Thus JFETs have better thermal stability.

The value of V_{GS} that will give drain current I_D for both N- and P-channel JFETs a zero temperature coefficient is given as $|V_{GS}| \quad |V_P| - 0.63\text{ V}$ ...(5.5)

5.8. JFET As An Amplifier

One of the applications of JFET is an amplification of input signal.

In FET the weak signal is applied between gate and source and an amplified output is taken across the drain-source circuit having load, as shown in Fig. 5.11, just as in a vacuum tube where signal is connected between grid and cathode and output is taken across a load connected in plate circuit.

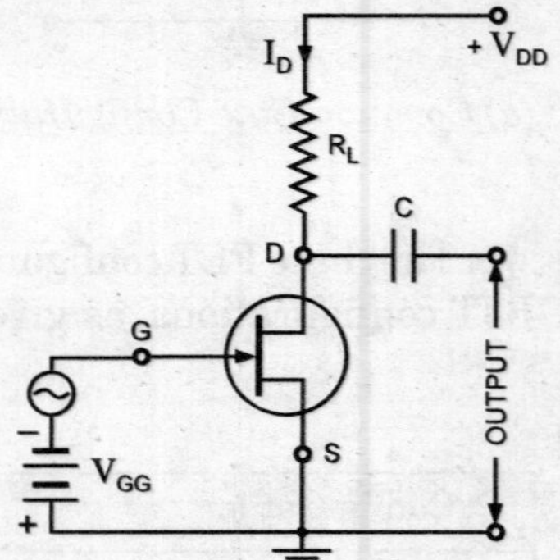

Fig. 5.11. ***JFET As An Amplifier***

For proper working of JFET it is necessary that the gate is made negative w.r.t. source (*i.e.* the input circuit is always reverse-biased). This is achieved either by inserting a battery V_{GG} in the gate circuit with the polarity shown in Fig. 5.11 or by providing a biasing circuit.

On application of an alternating weak signal, the fixed reverse bias of the gate changes and a small change in reverse bias on the gate causes a large change in drain current owing to change in width of the channel. This fact makes the FET capable of amplifying the strength of weak signals. During the positive half cycle of the input signal, the reverse bias on the gate decreases causing increase in channel width and hence the drain current. During the negative half cycle of the input signal, the reverse bias on the gate increases causing drain current to decrease. The result is that the small change in voltage at the gate produces a large change in the drain current. These large variations in drain current give large output across the load R_L and thus, JFET operates as an amplifier.

5.9. JFET Parameters

JFET has certain parameters which determine the performance. Such parameters are (*i*) ac drain resistance, (*ii*) transconductance, (*iii*) amplification factor, and (*iv*) dc drain resistance, as explained below :

1. *AC Drain Resistance.* It is defined as the ratio of change in drain-source voltage to change in drain current at constant gate-source voltage and is denoted by r_d.

$$\text{i.e. AC drain resistance, } r_d = \frac{\Delta V_{DS}}{\Delta I_D} \text{ at constant } V_{GS} \quad ...(5.6)$$

It is also called the ***dynamic drain resistance.***

Referring to the output or drain characteristic shown in Fig. 5.7 (*b*), it is clear that in the active region the change in drain current, I_D is very small for change in drain-source voltage, V_{DS} because the characteristic curves are almost flat. Hence ac drain resistance of a JFET is very large ranging from 10 kΩ to 1 MΩ.

2. *Transconductance.* The control that the gate-source voltage has over the drain current, I_D is measured by transconductance. It is denoted by g_{fs}. It may be defined as the ratio of change in drain current to the change in gate-source voltage at constant drain-source voltage.

$$i.e. \text{ Transconductance, } g_m = \frac{\Delta I_D}{\Delta V_{GS}} \text{ at constant } V_{DS} \qquad \text{...(5.7)}$$

It is also called the *forward transconductance* (g_{fs}) or forward transadmittance (Y_{fs}). It is measured in mA/volt or millisiemens.

The transconductance measured at I_{DSS} is denoted by g_{mo}.

Mathematical Expression For Transconductance

Differentiating both sides of Shockley equation $I_D = I_{DSS}\left(1 - \frac{V_{GS}}{V_P}\right)^2$, we get

$$d\,I_D = 2\,I_{DSS}\left(1 - \frac{V_{GS}}{V_P}\right)\left(-\frac{1}{V_P}\right) d\,V_{GS}$$

$$\text{or } \frac{d\,I_D}{d\,V_{GS}} = 2\,I_{DSS}\left(1 - \frac{V_{GS}}{V_P}\right)\left(-\frac{1}{V_P}\right)$$

$$\text{or } g_m = -\frac{2\,I_{DSS}}{V_P}\left(1 - \frac{V_{GS}}{V_P}\right) \qquad \text{...(5.8)}$$

Substituting $V_{GS} = 0$ in above expression, we get

$$g_{mo} = -\frac{2\,I_{DSS}}{V_P} \qquad \text{...(5.9)}$$

From Eqs. (5.8) and (5.9) we have

$$g_m = g_{mo}\left(1 - \frac{V_{GS}}{V_P}\right) \qquad \text{...(5.10)}$$

3. *Amplification Factor.* It is defined as the ratio of change in drain-source voltage to the change in gate-source voltage at constant drain current and is denoted by μ.

$$i.e. \text{ Amplification factor, } \mu = \frac{\Delta V_{DS}}{\Delta V_{GS}} \text{ at constant } I_D$$

Amplification factor of a JFET indicates how much more control the gate-source voltage has over drain current in comparison to the drain-source voltage.

$$\text{Amplification factor, } \mu = \frac{\Delta V_{DS}}{\Delta V_{GS}} = \frac{\Delta V_{DS}}{\Delta I_D} \times \frac{\Delta I_D}{\Delta V_{GS}}$$

$$= r_d \times g_m = \text{ac drain resistance} \times \text{transconductance} \qquad \text{...(5.11)}$$

Amplification factor μ of a FET may be as high as 100.

4. *DC Drain Resistance.* It is also called the static or ohmic resistance of the channel and is defined as ratio of drain-source voltage and drain current. It is denoted by R_{DS}

$$i.e., R_{DS} = \frac{V_{DS}}{I_D} \qquad \text{...(5.12)}$$

Example 5.1. When a reverse voltage of 10 V is applied between gate and source of JFET the gate current is 0.001 μ A. Determine resistance between gate and source.

Solution : Gate-source voltage, $V_{GS} = 10.0$ V

Gate current, $I_G = 0.001\ \mu A = 1 \times 10^{-9}$ A

Gate-to-source resistance, $R_{GS} = \frac{V_{GS}}{I_G} = \frac{10}{1 \times 10^{-9}} = 10{,}000$ MΩ **Ans.**

Example 5.2. When drain-source voltage is changed by 1.5 volts, the change in drain current is of 120 μA, the gate-source voltage remaining unchanged. Determine the ac drain resistance of the JFET.

Solution : Change in drain-source voltage, $\Delta V_{DS} = 1.5$ V

Change in drain-current, $\Delta I_D = 120 \times 10^{-6}$ A

AC drain resistance of the JFET, $r_d = \frac{\Delta V_{DS}}{\Delta I_D} = \frac{1.5}{120 \times 10^{-6}} = 12.5$ kΩ **Ans.**

Example 5.3. A JFET has $V_P = -4.5$ V, $I_{DSS} = 10$ mA and $I_{DS} = 2.5$ mA. Determine the transconductance.

Solution : Drain-source saturation current, $I_{DSS} = 10$ m A

Pinch-off voltage, $V_P = -4.5$ V

Drain-source current, $I_{DS} = 2.5$ m A

From Shockley's equation, drain-source current, $I_{DS} = I_{DSS}\left[1 - \frac{V_{GS}}{V_P}\right]^2$

$$\text{or } V_{GS} = V_P\left[1 - \sqrt{\frac{I_{DS}}{I_{DSS}}}\right] = -4.5\left[1 - \sqrt{\frac{2.5}{10}}\right] = -2.25 \text{ V}$$

From Eq. (5.8), transconductance,

$$g_m = \frac{-2I_{DSS}}{V_P}\left[1 - \frac{V_{GS}}{V_P}\right] = \frac{-2 \times 10 \times 10^{-3}}{-4.5}\left[1 - \frac{-2.25}{-4.5}\right] = 2.22 \text{ m A/V } \textbf{Ans.}$$

Example 5.4. A JFET has $g_m = 10$ mS, $I_{DSS} = 10$ μA. Calculate $V_{GS\,(OFF)}$.

Solution : Transconductance, $g_m = 10$ mS $= 10 \times 10^{-3}$ S

Drain-source saturation current, $I_{DSS} = 10\ \mu A = 10 \times 10^{-6}$ A

$V_{GS\,(OFF)} \rightarrow V_{GS} = V_P$

$$g_m = g_{mo} = \frac{-2\,I_{DSS}}{V_P} = \frac{-2\,I_{DSS}}{V_{GS(OFF)}}$$

$$\text{or } 10 \times 10^{-3} = \frac{-2 \times 10 \times 10^{-6}}{V_{GS(OFF)}}$$

$$\text{or } V_{GS\,(OFF)} = \frac{-2 \times 10 \times 10^{-6}}{10 \times 10^{-3}} = -2 \text{ mV } \textbf{Ans.}$$

Example 5.5. An N-channel JFET has $I_{DSS} = 10$ mA and $V_P = -4$ V. Determine the minimum value of V_{DS} for pinch-off region and drain current I_D for $V_{GS} = -2$ V in pinch-off region.

Solution : Pinch-off voltage, $V_P = -4$ V

Gate-source voltage, $V_{GS} = -2$ V

Drain-source saturation current, $I_{DSS} = 10$ m A $= 10 \times 10^{-3}$ A

$$\text{Drain current, } I_D = I_{DSS}\left[1 - \frac{V_{GS}}{V_P}\right]^2 = 10 \times 10^{-3}\left[1 - \frac{-2}{-4}\right]^2 = 2.5 \text{ m A } \textbf{Ans.}$$

The minimum value of V_{DS} for pinch-off region is equal to V_P. Thus the minimum value of V_{DS}

$$V_{DS\,(min)} = V_P = -4 \text{ V Ans.}$$

Example 5.6. For an N-channel JFET I_{DSS} = 8.7 mA; V_P = – 3 V; V_{GS} = – 1 V
Find the values of (i) I_D (ii) g_{mo} (iii) g_m *[Bombay Univ. Basic Electronics 1985]*

Solution : Drain-source saturation current,

$$I_{DSS} = 8.7 \text{ m A} = 0.0087 \text{ A}$$

Pinch-off voltage, $V_P = -3$ V

Gate-source voltage, $V_{GS} = -1$ V

(i) Drain current, $I_D = I_{DSS}\left(1-\dfrac{V_{GS}}{V_P}\right)^2 = 0.0087\left(1-\dfrac{-1}{-3}\right)^2 = 3.8667 \text{ mA}$ **Ans.**

(ii) Transconductance for $V_{GS} = 0$,

$$g_{mo} = \frac{-2\,I_{DSS}}{V_P} = \frac{-2\times 8.7\text{ m A}}{-3} = 5.8 \text{ m A/V or } 5.8 \text{ mS } \textbf{Ans.}$$

(iii) Transconductance, $g_m = g_{mo}\left(1-\dfrac{V_{GS}}{V_P}\right) = 5.8\left(1-\dfrac{-1}{-3}\right) = 3.867$ m A/V or 3.867 mS **Ans.**

Example 5.7. An FET follows the following relation:

$$I_D = I_{DSS}\left[1-\frac{V_{GS}}{V_P}\right]^2$$

I_{DSS} = 8.4 m A, V_P = – 3 V. What is the value of I_D for V_{GS} = – 1.5 V ? Find g_m at this point.

Solution : Drain-source saturation current, $I_{DSS} = 8.4$ m A

Pinch-off voltage, $V_P = -3$ V

Gate-source voltage, $V_{GS} = -1.5$ V

$$\text{Drain current, } I_D = I_{DSS}\left[1-\frac{V_{GS}}{V_P}\right]^2 = 8.4\left(1-\frac{-1.5}{-3}\right)^2 = 2.1 \text{ mA } \textbf{Ans.}$$

$$\text{Transconductance, } g_{mo} = \frac{-2\,I_{DSS}}{V_P} = \frac{-2\times 8.4}{-3} = 5.6 \text{ m A/V or } 5.6 \text{ mS}$$

$$\text{Transconductance, } g_m = g_{mo}\left(1-\frac{V_{GS}}{V_P}\right) = 5.6\left(1-\frac{-1.5}{-3}\right) = 2.8 \text{ mS } \textbf{Ans.}$$

Example 5.8. An N-channel JFET has a pinch-off voltage of – 4.5 V and I_{DSS} = 9 mA. At what value of V_{GS} will I_{DS} be equal to 3 m A? What is its g_m at this I_{DS} ?

Solution : Pinch-off voltage, $V_P = -4.5$ V

Drain-source saturation current, $I_{DSS} = 9 \text{ m A} = 9\times10^{-3}$ A

Drain-source current, $I_{DS} = 3 \text{ m A} = 3\times10^{-3}$ A

From Shockley's Eq. (5.1)

$$V_{GS} = V_P\left[1-\sqrt{\frac{I_{DS}}{I_{DSS}}}\right] = -4.5\times\left[1-\sqrt{\frac{3\times10^{-3}}{9\times10^{-3}}}\right] = -1.902 \text{ V } \textbf{Ans.}$$

Transconductance g_m for $I_{DS} = 3$ m A for which $V_{GS} = -1.902$ V

$$g_m = \frac{-2I_{DSS}}{V_P}\left[1-\frac{V_{GS}}{V_P}\right]$$

$$= \frac{-2\times9\times10^{-3}}{-4.5}\left(1-\frac{-1.902}{-4.5}\right) = 2.31 \text{ mA/V or } 2.31 \text{ mS } \textbf{Ans.}$$

5.10. Metal-Semiconductor Field-Effect Transistors (MESFETs)

The metal-semiconductor field-effect transistor (MESFET) was proposed by Mead in 1966 and subsequently fabricated by Hooper and Lehrer using a GaAs epitaxial layer on semi-insulating GaAs substrate. The operation of a MESFET is identical to that of a P-N junction JFET. In the MESFET, however, a metal-semiconductor rectifying contact instead of a P-N junction is used for the gate electrode. The MESFET have certain processing and performance advantages such as low-temperature formation of the metal–semiconductor barrier (in comparison with a P-N junction made by diffusion or grown processes), low resistance and low IR drop along the channel width, and good heat dissipation for power devices (the rectifying contact can also serve as an efficient thermal contact to heat sink). On the other hand, the P-N junction FET can have various gate configurations, such as a heterojunction or a buffered layer gate, that improve high-frequency performances.

Although MESFET can be fabricated in silicon, they are usually associated with gallium arsenide or other compound semiconductor materials, By using GaAs instead of Si, a higher electron mobility can be obtained, and furthermore GaAs can be operated at higher temperature (and so higher power levels).

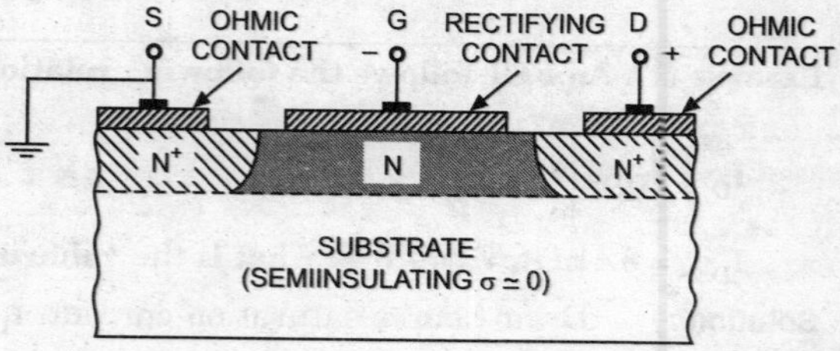

Fig. 5.12. *Cross-section of an N-Channel MESFET With a Semi-insulating Substrate*

Figure 5.12 depicts a simplified x-section of a GaAs MESFET. A thin epitaxial layer of a GaAs is used for the active region; the substrate is a very high resistivity GaAs material referred to as a semi-insulating substrate. The substrate is undoped or doped with chromium, which has energy level near the centre of GaAs band gap. In either case the Fermi level is near the centre of the gap, resulting in very high resistivity material (10^7 Ω-m), usually called semi-insulating GaAs. On this nonconducting substrate a thin layer of lightly-doped N-type GaAs is grown epitaxially, to form the channel region of the FET. The photolithographic processing consists of defining patterns in the metal layers for source and drain ohmic contacts (*e.g.* Au-Ge) and for the Schottky barrier gate (*e.g.* Al). By reverse biasing the Schottky gate, the channel can be depleted to the semi-insulating substrate, and the resulting ampere-volt characteristics are similar to the PN junction FET device. Since no diffusions are involved in the fabrication process, close geometrical tolerances can be had and the MESFET can be made very small in size. Gate lengths L ≤ 0.25 μm are common in such devices. This is very important at high frequencies, since drift time and capacitances must be kept to minimum.

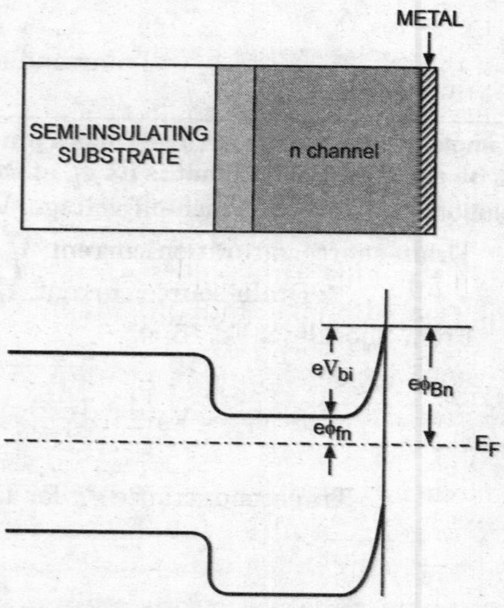

Fig. 5.13 *Idealized Energy-band Diagram of the Substrate Channel-Metal in The N-Channel MESFET*

It is possible to avoid epitaxial growth of the N-type layer and the etched isolation by using ion implantation. Starting with a semi-insulating GaAs substrate, a thin N-type layer at the surface of each transistor region can be formed by implanting silicon or a column VI donor impurity such as Se. This implantation needs an anneal for removing the radiation damage, but the epitaxial growth step is eliminated. In either the fully implanted device or the epitaxial device, the source and drain contacts may be improved by further N+ implantation in these regions. Owing to the relative simplicity of the implanted GaAs MESFET and the isolation between devices provided by the semi-insulating substrate, such structures are commonly employed in GaAs integrated circuits (ICs).

In the MESFET depicted in Fig. 5.12, a space charge region is induced by a reverse bias gate-to-source voltage V_{GS} under the metal gate and thus channel conductance is modulated as in the case of the PN junction FET. If the applied voltage is sufficiently large, the space charge will reach the substrate. This condition, again, is called the pinch-off. The device depicted in Fig. 5.12 is also a *depletion mode device,* because a gate voltage is required to be applied to pinch off the channel.

Treating the semi-insulating substrate as an intrinsic material, the energy-band diagram of the substrate-channel metal structure is as depicted in Fig. 5.13 with zero bias applied to the gate G. The majority carrier electrons are confined to the channel region due to a potential barrier between the channel and the substrate and between the channel and the metal.

Consider, now, MESFET of another type, in which the channel is pinched off even when gate-to-source voltage V_{GS} is zero. Fig. 5.14 (*a*) depicts this condition in which the channel thickness is smaller than the zero-biased space charge width. The depletion region is required to be reduced so as to open the channel which can be done by applying a forward-bias voltage to the gate-semiconductor junction. When a slightly forward-bias voltage is applied, the depletion region just extends through the channel—a condition called the *threshold,* illustrated in Fig. 5.14 (*b*). The threshold voltage V_{GST} is the gate-to-source voltage that is required to be applied to produce the pinch-off condition. The V_{GST} for such N-channel MESFET is positive, in contrast to the negative voltage for the N-channel depletion mode device. For V_{GS} exceeding V_{GST}, the channel region opens as illustrated in Fig. 5.14 (*c*). The applied forward-bias gate voltage is restricted to a few tenths of a volt before there is a significant gate current. This device is known as an N-channel enhancement mode MESFET. Enhancement mode P-channel MESFET and enhancement mode PN junction FETs are also available. The advantage of enhancement mode MESFETs is that circuits can be designed in which the voltage polarity on the gate and drain is same. However, the output voltage swing will be quite small with such devices.

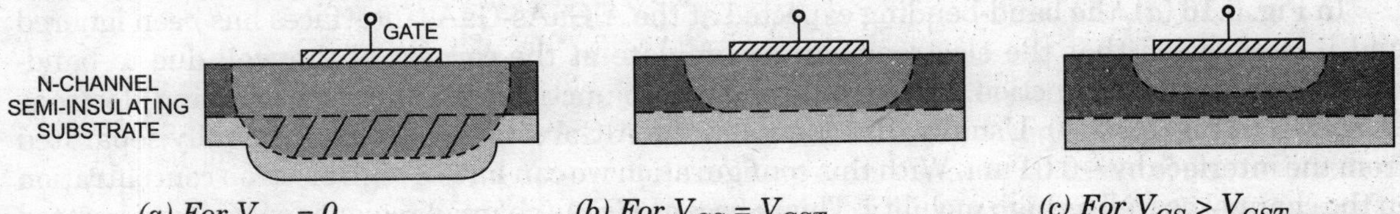

(a) For $V_{GS} = 0$ *(b) For* $V_{GS} = V_{GST}$ *(c) For* $V_{GS} > V_{GST}$

Fig. 5.14 ***Channel Space Charge Region of an Enhancement Mode MESFET***

5.11. High Electron Mobility Transistor (HEMT)

As frequency needs, power capacity, and low noise performance needs increase, the gallium arsenide MESFET is being pushed to its limit of design and performance. Such requirements imply a very small FET with a short channel length, large saturation current, and large transconductance. These requirements are usually met by increasing the channel doping under the gate. In all of the devices considered, the channel region is in a doped layer of bulk semiconductor with the majority carriers and doping impurities in the same region. The majority carriers experience ionized impurity scattering, which reduces carriers mobility and degrades the performance of the device.

What is required is a way of producing a high electron concentration in the channel of a MESFET by some means other than doping. An intelligent approach to this need is to grow a thin undoped well (e.g. GaAs) bounded by wider band gap, doped barriers (*e.g.* AlGaAs). This configuration, called *modulation doping,* results in conductive gallium arsenide when electrons from the doped AlGaAs barriers fall into the well and become trapped there, as illustrated in Fig. 5.16 (*a*). Since donors are in the AlGaAs rather than GaAs, there is no impurity scattering of electrons in the well. In case a MESFET is constructed with the channel along the GaAs well (perpendicular to the page in Fig. 5.16), we can have the advantage of this reduced scattering and resulting higher mobility. The effect is especially strong at low temperature where lattice (phonon) scattering is also low. The device is known by several names such as high electron mobility transistor (HEMT), modulation-doped field-effect transistor (MODFET), selectivity doped heterojunction field-effect transistor (SDHT), and two-dimensional electron gas field-effect transistor (TEGFET). A typical HEMT structure is depicted in Fig. 5.15.

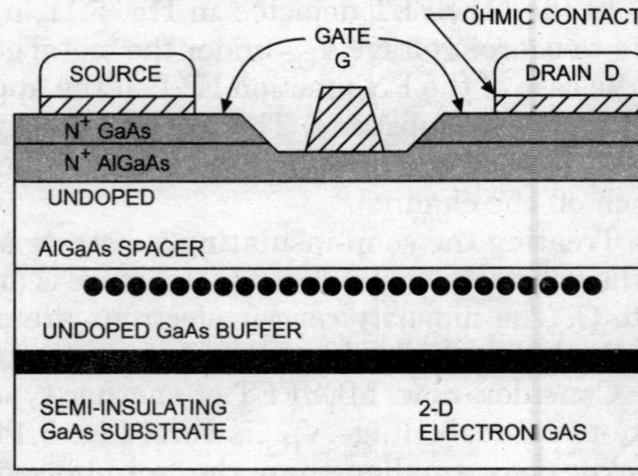

Fig. 5.15. ***A Metallic Crystal Structure***

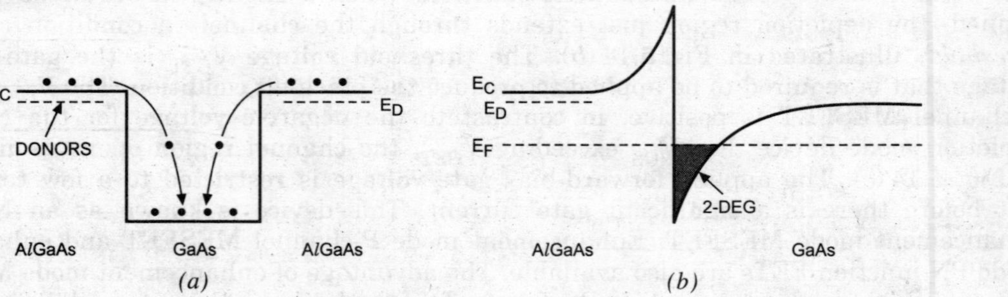

Fig. 5.16

In Fig. 5.16 (*a*), the band-bending expected at the AlGaAs-GaAs interfaces has been ignored and it is expected that the electrons will accumulate at the corners of the well due to band-bending at the heterojunction. In fact, only one heterojunction is required for trapping electrons as shown in Fig. 5.16 (*b*). Usually, the donors in the AlGaAs layers are intentionally separated from the interface by ~ 0.01 μm. With this configuration we can have a high electron concentration in the channel as well as high mobility. This is because GaAs channel region is spatially separated from the ionized impurities which provide the free carriers.

In Fig. 5.16 (*b*), mobile electrons produced by the donors in AlGaAs diffuse into the small band gap GaAs layer, and they are prevented from returning to the AlGaAs due to potential barrier at the AlGaAs-GaAs interface. The electrons in the (almost) triangular well form a two-dimensional electron gas, sometimes abbreviated as 2-DEG. Sheet carrier densities as large as 10^{16} m^{-2} can be had at a single interface such as that illustrated in Fig. 5.16 (*b*). Ionized impurity scattering is significantly reduced simply by separating the electrons from the donors. Also ionized impurity scattering can be reduced further by screening effects due to the extremely large density of the two-dimensional electron gas. In properly designed structures, the electron transport approaches that of bulk GaAs with no impurities, so that mobility is restricted by lattice scattering. This results in mobilities above 25 m^2/V-s at 77 K and 200 m^2/V-s at 4 K.

The advantages of this device are its ability to locate a large electron density (about 10^{16} m^{-2}) in a very thin layer (< 0.01 μm thick) very close to the gate and also to avoid ionized impurity scattering. The AlGaAs layer in this device is fully depleted under normal working conditions. The device behaviour closely resembles that of a MOSFET as the electrons are confined to the heterojunction. The advantages of the HEMT over the silicon MOSFET are the larger mobility and maximum electron velocity in GaAs in comparison to Si, and the smoother interfaces possible with an AlGaAs-GaAs heterojunction in comparison to $Si\text{-}SiO_2$ interface.

The high performance of the HEMT translates into an extremely high cut-off frequency, and devices with fast access times.

5.12. Short Channel Effects

Different modifications are required to be made to the simple theory of P-N junction FET and MESFET operation when the channel length is small (typically less than 1 μm). In the past, these short channel effects would be considered unusual, but now it is common to encounter FET devices in which these effects dominate the current-voltage characteristics. For instance, high-field effects occur when 1 V appears across a channel length of 1 μm, providing an electric field of 1,000 kV/m.

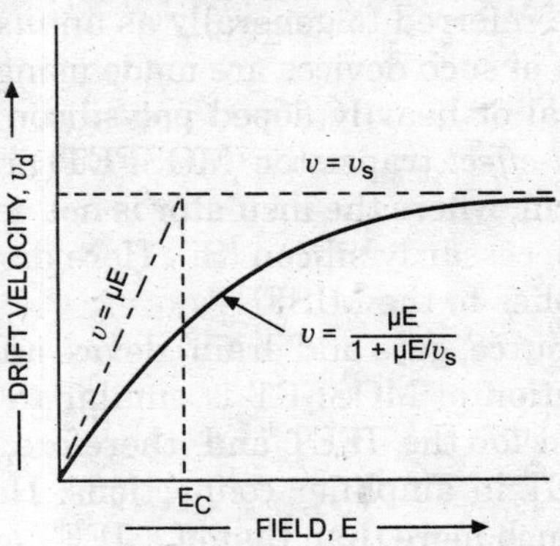

(a) *Approximations To The Saturation of Drift Velocity With Increasing Field*

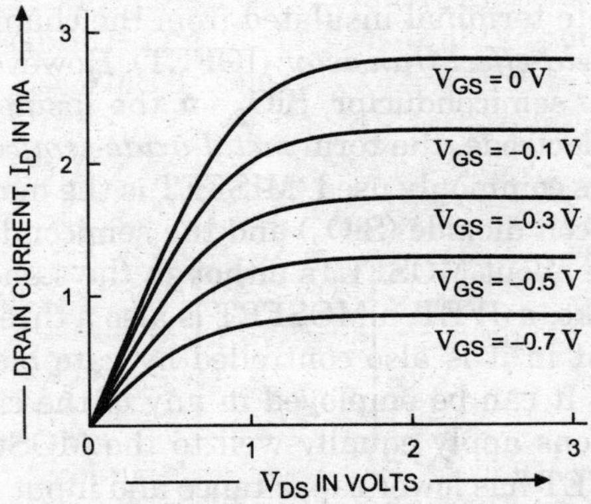

(b) *Drain Current-Voltage Characteristics For Saturated Velocity Case*

Fig. 5.17. ***Effects of Electron Velocity Saturation At High Electric Fields***

A simple piecewise-linear approximation to the velocity-field curve assumes a constant mobility (linear) dependence up to some critical field E_C and a constant saturation velocity v_s, for higher field. For silicon a better approximation is

$$\text{Drift velocity, } v_d = \frac{\mu E}{1 + \dfrac{\mu E}{v_s}} \qquad ...(5.13)$$

where μ is the low-field mobility. These two approximations are illustrated in Fig. 5.17(*a*). Assuming the electron movement through the channel drift with a constant saturation velocity v_s, the drain current takes a simple form

$$\text{Drain current, } I_D = env_s A = eN_D v_s Zh \qquad ...(5.14)$$

where h is a slow function of gate-to-source voltage V_{GS}. In this case saturated current follows the velocity saturation, and does not need a true pinch-off in the sense of depletion regions meeting at some point in the channel. In the saturated velocity case, the transconductance g_m is essentially constant, in contrast with the constant mobility case. The drain current-drain-to-source voltage ($I_D - V_{DS}$) curves are more evenly spaced, as illustrated in Fig. 5.17 (*b*), if constant saturation velocity dominates in comparison to the gate-to-source voltage V_{GS}—dependent spacing illustrated in Fig. 5.7 (*b*) for the long-channel constant mobility case.

Most of the devices operate with characteristics intermediate between the constant mobility and the constant velocity regimes. Depending upon the field distribution, it is possible to divide up the channel into regions dominated by the two extreme cases, or to use an approximation such as Eq. (5.13).

Another important effect of short channel is the reduction in effective channel length after pinch-off with the increase in drain-to-source voltage, as described in Art. 5.11. This effect does not play vital role in long channel devices because the change in length due to intrusion of the depletion region is a minor fraction of the channel length. But in short channel devices, the effective channel length may be substantially reduced, causing a slope in the saturated current-voltage characteristic that is analogous to the Early (base-width narrowing) effect in BJT discussed in Art. 4.12.

5.13. Metal-Insulator Semiconductor Field-Effect Transistor (MISFET)

The metal-insulator semiconductor field-effect transistor (MISFET) is one of the most widely used electronic devices, particularly in digital circuits, because of relatively small size, thousands of devices can be fabricated in a single integrated circuit (IC). Since it is constructed with the gate terminal insulated from the channel, it may be referred to generally as an *insulated-gate field-effect transistor* (IGFET). However, since most of such devices are made using silicon for the semiconductor, SiO_2 for the insulator, and metal or heavily doped polysilicon for the gate electrode, the term *metal-oxide-semiconductor field-effect* transistor (MOSFET) or simply MOS is commonly used. MISFET is the more general term, where the insulator is not necessarily silicon dioxide (SiO_2) and the semiconductor is not necessarily silicon (Si). Here discussion will be about MOSFETs although the same physics applies to the MISFETs.

Like, a JFET, a MOSFET is also a three terminal (source, gate and drain) device and drain current in it is also controlled by gate bias. The operation of MOSFET is similar to that of JFET. It can be employed in any of the circuits covered for the JFET and, therefore, all the equations apply equally well to the MOSFET and JFET in amplifier connections. However, MOSFET has lower capacitance and input impedance much more than that of a JFET owing to small leakage current. In case of a MOSFET the positive voltage may be applied to the gate and still the gate current remains zero.

MOSFETs are of two types namely (*i*) *enhancement type* MOSFET or E-MOSFET and (*ii*) depletion enhancement MOSFET or DE-MOSFET. In the depletion-mode construction a channel is physically constructed and a current between drain and source is due to voltage applied across the drain-source terminals. The enhancement MOSFET structure has no channel formed during its construction. Voltage is applied to the gate, in this case, to develop a channel of charge carriers so that a current results when a voltage is applied across the drain-source terminals.

The enhancement-type MOSFET (E-MOSFET) is widely used in both discrete and integrated circuits. In discrete circuits, the main use is in power switching, which means turning largest currents *on* and *off*. In ICs, the main use is in digital switching, the basic process behind modern computers. Although their use is declined, depletion-mode MOSFETs (DE-MOSFETs) are still found in high-frequency front-end communications circuits as RF amplifiers.

5.14. Depletion-Enhancement MOSFET (DE-MOSFET)

In Art. 5.3.5 we have seen that when the gate is biased negative w.r.t. the source in an N-channel JFET, the depletion-region widths are increased. The increase in the depletion regions reduces the channel thickness, which increases its resistance. The net result is that drain current I_D is reduced.

If the polarity of V_{GG} were reversed so as to apply a positive bias to the gate w.r.t. source, the P-N junctions between the gate and the channel would then be forward biased. Since a forward bias reduces the width of a depletion region, the thickness of channel would increase with a corresponding decrease in channel resistance. As a result, drain current I_D would increase beyond the JFET's I_{DSS} value.

The normal operation of a JFET is in its depletion mode of operation. However, as discussed above, it is also possible to *enhance* the conductivity of the JFET channel. However, the forward bias of the silicon P-N junction is usually restricted to a maximum of 0.5 V (more conservative limit is 0.2 V) so as to limit the gate current.

As we have seen in Art. 5.9, the greater the I_D in comparison to I_{DSS}, the greater transconductance g_m will be. In general, the higher the g_m, the better it is, because voltage gain is directly proportional to g_m. This is one of the advantages of being able to enhance the channel.

As its name suggests, the *depletion-enhancement* MOSFET (DE-MOSFET) was developed to be used in either or both the depletion and enhancement modes.

5.14.1. Construction

Figure 5.18 shows the construction of an N-channel depletion MOSFET. It consists of a highly doped P-type substrate into which two blocks of heavily doped N-type material are diffused forming the source and drain. An N-channel is formed by diffusion between the source and drain. The type of impurity for the channel is the same as for the source and drain. Now a thin layer of SiO_2 dielectric is grown over the entire surface and holes are cut through the SiO_2 (silicon-dioxide) layer to make contact with the N-type blocks (Source and Drain). Metal is deposited through the holes to provide drain and source terminals, and on the surface area between drain and source, a metal plate is deposited. This layer constitutes the gate. SiO_2 layer results in an extremely high input impedance of the order of 10^{10} to 10^{15} Ω for this area. The chip area of a MOSFET is typically 0.003 μm^2 or less which is about only 5% of the area required by a BJT.

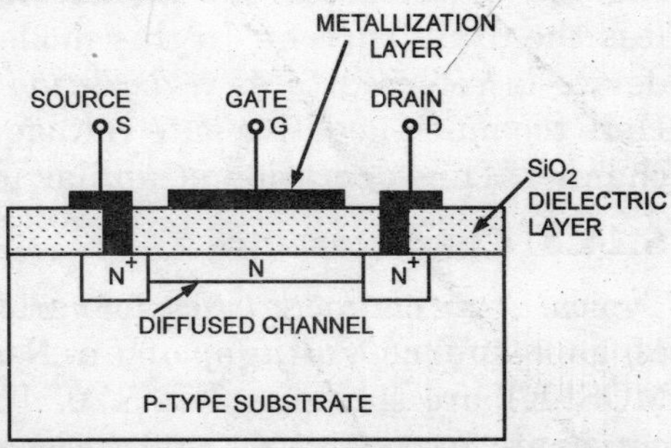

Fig. 5.18. *N-Channel DE-MOSFET Structure*

A P-channel DE-MOSFET is constructed like an N-channel DE-MOSFET, starting with an N-type substrate and diffusing P-type drain and source blocks and connecting them internally by a P-doped channel region.

In some cases the substrate is internally connected to the source terminal S and in others an additional terminal labelled SS is provided resulting in a four terminal device.

Silicon dioxide (SiO_2) is a type of insulator referred to as a *dielectric,* which develops opposing (as indicated by the prefix di-) electric fields within the dielectric when exposed to an externally applied field. The fact that the SiO_2 layer is an insulating layer means that there is no electrical connection between the gate terminal G and the channel of a MOSFET. It is the insulating layer of SiO_2 in the MOSFET construction that accounts for the very desirable high input impedance of the device.

In fact, the input resistance of a MOSFET is usually much higher than that of a typically JFET, even though the input impedance of most JFETs is sufficiently high for applications. Because of very high input impedance, the gate current I_G is essentially zero for dc-biased configurations.

5.14.2. Operation

It is obvious from Fig. 5.18 that there is no P-N junction between gate and channel. Here the diffused channel N, the insulating dielectric SiO_2 layer and the metal layer of the gate forms a parallel plate capacitor. DE-MOSFET can be operated with either a positive or a negative gate. When gate is positive with respect to the source it operates in the enhancement—or E-mode and when the gate is negative w.r.t. the source, as illustrated in Fig. 5.19, it operates in depletion-mode.

When the drain is made + ve with respect to source, a drain current will flow, even with zero gate potential and the MOSFET is said to be operating in E-mode. In this mode of operation gate attracts the – ve charge carriers from the P-substrate to the N-channel and thus reduces the channel resistance and increases the drain current. The more positive the gate is made, the more drain current flows.

On the other hand when the gate is made – ve with respect to the substrate, the gate repells some of the –ve charge carriers out of the N-channel. This creates a depletion region in the channel, as illustrated in Fig. 5.19, and, therefore, increases the channel resistance and reduces the drain current. The more – ve the gate, the less the drain current. In this mode of operation the device is referred to as a *depletion-mode MOSFET*. Here too much negative gate voltage can pinch-off the channel. Thus operation is similar to that of JFET.

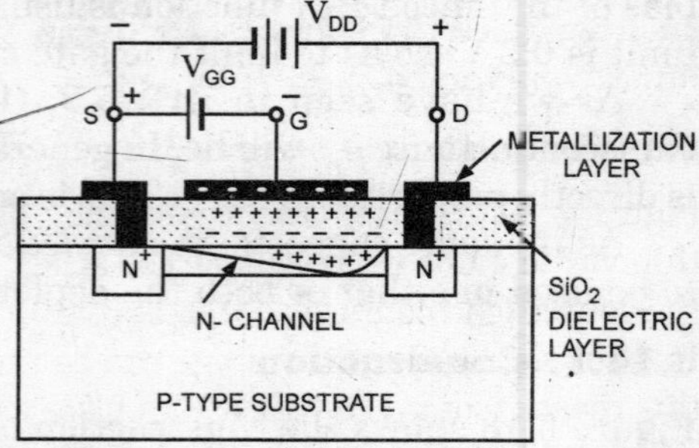

Fig. 5.19. *Depletion Mode Operation*

5.14.3. Characteristics

Typical *drain characteristics*, for various levels of gate-source voltage, of an N-channel MOSFET are shown in Fig. 5.20. The upper curves are for positive V_{GS} and the lower curves are for negative V_{GS}. The bottom drain curve is for $V_{GS} = V_{GS\,(OFF)}$. For a specified drain-source voltage V_{DS}, $V_{GS\,(OFF)}$ is the gate-source voltage at which drain current reduces to a certain specified negligibly small value, as shown in Fig. 5.20. This voltage corresponds to the pinch-off voltage V_P of JFET. For V_{GS} between $V_{GS\,(OFF)}$ and zero, the device operates in *depletion-mode* while for V_{GS} exceeding zero the device operates in *enhancement mode*. These drain curves again display an ohmic region, a constant-current source region and a cut-off region. MOSFET has two major applications: a constant current source and a voltage variable resistor.

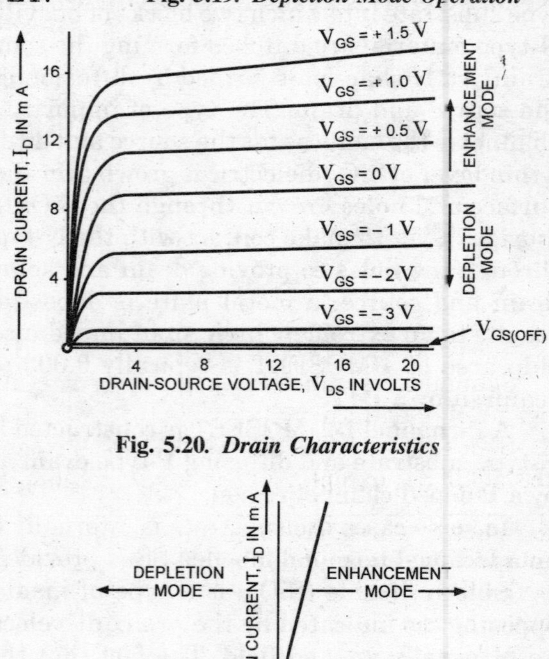

Fig. 5.20. *Drain Characteristics*

The *transfer* (or *transconductance*) characteristic for an N-channel DE-MOSFET is shown in Fig. 5.21. I_{DSS} is the drain current with a shorted gate. Since the curve extends to the right of the origin, I_{DSS} is no longer the maximum possible drain current. Mathematically, the curve is still part of a parabola and the same square-law relation exists as with a JFET. In fact, the depletion-mode MOSFET has a drain current given by the same transconductance equation as before, Eq. (5.1). Furthermore, it has the same equivalent circuits as a JFET. Because of this, the analysis of a depletion-mode MOSFET circuit is almost

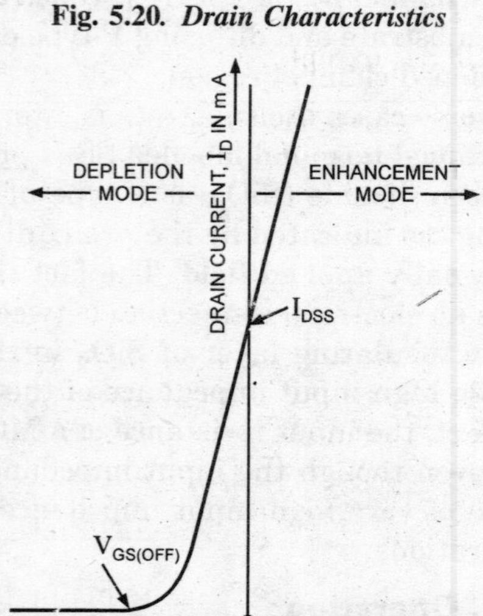

Fig. 5.21. *Transfer Characteristic*

identical to that of a JFET circuit. The only difference is the analysis for a positive gate, but even here the same basic formulas are used to determine the drain current I_D, gate-source voltage V_{GS} etc.

The foregoing discussion is applicable in principle also to the P-channel DE-MOSFET. For such a device the sign of all currents and voltages in the characteristics [Figs. 5.20 and 5.21] must be reversed.

5.14.4. Schematic Symbols

Figure 5.22 (*a*) shows the schematic symbol for a DE-MOSFET. Just to the right of the gate is the thin vertical line representing the channel. The drain lead comes out from the top of the channel and the source lead connects to the bottom. The arrow is on the P-substrate and points to the N-material. In some applications, a voltage can be applied to the substrate for added control of drain current. For this reason, some DE-MOSFETs have four terminal leads. But in most applications, the substrate is connected to the source. Usually the substrate is connected to the source internally by the manufacturer. This results in a three terminal device whose schematic symbol is shown in Fig. 5.22 (*b*).

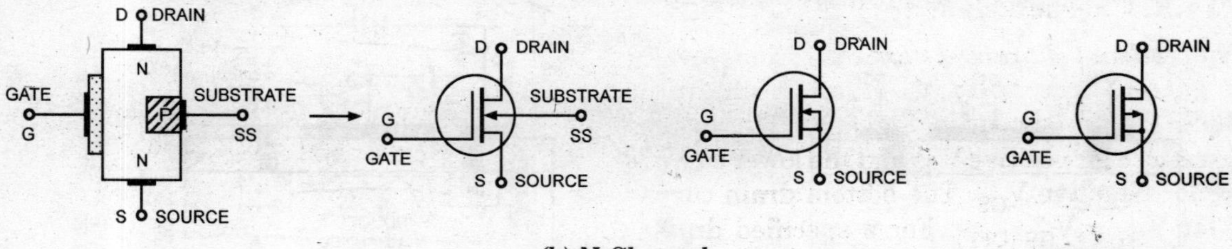

(*a*) (*b*) *N-Channel DE-MOSFET* (*c*) *P-Channel DE-MOSFET*

Fig. 5.22. ***Schematic Symbols For DE-MOSFETs***

Schematic symbol for a three terminal P-channel DE-MOSFET device is shown in Fig. 5.22 (*c*). The schematic symbol of a P-channel DE-MOSFET is similar to that of an N-channel DE-MOSFET, except that the arrow points outward.

Example 5.9. Sketch the transfer characteristic for an N-channel depletion type MOSFET with I_{DSS} = 10 mA and $V_P = -4$ V.

Solution : From Eq. (5.1)

At $V_{GS} = 0$, $\quad I_D = I_{DSS} = 10$ mA

At $V_{GS} = 1$, $\quad I_D = 10 \text{ mA}\left(1 - \frac{+1\text{ V}}{-4\text{ V}}\right)^2 = 15.625$ mA

At $V_{GS} = -1$ V, $\quad I_D = 10 \text{ mA}\left(1 - \frac{1}{4}\right)^2 = 5.625$ mA

At $V_{GS} = -2$ V, $\quad I_D = 10 \text{ mA}\left(1 - \frac{2}{4}\right)^2 = 2.5$ mA

At $V_{GS} = -4$ V $\quad I_D = 10 \text{ mA}\left(1 - \frac{4}{4}\right)^2 = 0$ mA

The transfer characteristic curve is drawn as shown in Fig. 5.23.

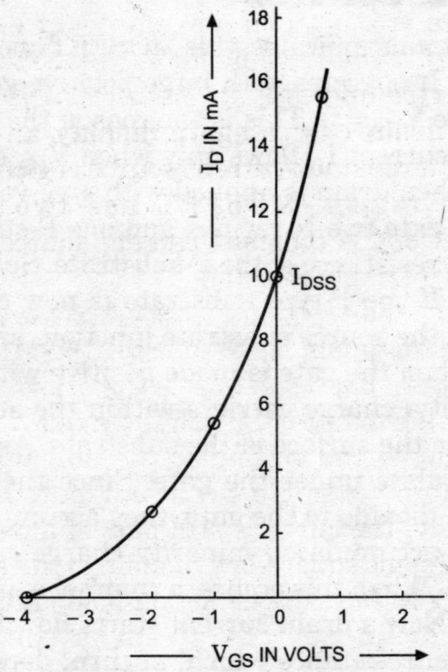

Fig. 5.23. ***Transfer Characteristic For an N-Channel Depletion Type MOSFET With I_{DSS} = 10 mA and $V_P = -4$ V***

5.15. Enhancement-Only MOSFET (E-MOSFET)

Although DE-MOSFET is useful in special applications, it does not enjoy widespread use. However, it played an important role in history because it was part of the evolution towards the E-mode MOSFET, a device that has revolutionized the electronic industry. E-MOSFET has become enormously important in digital electronics and computers. In the absence of E-MOSFETs, the personal computers (PCs), that are now so widespread, would not exist.

5.15.1. Construction

Figure 5.24 shows the construction of an N-channel E-MOSFET. The main difference between the construction of DE-MOSFET and that of E-MOSFET, as we see from Figs. 5.18 and 5.24 respectively, is that in E-MOSFET substrate extends all the way to the silicon dioxide (SiO_2) and no channels are doped between the source and the drain. Channels are electrically induced in these MOSFETs, when a positive gate-source voltage V_{GS} is applied to it.

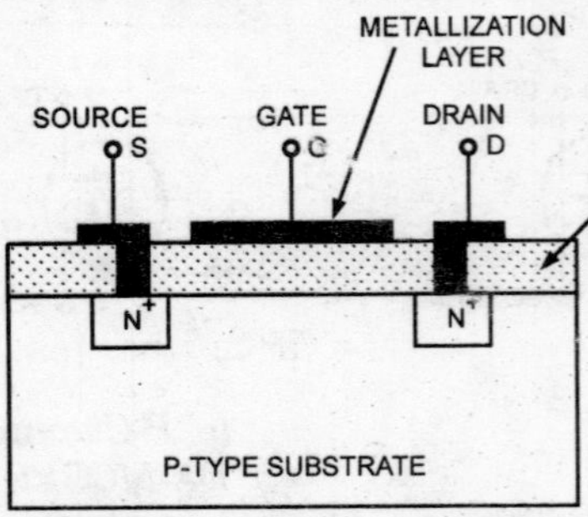

Fig. 5.24. *A Metallic Crystal Structure*

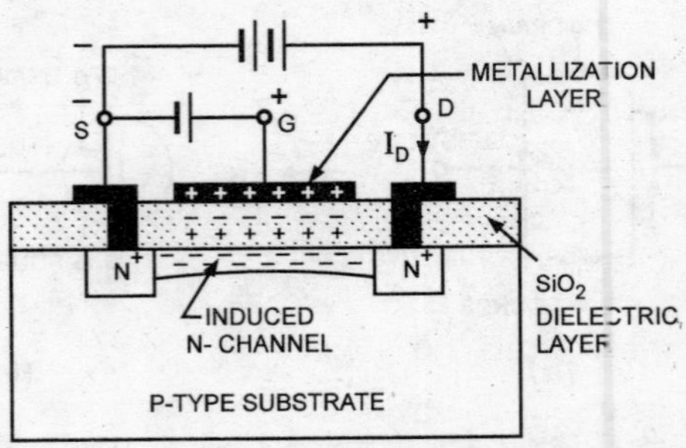

Fig. 5.25. *Operation of N-Channel E-MOSFET*

5.15.2. Operation

As its name indicates, this MOSFET operates only in the *enhancement mode* and has no depletion mode. It operates with large positive gate voltage only. It does not conduct when gate-source voltage $V_{GS} = 0$. This is the reason that it is called *normally-off* MOSFET. In these MOSFETs drain current I_D flows only when V_{GS} exceeds V_{GST} [gate-to-source threshold voltage].

When drain is applied with + ve voltage with respect to source and no potential is applied to the gate two N-regions and one P-substrate form two P-N junctions connected back to back with a resistance of the P-substrate. So a very small drain current *i.e.* a reverse leakage current flows. If the P-type substrate is now connected to the source terminal, there is zero voltage across the source substrate junction, and the drain-substrate junction remains reverse biased.

When the gate is made positive with respect to the source and the substrate, negative (*i.e.* minority) charge carriers within the substrate are attracted to the + ve gate and accumulate close to the surface of the substrate. As the gate voltage is increased, more and more electrons accumulate under the gate. Since these electrons cannot flow across the insulated layer of silicon dioxide to the gate, they accumulate at the surface of the substrate just below the gate. These accumulated minority charge carriers make N-type channel stretching from drain to source. When this occurs, a channel is induced by forming what is termed an *inversion layer* (N-type). Now a drain current starts flowing. The strength of the drain current depends upon the channel resistance which, in turn, depends on the number of charge carriers attracted to the positive gate. Thus drain current is controlled by the gate potential.

Since the conductivity of the channel is enhanced by the + ve bias on the gate, this device is also called the *enhancement MOSFET* or E-MOSFET.

The minimum value of gate-to-source voltage V_{GS} that is required to form the inversion layer (N-type) is termed the *gate-to-source threshold voltage* V_{GST}. For V_{GS} below V_{GST}, the drain current $I_D = 0$. But for V_{GS} exceeding V_{GST} an N-type inversion layer connects the source to drain and the drain current I_D is large. Depending upon the device being used, V_{GST} may vary from less than 1 V to more than 5 V.

JFETs and DE-MOSFETs are classified as the depletion-mode devices because their conductivity depends on the action of depletion layers. E-MOSFET is classified as an enhancement-mode device because its conductivity depends on the action of the inversion layer. Depletion-mode devices are normally ON when the gate-source voltage $V_{GS} = 0$, whereas the enhancement-mode devices are normally OFF when $V_{GS} = 0$.

5.15.3. Characteristics

Drain characteristics of an N-channel E-MOSFET are shown in Fig. 5.26. The lowest curve is the V_{GST} curve. When V_{GS} is lesser than V_{GST}, I_D is approximately zero. When V_{GS} is greater than V_{GST}, the device turns-on and the drain current I_D is controlled by the gate voltage. The characteristic curves have almost vertical and almost horizontal parts. The almost vertical components of the curves correspond to the ohmic region, and the horizontal components correspond to the constant current region. Thus E-MOSFET can be operated in either of these regions *i.e.* it can be used as a variable-voltage resistor (VVR) or as a constant current source.

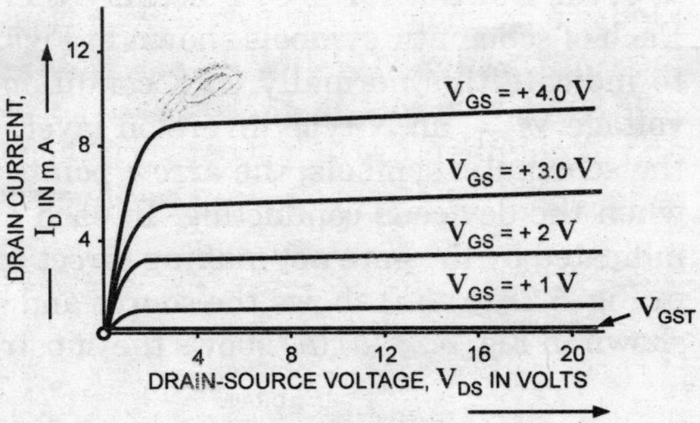

Fig. 5.26. ***Drain Characteristics***

Figure 5.27 shows a typical transconductance curve. The current I_{DSS} at $V_{GS} \leq 0$ is very small, being of the order of a few nanoamperes. When the V_{GS} is made positive, the drain current I_D increases slowly at first, and then much more rapidly with an increase in V_{GS}. The manufacturer sometimes indicates the *gate-source threshold voltage* V_{GST} at which the drain current I_D attains some defined small value, say 10 μA. A current $I_{D\,(ON)}$, corresponding approximately to the maximum value given on the drain characteristics and the values of V_{GS} required to give this current $V_{GS\,(ON)}$ are also usually given on the manufacturers data sheet.

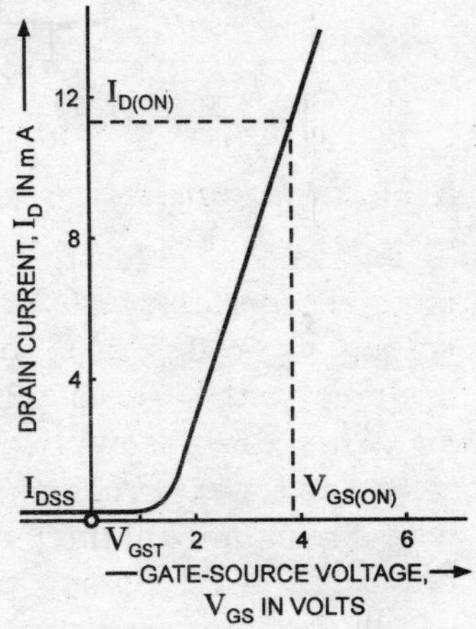

Fig. 5.27. ***Transfer Characteristic***

The equation for the transfer characteristic does not obey Eq. (5.1). However, it does follow a similar "square-law type" of relationship. The equation for the transfer characteristic of E-MOSFETs is given as

$$I_D = K\,(V_{GS} - V_{GST})^2 \qquad \text{...(5.15)}$$

where K, typically 0.3 mA/V^2, is a property of the device construction. Note that I_{DSS} cannot be associated with an E-MOSFET, because no drain current flows with $V_{GS} = 0$ volt.

With JFETs and DE-MOSFETs, the value of I_{DSS} and $V_{GS\,(OFF)}$ are the key quantities required for analysis. With E-MOSFETs, the key quantities are $I_{D\,(ON)}$, V_{GST}, and $V_{GS\,(ON)}$ indicated in Fig. 5.27. These three quantities are the first items to look for on a data sheet. By substituting these quantities in Eq. (5.15), the equation can be arranged in a more useful form:

$$I_D = K\, I_{D\,(ON)} \qquad ...(5.16)$$

$$\text{where } K = \left(\frac{V_{GS} - V_{GST}}{V_{GS\,(ON)} - V_{GST}} \right)^2 \qquad ...(5.17)$$

The input impedance of the E–MOSFET is very high (of the order of $10^{15}\ \Omega$) as the gate of the E-MOSFET is insulated from the channel. Transconductance for this device typically ranges from 1,000 μS to 2,000 μS (or 1 to 2 mA per volt).

5.15.4. Schematic Symbols

Schematic symbols for an N-channel E-MOSFET are shown in Fig. 5.28 (*a*). For zero value of V_{GS}, the E-MOSFET is OFF because there is no conducting channel between source and drain. Each of schematic symbols shown in Figs. 5.28 (*a*) (*ii*) and 5.28 (*a*) (*iii*), has broken channel line to indicate this normally OFF condition. As we know that for V_{GS} exceeding the threshold voltage V_{GST}, an N-type inversion layer, connecting the source to drain, is created. In each of the schematic symbols, the arrow points to this inversion layer, which acts like an N-channel when the device is conducting. In each case, the fact that the device has an insulated gate is indicated by the gate not making direct contact with the channel. The schematic symbol shown in Fig. 5.28 (*a*) (*iii*) shows the source and substrate internally connected, while the other symbol shown in Fig. 5.28 (*a*) (*ii*) shows the substrate connection brought out separately from the source.

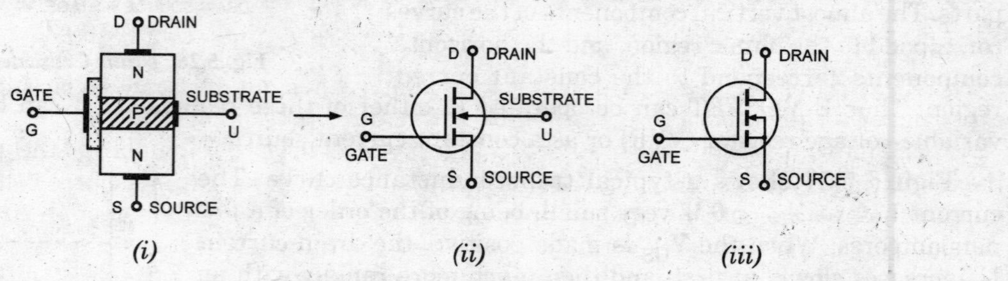

(*a*) *N-Channel E-MOSFET*

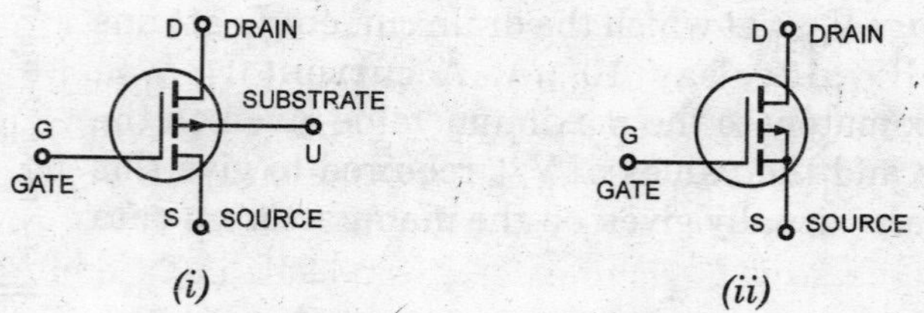

(*b*) *P-Channel E-MOSFET*

Fig. 5.28. ***Schematic Symbols For E-MOSFETs***

The schematic symbols for a P-channel E-MOSFET are shown in Fig. 5.28 (*b*). In these cases the arrow points outwards.

5.15.5. E-MOSFET Data Sheets

As already mentioned above, there are three key quantities [$I_{D\,(ON)}$, V_{GST} and $V_{GS\,(ON)}$] with E-MOSFETs. Data sheets may vary from one manufacturer to another one and one may contain some informations somewhat different from those given in another manufacturer's data sheet. But almost all data sheets include information regarding V_{GST}, $I_{D\,(ON)}$ and most include information regarding $V_{GS\,(ON)}$. Some also include information on the drain resistance R_{DS}. This resistance may be symbolised R_{DS} or it may appear as $r_{DS\,(ON)}$, $r_{ds\,(ON)}$ etc.

Example 5.10. An N-channel E-MOSFET has the following parameters :
$I_{D(ON)} = 5$ m A at $V_{GS} = 8$ V and $V_{GST} = 4$ V
Determine the drain current when $V_{GS} = 6$ V

Solution : From Eq. (5.15) $$K = \frac{I_{D(ON)}}{(V_{GS(ON)} - V_{GST})^2} = \frac{5 \text{ mA}}{(8-4)^2} = 0.3125 \text{ mA/V}^2$$

Now drain current, $I_D = K(V_{GS} - V_{GST})^2 = 0.3125\,(6-4)^2 = 1.25$ mA **Ans.**

5.16. Dual-Gate MOSFET

Cross-view of a dual-gate N-channel depletion type MOSFET is shown in Fig. 5.29. It acts as if two FETs are connected in series, as is obvious from Fig. 5.29. The middle block acts as drain for unit no. 1 and source for unit no. 2. Thus the current flow through the MOSFET is controlled by the voltage of both the gates. Drain current I_D can be cut-off by making either gate sufficiently negative.

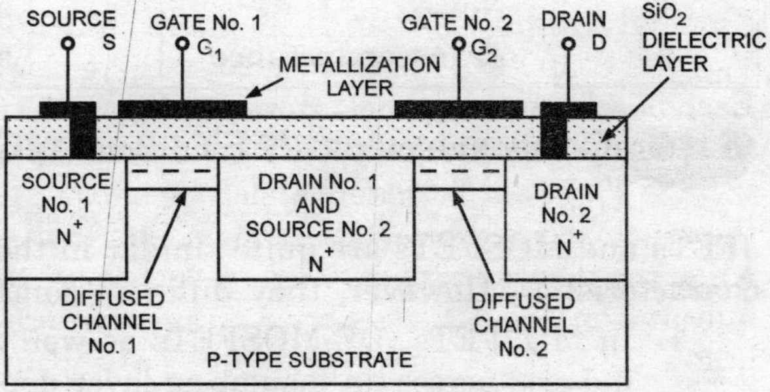

Fig. 5.29. ***Dual-Gate N-Channel Depletion Type MOSFET***

5.17. MOSFET Handling

MOSFET requires very careful handling particularly when out of circuit. In circuit a MOSFET is as rugged as any other solid-state device of similar construction and size.

MOSFETs have an ultra-thin silicon dioxide layer between the channel and the gate. Because the insulating layer is so thin, it is easily destroyed by excessive gate-to-source voltage V_{GS}. On application of large gate voltage, an open-circuit gate may accumulate enough charge so as to develop an electric field large enough to puncture the thin SiO_2 layer.

Aside due to direct application of an excessive gate-to-source voltage V_{GS}, the thin SiO_2 layer may get destroyed in more subtle ways. If a MOSFET is inserted or removed from a circuit while the power is ON, transient voltages caused by inductive kickback and other effects may exceed $V_{GS(max)}$ rating. This will wipe out the MOSFET. Even picking up a MOSFET may deposit enough static charge to exceed the $V_{GS(max)}$ rating. Generally, ground rings are used to short all leads of a MOSFET for avoiding any voltage build-up between gate and source. The grounding or shorting rings are removed after the MOSFET is connected in the circuit. Sometimes conducting foam is applied between the leads instead of using shorting or grounding rings.

Some MOSFETs are protected by built-in zener diodes in parallel with the gate and source, as shown in Fig. 5.30. With normal operating voltages, the zener diode remains open and does not influence the working of the circuit. In case of an extremely high V_{GS}, the zener diode breaks down thereby limiting the gate potential to a value equal to the zener diode breakdown voltage, much smaller than $V_{GS(max)}$ rating. The disadvantage of these internal zener diodes is that MOSFETs high input impedance is reduced. The trade-off is worth it in some applications because expensive MOSFETs get easily destroyed without zener protection.

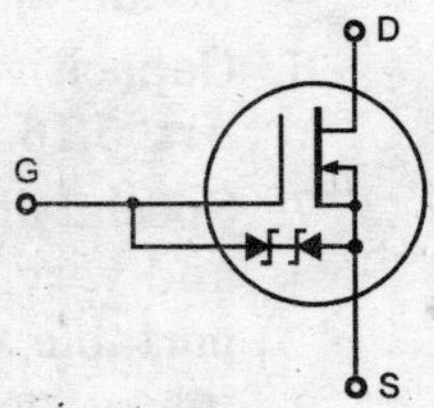

Fig. 5.30

Caution : *MOSFETs are delicate devices and can get easily destroyed. So they are to be handled carefully. Furthermore, they should never be connected or disconnected while the power is ON. Finally, before picking up a MOSFET device, get your body grounded by touching the chasis of the equipment you are working on.*

5.18. Comparison of N-Channel FETs With P-Channel FETs

S. No.	Particulars	N-Channel FETs	P-Channel FETs
1.	Current carriers	Electrons	Holes
2.	Mobility	Large	Poor
3.	Noise	Less	More
4.	Transconductance	Larger	Lesser

5.19. Comparison of JFETs and MOSFETs

JFETs and MOSFETs are quite similar in their operating principles and in their electrical characteristics. However, they differ in some aspects, as detailed below :

1. In MOSFETs (DE-MOSFETs as well as E-MOSFETs), the transverse electric field induced across an insulating layer deposited on the semiconductor material controls the conductivity of the channel whereas in the JFETs the transverse electric field across the reverse biased PN junction controls the conductivity of the channel.
2. JFETs can only be operated in the depletion mode whereas MOSFETs can be operated in either depletion or in enhancement mode. In a JFET, if the gate is forward biased, excess carrier injunction occurs and the gate current is substantial. Thus channel conductance is enhanced to some degree due to excess carriers but the device is never operated with gate forward biased because gate current is undesirable.
3. MOSFETs have input impedance much higher than that of JFETs because of small leakage currents. The input impedance of the MOSFET is of the order of 10^{10} to 10^{15} Ω whereas the input impedance of JFET is of the order of 10^8 Ω.
4. JFETs have characteristic curves more flatter than those of MOSFETs indicating a higher drain resistance.
   The drain resistance of a JFET is of the order of 10^5 to 10^6 Ω whereas the drain reistance of a JFET is of the order of 1 to 50 kΩ.
5. MOSFET is very susceptible to overload voltage and needs special handling during installation. It may get damaged easily if not properly handled.
6. Capacitive effects may be considerably lower in dual gate MOSFETs, as explained in Art. 5.16.
7. Special digital CMOS circuits are available which involve near zero power dissipation and very low voltage and current requirements. This makes them most suitable for portable systems.
8. When JFET is operated with a reverse bias on the junction, the gate current I_G is larger than it would be in a comparable MOSFET. The current caused by minority carrier extraction across a reverse-biased junction is greater, per unit area, than the leakage current that is supported by the oxide layer in a MOSFET. Thus MOSFET devices are more useful in electrometer applications than are the JFETs.

For the above reasons, and also because MOSFETs are somewhat easier to manufacture, they are more widely used than are the JFETs.

5.20. Comparison Between NMOS and PMOS

1. P-channel is much easier and cheaper to produce than the N-channel device.
2. The drain resistance of P-channel MOSFET is three times higher than that of an identical N-channel device.
3. The N-channel MOSFET is smaller for the same complexity than the P-channel MOSFET.
4. The N-channel MOSFET has the higher packing density which makes it faster in switching applications due to the smaller junction areas and low inherent capacitances.
5. A P-channel MOSFET occupies a larger area than the N-channel MOSFET for the given drain current rating. This is because the electron mobility is 2.5 times of that of an hole.
6. The N-channel MOSFET has higher false turn-on possibility than the P-channel device because of the positively charged contaminants.

Highlights

1. The FET is a majority carrier device, and is, therefore, often called a unipolar transistor. Like its bipolar counterpart, the FET is a three-terminal (namely drain, source and gate) semiconductor device in which current conduction is by one type of majority carriers (either electrons or holes).
2. Field effect devices are controlled by an electric field (or voltage) and give an extremely high input impedance—most important advantage over BJT devices. FETs are of several types such as JFETs, MESFETs and MISFETs.
3. The JFET is a voltage controlled device and has two diodes; the gate-source diode and the gate-drain diode for normal operation, the gate-source diode is reverse biased. Then the gate voltage controls the drain current.
4. The JFET has a high input impedance (of the order of 100 MΩ), a high degree of isolation between input and output, and a low noise figure. The main disadvantage of a JFET is its relatively low gain-bandwidth product, due to the junction capacitive effects.
5. The JFET can actually be used as a voltage-controlled resistor because of unique sensitivity of the drain-to-source resistance R_{DS} to the gate-to-source voltage V_{GS}.
6. The curve drawn between drain current I_D and drain-to-source voltage V_{DS} with gate-to-source voltage V_{GS} as the parameter is called the drain characteristic.
   Maximum drain current for any JFET labelled as I_{DSS} occurs when the gate-source voltage V_{GS} is zero. The minimum drain current for a JFET occurs at pinch-off defined by $V_{GS} = V_P$. The pinch-off voltage separates the ohmic and active regions for $V_{GS} = 0$. The gate-source cutoff voltage has the same magnitude as the pinch-off voltage $V_{GS(off)}$ turns the JFET off.
7. The relationship between the drain current I_D and the gate-to-source voltage V_{GS} of a JFET is a nonlinear defined by Shockley's equation

$$I_D = I_{DSS}\left[1 - \frac{V_{GS}}{V_P}\right]^2 = I_{DSS}\left[1 - \frac{V_{GS}}{V_{GS(off)}}\right]^2$$

   As the current level approaches I_{DSS}, the sensitivity of I_D to variations in V_{GS} increases significantly.
8. The curve drawn between drain current I_D and gate-to-source voltage V_{GS} for a given value of drain-to-source voltage V_{DS} is called the transfer characteristic. The transfer characteristics are the characteristics of the device itself and are not sensitive to the circuit

in which the JFET is used. The drain current increases more rapidly as V_{DS} approaches zero. Because the drain current equation contains a squared quantity, JFETs are referred to as square-law devices. The normalized transconductance curve indicates that I_D equals one-quarter of the maximum when V_{DS} equals half of the cut-off $\left(i.e.\ I_D = \frac{I_{DSS}}{4} \text{ when } V_{GS} = \frac{V_P}{2}\right)$ and at a point where $I_D = \frac{I_{DSS}}{2}$, $V_{GS} = 0.3$ V.

9. Maximum operating conditions are determined by the product of the V_{DS} and I_D.
10. Gate bias is used to bias a JFET in the ohmic region. When it operates in the ohmic region, a JFET is equivalent to a small resistance of R_{DS}. To ensure operation in the ohmic region, the JFET is driven into hard saturation by using $V_{GS} = 0$ and $I_{D(sat)} << I_{DSS}$.
11. The transconductance parameter g_m is determined by the ratio of the change in drain current ΔI_D associated with a particular change in gate-to-source voltage ΔV_{GS} in the region of interest. The steeper the slope of the I_D versus V_{GS} curve, the greater is the value of g_m. In addition, the closer the point or region of interest to the saturation current I_{DSS}, the greater is the transconductance parameter. The quantity g_m increases as V_{GS} approaches zero. Data sheets may list g_m or g_{fs} in siemens.
12. JFETs are mainly small-signal devices because most JFETs have a power rating of less than 1 W. When reading data sheets, start with the maximum ratings. Sometimes data sheets omit the minimum $V_{GS\ (off)}$ or other parameters. The large spread in JFET parameters justifies using ideal approximations for preliminary analysis and troubleshooting
13. MESFET is the same basic device (JFET) except that the PN junction is replaced by a Schottky barrier rectifying junction. The PN junction FET is fabricated with PN junction where as the MESFET is fabricated with a Schottky barrier rectifying junction.

    By using GaAs instead of silicon, a higher electron mobility can be achieved, and furthermore GaAs can be operated at higher temperature (and, therefore, higher power levels).

    GaAs MESFET devices have the advantages of higher electron mobility, therefore, smaller transit time and faster response and reduced parastatic capacitance and a simplified fabrication process, resulting from the semi-insulating GaAs substrate.
14. High-electron mobility transistor (HEMT) is also known as modulation-doped field-effect transistor (MODFET).

    The HEMT or MODFET, is also referred to as a two-dimensional electron gas FET (2-DEG FET, or TEGFET) to emphasize the fact that conduction along the channel occurs in a thin sheet of charge.

    HEMT is also referred to as a separately doped FET (SEDFET), to emphasize the fact that the doping occurs in a separate region from the channel.

    The advantages of HEMT over silicon MOSFET are the higher electron mobility and maximum electron velocity in AlGaAs/GaAs heterojunction compared with Si/SiO_2 interface. The high performance of the HEMT translates into an extremely high cut-off frequency, and devices with fast access times.
15. The operation of a metal-semiconductor field-effect transistor (MOSFET) is identical to that of a JFET. In the MOSFET, however, a metal-semiconductor rectifying contact, instead of a P-N junction, is used for the gate electrode.
16. The MOSFET is a field-effect transistor in which the metal gate is insulated by a very thin oxide layer from the semiconductor channel. It is acually a four-terminal device, with the substrate or body being the fourth terminal.

    There are two types of MOSFETs—the N-channel in which current is due to the flow of electrons in the inversion layer, and the P-channel in which current is due to the flow of holes in the inversion layer. Each of these devices can be either enhancement mode (E-mode) in which device is normally "off" and is turned on by applying a gate voltage, or depletion mode (DE-mode) in which the device is normally "on" and is turned off by applying a gate voltage.

17. The current in a MOSFET is due to the flow of carriers in the inversion layer between the source and drain terminals. The inversion layer charge density and channel conductance are controlled by the gate voltage, which means that the channel current is also controlled by the gate voltage.
18. The depletion mode MOSFET, has a source, gate, and drain. The gate is insulated from the channel. Because of this, the input resistance is very high. The DE-MOSFET has limited use, mainly in RF circuits.
19. The drain characteristic curves for a DE-MOSFET are similar to those of a JFET when operating in depletion mode. Unlike JFETs, DE-MOSFETs can also operate in the enhancement mode. When operating in enhancement mode, the drain current is greater than I_{DSS} (*i.e.* $I_D > I_{DSS}$).
DE-MOSFETs are mainly employed as RF amplifiers. They have good high frequency response, generate low levels of electrical noise, and maintain high input impedance values when V_{GS} is negative or positive.
20. The enhancement type MOSFET is normally off. When the gate voltage is equal to the threshold voltage (V_{GST}), an N-type inversion layer connects the source to the drain. When the gate voltage exceeds the V_{GST}, the device conducts heavily. Because of the thin insulating layer, MOSFETs are easily destroyed unless precautions are taken in their handling.
21. Since the E-MOSFET is primarily a switching device, it usually operates between cutoff and saturation. When it is biased in the ohmic region, it acts like a small resistance.
22. A dual-gate MOSFET, as the name implies provides a means of controlling I_D through the application of two input voltages, one on each gate, important applications of which include AGC, mixer circuits and demodulation.

Exercises

1. Explain the working of N-channel JFET. Also discuss its V-I characteristics.
[M.D. Univ. Third Semester May 2008]
2. Draw the symbol and characteristics of an N-channel JFET and mark linear region, saturation region and breakdown region.
3. Explaining the features of JFET, draw the drain characteristics.
[Pb. Univ. Applied Electronics I-Dec. 1995]
4. Sketch and explain the basic structure of an N-channel junction field-effect transistor.
[Pb. Univ. Applied Electronics-I, May 1994]
5. Draw the structure of a JFET and explain its principle of operation with neat diagrams along with its V-I characteristics. Define pinch-off voltage and mark it on the characteristics. Explain its significance in the operation of a JFET.
6. What is meant by depletion region in JFET ? Explain with suitable diagrams what are the basic differences between BJT and JFET.
7. What is pinch-off phenomenon in a JFET ? Mention the parameters that control the pinch-off voltage of a JFET.
8. Define pinch-off voltage of a JFET and explain its significance.
9. Why is a FET known as a unipolar device ? How do you compare this device with BJT ?
10. What are the relative merits and demerits of a FET amplifier over a transistor amplifier ?

11. Prove that the transconductance g_m of a JFET is given by

$$g_m = \frac{2}{|V_P|}\sqrt{I_{DS}\, I_{DSS}}$$

where V_P = pinch-off voltage, I_{DS} = drain current, I_{DSS} = maximum, I_{DS} corresponding to $V_{GS} = 0$ volt. **[U.P. Technical Univ. Fundamentals of Electronic Devices 2009-10]**

12. Explain the working of a JFET. Define the parameters of a JFET and develop its equivalent circuit.
13. How does MESFET differ from P-N junction FET ? Give the construction and operation of a MESFET.
14. Describe the short channel effects.
15. Enumerate the special feature of MESFET. Explain its working and discuss difference in its characteristics from the characteristics of MOSFET. **[U.P. Technical Univ. 2005-06]**
16. Draw the physical structure and equivalent circuit of the IGFET. Explain its special feature. Draw its I-V characteristic. **[U.P. Technical Univ. 2005-06]**
17. What is the advantage of the MESFET ? Draw its structure and explain its working. **[U.P. Technical Univ. 2007-08]**
18. What is HEMT ? Sketch the cross-section of a simple HEMT ? Describe its advantages over silicon MOSFET.
19. Describe briefly the construction and operation of a MOSFET in depletion region.
20. With the help of neat sketches explain the operation and construction of depletion and enhancement type of N-channel MOSFETs.
21. Explain the difference between the enhancement mode and depletion mode MOSFETs. **[U.P.S.C. I.E.S Elec. Engineering-II, 1997]**
22. Explain the construction and working of any two :
    (*i*) Depletion type MOSFET.
    (*ii*) Enhancement type MOSFET.
    (*iii*) JFET. **[M.D. Univ. Third Semester May 2007]**
23. What is a MOSFET ? How many types of MOSFET are there ? Point out the basic difference between the FET and BJT. **[U.P. Technical Univ. Fundamentals of Electronic Devices 2009-10]**
24. Explain the working of E-MOSFET in detail. Also discuss its V-I characteristics. **[M.D. Univ. Third Semester May 2008]**
25. Sketch the cross-sectional view of an enhancement mode MOSFET. Explain its operation and characteristics. **[U.P.S.C. I.E.S. Electrical Engineering II, 1998]**
26. Explain the basic construction of an enhancement type N-channel MOSFET. Draw and explain its static characteristics. How is the threshold voltage of the MOS-transistor adjusted ?
27. Draw the transfer characteristics of a MOSFET operation in enhancement mode.
28. Distinguish between JFET and MOSFET. **[U.P.S.C. I.E.S. Elec. Engineering-II, 1995]**
29. Distinguish the construction and operational features of a JFET and a MOSFET. **[U.P.S.C. I.E.S. Elec. Engineering-II, 1994]**
30. Write short notes on the following:
    (*i*) Characteristics of JFET.
    (*ii*) Gate control in JFET.
    (*iii*) Merits and demerits of JFETs over BJTs.
    (*iv*) MESFET.
    (*v*) HEMT.
    (*vi*) Short channel effects.
    (*vii*) MOSFET. **[M.D. Univ. Third Semester December 2005]**
    (*viii*) MOSFET characteristics.

Short Answer Type Questions With Answers

Q. 1. Why are ordinary transistors called bipolar transistors ?

Ans. In ordinary transistors, current conduction is by both majority carriers and minority carriers (both by holes as well as electrons) and, therefore, these are called bipolar transistors.

Q. 2. Why are field-effect transistors called unipolar transistors ?

Ans. In field-effect transistors current conduction is by only one type of majority carriers (either by electrons or by holes) and, therefore, these are called unipolar transistors.

Q. 3. Why the channel of a JFET is never completely closed at the drain end ?

Ans. If the channel is completely closed in JFET, then there will be no drain current, so there will be no voltage drop along the channel length and amount of reverse bias will become uniform and wedge-shaped depletion region will become rectangular one.

Q. 4. How is drain current controlled in a JFET ?

Ans. In a JFET drain current is controlled by controlling the reverse bias given to its gate (*i.e.* V_{GS}).

Q. 5. What is meant by drain characteristic of FETs ?

Ans. The curve drawn between drain current and drain-source voltage with gate-to-source voltage as the parameter is called the drain characteristic.

Q. 6. What is meant by transfer characteristic of FETs ?

Ans. The curve drawn between drain current and gate-source voltage for a given value of drain-source voltage is called the transfer characteristic.

Q. 7. No current flows through the channel of a JFET when $V_{GS} = 0$ and $V_{DS} = 0$, though channel is fully-open. Why ?

Ans. When drain-source voltage is zero, there is no attracting potential at the drain, so no current flows inspite of the fact that the channel is fully open.

Q. 8. Define pinch-off voltage for a JFET. **[U.P.S.C. I.E.S. 2002]**

Ans. The value of drain-source voltage at which channel is pinched-off (*i.e.* all the free charges from the channel get removed) is called the pinch-off voltage.

Q. 9. What are the parameters that control the pinch-off voltage of JFET ?

Ans. Parameters that control the pinch-off voltage of JFET are electron charge, donor/acceptor concentration density, permittivity of channel material and half width of channel bar.

Q. 10. Why FET is called a voltage controlled device ? **[Pb. Univ. April 1993]**

Ans. In a FET, drain current is controlled by the effect of the extension of the field associated with the depletion region developed by the reverse bias on the gate, so it is called a voltage controlled device.

Q. 11. How does the FET behave (*i*) for small values of $|V_{DS}|$ and (*ii*) for large values of $|V_{DS}|$? **[Pb. Univ. May 1994]**

Ans. (*i*) FET behaves as an ordinary resistor for small value of $|V_{DS}|$ *i.e.* in ohmic region.

(*ii*) FET behaves as a constant current source for large values of $|V_{DS}|$ till breakdown occurs.

Q. 12. How does the current vary with the gate voltage in the saturation region ? **[Pb. Univ. May 1994]**

Ans. Drain current decreases with the increase in gate-source bias $|V_{GS}|$. When $V_{GS} = 0$; drain current $I_D = I_{DSS}$, drain-source saturation current and when $V_{GS} = V_P$; drain current $I_D = 0$.

Q. 13. How does the transconductance vary with drain current ? **[Pb. Univ. May 1994]**

Ans. Transconductance g_m of a FET varies with the drain current I_D as given by the following equation

$$g_m = \frac{2}{|V_P|}\sqrt{I_{DS} \cdot I_{DSS}}$$

Q. 14. What is meant by gate-source cut-off voltage ?

Ans. The gate-source bias voltage required to reduce the drain current to zero is designated the gate-source cut-off voltage $V_{GS\,(OFF)}$. It is equal to pinch-off voltage V_P.

Q. 15. What is meant by saturation region ?

Ans. The region of drain characteristic of a FET in which drain current remains fairly constant is called the saturation or pinch-off region.

Q. 16. What is meant by drain-source saturation current I_{DSS} ?

Ans. The drain current in pinch-off or saturation region with zero gate-source voltage ($V_{GS} = 0$) is referred to the drain-source saturation current I_{DSS}.

Q. 17. Why is input impedance of the FET very high ?

Ans. FET has very high input impedance because its input circuit (gate-to-source) is reverse biased and the input gate current is very small (of the order of few nanoamperes).

Q. 18. What is the value of gate-source voltage V_{GS} that gives drain current of both N- and P-channel JFETs a zero temperature coefficient ?

Ans. $|V_{GS}| \quad |V_P| - 0.63$ V.

Q. 19. What is dynamic resistance of a JFET ?

Ans. The ratio of change in drain-source voltage to change in drain current at a given gate-source voltage is known as ac drain resistance or dynamic resistance r_d

$$i.e.\ r_d = \frac{\Delta V_{DS}}{\Delta I_D} \text{ at constant } V_{GS}.$$

Q. 20. What is meant by transconductance with reference to JFET ?

Ans. The control that gate-source voltage has over the drain current is measured by the transconductance of a JFET. It may be defined as the ratio of change in drain current to the change in gate-source voltage for a given value of drain-source voltage *i.e.*

$$g_m = \frac{\Delta I_D}{\Delta V_{GS}} \text{ at constant } V_{DS}.$$

Q. 21. What are the main disadvantages of JFETs ?

Ans. JFETs have the drawback of small gain, small gain-bandwidth product and gate leakage current as compared to MESFETs.

Q. 22. How does MESFET differ from P-N junction FET ?

Ans. MESFET is the same basic device (JFET) except that the PN junction is replaced by a Schottky barrier rectifying junction. The PN junction FET is fabricated with PN junction where as the MESFET is fabricated with a Schottky barrier rectifying junction.

Q. 23. Why MESFET is usually made from GaAs or other compound semiconductor materials rather than from silicon ?

Ans. GaAs or InP semiconductors are used instead of silicon in MESFET fabrication because of higher mobility and carrier drift velocity. Furthermore, GaAs can be operated at higher temperature (and, therefore higher power levels).

Q. 24. Enumerate the special features of MESFETs.

Ans. The MESFETs have special features of possibility of operation at high temperature (and, therefore, higher power levels); high operational speed, higher carrier drift velocities (about twice of silicon), higher electron mobility (about six times of silicon) and high frequency operation (because drift-time and capacitance may be reduced to minimum).

Q. 25. What is the main drawback of MESFET ?

Ans. The main drawback of MESFET is a limited gate voltage swing and that is due to small *turn-on* voltage of the Schottky gate. But this limitation is not of much significance in small-power circuits operating with a low supply voltage.

Q. 26. Enumerate the advantages of MESFETs.

Ans. MESFETs have the advantages of higher gain, higher gain-bandwidth product and relatively smaller gate leakage currents, lower parasitic capacitances; Good heat dissipation for power devices; Lower resistance and therefore, low voltage drop along the channel width; Lower parasitic resistances, larger transconductances and shorter electron transit time, and possibility of close geometrical tolerances and smaller channel lengths because MESFET fabrication does not require diffusion. Furthermore, the use of GaAs in MESFET fabrication provides higher mobility and higher carrier drift velocity. Higher electron mobility results in lower noise figures.

Q. 27. Enumerate the applications of MESFETs.

Ans. MESFETs are employed in high performance circuits of communication, computers and military systems, microwave amplifiers, oscillators, switches and mixers.

Q. 28. Give the advantages of HEMT over silicon MOSFET.

Ans. The advantages of HEMT over silicon MOSFET are the higher electron mobility and maximum electron velocity in AlGaAs/GaAs heterojunction compared with Si/SiO_2 interface. The high performance of the HEMT translates into an extremely high cut-off frequency, and devices with fast access times.

Q. 29. Why HEMT or MODFET, is also referred to as a two-dimensional electron gas FET ?

Ans. The HEMT or MODFET, is also referred to as a two-dimensional electron gas FET (2-DEG FET, or TEGFET) to emphasize the fact that conduction along the channel occurs in a thin sheet of charge.

Q. 30. Why HEMT is also referred to as SEDFET ?

Ans. HEMT is also referred to as a separately doped FET (SEDFET), to emphasize the fact that the doping occurs in a separate region from the channel.

Q. 31. Why is MOSFET called sometimes IGFET ?

Ans. MOSFET is constructed with the gate terminal insulated from the channel so it is sometimes called insulated gate FET (or IGFET).

Q. 32. How does MISFET differ from MOSFET ?

Ans. Metal-insulator semiconductor field-effect transistor (MISFET) is the more general term where insulator is not necessarily SiO_2 and semiconductor is not necessarily Si, where as in case of a MOSFET SiO_2 is necessarily an insulator and Si a semiconductor.

Q. 33. What is the significant difference between the construction of an enhancement type MOSFET and a depletion type MOSFET ?

Ans. In the depletion type MOSFET a channel is physically constructed and a current between drain and source is due to voltage applied across the drain-source terminals while in enhancement type construction no channel is formed during its construction. Voltage is applied to the gate in case of enhancement type MOSFET, to develop a channel of charge carriers so that a current results when a voltage is applied across the drain-source terminals.

Q. 34. Why E-MOSFET is called sometimes normally-off MOSFET ?

Ans. E-MOSFET operates with large positive gate voltages only and does not conduct when gate-source voltage $V_{GS} = 0$, so it is called normally-off MOSFET.

Q. 35. What is meant by gate-to-source threshold voltage in E-MOSFET ?

Ans. The minimum value of gate-to-source voltage V_{GS} that is required to form the inversion layer is termed the gate-to-source threshold voltage V_{GST}.

Q. 36. How is threshold voltage of the MOS transistor adjusted ?

Ans. Threshold voltage of a MOS transistor can be adjusted by the following three methods.

1. If the silicon crystal orientation is < 111 >, it results in high threshold voltage. If the crystal has orientation of < 100 >, the value of threshold will go to half as compared to the previous one.
2. Instead of SiO_2, a layer of $Si_3 N_4$ and SiO_2 is used. It makes the dielectric constant twice as that of SiO_2 alone. This causes V_{GST} to come down to half.
3. Instead of aluminium as gate electrode polycrystalline silicon doped with boron is used as the gate electrode. This causes reduction in contact potential between gate electrode and gate dielectric resulting in reduction of V_{GST}.

Q. 37. Why MOSFETs are never connected or disconnected in the circuit when power is ON ?

Ans. If a MOSFET is connected or disconnected in circuit while the power is ON, transient voltages caused by inductive kickback and other effects may exceed $V_{GS\ (max)}$ and thus wipe out the MOSFET.

Q. 38. Name the factors which make the JFET superior to BJT ?

Ans. The high input impedance, low output impedance and low noise level make JFET far superior to the BJT.

Problems

1. When gate-source voltage is – 12 V and gate current is 0.1 μ A, determine the resistance between the gate and source of a given JFET. **[Ans.** 120 MΩ]

2. In case of a JFET the variation in drain current is of 100 μ A when drain-source voltage is varied by 1.25 V, keeping gate-source voltage V_{GS} constant, determine the dynamic resistance of the JFET. **[Ans.** 12.5 kΩ]

3. In a JFET the drain current is changed by 0.25 m A when the gate-source voltage is changed by 0.125 V, keeping drain-source voltage constant. Calculate the transconductance of the given JFET. **[Ans.** 2 mS]

4. For an N-channel JFET, I_{DSS} = 8 mA ; V_P = – 4 V ; V_{GS} = – 1 V. Determine I_D, g_{mo} and g_m. **[Ans.** 4.5 mA; 4 m A/V; 3 m A/V]

❑❑❑

SECTION-D

Some Special Devices

Contents of This Section

SOME SPECIAL DEVICES : Photodiodes; Photodetectors; Solar Cells; Light-Emitting Diodes; Semiconductor Lasers; Light-Emitting Materials. Tunnel Diode; Degenerate Semiconductors; IMPATT Diode; The Transferred Electron Mechanism; The GUNN Diode P-N-P-N Diode; Silicon-Controlled Rectifier (SCR); Bilateral Devices : DIAC; TRIAC, IGBT.

CHAPTER 6

Photonic Devices

6.1. INTRODUCTION

Inside This Chapter

Photonic devices are those in which the basic particle of light (the photon) plays an important role. These devices can be categorised as follows:

1. Devices that detect optical signals through electronic processes, such as photodetectors.
2. Devices that convert optical radiation into electrical energy, such as photovoltaic device or solar cell.
3. Devices that convert electrical energy into optical radiation, such as light-emitting diodes (LEDs) and the diode LASER (light amplification by stimulated emission of radiation).

The characteristics of photodetectors and solar cells are dependent on optical energy, which is absorbed in a semiconductor and creates excess EHPs generating photocurrents. *Photodetectors* are the semiconductor devices that can be used to detect the presence of photons and convert optical signals into electrical signals. The *photoconductor*, in which photon-generated excess EHPs change the conductivity of a semiconductor, is perhaps the simplest type of photodetector: Photodiode is a two-terminal device designed to respond to photon absorption. It is a P-N junction diode operated with a reverse-bias voltage applied. The excess electrons and holes produced in the space charge region are separated very quickly by the electric field and thus photocurrent is produced. The output terminals of the *solar* or *photovoltaic cell* are connected to a resistive load in order to convert input optical power into electrical power.

Electroluminescence is the generation of light by an electric current passing through a material under an applied electric field. Excess carriers are created and then recombine, which may cause emission of photons in a forward-biased PN junction—such a device is called the *light-emitting diode (LED).* The LED photon emission is due to a spontaneous transition of an electron from the conduction to the valence band, resulting in a fairly wide spectral output bandwidth. LED spectral output has a relatively wide bandwidth of wavelengths but emission spectrum is narrow enough so that a particular colour (red, orange, yellow or green) will be visible to the eye, provided that the output is in the visible range. In case the light-emitting diode device structure and operating mode are modified, the device can be operated in a new mode in which the bandwidth is very narrow and the photon output is coherent. This new device is referred to as a *laser diode.* The laser diode relies on the concepts of stimulated emission and population inversion, and on the use of an optical resonant cavity.

6.2. PHOTODETECTORS

Photodetectors are the semiconductor devices that can detect optical signals through electronic processes. The extension of coherent and incoherent light sources into the far-infrared region on one hand and the ultraviolet (UV) region on the other has enhanced the requirement for high-speed, sensitive photodetectors. A general photodetector has basically three processes: (*i*) carrier generation by incident light, (*ii*) carrier transport and/or multiplication by whatever current-gain mechanism may be present, and (*iii*) interaction of current with external circuit to give the output signal.

Photodetectors play vital role in optical-fiber communication systems operated in the near-infrared region (0.8 to 1.6 μm). They demodulate optical signals (convert the optical variations into electrical variations), that are subsequently amplified and further processed. For these applications the stringent requirements that are to be met by the photodetectors include high speed response, high sensitivity at operating wavelengths and minimum noise. The photodetector must also be compact in size, use small biasing voltages/current, and be reliable under operating conditions.

The best known devices of these types are the light dependent resistor (LDR), the photodiode and phototransistor.

6.3. LIGHT DEPENDENT RESISTORS (LDRs)

Light dependent resistors (LDRs) are made from cadmium sulphide containing no or very few free electrons when not illuminated. Its resistance is then quite high. When it absorbs light, electrons are liberated and the conductivity of the material increases. Cadmium sulphide (CdS) is, therefore, a photoconductor. The approximate relationship between the resistance and illumination is

$$R = A E^{-\alpha} \qquad ...(6.1)$$

where E is illumination in lux, R is resistance in ohms, A and α are constants. The value of α depends on the cadmium sulphide used and on the manufacturing process. Values around 0.7 to 0.9 are quite common. The relationship between the resistance R and the illumination E is shown graphically in Fig. 6.1. The resistance rises to several megaohms under dark conditions.

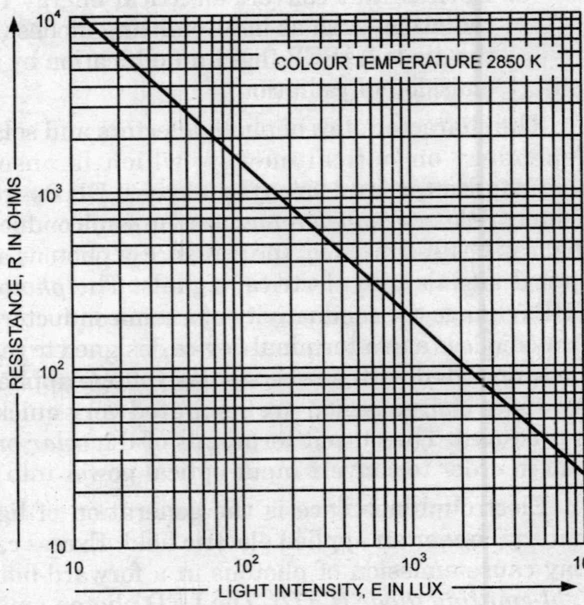

Fig. 6.1. ***LDR Characteristics***

The basic construction and symbol for LDR are shown in Figs. 6.2 (*a*) and 6.2 (*b*) respectively. The device consists of a pair of metal film contacts separated by a snake-like track of cadmium sulphide film, designed to provide the maximum possible contact area with the two metal films. The structure is housed in a clear plastic or resin case, to provide free access to external light. Practical LDRs are available in a variety of sizes and package styles, the most popular size having a face diameter of roughly 10 mm.

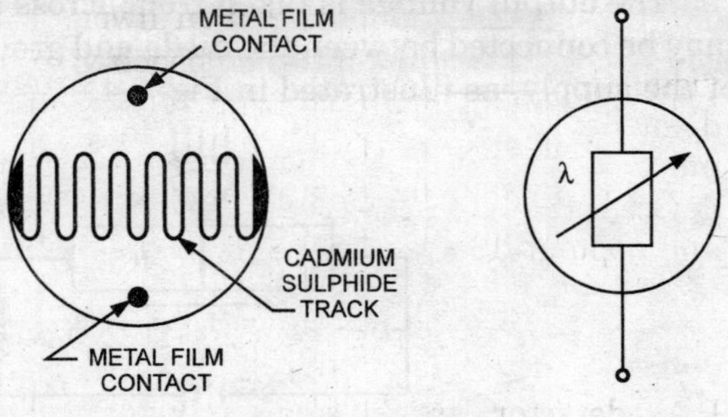

(*a*) *Basic Structure* (*b*) *Symbol*

Fig. 6.2. ***LDR***

Spectral Response. The resistors are only light dependent over a limited range of wavelengths. LDRs have their maximum response at about 680 nm.

Temperature Dependency. Electrons can be excited not only by photons but also by thermal agitation. The dark resistance is, therefore, not infinite at normal temperatures. It increases with the ambient temperature and can be reduced by cooling the device.

The temperature can also affect the resistance under illumination. At practical illumination levels and normal ambient temperatures the temperature coefficient is, however, very small and can be neglected.

Recovery Rate. When an LDR is brought from a certain illuminating level into total darkness, the resistance does not increase immediately to the dark value. The recovery rate is specified in kΩ/second and for current LDR types it is more than 200 kΩ/second (during the first 20 seconds starting at a light level of 1,000 lux). The recovery rate is much greater in the reverse direction, *e.g.* going from darkness to illumination level of 300 lux, it takes less than 10 ms to reach a resistance which corresponds with a light level of 400 lux.

LDRs are sensitive, inexpensive, and readily available devices. They have good power and voltage handling capabilities, similar to those of a conventional resistor. Their only significant defect is that they are fairly low acting, taking tens or hundreds of milliseconds to respond to sudden changes in light level. Useful practical applications of LDRs include light and dark-activated switches and alarms, light beam alarms and reflective smoke alarms etc.

An overlight detector is shown in Fig. 6.3. The LDR is a variable resistor whose resistance decreases with the increase in light intensity. When the light falling on an LDR has normal intensity, its resistance is large enough and the voltage across R_1 is insufficient to trigger the SCR. However, when light falling on LDR is of large intensity, the resistance of LDR falls and voltage drop across R_1 becomes large enough to trigger the SCR. Consequently, the buzzer sounds the alarm. It may be noted that if the strong light disappears, the buzzer continues to sound the alarm. It is because once the SCR is fired, the gate loses all control.

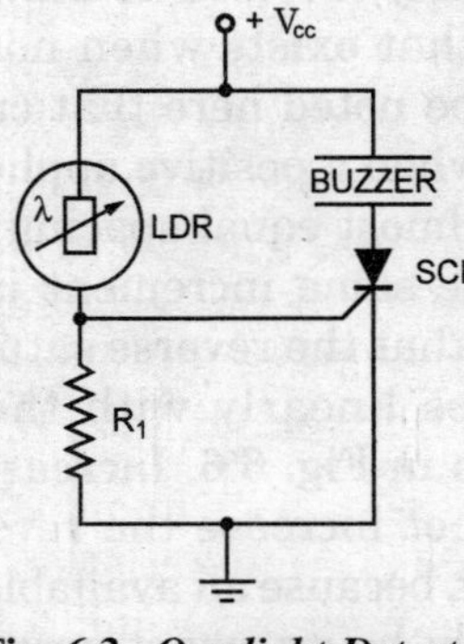

Fig. 6.3. ***Overlight Detector***

6.4. PHOTODIODE

Photodiode is a two-terminal semiconductor P-N junction device and is designed to operate with reverse bias. The basic biasing arrangement, construction and symbols for the device are given in Fig. 6.4. It is either mounted in translucent case or has its semiconductor junction mounted beneath an optical lens.

The output voltage is taken from across a series-connected load resistor R. This resistance may be connected between the diode and ground or between the diode and the positive terminal of the supply, as illustrated in Fig. 6.4.

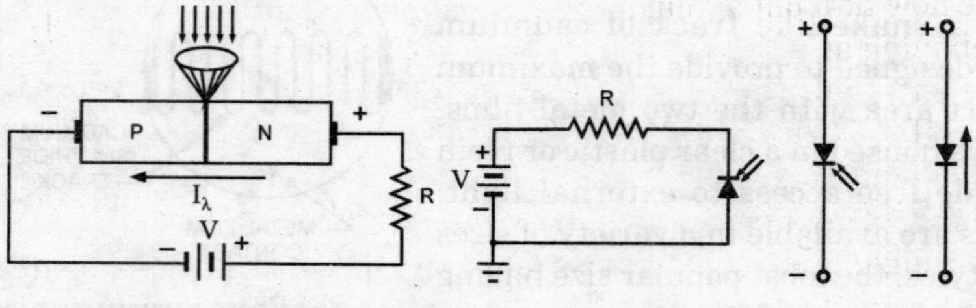

(a) Basic Biasing Arrangement and Construction *(b) Symbols*

Fig. 6.4. ***Photodiode***

When the P-N junction is reverse-biased, a reverse saturation current flows due to thermally generated holes and electrons being swept across the junction as the minority carriers. With the increase in temperature of the junction more and more hole-electron pairs are created and so the reverse saturation current I_0 increases. The same effect can be had by illuminating the junction. When light energy bombards a P-N junction, it dislodges valence electrons. The more light striking the junction the larger the reverse current in a diode. It is due to generation of more and more charge carriers with the increase in level of illumination. This is clearly shown in Fig. 6.5 for different intensity levels. The *dark current* is the current that exists when no light is incident. It is to be noted here that current becomes zero only when a positive applied bias equals to V_0. The almost equal spacing between the curves for the same increment in luminous flux reveals that the reverse saturation current I_0 increases linearly with the luminous flux as shown in Fig. 6.6. Increase in reverse voltage does not increase the reverse current significantly, because all available charge carriers are already being swept across the junction. For reducing the reverse saturation current I_0 to zero, it is necessary to forward bias the junction by an amount equal to barrier potential. Thus the photodiode can be used as a *photoconductive device*.

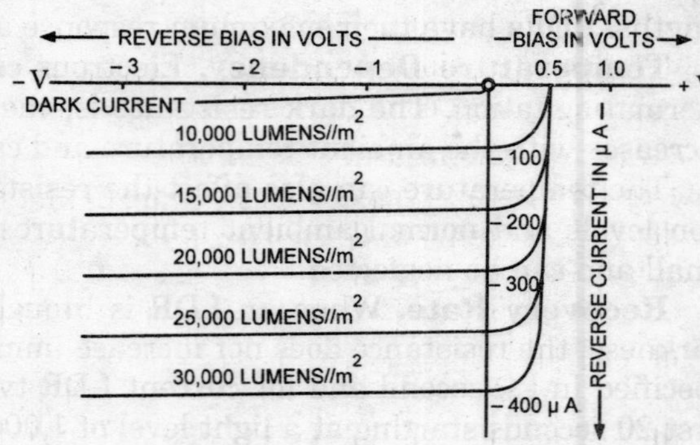

Fig. 6.5. ***Photodiode Characteristics***

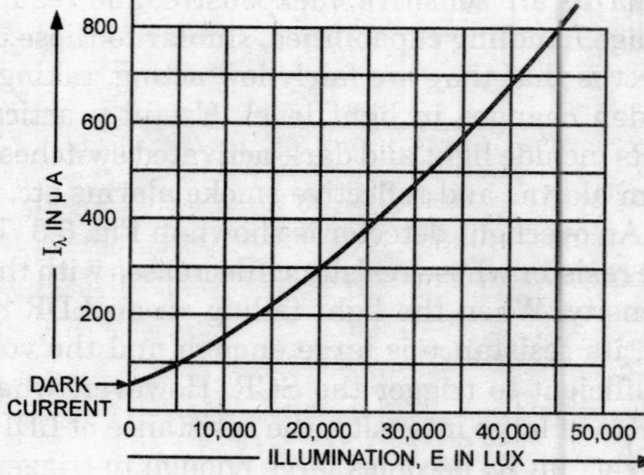

Fig. 6.6

On removal of reverse bias applied across the photodiode, minority charge carriers continue to be swept across the junction while the diode is illuminated. This has the effect of increasing the concentration of holes in the P-side and that of electrons in the N-side. But the barrier potential is negative on the P-side and positive on the N-side, and was created by holes flowing from P to N-side and electrons from N to P-side during fabrication of junction. Thus the flow of minority carriers tends to reduce the barrier potential.

When an external circuit is connected across the diode terminals, the minority carriers return to the original side via the external circuit. The electrons which crossed the junction from P to N-side now flow out through the N-terminal and into the P-terminal. This means that the device is behaving as a voltage cell with the N-side being the negative terminal and the P-side the positive terminal. Thus, the photodiode is a *photovoltaic* device as well as *photoconductive* device. When the device is operated with a reverse bias, it operates as a photoconductive device and when operated without the reverse bias, it operates as a photovoltaic device.

Since the rise and fall times (change-of-state parameters) are very small for this device (in the nanosecond range), it can be employed for high-speed counting or switching applications.

Photodiodes have a far lower light sensitivity than cadmium sulphide LDRs, but give a fair quicker response in light level. Generally LDRs are ideal for use in slow acting direct coupled light-level sensing applications, while photodiodes are ideal for use in fast acting ac coupled signalling applications. Typical photodiode applications include detection (both visible and invisible), demodulation, switching, logic circuits that need stability and high speed, character recognition, optical communication equipment, IR remote control circuits, encoders etc.

An alarm system using a photodiode is shown in Fig. 6.7. The reverse current I_λ will continue to flow as long as the light beam is not interrupted. When the light beam gets interrupted, I_λ falls to the dark current level and sounds the alarm.

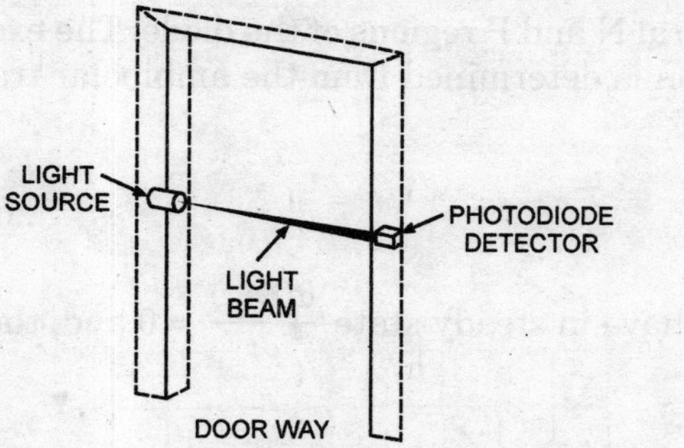

Fig. 6.7. ***Alarm System Using a Photodiode***

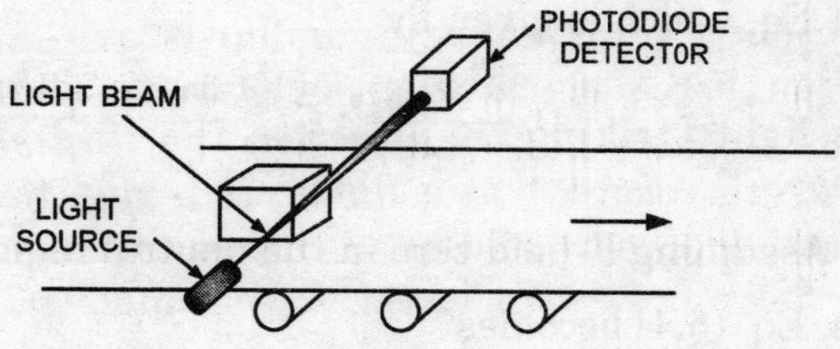

Fig. 6.8. ***Counter Operation Using a Photodiode***

In Fig. 6.8 a photodiode is used to count items on a conveyor belt. As each item passes the light beam is interrupted, I_λ falls to the dark current level and the counter is increased by one count.

Expression For Total Steady-State Diode Photocurrent For The Long Diode.

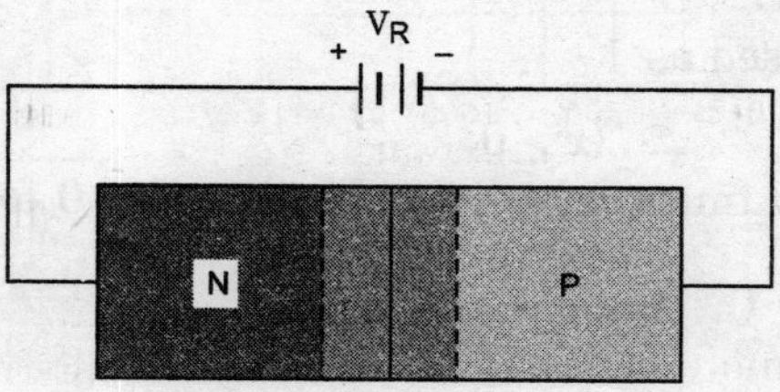

(a) Reverse-Biased P-N Junction

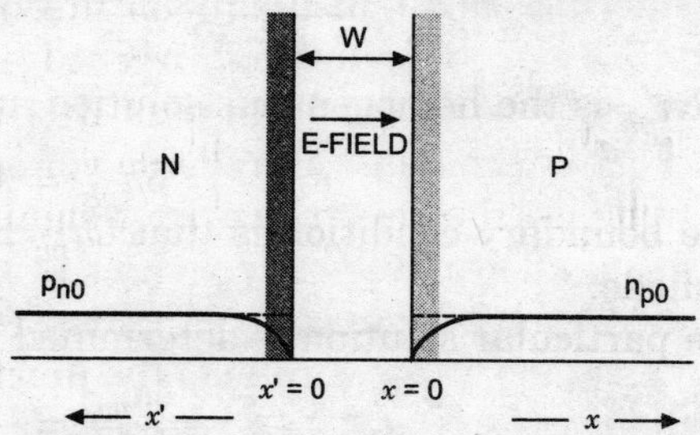

(b) Minority Carrier Concentration in The Reverse-Biased P-N Junction.

Fig. 6.9

Reverse-biased P-N junction and minority carrier distribution in the reverse-biased P-N junction (prior to photon illumination) is shown in Figs. 6.9(*a*) and 6.9(*b*) respectively.

If the generation rate of excess carriers is G_L, the photon-generated current density from the space charge region is given by

$$J_{L_1} = e\int G_L\, dx \qquad \text{...(6.2)}$$

where the integral is over the space charge region width. In case G_L is constant throughout the space charge volume, then

$$J_{L_1} = eG_L W \qquad \text{...(6.3)}$$

where W is the width of the space charge. This component of photocurrent responds very quickly to the photon illumination and is referred to as the *prompt photocurrent.*

Photoconductor gain is defined as the ratio of the rate at which charge is collected by the contacts to the rate at which charge is created within the photoconductor *i.e.*

$$\text{Photoconductor gain} = \frac{L}{eG_L AL}$$

By comparing above equation with Eq. (6.3) it may be noted that photodiode gain is unity. The photodiode speed is limited due to carrier transport through the space charge region. Assuming depletion width 2 μm and saturation drift velocity 10^5 m/s, the transit time $\tau_t = 20 \times 10^{-12}$ s. The ideal modulating frequency has a period of twice the transit time *i.e.* $2\tau_t$, hence the frequency is $\dfrac{1}{1\times 20\times 10^{-12}} = 25 \times 10^9$ Hz or 25 GHz. This frequency response is significantly higher than that of photoconductors.

Excess carriers are also created within the neutral N and P regions of the diode. The excess minority carrier electron distribution in the P-region is determined from the ambipolar transport Eq. which is given by

$$D_e \frac{\partial^2 (\delta n_p)}{\partial x^2} + G_L - \frac{\delta n_p}{\tau_{eo}} = \frac{\partial (\delta n_p)}{\partial t} \qquad \text{...(6.4)}$$

Assuming E-field zero in the neutral region we have in steady-state $\dfrac{\partial (\delta n_p)}{\partial t} = 0$ and, therefore, Eq. (6.4) becomes

$$\frac{d^2\, \delta n_p}{dx^2} - \frac{\delta n_p}{L_e^2} = -\frac{G_L}{D_e} \qquad \text{...(6.5) where } L_e = \sqrt{D_e\, \tau_{eo}}$$

The solution of above Eq. (6.5) can be taken as the sum of homogeneous and particular solutions. The homogeneous solution is found from the equation.

$$\frac{d^2\, \delta n_{ph}}{dx^2} - \frac{\delta n_{ph}}{L_e^2} = 0 \qquad \text{...(6.6)}$$

where δn_{ph} is the homogeneous solution and is expressed as

$$\delta n_{ph} = Ae^{-x/L_n} + Be^{+x/L_n} \qquad (x \geq 0) \qquad \text{...(6.7)}$$

One boundary condition is that δn_{ph} must remain finite, which means that $B \equiv 0$ for the "long" diode.

The particular solution is determined from

$$-\frac{\delta n_{pp}}{L_e^2} = -\frac{G_L}{D_e} \qquad \text{...(6.8)}$$

which gives

$$\delta n_{pp} = \frac{G_L L_e^2}{D_e} = \frac{G_L (D_e\, \tau_{eo})}{D_e} = G_L \tau_{eo} \qquad \text{...(6.9)}$$

The total steady-state solution for the excess minority carrier electron concentration in the P-region is then

$$\delta n_p = Ae^{-x/L_e} + G_L\tau_{eo} \quad ...(6.10)$$

The total electron concentration is zero at $x = 0$ for reverse biased junction. The excess electron concentration $x = 0$ is then

$$\delta n_p\,(x = 0) = n_{po} \quad ...(6.11)$$

Using the boundary condition from above Eq. (6.11), the electron concentration given by Eq. (6.10) becomes

$$\delta n_p = G_L\tau_{eo} - (G_L\tau_{eo} + n_{po})\,e^{-x/L_e} \quad ...(6.12)$$

Similarly the excess minority carrier hole concentration in the N-region can be determined. Using the x' notation depicted in Fig. 6.9, we can write

$$\delta p_n = G_L\tau_{ho} - (G_L\tau_{ho} + p_{no})\,e^{-x'/L_h} \quad ...(6.13)$$

The plot of Eqs. (6.12) and (6.13) is depicted in Fig. 6.10.

The gradient in the minority carrier concentration will generate diffusion currents in the P-N junction. The diffusion current density at $x = 0$ due to minority carrier electrons is

$$J_{e_1} = eD_e \left.\frac{d(\delta n_p)}{dx}\right|_{x=0} = eD_e\left[G_L\tau_{eo} - (G_L\tau_{eo} + n_{po})e^{-x/L_e}\right]_{x=0} = \frac{eD_e}{L_e}(G_L\tau_{eo} + n_{po}) \quad ...(6.14)$$

Above Eq. (6.14) can be written as

$$J_{e_1} = e\,G_L L_e + \frac{e\,D_e\,n_{po}}{L_e} \quad ...(6.15)$$

The first term in above Eq. (6.15) is the steady-state photocurrent density while the second term is the ideal reverse saturation current density due to the minority carrier electrons.

The diffusion current density (in the x-direction) at $x' = 0$ due to the minority carriers hole is

$$J_{h_1} = e\,G_L\,L_h + \frac{e\,D_h\,p_{no}}{L_h} \quad ...(6.16)$$

Similarly, the first term is the steady-state photocurrent density and the second term is the ideal reverse-saturation current density.

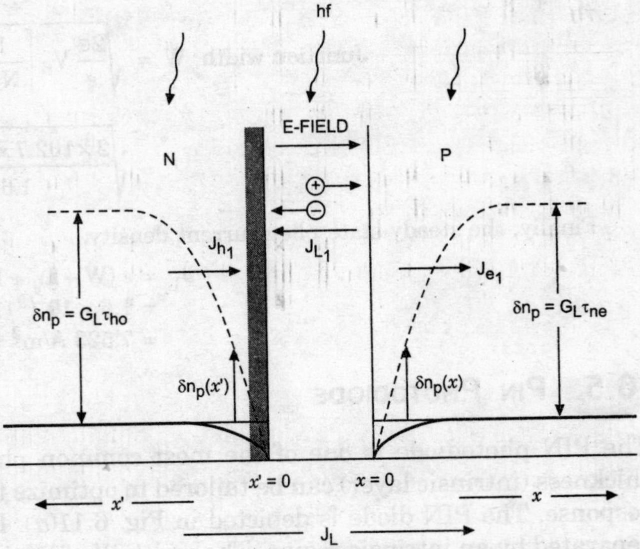

Fig. 6.10. Steady-State, Photoinduced Minority Carrier Concentrations and Photocurrents in a Long" Reverse Biased P-N Junction

The total steady-state diode photocurrent density for the long diode is

$$J_L = eG_LW + eG_LL_e + eG_LL_h = e(W + L_e + L_h)G_L \quad ...(6.17)$$

The photocurrent obtained from above Eq. (6.17) is the result of assuming uniform generation of excess carriers throughout the entire structure, a long diode, and steady state. The photocurrent is in the reverse-bias direction through the diode.

Example 6.1. Determine the steady-state photocurrent density in a reverse-biased, long P-N diode. The parameters of the P-N diode at 300 K are as follows:

$N_A = N_D = 10^{22}$ atoms/m^3 ; $D_e = 25 \times 10^{-4}$ m^2/s; $D_h = 1 \times 10^{-3}$ m^2/s

$\tau_{eo} = 500$ ns and $\tau_{ho} = 100$ ns; $n_i = 1.5 \times 10^{16}$ /m^3

Assume that external voltage applied voltage is – 10 V, $\epsilon_r = 11.6$; $\epsilon_0 = 8.854 \times 10^{-12}$ F/m $e = 1.6 \times 10^{-19}$; $V_T = 26$ mV; $G_L = 10^{27}$ m^{-3} s^{-1}.

Solution : Donor density, $N_D = 10^{22}$ atoms/m^3

Acceptor density, $N_A = 10^{22}$ atoms/m^3

$\tau_{eo} = 500 \text{ ns} = 500 \times 10^{-9}$ s

$\tau_{ho} = 100 \text{ ns} = 100 \times 10^{-9}$ s

Diffusion constant, $D_e = 25 \times 10^{-4}$ m^2/s

Diffusion constant, $D_h = 1 \times 10^{-3}$ m^2/s

Reverse bias voltage, $V_R = 10$ V

Electron charge, $e = 1.6 \times 10^{-19}$ C

Absolute permittivity for the material, $\in = \in_r \in_0 = 11.6 \times 8.854 \times 10^{-12} = 102.7 \times 10^{-12}$ F/m

Electron diffusion length, $L_e = \sqrt{D_e \tau_{eo}} = \sqrt{25 \times 10^{-4} \times 500 \times 10^{-9}} = 35.36$ μm

Hole diffusion length, $L_h = \sqrt{D_h \tau_{ho}} = \sqrt{1 \times 10^{-3} \times 100 \times 10^{-9}} = 10.0$ μm

Built-in potential barrier, V_{bi} or $V_0 = V_T \log_e \frac{N_A N_D}{n_i^2} = 26 \times 10^{-3} \log_e \frac{10^{22} \times 10^{22}}{(1.5 \times 10^{16})^2} = 0.597$ V

Junction voltage, $V_B = V_0 - V_R = 0.697 - (-10) = 10.697$ V

$$\text{Junction width, } W = \sqrt{\frac{2\in}{e} V_B \left[\frac{1}{N_A} + \frac{1}{N_D}\right]}$$

$$= \sqrt{\frac{2 \times 102.7 \times 10^{-12} \times 10.697}{1.6 \times 10^{-19}} \left[\frac{1}{10^{22}} + \frac{1}{10^{22}}\right]} = 1.657 \text{ μm}$$

Finally, the steady-state photocurrent density,

$$J_L = e\,(W + L_e + L_h)\,G_L$$
$$= 1.6 \times 10^{-19}\,(1.657 + 35.36 + 10) \times 10^{-6} \times 10^{27}$$
$$= 7{,}523 \text{ A/m}^2 = 0.7523 \text{ A/cm}^2$$ **Ans.**

6.5. PIN PHOTODIODE

The PIN photodiode is one of the most common photodetectors, since the depletion-region thickness (intrinsic layer) can be tailored to optimize the quantum efficiency and the frequency response. The PIN diode is depicted in Fig. 6.11(*a*). It consists of a P region and an N region separated by an intrinsic region. The width W of intrinsic region is much larger than the space charge width of an ordinary P-N junction. The 'I' region need not be truly intrinsic, as long as the resistivity is high. It can be grown epitaxially on the N-type substrate, and the P-region can be obtained by the implantation.

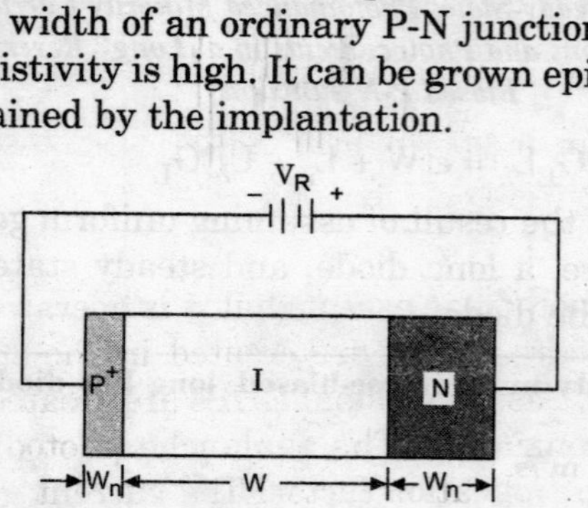

(a) A Reverse Biased PIN Photodiode

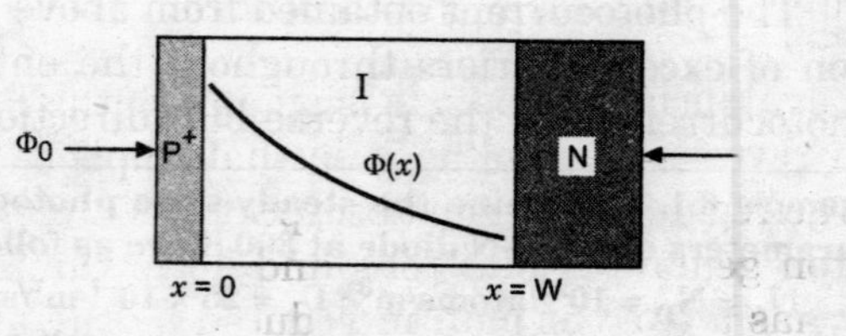

(b) Geometry Showing Non-uniform Photon Absorption

Fig. 6.11

When there is no bias applied to the diode, there will be diffusion of charge carriers due to concentration gradient across the junction. Diffusion holes and electrons across NI junction form a very thin depletion layer in the N-region and a relatively thicker depletion layer in I-region containing equal but opposite fixed charges. When a reverse bias is applied and gradually increased, the thickness of the depletion layer increases until the entire I-region is swept free of mobile carriers. Thus with a reverse bias applied to the PIN diode, the space charge region extends completely through the intrinsic region.

Assume that a photon flux ϕ_o is incident on the P^+ region and width W_P of P-region is very small. Now the photon flux, as a function of distance, in the intrinsic region is $\phi(x) = \phi_o e^{-\alpha x}$ where α is the photon absorption coefficient, as depicted in Fig. 6.11(*b*). The density of photocurrent produced in the intrinsic region can be given by

$$J_L = e\int_o^W G_L dx = e\int_o^W \phi_o e^{-\alpha x} dx = e\phi_o(1 - e^{-\alpha W}) \quad ...(6.18)$$

In derivation of above equation it has been assumed that there is no electron-hole recombination within the space charge region and also that each photon absorbed generates one EHP.

Although PN photodiode, discussed in Art. 6.4 provides a much high frequency response than CdS bulk-type photoconductor but even faster response may be achieved with the help of PIN photodiode. PIN photodiodes are relatively inexpensive and ultrafast having a switching speed of nanosecond or so and interface easily with conventional electronic circuitry. They have broad spectral response and generate very low noise. They are capable of processing very weak signals.

Hence, such devices are widely employed for detection of laser pulses and in ultrafast switching and logic circuits.

Example 6.2. Determine photocurrent density in a silicon PIN photodiode with an intrinsic region width of W = 25 μm assuming the photon flux of 10^{21} $m^{-2}s^{-1}$ and absorption coefficient α of 1×10^5 m^{-1}.

Solution: Photon flux, $\phi_o = 1 \times 10^{21}\ m^{-2}s^{-1}$

Absorption coefficient, $\alpha = 1 \times 10^5\ m^{-1}$

The generation rate of EHP at the front edge of the intrinsic region,

$$G_{L_1} = \alpha\,\phi_o = 1 \times 10^5 \times 1 \times 10^{21} = 1 \times 10^{26}\ m^{-3}s^{-1}$$

and the generation rate at the back edge of the intrinsic region,

$$G_{L_2} = \alpha\,\phi_o e^{-\alpha W} = 1 \times 10^5 \times 1 \times 10^{21}\ e^{-10^5 \times 25 \times 10^{-6}} = 0.082 \times 10^{26}\ m^{-3}s^{-1}$$

The generation rate is obviously non-uniform in the intrinsic region. The density of photocurrent is then

$$J_L = e\,\phi_o(1 - e^{-\alpha W}) = 1.6 \times 10^{-19} \times 1 \times 10^{21}\ (1 - e^{-10^5 \times 25 \times 10^{-6}})$$

$$= 146.9\ A/m^2 \text{ or } 14.69\ mA/cm^2 \textbf{ Ans.}$$

6.6. AVALANCHE PHOTODIODE

The avalanche photodiode is similar to the P-N or PIN photodiode except that it is operated at high reverse-bias voltages so that impact ionization occurs, EHPs are created in the space charge region by photon absorption, as already discussed. Also additional EHPs are created by photon generated electrons and holes through impact ionization. The avalanche photodiode now has a current gain introduced by the avalanche multiplication factor. The current gain-bandwidth product of an avalanche photodiode can be larger than 100 GHz, so the device can respond to light modulated at microwave frequencies.

The structure of an avalanche photodiode is shown in Fig. 6.12. It has four regions—N^+ and P^+ are heavily doped semiconductors and I region is very lightly doped.

For detection of low-level optical signals, it is often desirable to operate the photodiode in the avalanche region of its characteristic. In the avalanche photodiode junction must be uniform, and the guard ring is usually employed to ensure against edge breakdown. With proper design a silicon avalanche photodiode can have high sensitivity to low-level optical signals, and the response time is in the neighbourhood of 1 ns. These devices are particularly useful in fibre optic communication systems.

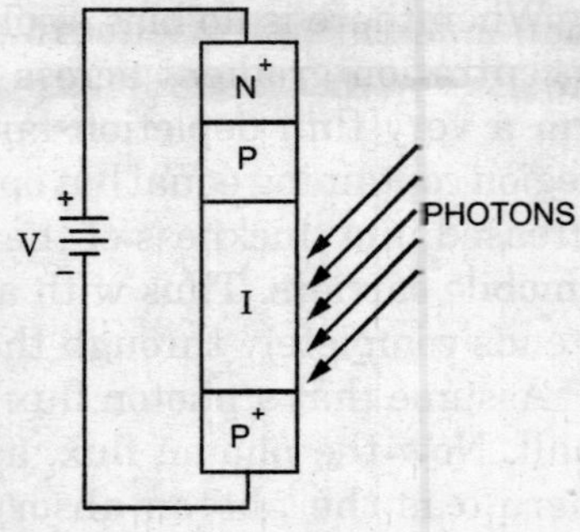

Fig. 6.12. ***Structure of Avalanche Photodiode***

6.7. PHOTOTRANSISTOR

A phototransistor is similar to an ordinary BJT, except that no base terminal is provided. Instead of a base current, the input to the transistor is in the form of light. The current induced by photoelectric effects is the base current of the transistor. Let it be I_λ. The resulting collector current, on an approximate basis, is

$$I_C = (1 + \beta_{dc})\, I_\lambda \cong \beta_{dc} I_\lambda \qquad ...(6.19)$$

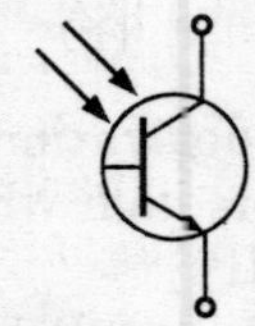

Fig. 6.13. ***Standard Symbol of Phototransistor***

Figure 6.13 shows the standard symbol of a phototransistor, which can be regarded as a conventional transistor housed in a case that enables its collector-base junction to be exposed to external light. The device is usually packaged in a metal can with a lens on top. Clear plastic encapsulated phototransistors are also available. A sketch of the phototransistor with the terminal identification and the angular alignment is given in Fig. 6.14.

The device is normally used with its base open, in either of the configuration shown in Fig. 6.15 and functions as follows.

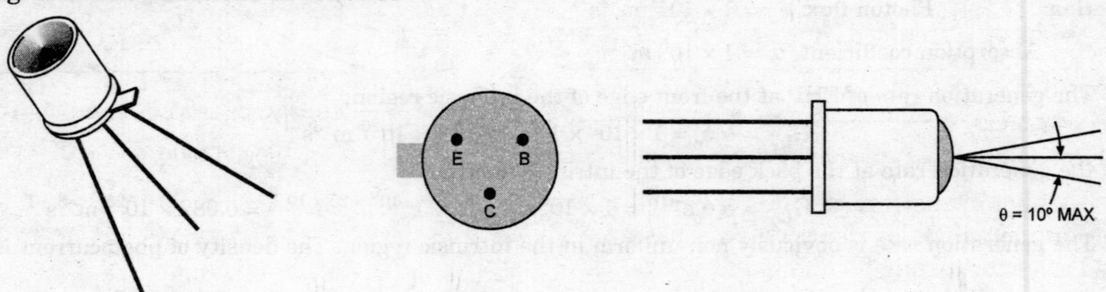

Fig. 6.14

In Fig. 6.15 (*a*) the base-collector junction of the transistor is effectively reverse biased and thus acts as a photodiode. The photo-generated currents of the base collector junction feed directly into the base of the device, and the nominal current-amplifying transistor action causes the output current to appear (β_{dc} times base current), as collector current and resistor R causes this current to provide an output voltage as illustrated.

In practice, the collector and emitter currents of a transistor are virtually identical and, since the base is open, the device is not subjected to significant negative feedback. Consequently, the alternative circuit shown in Fig. 6.15 (b), in which resistor R is connected to emitter of the transistor, gives a virtually identical performance to that of circuit shown in Fig. 6.15 (*a*).

Typical output characteristics are shown in Fig. 6.16(*a*), which are similar to those of an ordinary transistor except that illumination level is substituted for base current.

A curve of base current versus illumination level is shown in Fig. 6.16(*b*).

The main difference between a phototransistor and a photodiode is the current gain h_{fe} or β_{dc}. The same amount of light striking both devices produces β_{dc} times more current in a phototransistor than in a photodiode. The increased sensitivity of a phototransistor is a big advantage over a photodiode. However, photodiodes are the faster of the two. A photodiode has typical output currents in microamperes and can switch *on* or *off* in nanoseconds while the phototransistor has typical output currents in milliamperes but switches *on* or *off* in microseconds.

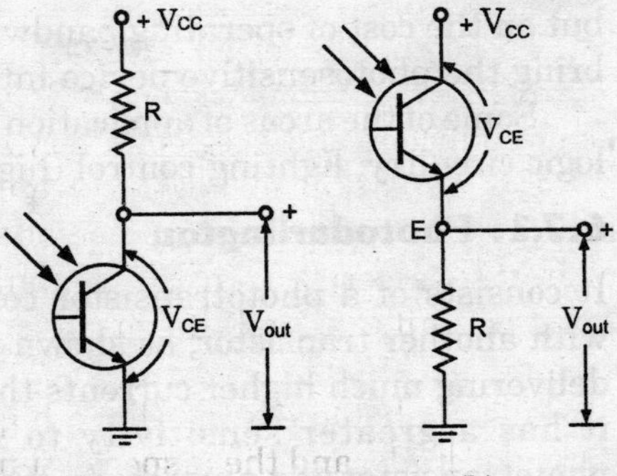

(*a*) (*b*)

Fig. 6.15. ***Alternative Phototransistor Configurations***

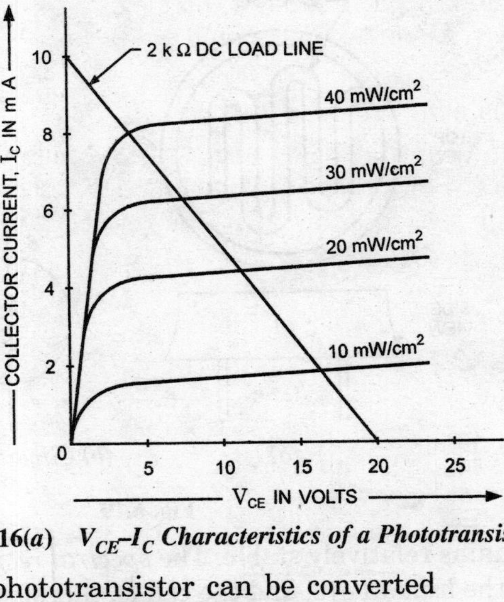

Fig. 6.16(*a*) ***V_{CE}–I_C Characteristics of a Phototransistor***

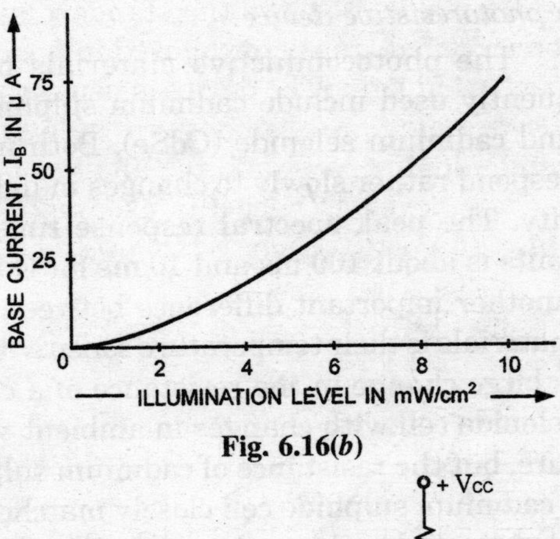

Fig. 6.16(*b*)

A phototransistor can be converted into a photodiode by using only its base and collector terminals and ignoring the emitter, as illustrated in Fig. 6.17 (*a*).

Alternatively, the sensitivity (and operating speed) can be controlled with a variable base return resistor, as shown in Fig. 6.17(*b*). With R_B open-circuit phototransistor operation is obtained while with R_B short-circuited, photodiode operation is available.

In practice the value of resistor R in the photodiode or phototransistor circuit is usually chosen on a compromise basis, since the circuit voltage gain increases

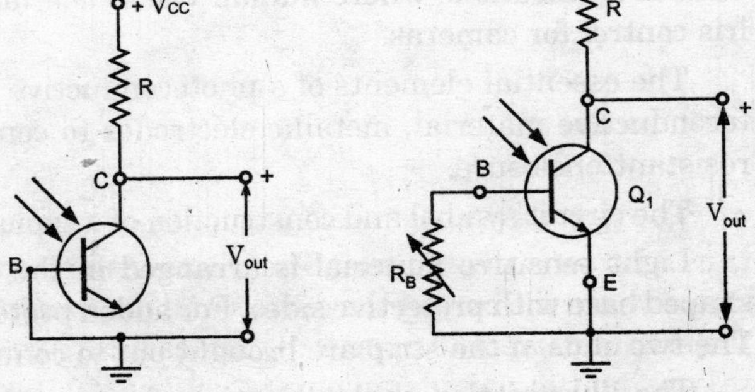

(a) Phototransistor Used as a Photodiode *(b) Variable-Sensitivity Phototransistor Circuit*

Fig. 6.17

but on the cost of operating bandwidth. Also, the value of R, in many applications, is chosen to bring the photosensitive device into its linear operating region.

Some of the areas of application for the phototransistor include punch-card readers, computer logic circuitry, lighting control (highways etc.), level indication, relays and counting system.

6.7.1. Photodarlington

It consists of a phototransistor connected in Darlington arrangement with another transistor, as shown in Fig. 6.18. The device is capable of delivering much higher currents than a phototransistor, and, therefore, it has a greater sensitivity to illumination levels than either a phototransistor or a photodiode. With the additional transistor involved, the photodarlington has a considerable longer switching time than a phototransistor. The device has limited useful maximum operating frequency (only tens of kilohertz). Devices are also available with integral amplifiers, and in the form of photosensitive SCRs and triacs, etc.

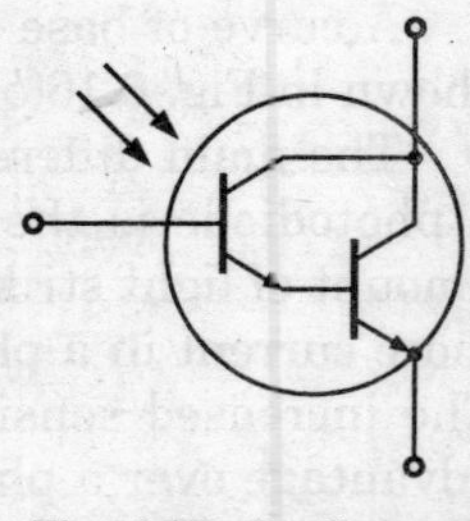

Fig. 6.18. ***Darlington Phototransistor***

6.8. PHOTOCONDUCTIVE CELLS

The photoconductive cell is a two terminal semiconductor device whose terminal resistance will vary (linearly) with the intensity of the incident light. For obvious reasons, it is frequently called a *photoresistive device*.

The photoconductive materials most frequently used include cadmium sulphide (CdS) and cadmium selenide (CdSe). Both materials respond rather slowly to changes in light intensity. The peak spectral response time of CdS units is about 100 ms and 10 ms for CdSe cells. Another important difference between the two materials is their temperature sensitivity. There is large change in the resistance of a cadmium selenide cell with changes in ambient temperature, but the resistance of cadmium sulphide remains relatively stable. The *spectral response* of a cadmium sulphide cell closely matches that of the human eye, and the cell is, therefore, often used in applications where human vision is a factor, such as street light control or automatic iris control for cameras.

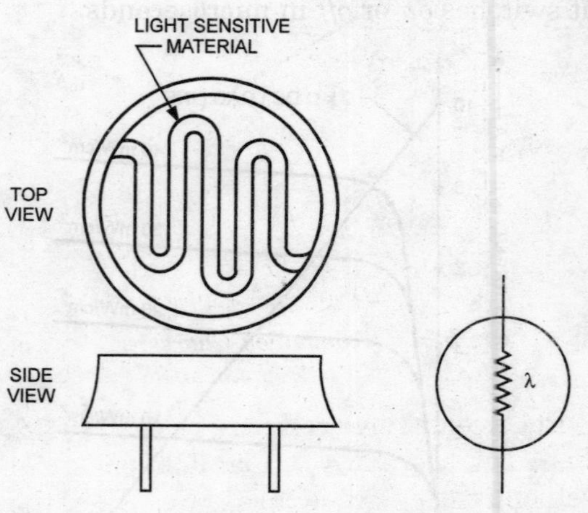

(a) *(b) Circuit Symbol*

Fig. 6.19

The essential elements of a photoconductive cell are the ceramic substrate, a layer of photoconductive material, metallic electrodes to connect the device into a circuit and a moisture resistant enclosure.

The circuit symbol and construction of a typical photoconductive cell are shown in Fig. 6.19.

Light sensitive material is arranged in the form of a long strip, zigzagged across a disc shaped base with protective sides. For added protection, a glass or plastic cover may be included. The two ends of the strip are brought out to connecting pins below the base.

The illumination characteristics of a typical photoconductive cell are shown in Fig. 6.20 from which it is obvious that when the cell is not illuminated its resistance may be more than 100 kΩ. This resistance is called the *dark resistance*. When the cell is illuminated, the resistance may fall to a few hundred ohms. Note that the scales on the illumination characteristic are logarithmic to cover a wide ranges of resistance and illumination that are possible. Cell sensitivity may be expressed in terms of the cell current for a given voltage and given level of illumination.

The photoconductive cells have the advantages of (*i*) high sensitivity (*ii*) low cost and (*iii*) high dark-to-light resistance ratio exceeding 100 : 1. The drawbacks of photoconductive cells are large response time and relatively narrow spectral response.

The major drawback of the photoconductive cells is that temperature variations cause substantial variations in resistance for a particular light intensity. Therefore, such a cell is unsuitable for analog applications.

Such cells find wide use in industrial and laboratory control applications.

The photoconductive cell used for relay control is shown in Fig. 6.21. When the cell is illuminated, its resistance is low and the relay current is at its maximum. When the cell is dark, its high resistance reduces the current down to a level too low to energize the relay. Resistance R is included to limit the relay current to the desired level when the resistance of the cell is low.

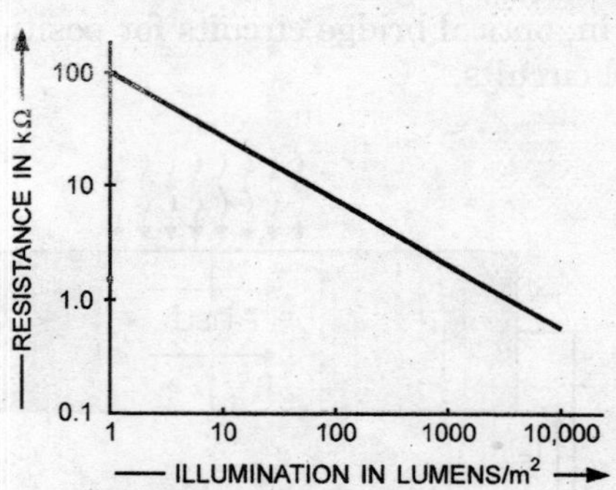

Fig. 6.20 ***Illumination Characteristic of a Typical Photoconductive Cell***

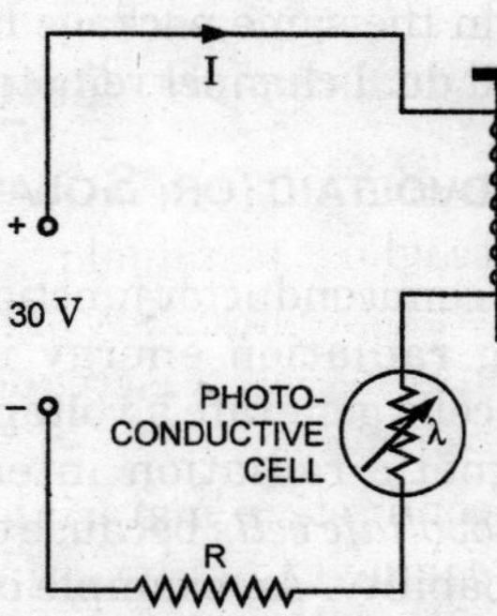

Fig. 6.21

Photoconductive cells are used to switch transistors *on* and *off*, as illustrated in Fig. 6.22. When the cell shown in Fig. 6.22 (*a*) is dark, the transistor base is biased above its emitter level, and the device is turned *on*. When the cell is illuminated, the lower resistance of the cell in series with R biases the transistor base voltage below its emitter level. Thus, the device is turned *off*. Fig. 6.22 (*b*) shows a circuit in which the transistor is biased *on* when the cell is illuminated, and *off* when dark.

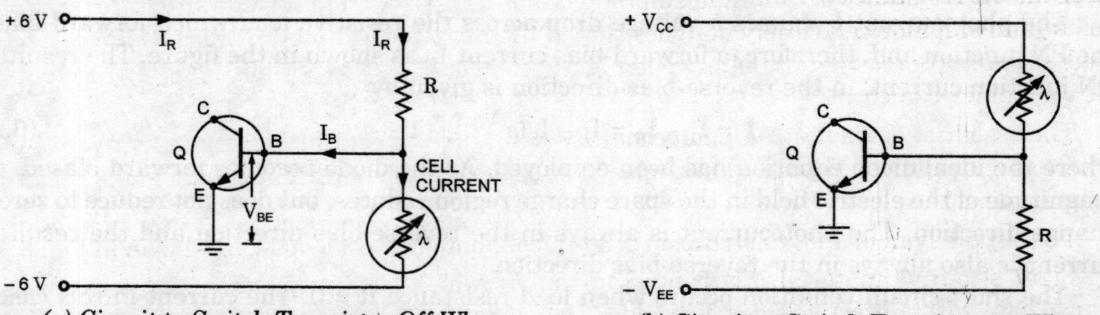

(a) Circuit to Switch Transistor Off When Photoconductive Cell is Illuminated

(b) Circuit to Switch Transistor on When Photoconductive Cell is Illuminated

Fig. 6.22

Voltage regulator employing a photoconductive cell is shown in Fig. 6.23. This system is formed by a photoconductive cell, lamp and a resistor R_1. If input voltage V_{in} falls in magnitude for any reason, the brightness of the lamp would also decrease. The decrease in illumination would cause increase in resistance R_λ of the photoconductive cell to maintain output voltage V_{out} at its rated level as determined by the voltage divider rule; that is

$$V_{out} = \frac{R_\lambda V_{in}}{R_1 + R_\lambda} \quad ...(6.20)$$

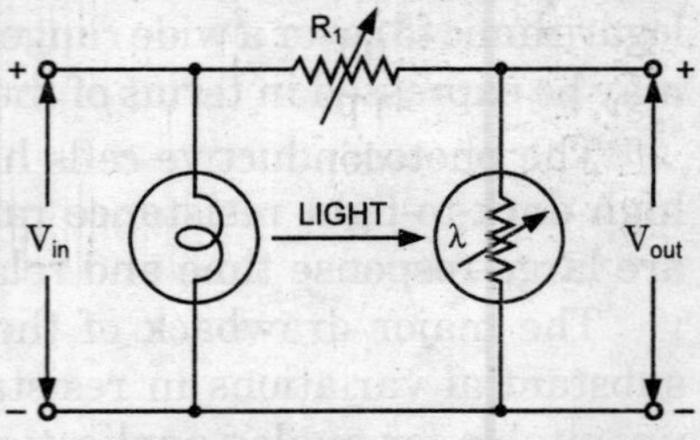

Fig. 6.23. *Voltage Regulator Using a Photoconductive Cell*

Note : As with all other devices, care must be taken to ensure that the power dissipation is not excessive.

It is used in counting system where the objects on a conveyor belt interrupt a light beam to generate a series of pulses which operate a counter.

It is used in twilight circuits. When the day light has faded to a given level, the corresponding resistance of the detector makes another circuit to switch **on** the required lights.

It is widely used in cameras to control shutter opening during the flash. Twin photoconductive cells mounted in the same package have been used in, optical bridge circuits for position control mechanism and dual channel remote volume control circuits.

6.9. Photovoltaic or Solar Cells

These cells are semiconductor junction devices used for converting radiation energy into electrical energy. These cells generate a voltage proportional to electromagnetic radiation intensity and are called the *photovoltaic cells* because of their voltage generating capability. An example of such a cell is a silicon solar cell, which converts radiant energy of the Sun into electrical energy.

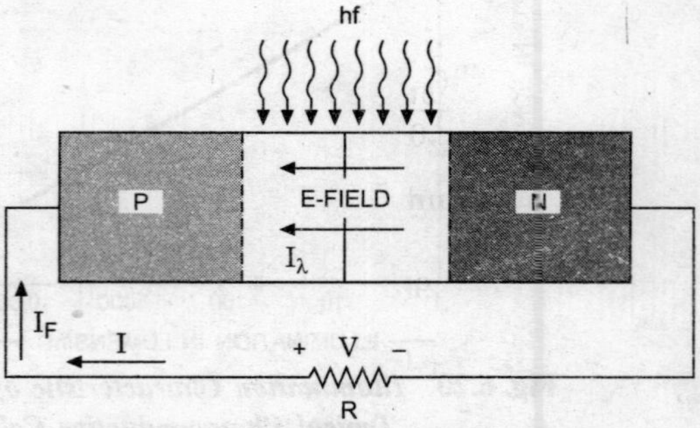

Fig. 6.24. *P-N Junction Solar Cell With Resistive Load*

Selenium and silicon are the most widely used materials for solar cells, though gallium arsenide, indium arsenide and cadmium sulphide are also used.

Consider a P-N junction depicted in Fig. 6.24 with a resistive load. Even when there is no bias applied to the diode, an electric field exists in the space charge region as illustrated in the figure. Incident photon illumination can generate EHPs in the space charge region that will be swept out generating photocurrent I_λ in the reverse-bias direction as illustrated.

The photocurrent I_λ causes a voltage drop across the resistive load which forward biases the PN junction and, therefore, a forward-bias current I_F as shown in the figure. The resultant PN junction current, in the reverse-bias direction is given by

$$I = I_\lambda - I_F = I_\lambda - I_S(e^{eV/kT} - 1) \quad ...(6.21)$$

where the ideal diode equation has been employed. As the diode becomes forward biased, the magnitude of the electric field in the space charge region reduces, but does not reduce to zero or change direction. The photocurrent is always in the reverse-bias direction and the resultant current is also always in the reverse-bias direction.

The short-circuit condition occurs when load resistance R = 0. The current in this case is called the *short-circuit current,* or

$$I = I_{SC} = I_\lambda \quad ...(6.22)$$

The open-circuit condition occurs when load resistance $R \to \infty$. The resultant current is zero and the voltage developed is the *open-circuit voltage.* The photocurrent is just balanced by the forward-biased junction current and, therefore, we have

$$I = 0 = I_\lambda - I_S\left[e^{eVoc/kT} - 1\right] \quad ...(6.23)$$

The open-circuit voltage V_{OC} can, therefore, be determined by equation

$$V_{OC} = V_T \log_e\left(1 + \frac{I_\lambda}{I_S}\right) \quad ...(6.24)$$

I-V characteristic is shown in Fig. 6.25.

The open-circuit voltage V_{OC} and short-circuit current I_{SC} are determined for a given level of light by the cell properties.

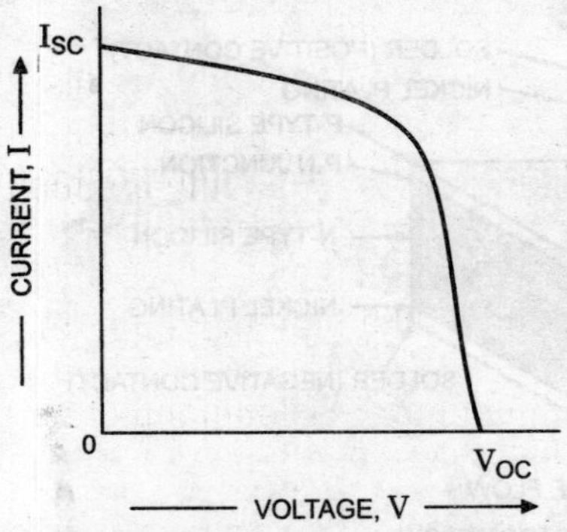

Fig. 6.25. ***Current-Voltage Characteristic of a PN Junction Solar Cell***

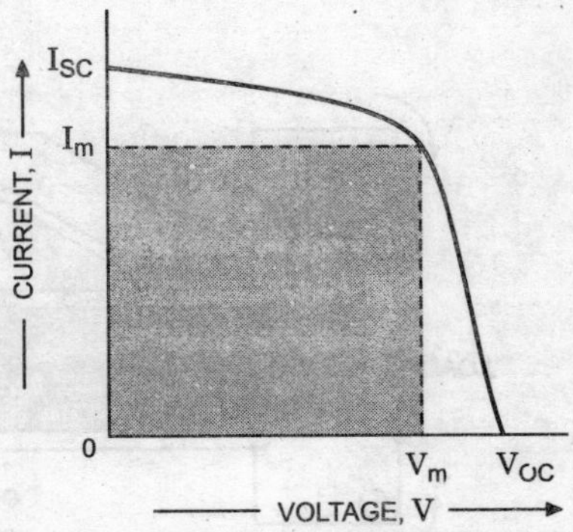

Fig. 6.26. ***Maximum Power Rectangle of Solar Cell I-V Characteristic***

The power delivered to the load is given by the product of voltage and current, therefore,

$$P = VI = V\left[I_\lambda - I_S\left(e^{eV/kT} - 1\right)\right] \quad ...(6.25)$$

The power delivered to the load will be maximum if its derivative is zero *i.e.* $\frac{dP}{dV} = 0$.

$$\text{or } I_\lambda - I_S\left(e^{eV_m/kT} - 1\right) - I_S V_m \frac{e}{kT} e^{eV_m/kT} = 0 \quad ...(6.26)$$

where V_m is the voltage that makes the power delivered to the load maximum. The above Eq. (6.26) may be rewritten as

$$\left(1 + \frac{V_m}{V_T}\right) e^{V_m/kT} = 1 + \frac{I_\lambda}{I_S} \quad ...(6.27)$$

The value of V_m can be found by trial and error. It is to be noted that the maximum power delivered to the load illustrated by the shaded rectangle in Fig. 6.26 is less than the $V_{OC}\, I_{SC}$ product. The ratio $\frac{V_m I_m}{V_{OC} I_{SC}}$ is called the *fill factor,* and is a figure of merit for solar cell design.

The worthnoting point is that J_S is a function of the semiconductor doping concentrations. With the increase in doping concentrations, J_S decreases, which results in increase of open-circuit voltage V_{OC}. However, as the open-circuit voltage V_{OC} is a function of log of I_λ and I_S, the V_{OC} is not a strong function of these parameters.

The construction and cross-section of a typical power solar cell for use as an energy converter are given in Fig. 6.27. The surface layer of P-type material is extremely thin so that light can penetrate to the junction. The nickel-plated ring around the P-type material is the positive output terminal, and the plating at the bottom of the N-type material is the negative output terminal. Power solar cells are also available in flat strip form for efficient coverage of available surface areas. The circuit symbol often used for a photovoltaic cell is given in Fig. 6.28 (*a*).

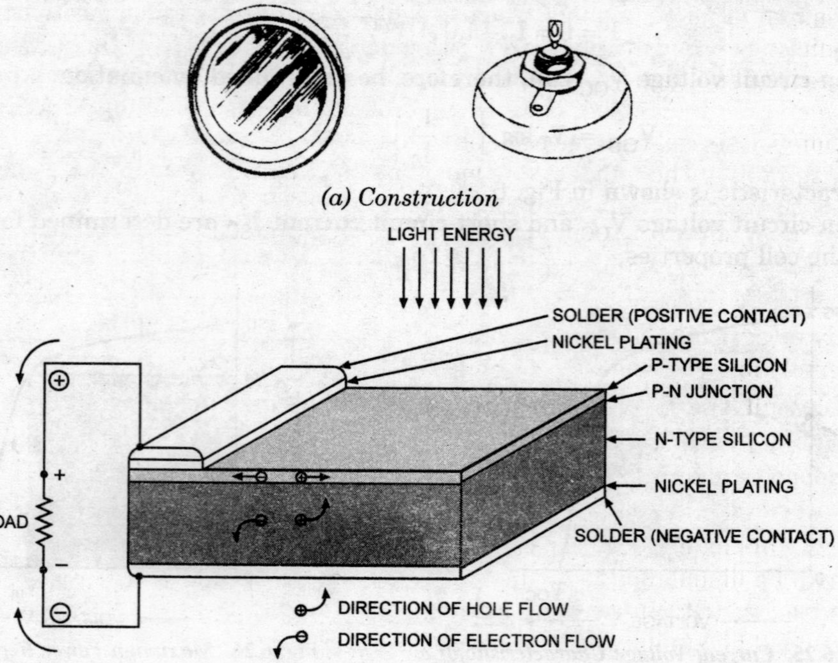

(a) Construction

(b) Cross-section

Fig. 6.27. *Solar Cell*

The open-circuit output voltage characteristic of a typical photovoltaic cell is given in Fig. 6.28 (*b*), the graph is *logarithmic* on *light intensity axis*. This characteristic shows that the cell is more sensitive for low light levels, because a small change in light intensity (say from 10 to 100 lux) can produce the same increase in output voltage as a large change in light intensity (say from 100 to 1,000 lux) at a higher light intensity level.

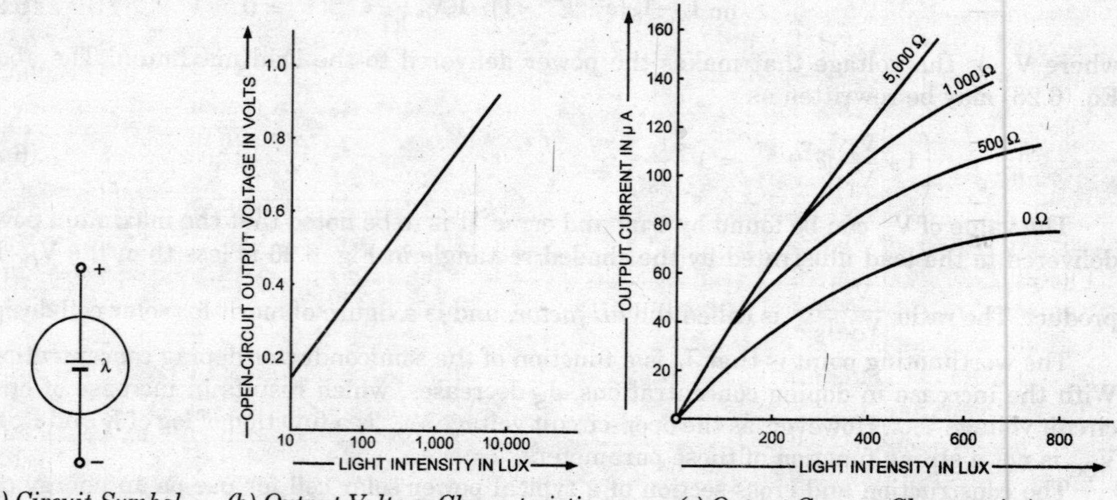

(a) Circuit Symbol *(b) Output Voltage Characteristic* *(c) Output Current Characteristics*

Fig. 6.28. *Photovoltaic Cell*

The output current characteristics of a typical photovoltaic cell for various load resistances are given in Fig. 6.28 (*c*). These characteristics show that the output current of a photovoltaic cell increases with the increase in light intensity. The output current of such a cell is very low and is measured in microamperes. The available output current depends upon the light intensity, cell efficiency (typically only a few per cent), and on the size of the active area of the cell face. The conversion efficiency depends upon the spectral content and the illumination. Photocells can be stacked in parallel, however, in order to increase their output current capability.

The photovoltaic cell can be operated satisfactorily over a wide range of temperature (say from – 100 to 125°C). The temperature variations have little effect on short-circuit current but affect the open-circuit output voltage considerably. These variations may be of the order of a few mV per °C in output voltage. An individual solar cell generates an open-circuit voltage of about 500 mV (depending on light intensity) when active. However, photo cells can be connected in series to increase the available terminal voltage.

The advantages of such devices are their ability to generate a voltage without any bias and their extremely fast response *i.e.* these devices can be used as energy convertors directly.

Multiple-unit silicon photovoltaic devices may be used for sensing light in applications such as reading punched cards in the data processing industry.

Gold-doped germanium cells with controlled spectral response characteristic act as photovoltaic devices in the infrared region of the spectrum and may be used as infrared detectors.

Solar cells are extensively used as a source of power for many space satellites. The array of junctions can be distributed over the surface of the satellite or can be contained in solar cell "paddles" attached to the main body of the satellite. Applications of solar cells are not restricted to outer space. It is possible to have useful power from the sun in terrestrial applications using solar cells, even though the solar intensity is reduced by the atmosphere. About 1 kW/m^2 is available in a particularly sunny location but not all of this solar power can be converted into electrical power. Much of the photon flux is at energies less than the E_G of the cell, and is not absorbed. High-energy photons are strongly absorbed, and the resulting electron-hole pairs may recombine at the surface. A well made silicon cell can have 10 per cent efficiency for solar energy conversion, providing approximately 100 W/m^2 of electrical power under full illumination. This is the modest amount of power per unit solar cell area, considering the effort involved in fabricating a large area of silicon cells. One approach to have more power per cell is to focus considerable light onto the cell using mirrors. Although Si cells lose efficiency at the resulting high temperatures, GaAs and related compounds can be employed at 100°C or higher. In such solar concentrator systems more effort and expense can be put into the fabrication of solar cell, as fewer cells are required. For instance, a GaAs-AlGaAs heterojunction cell provides good conversion efficiency and operates at the elevated temperatures common in solar concentration systems.

A group of series-parallel connected solar cells operating as a battery charger is shown in Fig. 6.29. Several cells are required to be connected in series to give the required output voltage and several of such series-connected groups are to be connected in parallel to provide the necessary output current.

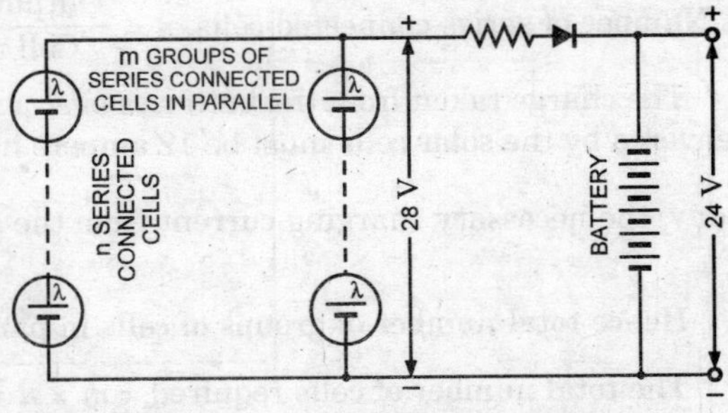

Fig. 6.29. ***Group of Series-parallel Connected Solar Cells Operating as a Battery Charger***

Though photoconductive cells can be used for many of the same purposes that photovoltaic cells are used, but they cannot be employed as energy sources. Photoconductive cells are preferred to photovoltaic cells when very sensitive response to changing light conditions is required.

When fast response is necessary, photovoltaic cells are preferred to photoconductive cells. Likewise, if a photocell is to be rapidly switched *on* and *off*, photovoltaic cells are preferred because they can be switched at higher frequencies than photoconductive cells. As a rough rule of thumb, photoconductive cells cannot be successfully switched at frequencies higher than about 1 kHz, whereas photovoltaic cells can be switched successfully at frequencies up to about 100 kHz and sometimes higher even.

Example 6.3. Determine the open-circuit voltage of a silicon P-N junction solar cell at 300 K with the following parameters

$N_A = 7.5 \times 10^{24}$ atoms/m³, $N_D = 1.5 \times 10^{22}$ atoms/m³, $D_e = 25 \times 10^{-4}$ m²/s;
$D_h = 1 \times 10^{-3}$ m²/s, $\tau_{eo} = 500$ ns, $T_{ho} = 100$ ns ; $n_i = 1.5 \times 10^{16}$/m³
$e = 1.6 \times 10^{-19}$C. Assume photon current density to be 12.5 mA/cm².

Solution : Electron diffusion length, $L_e = \sqrt{D_e \tau_{eo}} = \sqrt{25 \times 10^{-4} \times 500 \times 10^{-9}} = 35.36$ μm

Hole diffusion length, $L_h = D_h \tau_{ho} = \sqrt{1 \times 10^{-3} \times 100 \times 10^{-9}} = 10.0$ μm

$$J_S = e \times n_i^2 \left(\frac{D_e}{L_e N_A} + \frac{D_h}{L_h N_D} \right)$$

$$= 1.6 \times 10^{-19} \times (1.5 \times 10^{16})^2 \left[\frac{25 \times 10^{-4}}{35.36 \times 10^{-6} \times 7.5 \times 10^{24}} + \frac{1 \times 10^{-3}}{10 \times 10^{-6} \times 1.5 \times 10^{22}} \right]$$

$$= 2.4 \times 10^{-7} \text{ A/m}^2 = 2.4 \times 10^{-11} \text{ A/cm}^2$$

$J_L = 12.5$ mA/cm² ...(given)

From Eq. (6.24) we have

$$\text{Open-circuit voltage, } V_{OC} = V_T \log_e \left(1 + \frac{I_\lambda}{I_S} \right)$$

$$= V_T \log_e \left(1 + \frac{I_\lambda}{I_S} \right) = 26 \text{ mV} \log_e \left(1 + \frac{12.5 \times 10^{-3}}{2.4 \times 10^{-11}} \right) = 521.84 \text{ mV or } 0.522 \text{ V}$$ **Ans.**

Example 6.4. An earth satellite has 24 V batteries which supply a continuous current of 0.5 A. Solar cells (each cell supplying 50 mA at 0.45 V) are employed to keep the battery fully charged. If the illumination from the sun for 12 hours in every 24 hours is 125 mW/cm², determine the approximate number of solar cells required.

Solution : The circuit for the solar cell battery charger is given in Fig. 6.29. Each cell supplies 50 mA at 0.45 V. Allowing for the voltage drop across the rectifier and series resistor, a maximum output voltage of approximately 28 V is required. Thus,

$$\text{Number of series-connected cells, } n = \frac{\text{Output voltage}}{\text{Cell voltage}} = \frac{28 \text{ V}}{0.45 \text{ V}} \simeq 64$$

The charge taken from the batteries over a period of 24 hours is 24 × 0.5 = 12 Ah. Thus, the charge delivered by the solar cells must be 12 ampere hours. Since the solar cells are illuminated only 12 hours a day, the necessary charging current from the solar cells is $\frac{\text{12 ampere-hours}}{\text{12 hours}}$ *i.e.* 1 A.

$$\text{Hence total number of groups of cells in parallel, } m = \frac{\text{Output current}}{\text{Current supplied per cell}} = \frac{1 \text{ A}}{50 \text{ mA}} = 20$$

The total number of cells required = $m \times n$ = 20 × 64 = 1,280 **Ans.**

6.10. LIGHT-EMITTING DIODES (LEDs)

The best known of all optoelectronic devices is the so called LED (light-emitting diode), which emits a fairly narrow bandwidth of visible (usually red, orange, yellow or green) or invisible (infrared) light when its internal diode junction is stimulated by a forward electric current/

voltage (power). LEDs have typical power-to-light energy conversion efficiencies some 10 to 50 times greater than of a simple tungsten lamp and have very fast response times (approximately 0.1 μs compared with tens or hundreds of milliseconds for a tungsten lamp), and are thus widely employed as visual indicators and as moving-light displays.

As already discussed in the foregoing article that a P-N junction can absorb light energy and produce electric current. The reverse is also true *i.e.* P-N junction emits light when it is energized.

The operation of light-emitting diode (LED) is based on the phenomenon of electroluminance, which is the emission of light from a semiconductor under the influence of an electrical field. As already discussed in previous articles, the recombination of charge carriers takes place in a forward P-N junction as the electrons cross from the N-region and recombine with holes existing in P-region. Free electrons are in the conduction band of energy levels, while holes are in the valence energy band. Therefore, the electrons are at high energy levels than the holes. For the electrons to recombine with the holes, they must give some of their energy. Typically, these electrons give up energy in the form of heat and light.

In silicon and germanium diodes, most of the electrons give up their energy in the form of heat. However, with GaAsP, and GaP semiconductors, the electrons give up their energy by emitting photons. If the semiconductor is transluscent, the light will be emitted and the junction becomes a source of light *i.e.* a *light-emitting diode* (LED). LEDs emit no light when *reverse-biased*. In fact, operating LEDs in reverse bias mode will quickly destroy them.

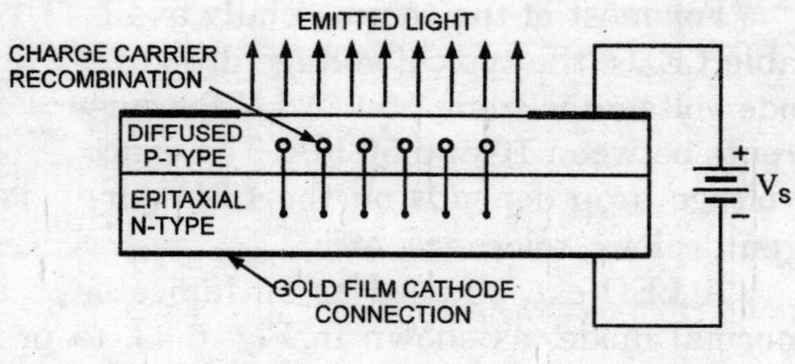

Fig. 6.30

A cross-sectional view of a typically diffused LED is given in Fig. 6.30. The semiconductor material used is gallium arsenide (GaAs), gallium arsenide phosphide (GaAsP), or gallium phosphide (GaP). An N-type epitaxial layer is grown upon a substrate, and the P-region is produced by diffusion. Recombination of charge carriers occur in the P-region so this region is required to be kept at the top. Thus, the P-region becomes the device surface. The metal anode connections are made at the outer edges of P-layer so as to allow more surface area for the light to come out. A gold film is applied to the substrate bottom to reflect as much as possible of the light toward the surface of the device and to provide a cathode connection. LEDs are manufactured with domed lenses so as to reduce the re-absorption problem. LEDs are always encased in order to protect their delicate wires. LEDs made from GaAs emit invisible infrared light, LEDs constructed of GaAsP tend to emit either red or yellow light. The GaP LEDs give either red or green light.

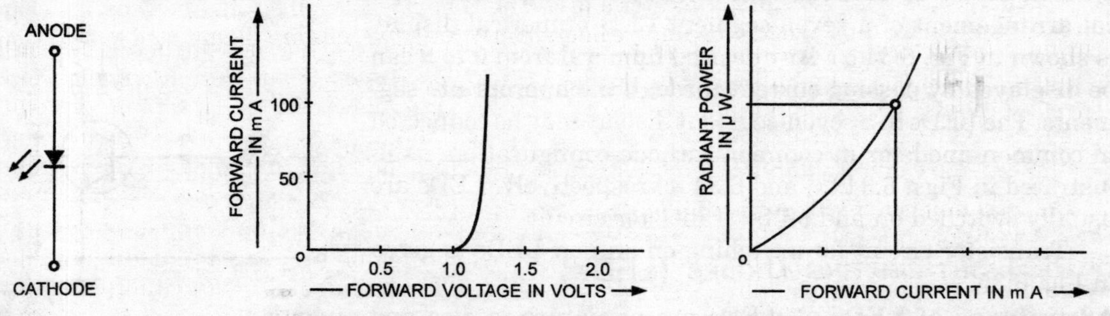

(a) *(b) Forward Bias V–I Curve* *(c) Output Characteristic Curve*

Fig. 6.31. ***LED***

Figure 6.31 (*a*) shows a circuit symbol for a LED. The arrows indicate radiation emitted by the diode. Figures 6.31 (*b*) and 6.31 (*c*) give two curves used to determine LED operating characteristics. Fig. 6.31 (*b*) is forward bias V-I curve for a typical LED employed in burglar alarms. Forward bias of approximately one volt is required to give significant forward current. Figure 6.31 (*c*) provides radiant power-forward current curve. The radiant output power is rather small and indicates a very low efficiency of electrical to radiant energy conversion.

Figure 6.32 shows a source connected to a series resistor R_S and a LED. The outward arrows symbolise the radiated light. The forward resistance of LEDs is very low which means that once the forward bias exceeds, the current through it will increase rapidly for only a very small increase in voltage. Thus it becomes necessary to use an external series current limiting resistor. The value of series resistor R_s can be determined from the following equation.

$$R_s = \frac{V_s - V}{I} \qquad ...(6.28)$$

where V_s is the supply voltage, V is LED forward bias voltage and I is the desired current.

For most of the commercially available LEDs, the typical voltage drop (diode voltage) is from 1.5 – 2.5 V for currents between 10 and 50 mA. The exact voltage drop depends on the LED current, colour, tolerance, etc.

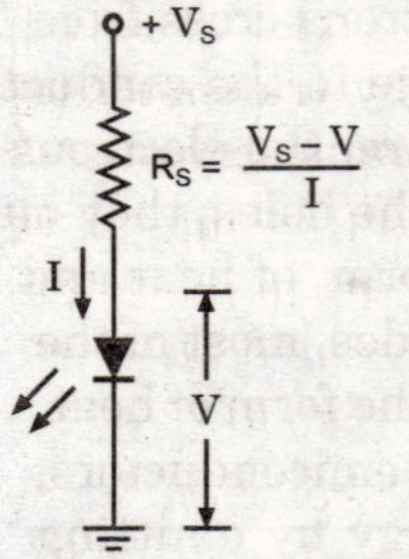

Fig. 6.32. *LED Circuit*

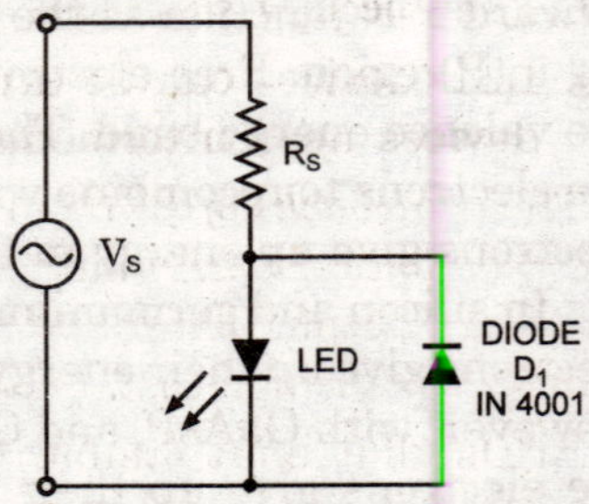

Fig. 6.33. *LED As An Indicator in AC Circuit*

A LED can be used as an indicator in ac circuit by wiring it in inverse parallel with a normal diode, as shown in Fig. 6.33, to prevent the LED being reverse biased; for a given brightness, the value of R_s should be halved relative to that of the dc circuit.

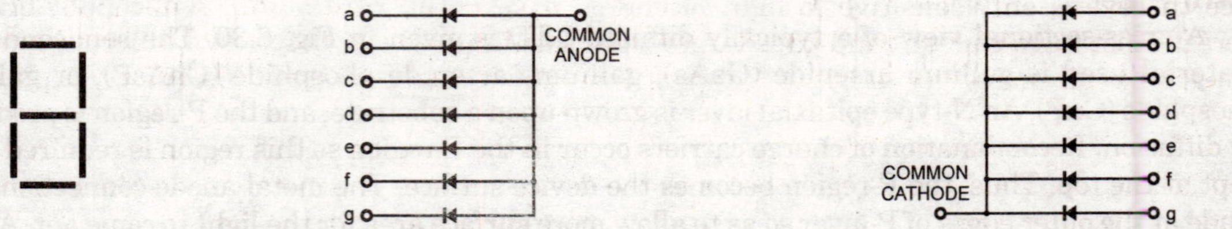

(a) Display Format *(b) Common Anode Configuration* *(c) Common Cathode Configuration*

Fig. 6.34

LED displays are available in many different sizes and shapes. The light-emitting region is available in lengths from 2.5 to 25 mm. Numbers can be created by such segments. Typical arrangement of a seven-segment LED numerical display is shown in Fig. 6.34(*a*). Any desired numeral from 0 to 9 can be displayed by passing current through the appropriate segments. The LEDs in a seven-segment display may be connected in common anode or in common-cathode configuration, as illustrated in Figs. 6.34 (*b*) and 6.34 (*c*) respectively. LEDs are usually switched on and off by transistor circuit.

Transistor circuit for switching on and off LEDs is given in Fig. 6.35.

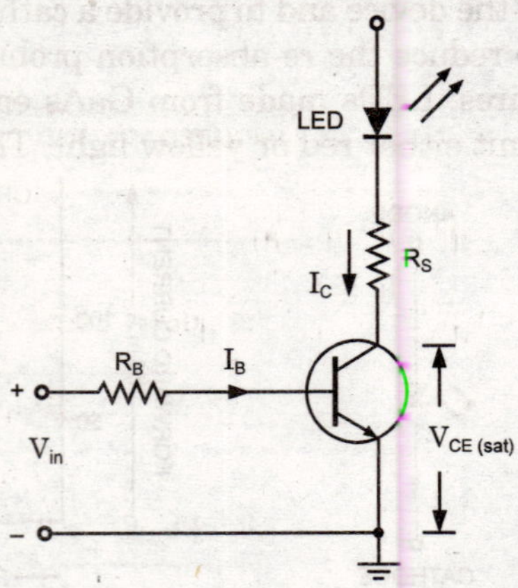

Fig. 6.35. *Transistor Circuit For Switching On and OFF a LED*

Advantages of LEDs. 1. LEDs are miniature in size and can be stacked together to form numeric and alphanumeric displays in high density matrix.

2. Smooth control of light intensity from a LED because light output from a LED is a function of current flowing through it.

3. LEDs are economical and have a high degree of reliability because they are manufactured with the same type of technology as is used for transistors and ICs.
4. LEDs are rugged and can, therefore, withstand shocks and vibrations.
5. LEDs can be operated over a wide temperature range (0–70ºC).
6. The switching time (both *on* and *off*) is less than 1 ns. So they are very suitable where dynamic operation of large number of arrays is involved.
7. LEDs are available in different colours like red, yellow, green and amber.
8. They need moderate power. They are used where low dc power is available.
9. They occupy small area.
10. LED devices can be driven by transistor-transistor logic (TTL) whereas gas discharge devices need intermediate transistor stages to allow them to be driven by 5 V TTL.

The *drawbacks of LEDs* are that they may get damaged by overvoltage or overcurrent. They have wide optical bandwidth compared to the LASER ($\simeq$ 10 nm). Their temperature depends on the radiant output power and wavelength. LEDs are not suited for large area displays, primarily because of their high cost. For large displays, devices using gas filled plasma are used.

LEDs that emit visible light are widely used in instrument display panel indicator, digital watches, calculators, multimeters, intercom, telephone switch boards, etc. LEDs that emit invisible infrared light find application in remote control schemes, object detectors, burglar alarm systems; and other areas requiring invisible radiation.

6.11. LIGHT-EMITTING MATERIALS

Since the eye is only sensitive to light of energy $h_f \geq 1.8$ eV (~ 0.7 μm), semiconductors of interest must have energy band gaps larger than this limit. The band gaps of different binary compound semiconductors are illustrated in Fig. 6.36(*a*). A large variation exists in photon

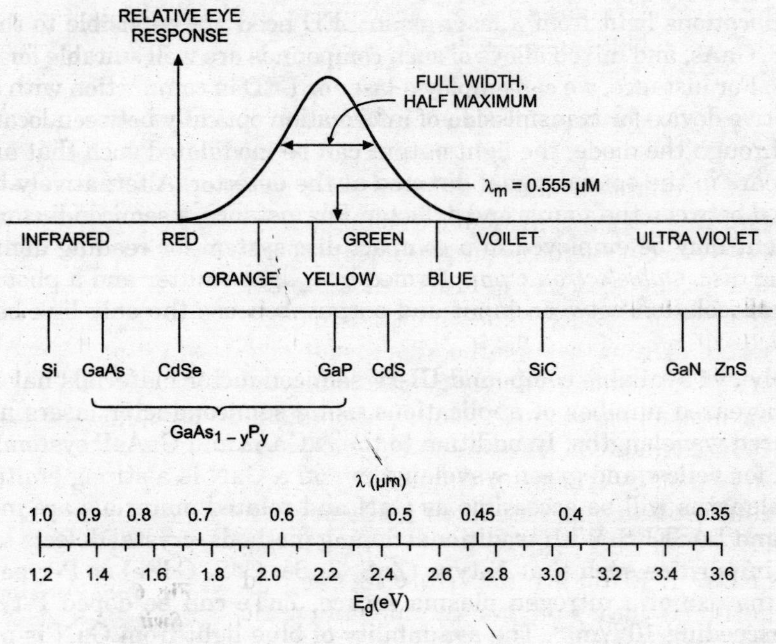

Fig. 6.36. *(a) Light Spectrum Vs Wavelength and Energy*

energy between the infrared band gaps (InSb of 0.18 eV) to the ultraviolet band gap (GaN of 3.4 eV). In fact, use of ternary and quatenary compounds can increase number of available energy significantly. A good example, of the variation in photon energy available from the compound semiconductors is the ternary alloy gallium arsenide-phosphide, as illustrated in Fig. 6.36(*b*). With the decrease in percentage of As and increase in percentage of P, the resulting band gap varies from the indirect 1.43 eV gap of GaAs (infrared) to the indirect 2.26 eV gap of GaP (green). The band gap of $GaAs_{1-x}P_x$ varies almost linearly with x until the 0.45 composition is reached, and electron-hole recombination is direct over this range. The most common alloy composition employed in LED displays is x 0.4. For this composition the band gap is direct. This results in efficient radiative recombination, and the emitted photons (~1.9 eV) are in the red portion of the spectrum.

For $GaAs_{1-x}P_x$ with P concentrations exceeding 45 per cent, the band gap is due to the indirect X minimum. Radiative recombination in such indirect materials is generally unlikely, since electrons in the conduction band have different momentum from holes in the valence band.

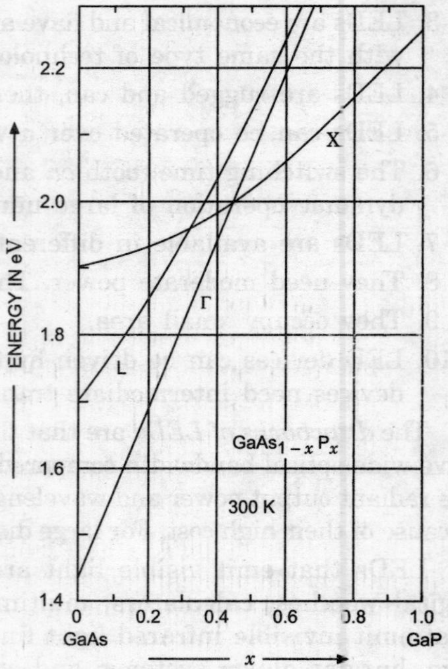

Fig. 6.36. (*b*) *Conduction Band Energies as a Function of Alloy Composition For* $GaAs_{1-x}P_x$

However, indirect $GaAs_{1-x}P_x$ (including GaP, $x = 1$) doped with nitrogen can be employed in LEDs to provide light in the yellow to green spectrum. This is possible because the nitrogen impurity binds an electron very tightly.

In many applications light from a laser or an LED need not be visible to the eye. Infrared emitters like InP, GaAs, and mixed alloys of such compounds are well suitable for optical communication systems. For instance, we can employ a laser or LED in conjunction with a photodiode or other photosensitive device for transmission of information optically between locations. By variation of current through the diode, the light output can be modulated such that analog or digital information appears in the optical signal directed at the detector. Alternatively the information may be introduced between the source and detector. For instance, a semiconductor laser-photodetector arrangement may be employed in a compact disc system for reading digital information from the spinning disc. *Optoelectronic pair* formed by a light emitter and a photodiode provides complete electrical isolation between input and output, because the only link between the two devices is optical.

A large number of available compound III-IV semiconductor materials have proved useful to provide for increased number of applications using semiconductor lasers and LEDs with visible and infrared wavelengths. In addition to the AlGaAs and GaAsP systems, the InAlGaP system is useful for yellow and green wavelengths and a GaN is a strong emitter in the blue. Even more wavelengths will be accessible as GaN and related materials are increasingly employed in LEDs and LASERS. With traditional doping methods, crystal defects tend to compensate the doping impurities such that N-type (ZnS, ZnSe, CdS, CdSe) or P-type (ZnTe) can be obtained. With the use of a nitrogen plasma source, ZnTe can be doped P-type to acceptor concentrations exceeding $10^{25}/m^{-3}$. The availability of blue light from GaN is particularly important in extending III-V light emission across the entire visible spectrum.

6.12. Semiconductor Lasers

The word LASER stands for *light amplification by stimulated emission of radiation,* which sums up the operation of an important optical and electronic device. The laser is a source of highly directional, monochromatic, coherent light, and as such it has revolutionized some longstanding optical problems and has produced some new fields of basic and applied optics. The light from a laser, depending on the type, can be continuous beam of low or medium power, or it can be a short burst of intense light delivering millions of watts. Light has always been a primary communication link between humans and the environment, but until the laser was invented, the light sources available for transmission of informations, and performance of experiments were generally neither monochromatic nor coherent, and were of relatively low intensity. Thus the laser is of great interest in optics, but it is equally important in optoelectronics, particularly in fibre optic communications. The last three letters in the word LASER are intended to imply how the device operates by the *stimulated emission of radiation.*

The laser was used in 1962 when the first P-N junction lasers were fabricated in GaAs infrared wavelength and the GaAsP visible wavelength. For the P-N junction laser, the laser action is produced by simply passing a forward current through the diode itself. The result is a very efficient overall system that can be modulated easily by modulating the current. Since semiconductor lasers have very short photon lifetimes, modulation at high frequencies can be achieved.

A semiconductor laser is very compact in size (of the order of 0.1 mm long). In semiconductor lasers, the transitions are associated with the band properties of materials. They operate at low power inputs and their output is easily adjusted/varied by controlling junction current. Thus the function of semiconductor laser is to provide portable and easily controlled source of low power coherent radiation.

Because of its compact size and capability for high-frequency modulation, the semiconductor laser is one of the most important light sources for optical fibre communication.

In choosing a material for junction laser fabrication, it is necessary that electron-hole recombination occur directly, rather than through trapping processes such as are dominant in silicon and germanium. GaAs is an example of such a "direct" semiconductor. Also, we must be able to dope the material N-type or P-type to form a junction. In case an appropriate resonant cavity can be made in the junction region, a laser results in which population inversion is done by the bias current applied to the junction (Fig. 6.37).

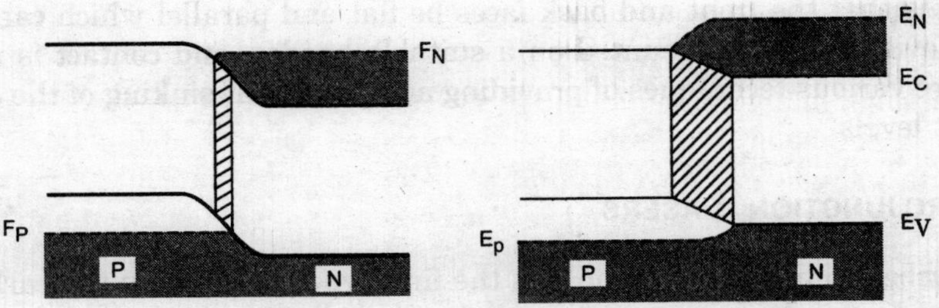

Fig. 6.37. ***Variation of Inversion Region Width With Forward Bias V(a) < V(b)***

The light intensity against photon energy (hf) for a laser junction is shown in Fig. 6.38, in three different states.

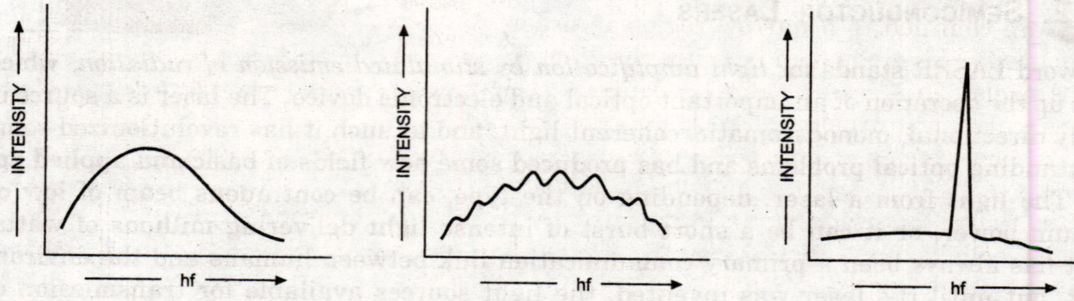

(a) Incoherent Emission Below Threshold *(b) Laser Modes At Threshold* *(c) Dominant Laser Mode Above Threshold*

Fig. 6.38. ***Light Intensity Vs Photon Energy (hf) For a Junction Laser***

At low current levels [Fig. 6.38(*a*)], spontaneous emission spectrum containing energies in the range $E_G < hf < (F_N - F_P)$ is obtained. With the increase in current to the point that significant population inversion exists, stimulated emission occurs at frequencies corresponding to the cavity modes, as depicted in Fig. 6.38(*b*). Finally, at a still higher current level, a most preferred mode or set of modes will dominate the spectral output, as depicted in Fig. 6.38 (*c*). This very intense mode represents the main laser output of the device; the output light will contain almost monochromatic radiation superimposed on a relatively weak radiation background, due primarily to spontaneous emission.

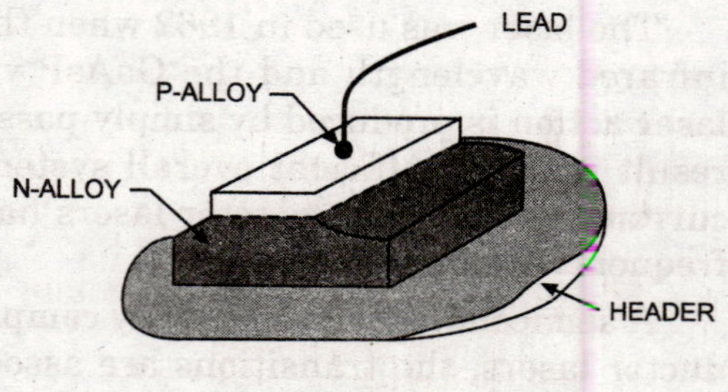

Fig. 6.39. ***Mounted Junction Laser Structure***

To build a PN junction laser, we are required to form a junction in a highly doped, direct semiconductor (say GaAs). For it a resonant cavity is made in the proper geometrical relationship to the junction and then contact is made to the junction in a mounting which is required for efficient heat transfer. Starting with a degenerate N-type sample, a P region is formed on one side, say, by diffusing Zn into the N-type GaAs. Zn serves as an acceptor in GaAs and therefore, the heavily doped Zn diffused layer constitutes a P^+ region. Thus we have a larger-area planar P-N junction. Now grooves are cut or etched along the sample length leaving a series of long P-regions isolated from each other. Such P-N junctions can be cut or broken apart and then cleaved into devices of the required length. The very important requirement of a resonant cavity is that the front and back faces be flat and parallel which can be fulfilled by cleaving. The device is finally mounted on a suitable header, and contact is made to the P-region. There are various techniques of providing adequate heat sinking of the device for large forward current levels.

6.13. Heterojunction Lasers

The device described in previous section was the first type employed in the early development of semiconductor lasers. It is known as an *homojunction laser* because it contains only one junction in a single type of material. Multiple layers have to be built-up in the laser structure to develop more efficient lasers operating at room temperature. These devices are called *heterojunction lasers,* and can be operated continuously at room temperature to meet with the optical communication needs.

In the structure shown in Fig. 6.40, the injected carriers are confined to a narrow region in order that population inversion can be built-up at lower current levels. This results in a lowering of the threshold current at which the laser action begins.

Fig. 6.40. ***AlGaAs Heterojunction Grown on Thin P-Type GaAs Layer***

A further improvement can be had by sandwiching the active GaAs layer between two AlGaAs layers. This *double-heterojunction structure* further confines injected carriers to the active region, and the change in refractive index at the GaAs — AlGaAs boundaries assists in confining the produced light waves. In the double-heterojunction laser depicted in Fig. 6.41 (*b*), the injected current is restricted to a narrow-stripe along the lasing direction, in order that the total current required to drive the device is reduced. This type of laser was a major step forward in the development of lasers for fibre optic communications.

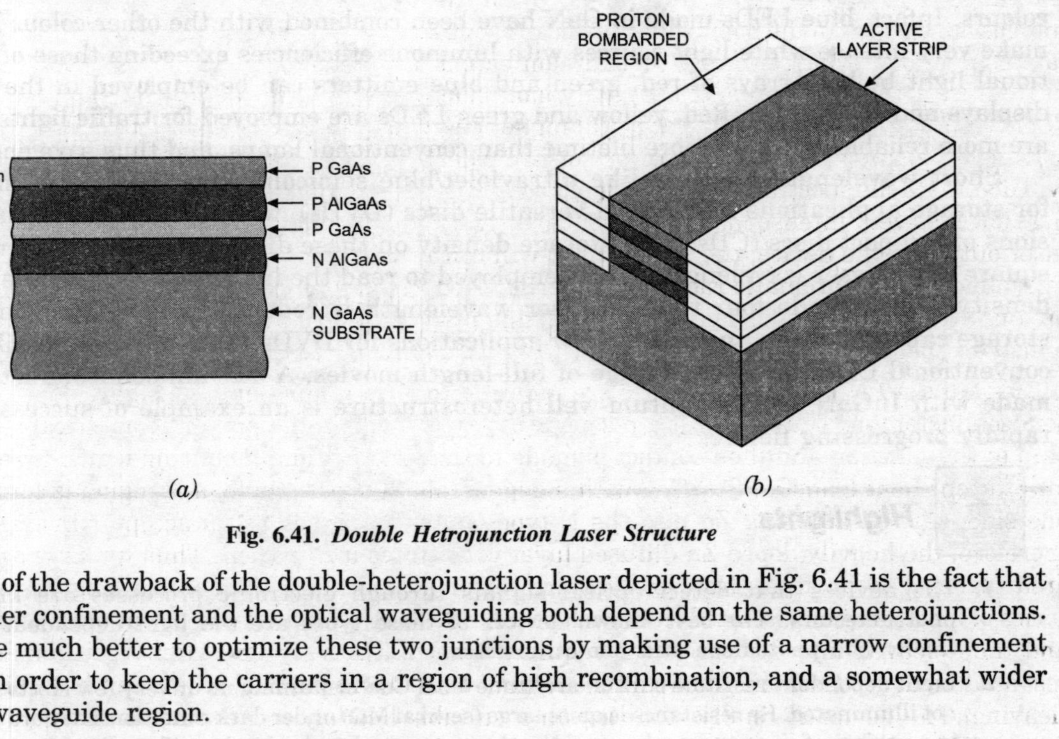

Fig. 6.41. ***Double Hetrojunction Laser Structure***

One of the drawback of the double-heterojunction laser depicted in Fig. 6.41 is the fact that the carrier confinement and the optical waveguiding both depend on the same heterojunctions. It will be much better to optimize these two junctions by making use of a narrow confinement region in order to keep the carriers in a region of high recombination, and a somewhat wider optical waveguide region.

6.14. MATERIALS FOR SEMICONDUCTOR LASERS

The InGaAsP/InP system is particularly very suitable for the type of lasers employed in fibre-optic communication systems. The quaternary alloy InGaAsP is particularly versatile in the fabrication of laser diodes, permitting considerable choice of wavelength and flexibility in lattice matching, which is important in producing heterostructures by epitaxial growth. By choice of composition, lasers can be made in the infrared range 1.3-1.55 μm required for fibre optics. In many applications, however, other wavelength ranges are required for laser output. For instance, the use of lasers in pollution diagnostics needs wavelengths farther in the infrared than are available from InGaAsP and AlGaAs. For such an application, the ternary alloy

PbSnTe provides laser output wavelengths from about 7 μm to more than 30 μm at low temperatures, depending on the composition of material. For intermediate wavelengths, the InGaSb system can be employed.

Materials chosen for the fabrication of semiconductor lasers must be efficient light emitters and also be amenable to the formation of P-N junctions and in most cases the formation of heterojunction barriers. By modern crystal growth techniques it is possible to grow junctions in ZnS, ZnSe, ZnTe, and alloys of these materials using N as the acceptor. Lasers can be made in these materials which emit in the green and blue-green regions of the spectrum. Much progress has been made in the growth of large band gap semiconductors making use of GaN, and its alloys with InN and AlN. The InAlGaN system has direct band gaps over the entire alloy composition range and, therefore, offers very efficient light emission. Band gaps range from about 2 eV for InN, to 3.4 eV for GaN and 5 eV for AlN. This covers the wavelength range from about 620 nm to about 248 nm (from blue to ultraviolet).

It has been a major goal of the optoelectronics community to have high efficiency blue emitters, since along with red and green, blue completes the list of three primary additive colours. Infact, blue LEDs made in GaN have been combined with the other colour LEDs to make very intense white light sources with luminous efficiencies exceeding those of conventional light bulbs. Arrays of red, green and blue emitters can be employed in the outdoor displays and TV screens. Red, yellow and green LEDs are employed for traffic lights as they are more reliable and have more lifetime than conventional lamps, and thus save energy.

Short wavelength emitters like ultraviolet/blue semiconductor lasers are important for storage applications like digital versatile discs (DVDs), which are higher density versions of compact discs (CDs). The storage density on these discs varies inversely with the square of the laser wavelength that is employed to read the information. Thus the storage density can be made four times if laser wavelength is reduced to half. Such increased storage capacity opens up entirely new applications for DVDs that were not possible with conventional CDs, say, the storage of full-length movies. A 417 nm semiconductor laser made with InGaN multi-quantum-well heterostructure is an example of success in this rapidly progressing field.

Highlights

1. The devices that detect optical signals through electronic processes are known as photodetectors. The best known devices of these types are the light dependent resistor (LDR), the photodiode and phototransistor.
2. Light dependent resistors (LDRs) are made from CdS containing no or very few electrons when not illuminated. Its resistance is quite large (several MΩ) under dark condition They are available in a variety of sizes and package styles, the most popular size having a face diameter of roughly 10 mm. They are sensitive, inexpensive, and readily available devices. Usual practical applications of LDRs include light and dark-activated switches and alarms, light beam alarms and reflective smoke alarms etc.
3. Photodiode is a two terminal semiconductor P-N junction and is designed to operate with reverse bias. The photodiode is a photovoltaic device as well as photoconductive device. When the device is operated with a reverse bias, it operates as a photoconductive device and when operated without the reverse bias, it operates as a photovoltaic device.
4. Although PN photodiode provides a much high frequency response than CdS bulk-type photoconductor but even faster response may be achieved with the help of PIN photodiode. PIN photodiode consists of a P region and an N region separated by an intrinsic region. PIN

photodiodes are relatively inexpensive and ultrafast having a switching speed of nanosecond or so and interface easily with conventional electronic circuitry.

5. The avalanche photodiode is similar to PN or PIN photodiode except that it is operated at high reverse-bias voltages to that impact ionization occurs, EHPs are created in the space-charge region by photon absorption. With proper design a silicon avalanche photodiode can have high sensitivity to low level optical signals, and the response time is about 1 ns. These devices are particularly useful in fiber optic communication systems.

6. The phototransistor is a three-terminal device having characteristics very similar to those of a BJT with a base and collector current sensitive to the incident light intensity. The base current that results is essentially linearly related to the applied light with a level almost independent of the voltage across the device until breakdown results.

   Photodiodes and phototransistors may be operated as switching devices in light operated relays, shaft encoders, paper tape readers, or used to provide an output proportional to the light input as in the accelerometer.

7. The photoconductive cell is a two-terminal semiconductor device whose terminal resistance will vary (linearly) with the intensity of the incident light. It is frequently called a photoconductive device.

   Photoconductive cells are constructed from cadmium sulphide or cadmium selenide semiconductors. They have their conductivity increased by incident photons of sufficient energy raising valence electrons to the conduction band. The photoconductive cells have the advantages of (*i*) high sensitivity (*ii*) low cost and (*iii*) high dark-to-light resistance exceeding 100 : 1. The drawbacks of photoconductive cells are large response time and relatively narrow spectral response.

8. Photovoltaic cells are semiconductor junction devices used for converting radiation energy into electrical energy. These cells generate a voltage proportional to electromagnetic radiation intensity and are called the *photovoltaic cells* because of their voltage generating capability. An example of such a cell is a silicon solar cell, which converts radiant energy of the Sun into electrical energy. Selenium and silicon are the most widely used materials for solar cells, though gallium arsenide, indium arsenide and cadmium sulphide are also used. Photovoltaic cells may be series-parallel interconnected to provide large area solar power cells.

9. Light-emitting diodes (LEDs) emit light under forward-bias conditions but needs 2 V to 4 V for good emission.

10. A large number of available compound III-IV semiconductor materials have proved useful to provide for increased number of applications using semiconductor lasers and LEDs with visible and infrared wavelengths. In addition to the AlGaAs and GaAsP systems, the InAlGaP system is useful for yellow and green wavelengths and a GaN is a strong emitter in the blue. Even more wavelengths will be accessible as GaN and related materials are increasingly employed in LEDs and LASERS. With traditional doping methods, crystal defects tend to compensate the doping impurities such that N-type (ZnS, ZnSe, CdS, CdSe) or P-type (ZnTe) can be obtained. With the use of a nitrogen plasma source, ZnTe can be doped P-type to acceptor concentrations exceeding $10^{25}/m^{-3}$. The availability of blue light from GaN is particularly important in extending III-V light emission across the entire visible spectrum.

11. LASER stands for light amplification by stimulated emission of radiation, which sums up the operation of an important optical and electronic devices.

    A semiconductor laser is very compact in size (of the order of 0.1 mm long). In semiconductor lasers, the transitions are associated with the band properties of materials. They operate at low power inputs and their output is easily adjusted/varied by controlling junction current. Thus the function of semiconductor laser is to provide portable and easily controlled source of low power coherent radiation.

    Because of its compact size and capability for high-frequency modulation, the semiconductor laser is one of the most important light sources for optical fibre communication.

12. Heterojunction lasers are the multiple layer PN junction lasers with more than one junction and can provide still greater efficiencies and more power.

Exercises

1. Define photonic devices and give their classification with categories in each category.
2. What is a photodetector ? List the stringent requirements that are to be met by photodetectors. Name such few devices.
3. What is an LDR ? Give its basic construction, symbol and characteristics. How an LDR can be used as an overlight detector ?
4. What is a photodiode ? Give its biasing arrangement and construction. Give its one or two applications.
5. Derive an expression for total steady-state diode photocurrent for a long diode.
6. Why PIN photodiode is considered one of the most common photodetectors ? Give its construction and operation. Give the expression for photocurrent density along with the assumptions made in derivation of this equation.
7. Why PIN diode detector does not provide gain ? How can it be made more sensitive to low intensity of light ? What material should be used to detect light of wavelength $\lambda = 0.6$ μm ? **[U.P. Technical Univ. 2005-06]**
8. What is a phototransistor ? How does it differ from an ordinary transistor and photodiode ? Give its standard symbol and characteristics. How can it be converted into a photodiode ? How can its sensitivity and operating speed be controlled ?
9. Compare a phototransistor and photodiode. Name some of the applications of a phototransistor.
10. What is photodarlington ? Give its symbol and special features.
11. Explain the construction and working principle of a photoconductive cell. What are its major drawbacks ? Give its characteristics and one or two applications.
12. Explain the construction and working principle of a solar cell.
13. Describe the construction and working of a photovoltaic cell. Give its circuit symbol, voltage and current characteristics. Give also its merits and applications.
14. Sketch the current-voltage characteristics of a solar cell. Define what is meant by the maximum power rectangle of a solar cell.
15. Explain the operation of solar cell and photodetectors in detail. **[M.D. Univ. Third Semester May 2008]**
16. What is a photodetector ? Describe the working of solar cell. What are the applications of these devices ? **[U.P. Technical Univ. Fundamentals of Electronic Devices 2009-10]**
17. What is LED ? Explain. **[M.D. Univ. Third Semester May 2007]**
18. Discuss LED. **[M.D. Univ. Third Semester December 2007]**
19. How do you define light-emitting materials ? What are the voltage drop and current limits of a general LED ? What determines the emission of colour of the LED ? **[U.P. Technical Univ. 2007-08]**
20. What is LED ? Give its principle of working, construction, circuit symbol, merits, demerits and applications.
21. Briefly discuss the use of LEDs as display devices in instrumentation.
22. How do heterojunction lasers differ from semiconductor lasers. Describe semiconductor and heterojunction lasers.

23. Write short notes on the following:

(*i*) Photonic devices.

(*ii*) Photodetector.

(*iii*) LDR.

(*iv*) PN junction photodiode.

(*v*) PIN photodiode. **[U.P. Technical Univ. 2006-07]**

(*vi*) Solar cells. **[M.D. Univ. May 2005; U.P. Technical Univ. 2006-07]**

(*vii*) LEDs. **[M.D. Univ. December 2005; December 2006]**

(*viii*) Light-emitting materials.

(*ix*) Semiconductor lasers.

(*x*) Materials for semiconductor lasers.

Short Answer Type Questions With Answers

Q. 1. What are photonic devices ?

Ans. Photonic devices are those in which the basic particle of light (the photone) plays an important role. Photodetectors, solar cells and LASERS etc. are photonic devices.

Q. 2. What are photodetectors ?

Ans. Photodetectors are the semiconductor devices that can detect optical signals through electronic processes.

Q. 3. Enlist the three electronic processes involved in photodetection.

Ans. Three electronic processes involved in photodetection are (*i*) carrier generation by incident light (*ii*) carrier transport and/or multiplication and (*iii*) interaction of current with the external circuit to give the output signal.

Q. 4. Name some devices used for photodetection.

Ans. The best known devices used for photodetection are LDRs, photodiodes and phototransistors.

Q. 5. What is photodiode ?

Ans. Photodiode is a two-terminal semiconductor P-N junction device and is designed to operate with reverse bias.

Q. 6. How the current is reduced to zero in a photodiode ?

Ans. The reverse saturation current I_0 flowing through a photodiode is reduced to zero by applying a forward bias voltage of magnitude equal to barrier potential.

Q. 7. Photodiode is a photovoltaic device or a photoconductive device or both ?

Ans. Photodiode is a photovoltaic device as well as photoconductive device. When it is operated with a reverse bias, it is photoconductive device and when operated without the reverse bias, it is a photovoltaic device.

Q. 8. What is PIN photodiode ?

Ans. PIN photodiode consists of a P-region and N-region separated by an intrinsic region where width of intrinsic region is much larger than the space charge width of an ordinary PN junction.

Q. 9. How does avalanche photodiode differ from PIN photodiode ?

Ans. The avalanche photodiode is similar to PIN photodiode except that it is operated at high reverse-bias voltages so that impact ionization occurs.

Q. 10. How does a phototransistor differ from an ordinary BJT ?

Ans. Phototransistor is similar to an ordinary BJT except that no base terminal is provided. Instead of a base current, the input to the transistor is in the form of light.

Q. 11. What are photovoltaic or solar cells ?

Ans. Photovoltaic or solar cells are semiconductor junction devices used to convert radiation energy into electrical energy. These cells generate a voltage proportional to electromagnetic radiation intensity and are called the photovoltaic cells because of their voltage generating capability.

Q. 12. What is meant by LED ?

Ans. LED stands for light-emitting diode.

Q. 13. What precautions are required to be observed in the use of LEDs ?

Ans. LEDs should neither be reverse-biased nor operated near their maximum current rating. The leads of a LED should never be bent closer than about 2 mm from the encapsulation.

Q. 14. What for does LASER stand ?

Ans. LASER stands for light amplification by stimulated emission of radiation.

Q. 15. List the advantages of laser devices.

Ans. Laser devices are small in size, highly efficient, operate at low power inputs and their outputs are adjustable by varying the junction currents.

Q. 16. What are heterojunction lasers ?

Ans. Heterojunction lasers are the multiple layer PN junction lasers with more than one junction and can provide still greater efficiencies and more power.

Q. 17. Why II-VI compounds are not used for the fabrication of semiconductor lasers though they are generally very efficient light emitters ?

Ans. Because junctions are difficult to form.

Q. 18. Why semiconductors with indirect band gaps are not used for the fabrication of semiconductor lasers ?

Ans. Because semiconductors with indirect band gaps are not sufficiently efficient light emitters for practical laser fabrication.

□□□

CHAPTER 7

Negative Conductance Microwave Devices

Inside This Chapter

7.1. Introduction

Semiconductor instability is affected by negative conductance and this instability is effectively used in semiconductors for high-frequency applications. There are a number of devices that are useful in microwave circuits like the Varactor and the specially designed high-frequency transistors, but they can provide amplification and other functions at microwave frequencies up to 10^5 MHz. This is due to transit time and other effects. For performing electronic functions like switching and conversion of dc power to microwave power, other devices are required to be specially designed. The most commonly used negative conductance devices are:

1. Tunnel or Esaki diodes.
2. Transit time diodes.
3. Gunn diodes.

As all of the above devices are diodes, they have two terminals and can be designed to operate in a negative conductance mode to provide amplification or oscillation at microwave frequencies in a proper circuit.

A number of devices which are extensively employed in pulse and switching circuitry are most conveniently characterised in terms of volt-ampere curve which displays, over a limited range, a negative incremental resistance. Negative resistance may result in two-ways: Some devices exhibit a characteristic in part of which as the voltage increases, the current drops thereby resulting in a negative resistance while some other devices may exhibit a characteristic in part of which decrease in voltage causes increase in current thereby resulting in a negative resistance. Negative-resistance devices may be voltage controllable type or current controllable type.

7.2. Tunnel Diode

A tunnel diode (sometimes called an *Esaki diode* after its inventor, Dr. Leo Esaki), is a high conductivity two-terminal P-N junction doped heavily–about 1,000 times higher than a conventional junction diode. Because of heavy doping depletion layer width is reduced* to an extremely small value of the order of 10^{-5} mm : reverse breakdown voltage is also reduced to a very small value (approaching zero) resulting in appearance of the diode to be broken for any reverse voltage, and a –ve resistance section is produced in volt-ampere characteristic of the diode. It is found that the reduced depletion layer can result in carriers *'punching through'* the junction with the velocity of light even when they do not possess enough energy to overcome the potential barrier. The result is that large forward current is produced at relatively low forward voltage (less than 100 mV). Such a mechanism of conduction in which charge carriers (possessing very little energy) punch through a barrier directly instead of climbing over it is called *tunneling*. That is why such diodes are called the *tunnel diodes*. Because of heavy doping the tunnel diode can conduct in reverse as well as in forward direction but it is usually used in forward biased mode.

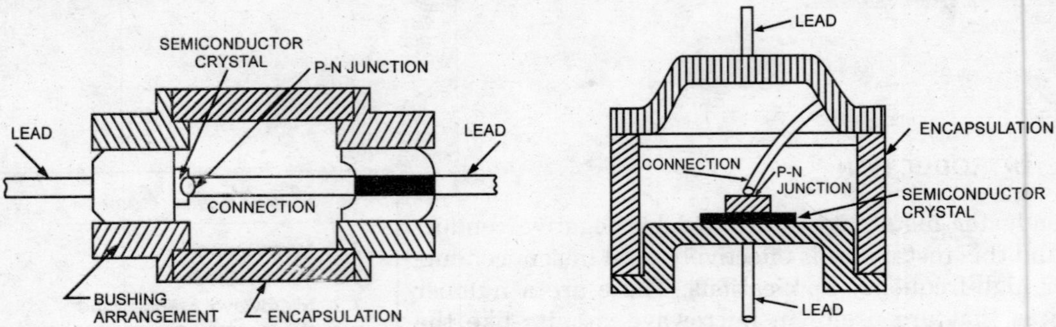

(*a*) (*b*) *Pellet Tunnel Diode*

Fig. 7.1

Construction. Tunnel diodes are usually fabricated from germanium, gallium arsenide or gallium antimonide. Silicon is not used in the construction of tunnel diodes because the ratio of peak value of forward current to the value of valley current (*i.e.* I_P/I_V) is maximum in case of gallium arsenide (approximately 15) and comparatively smaller for germanium (roughly 8) but very small in case of silicon (about 3).

Tunnel diodes are ordinarily manufactured by alloying from gallium arsenide. The source materials are highly doped semiconductor crystals with an impurity concentration of the order of 10^{25} per cubic metre or more. Arrangement is shown in Fig. 7.1 (*a*).

An alternative arrangement is shown in Fig. 7.1 (*b*). Such a tunnel diode is called the *pellet tunnel diode*.

It is to be noted that the tunnel diode, being a low power device, can be easily damaged by heat and static electricity.

The commonly used schematic symbols for the tunnel diodes are shown in Fig. 7.2.

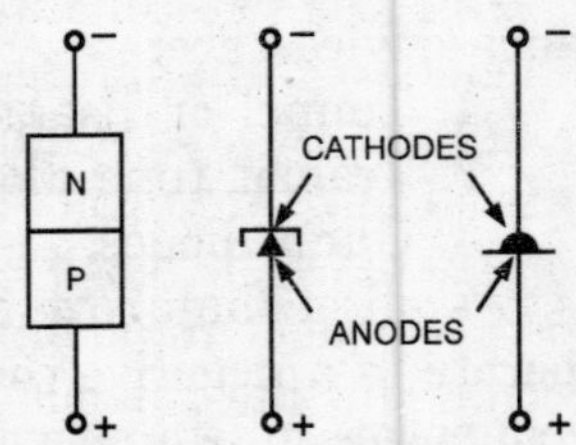

Fig. 7.2. ***Tunnel Diode Symbols***

Volt-Ampere Characteristic. Forward V-I characteristic of a tunnel diode is illustrated in Fig. 7.3. Forward bias produces immediate conduction—as soon as forward bias is applied, significant current is produced. The current attains quickly its peak value, I_P when the applied forward bias attains a value of V_P volts. The current variation in the vicinity of origin is due to quantum mechanical tunneling of electrons through narrow space charge region of the junction. With the further increase in forward voltage, the diode current starts decreasing till it attains its minimum value, called the *valley current* I_V corresponding to *valley voltage* V_V. Thus from

* The width of junction barrier varies inversely as the square root of impurity concentration.

peak point A to valley point B the diode current falls with the increase in voltage resulting in *negative resistance* in this region. In fact this portion AB of the characteristic constitutes the most useful property of the diode. In this region the diode, instead of absorbing power, produces power. So the tunnel diode can be used as a very high frequency oscillator. For voltages higher than valley voltage V_V current starts increasing as in any normal diode.

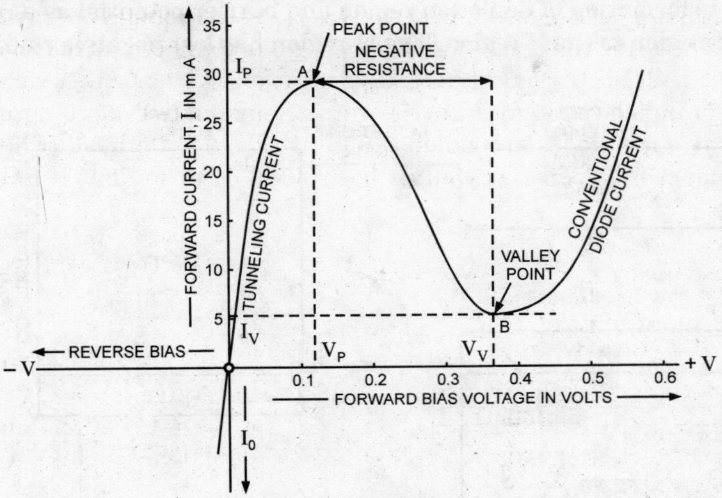

Fig. 7.3. *V-I Characteristic of Tunnel Diode*

Depletion Region and Energy Band Diagrams. The tunneling phenomenon can be explained with the help of energy band diagrams of P- and N-type semiconductor materials.

Because of heavy doping in case of tunnel diode, the depletion region is extremely narrow and, therefore, it does not constitute much of a barrier to the electron flow. As a result, a small forward or reverse bias (not large enough to overcome the barrier potential) can provide charge carriers sufficient energy to cross the depletion region. When this occurs, the charge carriers are said to be *tunneling* through the barrier.

When the semiconductor material is normally doped (either N-type or P-type), electrons fill all the holes in the valence band of energy level and conduction band is empty as illustrated in Fig. 7.4 (*a*). In case of a heavily doped P-type semiconductor material there is a deficiency of electrons and the valence band cannot be taken as filled. The result is that at the top of valence band there is a layer of empty energy levels, as illustrated in Fig. 7.4 (*b*). With very heavily doped N-type semiconductor material, there is an abundance of electrons. The result is that electrons fill the valence band and create a layer of filled energy level at the bottom of the conduction band, as illustrated in Fig. 7.4(*c*).

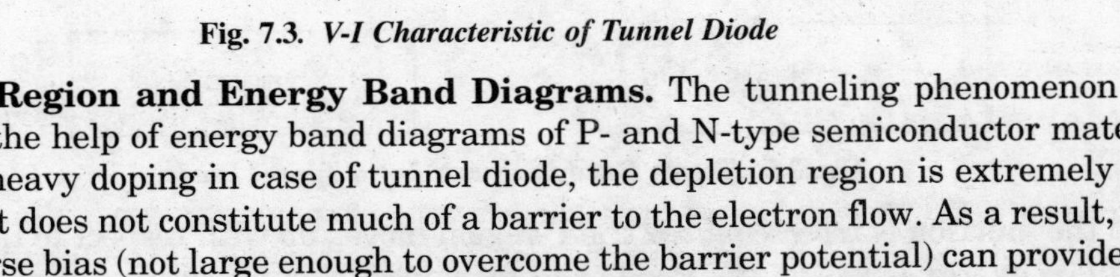

(*a*) *Normally Doped P and N-Type Materials* (*b*) *Very Heavily Doped P-Type Material* (*c*) *Very Heavily Doped N-Type Material*

Fig. 7.4. ***Energy Band Diagrams For Normally Doped and For very Heavily Doped Semiconductor Materials***

The energy band diagram for a heavily doped unbiased P-N junction is shown in Fig. 7.5. Note that the depletion region is very narrow and the filled levels on the N-side are exactly opposite those on the P-side. In this condition, no tunneling occurs as there are no empty lower energy levels to which electrons from either side might cross the depletion region. It is also to be noted that the conduction and valence bands on P-side are (negatively) higher than those on the N-side. This is due to formation of depletion region and barrier potential as a result of electrons crossing from the N-region to the P-region. The N-region has lost negative charges and P-region has gained them.

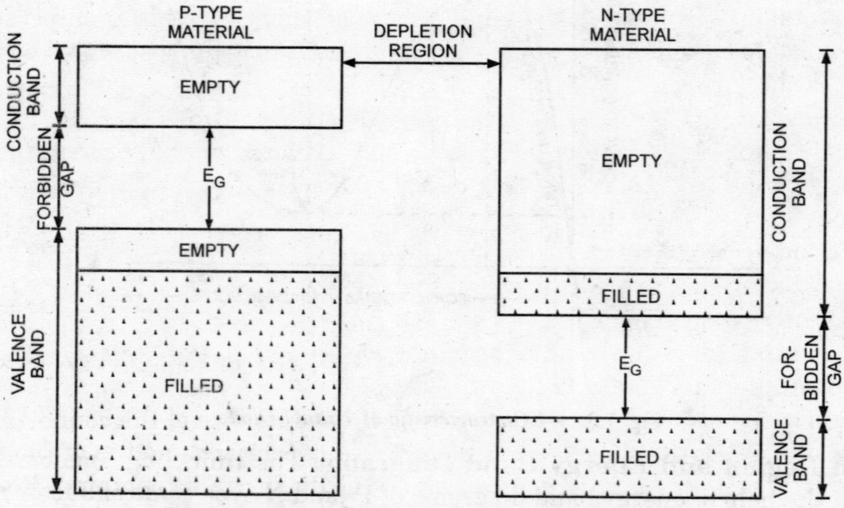

Fig. 7.5. ***Energy Band Diagrams For a Tunnel Diode***

When the junction is *reverse biased* the P-region moves up with respect to the N-region. As a result, filled energy levels on the P-side become opposite empty energy levels on the N-side. At this stage electrons tunnel through the narrow space charge region from the higher energy levels on the P-side to the lower energy levels on the N-side. Despite the fact that the junction is reverse biased, significant current flows. With the increase in reverse bias, more and more electrons tunnel from the P-side to N-side and a larger current flows. Thus the reverse characteristic of a tunnel diode is linear, just like that of a resistor (Fig. 7.3).

When the tunnel diode is forward biased, its initial behaviour is similar to that when it is reverse biased. Now, some of the filled energy levels on the N-side shift to a high energy level than empty level on the P-side (Fig. 7.5). Electron tunneling now occurs from the N-side to the P-side. With the increase in forward bias, more and more electrons tunnel from the N-side to the P-side. Eventually, however, a maximum level of tunneling is reached when a band of filled energy levels at the bottom of conduction band on the N-side is directly opposite the band of empty energy levels at the top of the valence band on the P-side. At this stage diode current attains its peak value I_P and the forward bias is V_P ($\simeq$ 100 m V). With the further increase in forward bias, part of the band of filled energy levels on the N-side is raised to an energy level corresponding to forbidden energy gap on the P-side. Electrons cannot tunnel to a forbidden energy level; thus the current flow due to tunneling is reduced. With continued increase in forward bias, tunneling goes on reducing. When all of the band of filled levels at the bottom of the conduction band on the N-side is raised to a level corresponding to the forbidden energy gap on the P-side, the current flow due to tunneling is reduced to the minimum level, called the *valley current* I_V. Now, however, the normal process of current flow across the forward biased junction begins to take over, as the bias becomes large enough to overcome the barrier potential. For voltages exceeding valley voltage V_V, the current starts increasing as in a normal diode.

Diode Parameters. We have seen that between I_P and I_V (points A and B on the V–I characteristic of tunnel diode illustrated in Fig. 7.3) the diode current decreases with the increase in voltage applied across the diode. This region of the characteristic is called the *negative resistance region*, and the negative resistance R_N of the tunnel diode is its most important property. Its value may be determined by computing the slope of the characteristic in the negative resistance region. The value of negative resistance slightly differs from point to point on the negative resistance portion of the characteristic. So R_N is usually specified at the centre of the negative resistance region.

Parameters for a typical tunnel diode are given below :

Peak current I_P : 1 – 100 mA; Peak voltage V_P : 50 – 200 mV; Valley current I_V : 0.1 – 5 mA; Valley voltage V_V : 350 – 500 mV; Forward bias V : 0.5 – 1 V; Negative resistance R_N : – 10 to – 200 Ω.

Tunnel Diode Equivalent Circuit. The equivalent circuit for a tunnel diode biased in the negative resistance region is given in Fig. 7.6. R_S represents the resistance of the connecting leads to the tunnel diode and the semiconductor material, and is of the order of 1 Ω; L_S represents the inductance of the connecting leads and is of the order of 0.5 nH; and C_D is junction diffusion capacitance and is of the order of 5 to 100 pF. These factors limit the frequency at which the diode can be operated. Frequency may range from 100 MHz to 10,000 MHz.

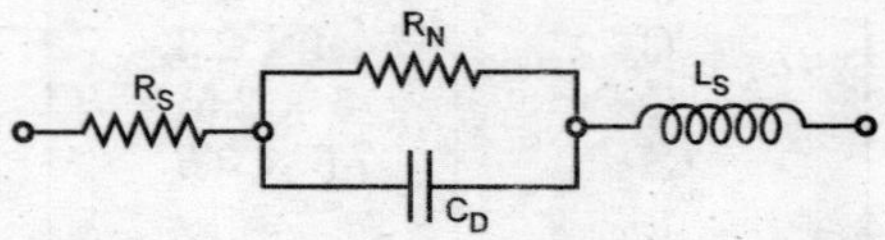

Fig. 7.6. ***Equivalent Circuit For a Tunnel Diode***

Advantages and Disadvantages of Tunnel Diodes. The advantages of tunnel diodes are low cost, low noise, simplicity, high speed of operation, ease of operation, low power and environmental immunity.

The disadvantages of the tunnel diodes are its low output-voltage swing and being a two-terminal device, it cannot provide isolation between the input and output circuits.

Applications. The tunnel diode can be used as an amplifier, an oscillator, or a switch, Because of very fast response to inputs, it is almost exclusively a high frequency component. The tunnel diode is, however, not widely used because it is a low current device and also that other better devices are available for the operations listed above.

7.3. Microwave Frequencies

The microwave frequencies are grouped both in frequency range and band sizes. Both of these factors are specified in all the cases to avoid any confusion. Table 7.1 provides frequencies from 0.1 GHz to 1,000 GHz for wavelengths of 0.3 mm to 3.0 m. In 1970, a new band designation for microwave frequencies has been adopted, as listed in Table 7.2. Many devices discussed earlier can be made to operate in the microwave bands.

7.4. Degenerate Semiconductors

In the devices discussed so far, the properties of relatively pure semiconductors (pure semiconductors with a very small fraction of doping impurities) have been studied. In such semiconductors no charge transport can occur within the donor or acceptor levels themselves because the few impurity atoms are so widely spaced throughout the sample. But when doping levels are increased, the impurities are very close to each other in the structure leading to

TABLE 7.1
Bands and Frequency Range

Band	*Waveband Size in Cm*	*Frequency Range in GHz*
Submillimeter	—	> 300
Millimeter	—	> 30-300
R	0.9 × 0.56	26.5-40
K	1.2 × 0.65	180-26.5
Ku	—	15.3-18.0
N	1.5 × 0.85	15.0-22.0
P	1.8 × 1.0	12.4-18.0
M	2.1 × 1.2	10.0-15.0
X	2.5 × 1.25	8.2-12.4
H	3.2 × 1.3	7.05-10.0
J	3.8 × 1.9	5.3-8.2
C	4.4 × 2.3	4.9-7.05
G	5 × 2.5	3.95-5.85
S	7.6 × 3.8	2.6-3.95
L	—	1.0-2.6

TABLE 7.2
New Bands and Frequency Range

Band	*Frequency Range in GHz*
Submillimeter	> 300
Millimeter	> 30-300
M	60.000-100.000
L	40.000-60.000
K	20.000-40.000
J	10.000-20.000
I	8.000-10.000
H	6.000-8.000
G	4.000-6.000
F	3.000-4.000
E	2.000-3.000
D	1.000-2.000
C	0.500-1.000
B	0.250-0.500
A	0.100-0.250

interaction within the donors or acceptors. In case the concentration of electrons in the conduction band is greater than the effective density of states, the Fermi level is no longer within the band gap but extends into the conduction band and when this occurs, the material is referred to as *degenerate N-type*. A similar reaction occurs in a degenerate P-type material where the concentration of acceptors is high and the Fermi level extends to the valence band.

7.5. Transit Time Devices

The impact ionization and transit-time properties of the semiconductor structures are used to develop negative resistance at microwave frequencies.

The negative conductance occurs because of two processes, causing the current to lag behind the voltage in time:

(*i*) delay due to the avalanche process and (*ii*) a further delay due to the transit time of the carriers across the drift region.

If the sum of above delay times is approximately one-half cycle of the operating frequency, negative conductance occurs and the device can be used for oscillation and amplification.

The advantage of a simple structure, as in two-terminal devices in comparison to three-terminal transistors, was considered to produce negative resistance from transit time in semiconductor diodes by Shockley in 1954. The devices used are:

1. IMPATT (Impact ionization avalanche transit time) diodes.
2. BARRIT (Barrier injection and transit time) diodes.
3. DOVETT (Double velocity transit time) diodes.
4. TRAPATT (Trapped plasma avalanche triggered transit time) diodes.

Transit time devices can convert dc to microwave ac signals with high efficiency and are very useful in the generation of microwave power for many applications. Here we will be discussing IMPATT diode only.

7.6. Impatt Diode

IMPATT diode is the impact ionization avalanche transit time diode and the Read diode proposed by WT Read in 1958 is best suited for illustration of its basic principles. The Read diode consists of (*i*) one-sided abrupt P-N junction (*ii*) two sided (double-drift) diode (*iii*) Modified Read diode and (*iv*) P-I-N diode. Figure 7.7 shows the static characteristics of the Read diode (P^+–N–I–N^+) or its dual (N^+–P–I–P^+).

The device consists essentially of two regions P^+–N region at which avalanche multiplication takes place and (*ii*) the I (essentially intrinsic) region through which generated electrons, taking the advantage of the higher mobility of electrons compared to holes, must drift in moving to the N^+ contact.

Similarly devices can be built-up in the N^+-P-I-P^+ configuration, in which holes resulting from the avalanche multiplication drift through the I-region.

In the Read diode shown in Fig. 7.7 (*a*), the avalanche region is very narrow and the maximum multiplication occurs in this narrow region between O and A in Fig. 7.7 (*c*), the avalanche region. Drift region is outside the avalanche region. The width of the avalanche region and the voltage drop across this region strongly affect optimum current density and the maximum efficiency of the IMPATT diode.

There are two limiting cases of the Read doping profiles. When the N_2 region becomes zero, we have a one-sided abrupt P^+-N junction. On the other hand, when the N_1 region becomes zero we have a P-I-N diode, also called the Misawa diode.

(a) Read Diode

(b) Doping Profile

(c) Electric Field Distribution

(d) Ionization Integrand at Avalanche Breakdown

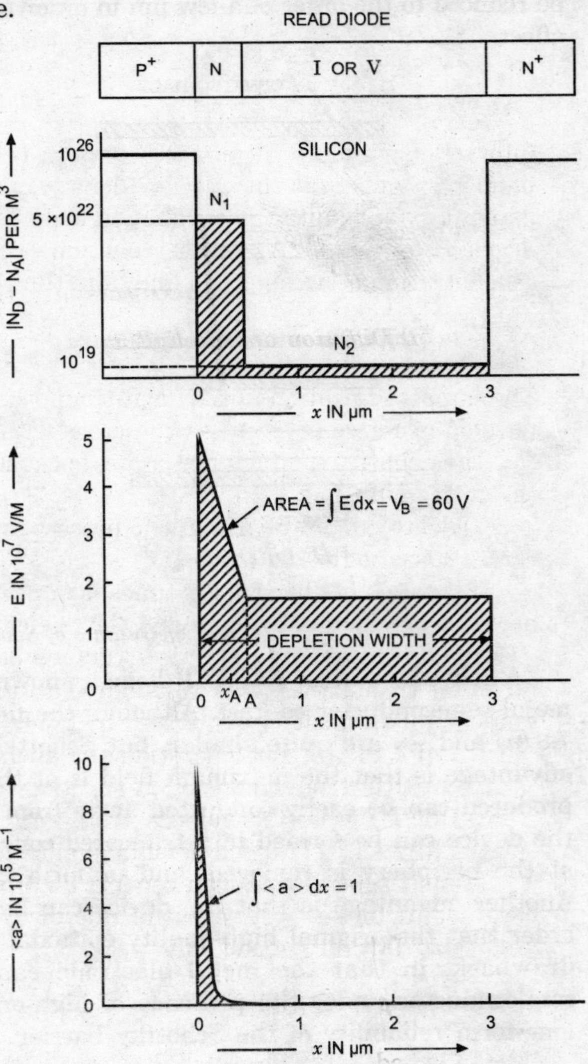

Fig. 7.7 *Read Diode*

Although the Read diode depicted in Fig. 7.7 displays most directly the operation of IMPATT devices, simpler structures can be used, and in some cases they may be more efficient. Negative conductance can be had in simple P-N junctions or in P-I-N devices. In the case of the P-I-N diode, most of the applied voltage occurs across the I region, which serves as a uniform avalanche region and also as a drift region. Thus the two processes of delay due to avalanche and drift, which are separate in the case of Read diode, are distributed within the I region of the P-I-N diode. This means that both electron and holes participate in the avalanche and drift processes.

7.6.1. Diode Fabrication

IMPATT diodes are usually designed to maximize power output and efficiency. Some typical high-power IMPATT diode designs are depicted in Fig. 7.8.

The configuration given in Fig. 7.8 (*a*) is formed by using a double epitaxial process or by diffusion into an epitaxial layer. Series resistance is reduced by using N^+ substrate. The thickness of epitaxial layer is also controlled in order that at breakdown essentially no unswept epitaxial layer is left. For high-frequency operation, even the thickness of N^+ substrate should be reduced to the order of a few μm in order to reduce losses and non-uniformity from the skin effect.

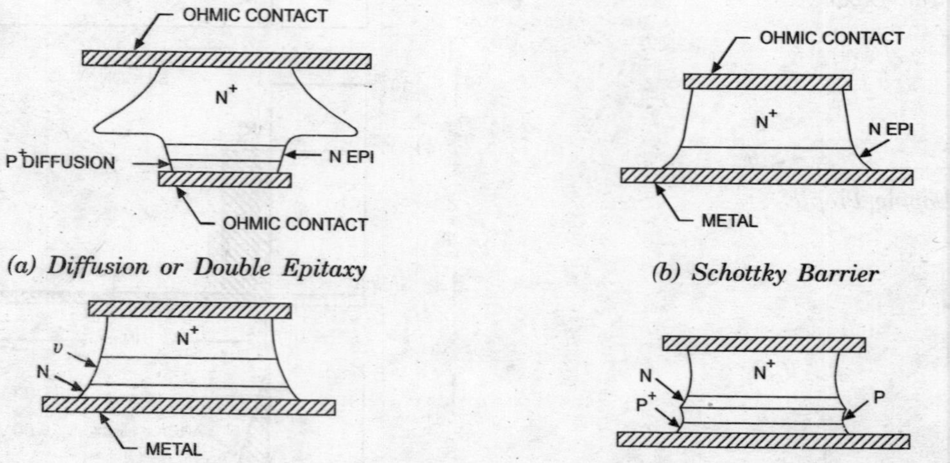

(a) Diffusion or Double Epitaxy *(b) Schottky Barrier*

(c) Hi-Lo Diode *(d) Double Drift IMPATT Diode Using Ion Implantation*

Fig. 7.8. ***Outlines of Some IMPATT Diode Designs***

A Schottky-barrier IMPATT diode is shown in Fig. 7.8(*b*). A Schottky barrier is a rectifying metal-semiconductor contact. Although the field distributions for the devices shown in Figs. 7.8 (*a*) and (*b*) are quite similar, but Schottky barrier offers some unique advantages. One advantage is that the maximum field is at the metal-semiconductor interface, and the heat produced can be easily conducted away from the metal contact. As depicted in Fig. 7.8 (*b*), the device can be formed in a truncated cone shape in order that the high-field singularity at the periphery is removed, and uniform breakdown can be achieved inside the device. Another advantage is that the device can be fabricated at relatively low temperatures, in order that the original high-quality epitaxial layer can be preserved. There is also a major drawback, in that the metal electrode can sometimes be attacked chemically by the semiconductor under the presence of high-energy electrons and holes, thus impairing the long-term reliability of the Schottky barrier.

The Schottky-barrier approach can be employed in modified Read diode structures, as illustrated in Fig. 7.8 (*c*), by replacing P^+ layer by metal contact. Since the Schottky barrier is basically a majority-carrier device, the minority-carrier storage effect that occurs in the original Read structure, is eliminated and efficiency is improved. Use of the modified Read diodes has resulted in a dramatic increase in microwave frequency in comparison to the diode with constant doping profile. However, the modified Read diode requires a more stringent control of doping profiles to have a specific-frequency device.

A new device configuration is shown in Fig. 7.8(*d*). It uses an ion implantation technique. For the most common dopants, for example, B and P, the ion penetration depth is of the order of 0.5 μm per 100 keV, so that 1 μm depth can easily be obtained from ion implantation equipment with energy sources a few hundred keV. These energy ranges are useful in making millimetre-wave IMPATT diodes. The possibility of making the double-drift structures depicted in Fig. 7.8 (*d*) is of particular importance. Power output per unit area and impedance per unit area are both approximately doubled. Thus from such structures increased power output with improved efficiency can be had.

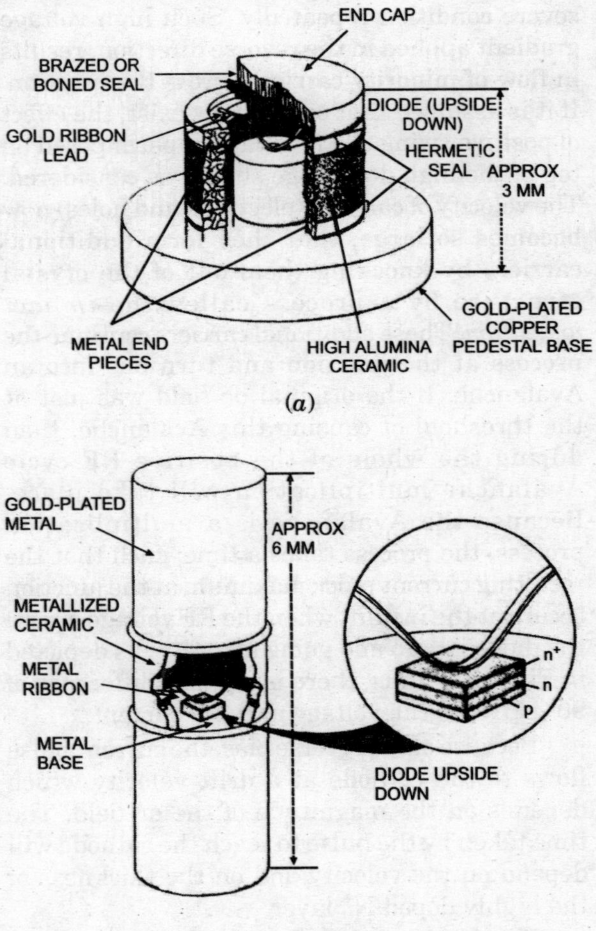

Fig. 7.9. ***Two Microwave Packages With IMPATT Diodes Mounted***

IMPATT diodes can also be fabricated by the molecular-beam epitaxy (MBE) method. Since the doping and thickness of layer down to atomic dimensions can be controlled in this method, this method is very useful for millimetre and submillimetre IMPATT diodes.

The fabricated diode is usually mounted in a microwave package. Two typical packages are illustrated in Fig. 7.9. In both cases the IMPATT diode is mounted with its diffused side or metal-electrode side in contact with a copper or diamond heat sink so as to conduct away the heat produced at the junction easily.

A typical commercial IMPATT diode for use below about 50 GHz is depicted in Fig. 7.10.

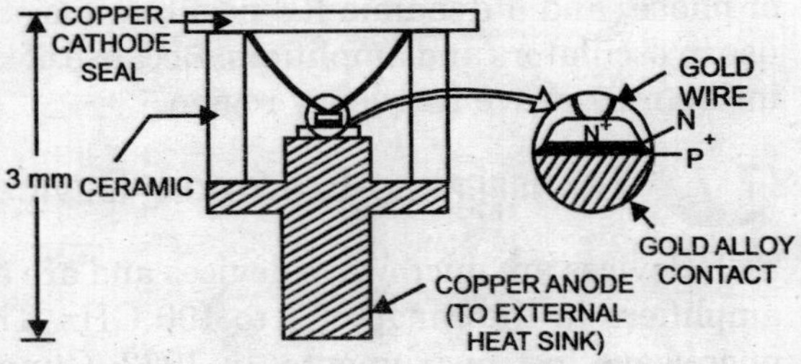

Fig. 7.10. ***Structure of IMPATT Diode***

IMPATT diode makes use of a combination of delay involved in producing Avalanche current amplification together with the delay due to transit time through a drift space to provide the required 180° phase difference between applied voltage and the resulting current. The x-section of the active region of the IMPATT diode is depicted in

Fig. 7.11. From the figure, it is to be noted that it is a diode in which the junction is between the P^+ and N layers.

An extremely high voltage gradient of the order of 40 kV/mm is applied across the IMPATT diode, eventually causing a very high current. The device is constructed to withstand such severe condition repeatedly. Such high voltage gradient applied in the reverse direction, results in flow of minority carriers across the junction. If it is assumed that oscillations exist, the effect of positive swing of RF voltage superimposed on top of the high dc voltage should be considered. The velocity of carriers (electrons and holes) now becomes so large, that they form additional carriers by knocking them out of the crystal structure by a process called the *impact ionization*. These additional carriers continue the process at the junction and turn out into an Avalanche. If the original dc field was just at the threshold of causing this Avalanche, then during the whole of the positive RF cycle Avalanche multiplication will take place. Because the Avalanche is a multiplication process, the process takes a time, such that the resulting current pulse maximum at the junction occurs at the instant when the RF voltage across the diode is zero and going negative, as depicted in Fig. 7.12. Thus, there is a phase difference of 90° between the voltage and the current.

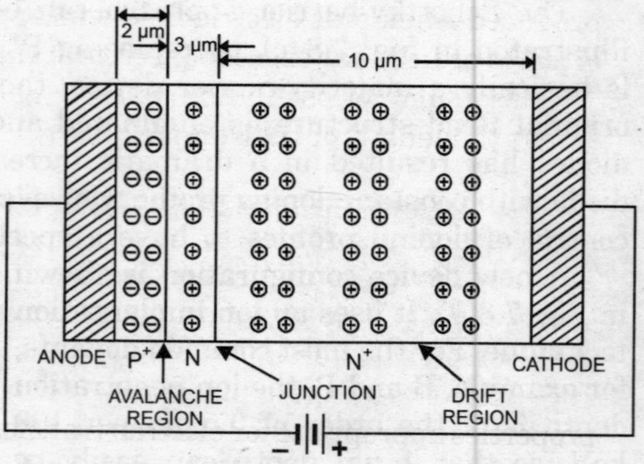

Fig. 7.11. ***Active Region of IMPATT Diode***

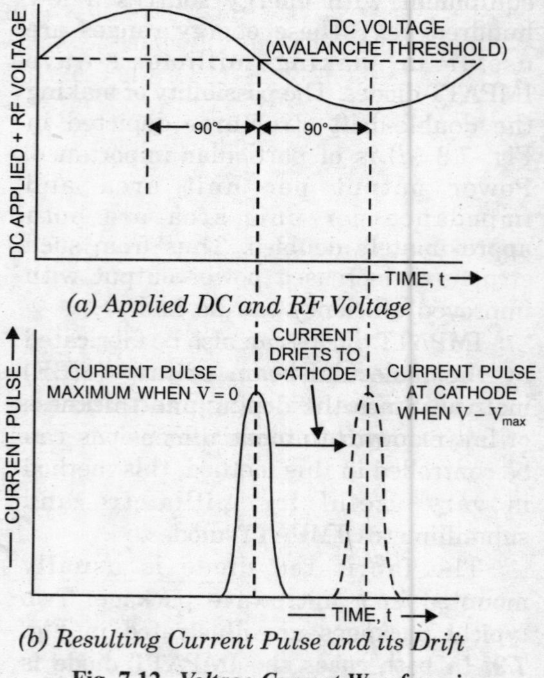

Fig. 7.12. ***Voltage-Current Waveform in an IMPATT Diode***

Because of the reverse bias, the current pulse flows to the cathode at a drift velocity which depends on the magnitude of the dc field. The time taken by the pulse to reach the cathode will depend on the velocity and on the thickness of the highly doped N^+ layer.

The drift region thickness is taken such that the time taken for the current pulse to reach the cathode corresponds to a further 90° phase difference. Thus, when the current pulse actually arrives at the cathode terminal the RF voltage is at its negative peak (Fig. 7.12). Thus, voltage and current in the IMPATT diode are 180° out of phase, and a dynamic RF negative resistance exists. Such a negative resistance find wide use in oscillators and amplifiers. Because of the short transit time involved they can be operated in the microwave frequency region.

7.7. Transferred-Electron Devices (TEDs)

TED devices are microwave devices and are extensively employed as local oscillators and power amplifiers in the range of 1 to 100 GHz. These are used in RADARs, intrusion alarms, and microwave test instruments. In 1963, Gunn discovered that coherent microwave output was produced when a dc electric field was applied across a randomly oriented, short N-type sample of GaAs or InP that exceeded a critical threshold value of several thousand volts per centimetre. The frequency of oscillation was roughly equal to the reciprocal of the carrier transit time across the length of the sample.

7.8. TRANSFERRED-ELECTRON EFFECT

The transfer of conduction electrons from a high-mobility energy valley to low-mobility, higher-energy satellite valleys is called the *transferred-electron effect*. Although GaAs and InP are the most common materials for TEDs, transferred-electron effects have also been observed in CdTe, ZnSe, GaAsP, and other materials. The band structure of some materials can be changed to exhibit properties appropriate for electron transfer. For instance, the energy bands of InAs can be distorted by applying pressure to the crystal, so that a set of satellite valleys becomes available for electron transfer, although these upper valleys are too far above the lower valley at normal pressures.

As observed from Fig. 7.13, the energy band structures of both GaAs and InP at 300 K are quite similar. The energy separation between the upper and lower valleys, ΔE is approximately 0.31 eV for GaAs and 0.53 eV for InP.

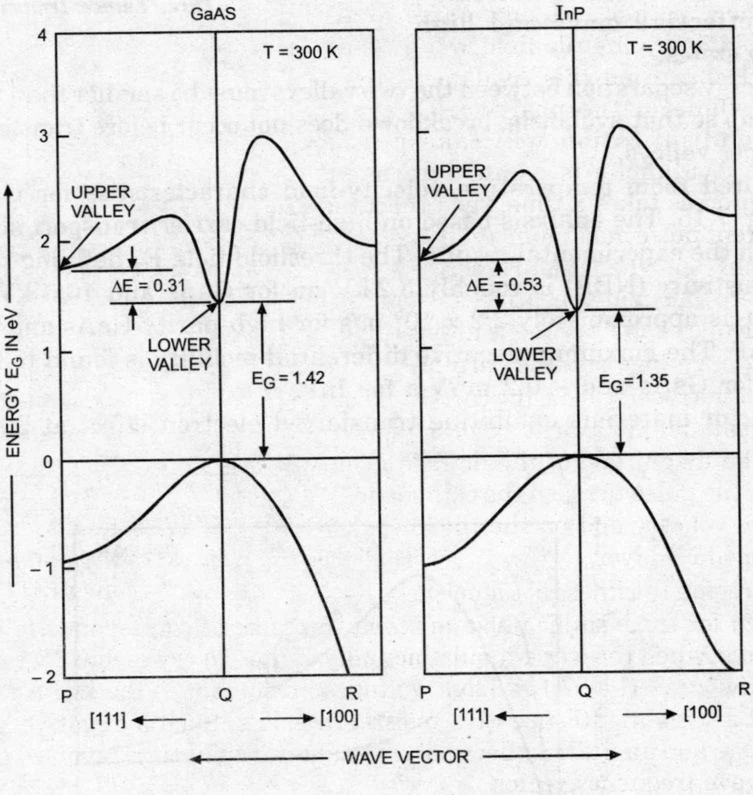

Fig. 7.13. ***Energy-Band Structures of GaAs and InP***

The general velocity-field curves are shown in Fig. 7.14, from which we observe that

(*i*) There is a well defined threshold field E_T for the onset of negative differential resistivity (NDR) or negative differential mobility.

(*ii*) The threshold field E_T increases with the increase in lattice temperature.

(*iii*) The negative mobility can be destroyed by having the lattice temperature too high or the energy difference ΔE too small.

Thus it becomes necessary to meet certain requirements for the electron transfer mechanism to give rise to bulk negative differential resistivity (NDR). These requirements are:

1. The lattice temperature must be low enough that, in the absence of a bias electric field, most electrons are in the lower conduction-band minimum, or $kT < \Delta E$.
2. In the lower conduction-band minimum, the electrons must have high mobility, small effective mass, and low density of states, whereas in upper satellite valleys, the electrons must have low mobility, large effective mass and high density states.
3. The energy separation between the two valleys must be smaller than the semiconductor band gap so that avalanche breakdown does not occur before transfer of electrons into the upper valleys.

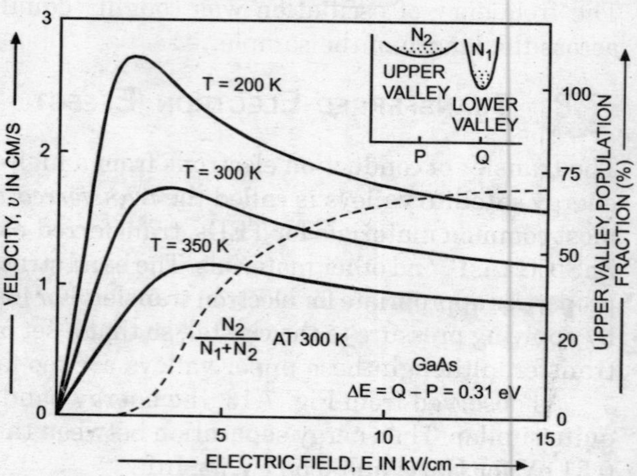

Fig. 7.14. ***Velocity-Field Characteristics For GaAs at Three Lattice Temperatures***

The measured room temperature velocity-field characteristics for GaAs and InP are depicted in Fig. 7.15. The analysis based on high-field carrier transport studies are in good agreement with the experimental results. The threshold field E_T defining the onset negative differential resistivity (NDR) is roughly 3.2 kV/cm for GaAs and 10.5 kV/cm for InP. The peak velocity v_p is approximately 2.2×10^5 m/s for high-purity GaAs and 2.5×10^5 m/s for high- purity InP. The maximum negative differential mobility is found to be approximately $-$ 0.24 m^2/V-s for GaAs and $-$ 0.2 m^2/V-s for InP.

Semiconductor materials exhibiting transferred electron effect at 300 K are listed in Table 7.3.

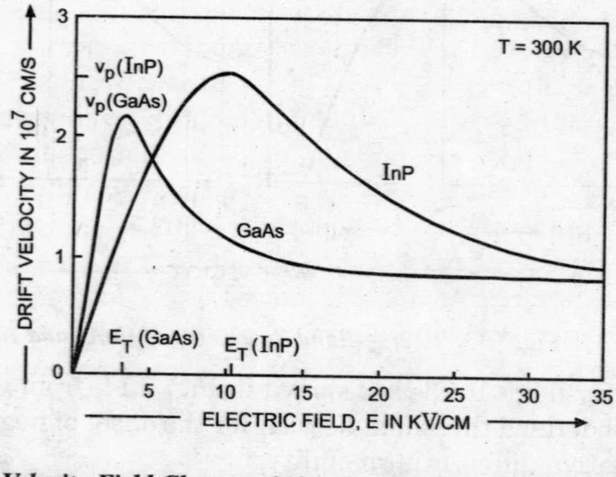

Fig. 7.15. ***Measured Velocity-Field Characteristics For GaAs and In P at Room Temperature***

TABLE 7.3. Semiconductor Materials, Exhibiting Transferred Electron Effect at 300 K

Semiconductor	Energy in eV	Valley Separation		Threshold Field E_T in kV/cm	Velocity of Drift in 10^5 m/s
		Between	ΔE in eV		
GaAs	1.42	PQ	0.31	3.2	2.2
InP	1.35	—	0.53	10.5	2.5
Ge[1]	0.74	Q-P	0.18	2.3	1.4
InAs[2]	0.36	Q-P	1.28	1.6	3.6
InSb[3]	0.28	Q-P	0.41	0.6	5.0
CdTe	1.5	Q-P	0.51	11.0	1.5
ZnSe	2.6	Q-P	—	38.0	1.5
$Ga_{0.5}In_{0.5}Sb$	0.36	Q-P	0.36	0.6	2.5
$Ga_{0.3}In_{0.7}Sb$	0.24	Q-P	—	0.6	2.9
$InAs_{0.2}P_{0.8}$	1.10	Q-P	0.95	5.7	2.7
$Ga_{0.13}In_{0.87}As_{0.37}P_{0.63}$	1.05	—	—	5.5-8.6	1.2

1. At 77 K, (100) -or (110) -oriented. 2. Under 14-k bar pressure. 3. At 77 K under 8-k bar pressure.

7.9. Gunn Diodes

The Gunn diodes are named after J.B. Gunn, who first demonstrated one of the forms of oscillation, are microwave devices which operate by the transferred electron mechanism. These devices are called diodes only because they are two-terminal devices, though no P-N junction is involved. However, the Gunn diode and other related devices use bulk instabilities which do not require junctions.

7.9.1. Fabrication of Gunn Diode

These are simple homogeneous structures which have ohmic contacts on both ends. In practice, however, considerable care is required to be taken in fabricating and mounting workable devices. In addition to the obvious requirements on doping, carrier mobility and sample length, there are important problems with contacts, heat sinking and parasitic reactances of the packaged device. The samples must have high mobility, few lattice defects, and homogeneous doping in the range giving carrier concentrations $n_0 \simeq 10^{19} - 10^{22}$ m^{-3}.

Devices can be fabricated from GaAs or InP bulk samples cut from an ingot, but ingot material is commonly employed as a substrate for an epitaxial layer, which serves as the active region of the device. The material properties of epitaxial layers are usually superior to bulk samples, and the precise control of thickness of layers is helpful in such devices, which need exact lengths of samples. In a typical configuration, an N-type epitaxial layer of approximately 10 μm thickness is grown on an N^+ substrate wafer, which is perhaps of 100 μm thickness. The substrate serves as one of the contacts to the active region. A thin N^{++} (very heavily doped) layer is grown on top of the N-region, so that an N^+-N-N^{++} sandwiched structure is formed. External contacts can be made by evaporation of a thin layer of Au-Sn or Au-Ge on each surface, followed by a brief alloying step in a hydrogen atmosphere. The wafer is divided into individual devices by cutting or cleaving (providing a cube structure) or by selective etching (providing a mesa structure). Each device is mounted with the N^{++} side down on a copper stud or other heat sink, so that the heat generated in the active region can be dissipated to the mount in one direction and to the substrate layer in the other direction. Then the substrate side can be contacted by a wire or pressure contact. In other configurations, planar fabrication techniques can be employed to make lateral devices in an N-type epitaxial layer grown on a high-resistivity substrate.

Power dissipation is very important in these devices because considerable heating (roughly 10^7 W/cm^3) is involved. With the increase in temperature, the characteristics of the device change due to variations in n_0 and mobility. Because of such heating effects, such devices seldom attain their theoretical maximum efficiency. Pulsed operation permits better control of heat dissipation than does continuous operation. In the pulse mode, the efficiencies near theoretical limits can sometimes be achieved. In case continuous operation is not required, peak powers of hundreds of watts can be achieved in pulses of microwave oscillation.

7.9.2. Operation of Gunn Diode

If a sample of GaAs is so biased that the field falls in the negative conductivity region, space charge instabilities are caused, and the device cannot be maintained in a dc stable condition. To understand the formation of such instabilities, we will first consider the dissipation of space charge in the conventional semiconductor. It can be shown that a localized space charge dies out exponentially with time in a homogeneous sample with positive resistance. The instantaneous charge is given as

$$q = Q_o e^{-t/\tau_d} \qquad ...(7.1)$$

where Q_o is the initial space charge and $\tau_d = \in/\sigma$ is known as the dielectric relaxation time. Because of this process, random fluctuations in carrier concentration are quickly neutralized, and space charge neutrality is a good approximation for most of the semiconductors in the usual range of conductivities. For instance, the dielectric relaxation time for a 1.0 Ω-cm Si or GaAs sample is about 10^{-12} s.

Equation (7.1) provides a rather remarkable result in cases in which the conductivity is negative. For such cases τ_d is negative also, and space charge fluctuations build up exponentially in time rather than dying out. This means that normal random fluctuations in the carrier distribution can grow into large space charge regions in the sample. For understanding the occurrence of this phenomenon in a GaAs sample biased in the negative conductivity regime consider velocity-field diagram for N-type GaAs illustrated in Fig. 7.16 (*a*). If we assume a small shift of electron concentration in some region of the device, a dipole layer can form as illustrated in Fig. 7.16 (*b*). Under normal conditions this dipole would die out quickly. However, under conditions of negative conductivity the charge within the dipole, and so the local electric field grows, as depicted in Fig. 7.16 (*c*). No doubt this growth of electric field occurs in a stream of electrons drifting from cathode to anode, and the dipole, now may be called a domain, drifts along with the stream as it grows. Finally the drifting domain will reach the anode, where it gives up its energy as a pulse of current in the external circuit.

During the initial growth of the domain, an increasing fraction of the applied voltage appears across it, at the expense of electric field in the rest of the bar. Consequently, it is not likely that more than one domain will be present in the bar at a time, after the formation of one domain, the electric field in the rest of the bar falls below the threshold value for negative conductivity. In case of a dc bias, the field outside the moving domain will stabilize at a positive conductivity point such as A in Fig. 7.16 (*a*), and the field in the domain will stabilize at the high-field value B.

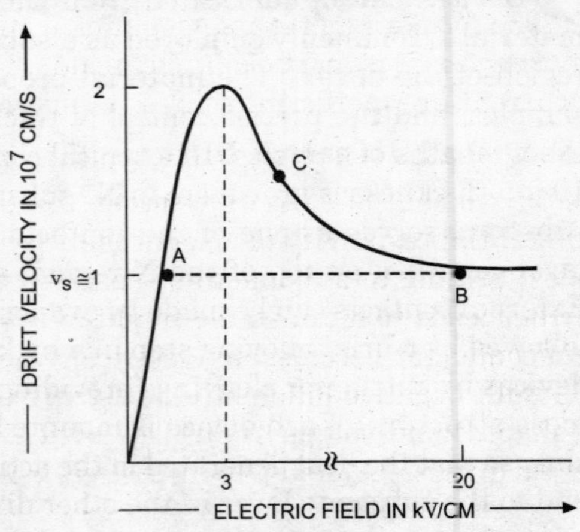

(a) Velocity-Field Characteristic For N-Type GaAs

A small dipole forms from a random noise fluctuation (or more likely at a permanent nucleation site such as a crystal defect, a doping inhomogeneity or the cathode itself), and this dipole grows and drifts down the bar as a domain. During the initial stages of growth of the domain, it can be assumed that there is a uniform electric field in the bar, except just at the small dipole layer. When the field is in the region of negative mobility, such as point C in Fig. 7.16 (*a*), the slightly higher field within the dipole results in a lower value of electron drift velocity inside the dipole in comparison to that outside. The result is that electrons on the right (downstream) of the domain drift away, while electrons pile up on the left-(up-stream) side. This makes the accumulation and depletion layers of the dipole to grow, thereby further strengthening the electric field in the domain. This is obviously a runaway process, in which electric field within the domain builds up while that outside the domain weakened. A stable condition is realized when the field attains the value indicated by point B in Fig. 7.16 (*a*), and the field outside drops to point A. When this condition is fulfilled, the electrons drift at a constant velocity v_s everywhere and the domain moves down the bar without further growth.

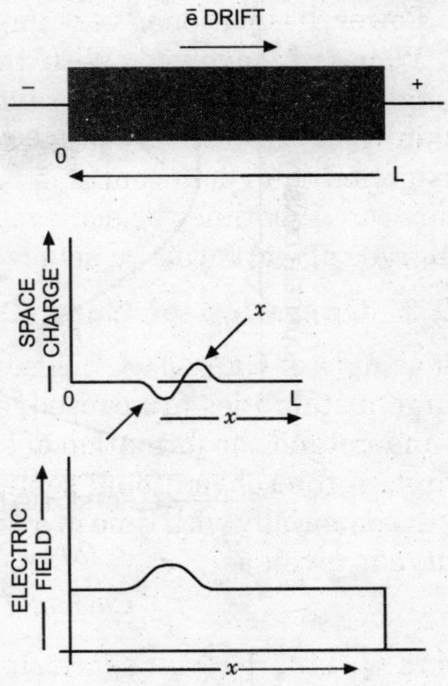

(b) Formation of a Dipole

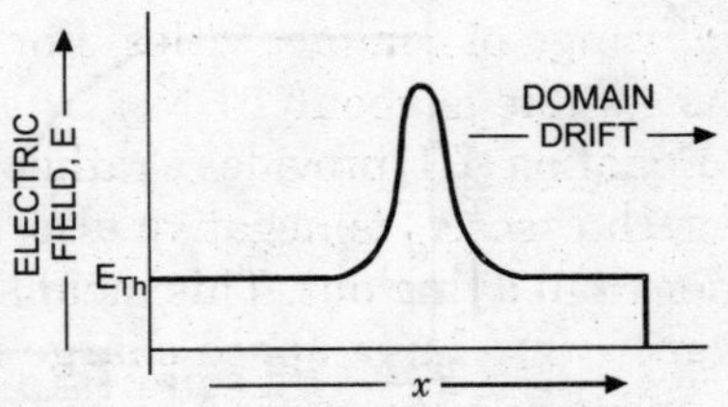

(c) Growth and Drift of a Dipole Under Conditions of Negative Conductivity

Fig. 7.16

In the above discussion it was assumed that the domain has time to build up to its stable condition before it drifts outside the bar. This is not always true, for instance in a short bar with a low concentration of electrons, a dipole can drift the length of the bar before it grows into a domain. Limits can be specified on the electron concentration n_o and sample length L for successful formation of domain by requiring the transit times (L/v_s) exceeding the dielectric relaxation time (absolute value) in the negative mobility region. Thus the requirement is

$$\frac{L}{v_s} > \frac{\epsilon}{e\mu n_o}$$

$$\text{or} \quad Ln_o > \frac{\epsilon v_s}{e\mu} = 10^{16} \text{ m}^{-2} \qquad \ldots(7.2)$$

for N-type GaAs, where the average differential mobility is taken to be – 0.01 m^2/V-s. Thus for successful domain formation, there is a critical product of electron concentration and sample length.

The formation of stable domains is not the only mode of operation for transferred electron devices. Nor it is most desirable mode for most applications, because the resulting short pulses of current are inefficient sources of microwave power.

7.9.3. Bulk Negative Differential Resistivity (Bulk NDR)

The movement of carriers between the upper and lower valleys gives rise to a bulk negative differential resistivity of a semiconductor. This results in a random fluctuation of electrons (or holes) leading to a momentary space charge growing exponentially with respect to time. This further leads to an unstable condition. Due to negative differential resistivity, a homogeneous semiconductor becomes an electrically heterogeneous semiconductor to attain stability.

Bulk negative differential resistivity devices are of two types *viz.* (*i*) current-controlled NDR (S-shaped) and voltage controlled NDR (N-shaped). The characteristics of these two NDRs are shown in Figs. 7.17 and 7.18 respectively. Figures 7.17 (*a*), 7.17(*b*) depict the general current density versus electric field characteristics for current controlled and voltage-controlled NDRs respectively and Figs. 7.17 (*c*) and 7.17(*d*) show the differential resistivities $\left(\frac{\partial E}{\partial J}\right)$ for the two groups.

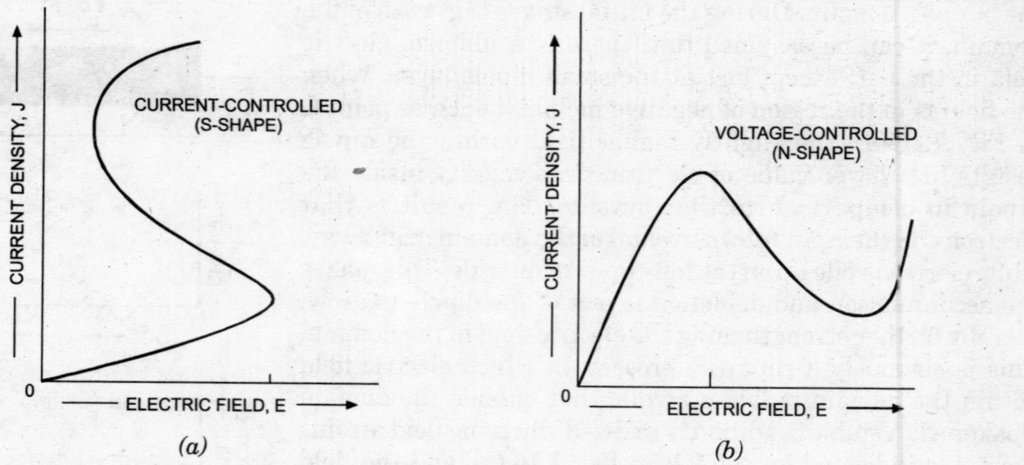

(a) *(b)*

Current Density Versus Electric Field Characteristics

(c) For Current Controlled NDR *(d) For Voltage-Controlled NDR*

Fig. 7.17. ***Differential Resistivities***

For current-controlled NDR, the initial positive differential resistivity decreases with increasing field; *i.e.*, $\frac{d\rho}{dE} < 0$. If a region has a sightly higher field, as depicted in Fig. 7.18 (*a*),

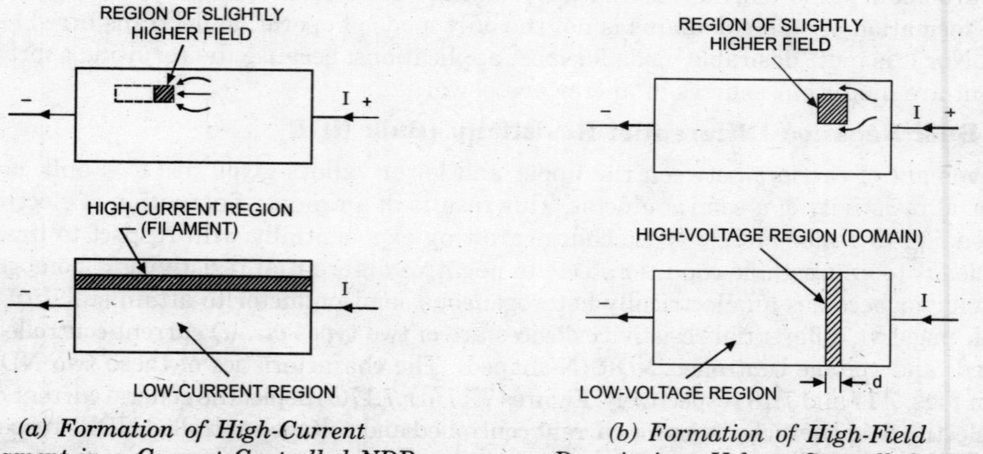

(a) Formation of High-Current Filament in a Current Controlled NDR

(b) Formation of High-Field Domain in a Voltage-Controlled NDR

Fig. 7.18

the resistance reduces. Thus more current will flow into it. This causes an elongation of the region along the current path, and finally in the formation of a high-current filament operating along the field direction.

For the voltage-controlled NDR, the positive differential resistivity increases with the increasing field and for a slightly higher field the resistance is large. Thus less current will flow through it. This causes an elongation of the region, and a high-field domain is formed separating regions of low field. The interfaces separating low- and high-field domains lie along equipotentials, so that they are in planes perpendicular to the current direction.

Highlights

1. A number of devices which are extensively employed in pulse and switching circuitry are most conveniently characterised in terms of volt-ampere curve which displays, over a limited range, a negative incremental resistance. Negative resistance may result in two-ways: Some devices exhibit a characteristic in part of which as the voltage increases, the current drops thereby resulting in a negative resistance while some other devices may exhibit a characteristic in part of which decrease in voltage causes increase in current thereby resulting in a negative resistance. Negative-resistance devices may be voltage controllable type or current controllable type.
2. A tunnel diode is a high-conductivity two-terminal heavily (1000 times higher than a conventional junction diode) doped P-N junction diode. It can conduct in reverse as well as forward direction but is usually used in forward biased mode.

   A tunnel diode's characteristics exhibit a negative resistance region and it may be used as an ultra-fast switching device or in ultra-high frequency oscillators.
3. In relatively pure semiconductors, no charge transport can occur within the donor or acceptor themselves because the few impurity atoms are so widely spaced throughout the sample. But when doping levels are increased, the impurities are very close to each other in the structure leading to interaction within the donors or acceptors. In case the concentration of electrons in the conduction band is greater than the effective density of states, the Fermi level is no longer within the band gap but extends into the conduction band and when this occurs, the material is referred to as *degenerate N-type.* A similar reaction occurs in a degenerate P-type material where the concentration of acceptors is high and the Fermi level extends to the valence band.
4. The impact ionization and transit-time properties of the semiconductor structures are used to develop negative resistance at microwave frequencies.

   Transit time devices can convert dc to microwave ac signals with high efficiency and are very useful in the generation of microwave power for many applications.
5. TED devices are microwave devices and are extensively employed as local oscillators and power amplifiers in the range of 1 to 100 GHz. These are used in RADARs, intrusion alarms, and microwave test instruments.
6. IMPATT diode stands for IMPact Avalanche and Transit Time diode. Any device exhibiting dynamic negative resistance for dc will also exhibit it for ac. When alternating voltage is applied, the current will increase with the decrease in applied voltage. Thus, negative resistance may be redefined as that property of the device which causes the current through it to be 180° out of phase with the voltage applied across it.
7. The transfer of conduction electrons from a high-mobility energy valley to low-mobility, higher-energy satellite valleys is called the *transferred-electron effect*. Although GaAs and InP are the most common materials for TEDs, transferred-electron effects have also been observed in CdTe, ZnSe, GaAsP, and other materials.
8. The Gunn diodes are named after J.B. Gunn, who first demonstrated one of the forms of oscillation, are microwave devices which operate by the transferred electron mechanism. These devices are called diodes only because they are two-terminal devices, though no P-N junction is involved.

However, the Gunn diode and other related devices use bulk instabilities which do not require junctions.

Power dissipation is very important in Gunn devices because considerable heating (roughly 10,000 kW/cm^3) and the increase in temperature results in change of device characteristics.

Exercises

1. What are the basic constructional differences between a conventional P-N junction diode and a tunnel diode ? Explain briefly the operation of a tunnel diode. **[U.P. Technical Univ. 2006-07]**
2. How does a tunnel diode differ from a conventional P-N junction diode ? With the help of its energy-band diagram explain the principle of operation in forward and reverse bias conditions. Sketch its V-I characteristics and mark negative resistance on it.
3. Draw the V-I characteristics of tunnel diode. Briefly describe the mechanism of junction breakdown.
4. What is tunneling ? In what mode the tunnel diodes are usually operated ? Give the construction and V-I characteristics of tunnel diodes.
5. Obtain I-V characteristics of a tunnel diode with special reference to its negative resistance starting from equilibrium condition to forward and reverse bias conditions. **[U.P. Technical Univ. 2005-06]**
6. Explain the tunnel diode in detail. What are its applications ? **[U.P. Technical Univ. Fundamentals of Electronic Devices 2009-10]**
7. Describe briefly the principle of operation of an IMPATT diode. **[U.P. Technical Univ. 2006-07]**
8. What is IMPATT diode ? Give its construction. Give outlines of some IMPATT diode designs.
9. What are transferred electron devices and where they are used ? What is transferred electron effect ? Name the materials in which transferred-electron effect can be observed. Draw the velocity-field characteristics for GaAs at different temperatures.
10. What is Gunn diode ? Why is it called a diode ? How it can be fabricated ? Explain formation and drift of space charge domains.
11. What is the basic difference between GUNN diode and the IMPATT diode ? **[U.P. Technical Univ. 2005-06]**
12. Write short notes on the following:
    (*i*) Tunnel diode.
    (*ii*) Degenerate semiconductors.
    (*iii*) IMPATT diode.
    (*iv*) Transferred electron devices.
    (*v*) Transit time device.
    (*vi*) Gunn diode. **[U.P. Technical Univ. 2006-07]**

Short Answer Type Questions With Answers

Q. 1. What is tunnel diode ?

Ans. Tunnel diode is a highly conductivity two-terminal P-N junction doped heavily (about 1,000 times higher than a normal diode).

Q. 2. What is tunneling ?

Ans. The mechanism of conduction in a semiconductor diode in which charge carriers (possessing very little energy) punch through a barrier directly instead of climbing over it is called tunneling.

Q. 3. What are the applications of tunnel diodes ?

Ans. Tunnel diodes can be used as amplifiers, oscillators or switching devices, being an exclusive high-frequency component because of its very fast response of inputs.

Q. 4. What are degenerate devices ?

Ans. Degenerate devices are the devices in which the doping levels are increased largely, the impurities are close to each other leading to interaction between donors and acceptors.

Q. 5. How does negative resistance develop in transit time devices ?

Ans. The negative resistance occurs because of two processes, causing the current to lag behind the voltage in time:

(*i*) delay due to the avalanche process and (*ii*) a further delay due to the transit time of the carriers across the drift region.

If the sum of above delay times is approximately one half cycle of the operating frequency, negative resistance occurs.

Q. 6. Name a few transit time devices.

Ans. Transit time devices include (*i*) IMPATT (*ii*) BARITT (*iii*) DOVETT and (*iv*) TRAPATT diodes.

Q. 7. What are the applications of transit time devices ?

Ans. Transit time devices can convert dc to microwave signals with high efficiency and are very useful in the generation of microwave power for many applications.

Q. 8. What is transferred-electron effect ?

Ans. The transfer of conduction electrons from a high-mobility energy valley to low-mobility higher energy satellite valleys is called the transferred electron effect.

Q. 9. Name the materials most commonly used for TEDs ?

Ans. GaAs and InP are the materials most commonly used for TEDs.

Q. 10. What are Gunn diode ?

Ans. The Gunn diodes are named after J.B. Gunn, who first demonstrated one of the forms of oscillation, are microwave devices which operate by the transferred electron mechanism. These devices are called diodes only because they are two-terminal devices, though no P-N junction is involved. However, the Gunn diode and other related devices use bulk instabilities which do not require junctions.

Q. 11. Why power dissipation is very important in Gunn devices ?

Ans. Power dissipation is very important in Gunn devices because considerable heating (roughly 10,000 kW/cm^3) and the increase in temperature results in change of device characteristics.

Q. 12. Name the bulk NDR devices.

Ans. Bulk NDR devices are of two types: *viz.* (*i*) current controlled NDR (S-shaped) and voltage-controlled NDR (N-shaped).

❑❑❑

CHAPTER 8

Power Devices

8.1. THYRISTORS AND THYRISTOR FAMILY

Inside This Chapter

Thyristor is the general name given to a family of semiconductor devices having four layers with a control mechanism, although this term is most commonly applied to the SCR (silicon-controlled rectifier). This term is derived from *thyratron* and *transistor* because the device combines the rectification action of thyratron and control action of transistor. Thyristor stands for *THYRatron transISTOR.*

The device was first developed at the Bell Laboratories in the USA and was commercially introduced by the General Electrical Company in the USA in 1957. The device appeared in the market under different names such as SCR, thyristor, thyrode etc. Since its inception the thyristor has come to stay as a basic building block in many industrial and power system applications. Its ability to be controlled, compactness, fast response, high reliability, better efficiency, large power handling capacity, high voltage and current ratings, good trigger sensitivity, static operation, large power gain, sturdy construction, long life, very little maintenance and low cost of fabrication—due to advancement in the field of fabrication—have given the thyristor a colourful reception in every field. Today thyristors are finding applications in the control of dc/ac motors; for the improvement of power factor; and as switching devices. They have become an integral part of HVDC transmission. They are so advantageous that our major steel plants had decided to modernise their plants by replacing MG sets and mercury-arc rectifiers by thyristor converters. Thyristors have helped in further cost reduction and in the development of drive system by changing the emphasis from dc motors to ac motors. With cycle converters and inverters, the speed of an ac motor can also be controlled with ease and reliability. Apart from these main applications it finds use as a switching device, particularly in the improvement of power factor of transmission lines and mains. Thyristor can be used as a power switching device with a power handling capacity ranging from a few watts to as high as 4 MW (2,500 A at 1,600 V). Some thyristors have a rating as high as 400 A, 10,000 V for use in hv ac transmission lines.

Thyristors, with its large number of advantages and tremendous control capabilities, have numerous applications and have completely replaced the electromagnetic control systems.

SCRs and triacs having high voltage and current ratings are widely employed for power control applications whereas other members of thyristor family are employed for small power applications and for switching in control and digital circuits.

8.2. Comparison Between Transistors and Thyristors

As already mentioned, transistors and thyristors are both semiconductor devices. They are now widely employed in switching operations because of their numerous advantages such as noiseless operation owing to absence of moving parts, very high switching speed (say 10^9 operations per second), high efficiency, low maintenance, small size, little weight and trouble free service for long period, large control current range (say from 30 A to 100 A) with small gate current of few mA over mechanical switches or electromechanical relays. However, transistors and thyristors both have their own areas of applications. Thyristors have some advantages over transistors, as enumerated below :

1. Thyristor is a four-layer device while the transistor is a three-layer device.
2. Due to difference in fabrication and operation it is possible to have thyristors with higher voltage and current ratings.
3. Rating of a transistor is always in watts while that of a thyristor is in kWs *i.e.* thyristors have better power handling capacity.
4. A thyristor needs only a pulse to make it conducting and thereafter it remains conducting. On the other hand a transistor needs a continuous current for keeping it in a conducting state.
5. Internal power losses in a thyristor are much smaller than those in a transistor.

Transistors, however, have lower voltage drop, and need no turn-off circuit. Power transistors are linear devices better suited for some linear power applications such as audio and high frequency power amplifiers, regulator systems etc.

8.3. Silicon Controlled Rectifier (SCR)

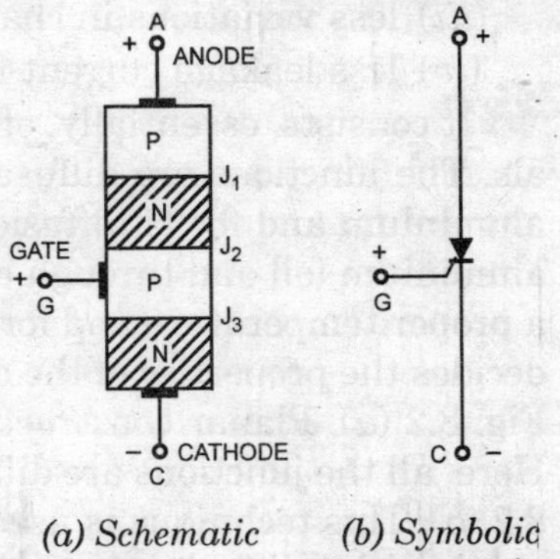

(a) Schematic Diagram *(b) Symbolic Representation*

Fig. 8.1. ***Silicon Controlled Rectifier***

As the terminology indicates, the SCR is a controlled rectifier constructed of a silicon semiconductor material with a third terminal for control purposes. Silicon was chosen because of its high temperature and power capabilities. The basic operation of the SCR is different from that of an ordinary two-layer semiconductor diode in that a third terminal, called a *gate*, determines when the rectifier switches from the open-circuit to short-circuit state. It is not enough simply to forward-bias the anode-to-cathode region of the device. In the conduction state the dynamic resistance of the SCR is typically 0.01 to 0.1 Ω and reverse resistance is typically 100 k Ω or more. It is widely used as a switching device in power control applications. It can control loads by switching *on* and *off* up to many thousand times a second. It can switch *on* for a variable length of time duration, thereby delivering desired amount of power to the load. Thus, it possesses the advantage of a rheostat as well as a switch with none of their drawback. A schematic diagram and symbolic representation of an SCR are shown in Figs. 8.1 (*a*) and 8.1 (*b*) respectively. As illustrated in Fig. 8.1 (*a*), SCR is a three-terminal four-layer semiconductor device, the layers being alternately of P-type and N-type. The junctions are marked J_1, J_2 and J_3 (junctions J_1 and J_3 operate in forward direction while middle junction J_2 operates in the reverse direction) whereas the three terminals are anode (A), cathode (C) and gate (G) which is connected to the inner P-type layer. The function of the gate is to control the firing of SCR. In normal operating conditions, anode is positive with respect to cathode.

8.3.1. Construction

From Fig. 8.2 (*a*) it is clear that SCR is essentially an ordinary rectifier (PN) and a junction transistor (N-P-N) combined in one unit to form PNPN device. Three terminals are taken: one from the outer P-type material, known as *anode*, second from the outer N-type material, known as *cathode* and the third from the base of transistor section known as the *gate*.

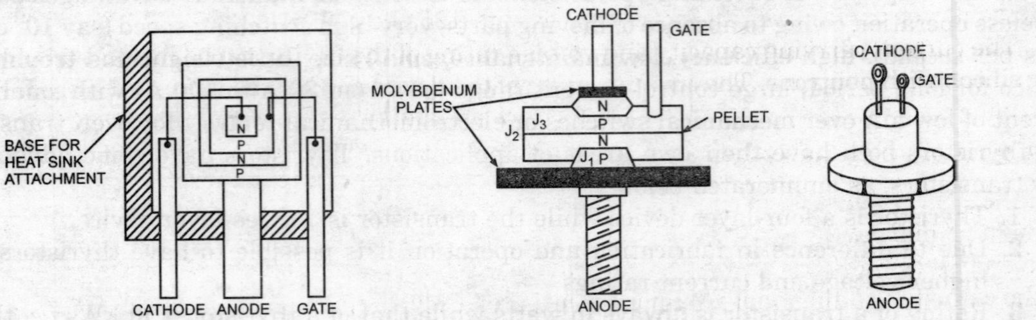

(*a*) *Planar Type Construction* (*b*) *Mesa Type Construction* (*c*) *External Construction*

Fig. 8.2. ***Silicon Controlled Rectifier***

The basic material used for fabrication of an SCR is N-type silicon. It has a specific resistance of about 6 Ω-mm. Silicon is the natural choice as base material because of the following advantages.

(*i*) ability to withstand high junction temperature of the order of 150° C;

(*ii*) high thermal conductivity;

(*iii*) less variations in characteristics with temperature; and

(*iv*) less leakage current in P-N junction.

It consists, essentially, of a four layer pellet of P and N type silicon semiconductor materials. The junctions are diffused or alloyed. The material which may be used for P diffusion is aluminium and for N diffusion is phosphorous. The contact with anode can be made with an aluminium foil and through cathode and gate by metal sheet. Diffusion must be carried out at a proper temperature and for necessary duration to provide correct concentration because this decides the properties of the device. Low power SCRs employ the planar construction shown in Fig. 8.2 (*a*). Planar construction is useful for making a number of units from a silicon wafer. Here, all the junctions are diffused. The other technique is the mesa construction shown in Fig. 8.2 (*b*). This technique is used for high power SCRs. In this technique, the inner junction J_2 is obtained by diffusion, and then the outer two layers are alloyed to it. The PNPN pellet is properly braced with tungsten or molybdenum plates to provide greater mechanical strength and make it capable of handling large currents. One of these plates is hard soldered to a copper or an aluminium stud, which is threaded for attachment to a heat sink. This provides an efficient thermal path for conducting the internal losses to the surrounding medium. The uses of hard solder between the pellet and back-up plates minimises thermal fatigue, when the SCRs are subjected to temperature induced stresses. For medium and low power SCRs, the pellet is mounted directly on the copper stud or casing, using a soft solder which absorbs the thermal stresses set up by differential expansion and provides a good thermal path for heat transfer. For a larger cooling arrangement, which is required for high power SCRs, the press-pack or hockey-puck construction is employed, which provides for double-sided air for cooling.

The salient features to be considered, while designing an SCR, are the diameter and thickness of wafer, composition of the base material, type and amount of the material to be diffused into the wafer, shape, position and contact area of the gate, shape and size of the SCR, type of heat sink etc.

Fabrication technology determines various properties of the device. The voltage rating of a device can be increased by lightly doping the inner two layers and increasing their thickness. But due to this increased resistance, forward voltage drop increases and large triggering currents are required causing greater power dissipation accompanied by smaller current ratings. The heat dissipation of silicon falls from 1.5 W/cm^2 at 25°C to 1.25 W per cm^2 at 125°C. A high voltage power device can seldom be used beyond 125°C.

The current carrying capacity and voltage rating of the device can be increased by irradiating silicon with neutrons. The current rating of the device can also be increased by reducing the current density at the junction but this results in a bulky device with large *turn-on* time.

8.3.2. Principle of Operation

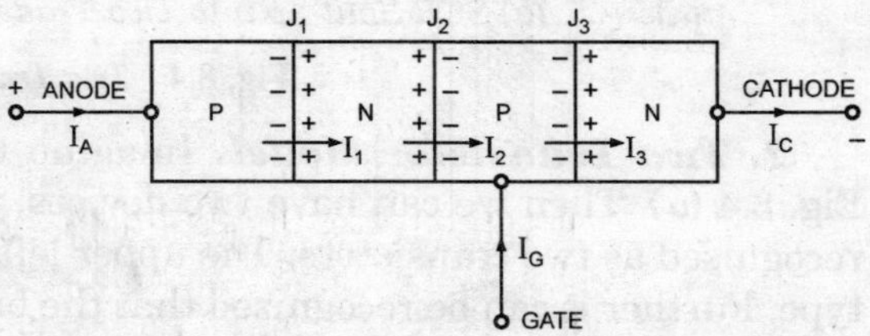

Fig. 8.3. *Diagrammatic Representation Showing Current Flow and Voltage Bias in an SCR*

The SCR is a four-layer, three-junction and a three-terminal device and is shown in Fig. 8.1 (*a*). The end P-region is the anode, the end N-region is the cathode and the inner P-region is the gate. The anode to cathode is connected in series with the load circuit. Essentially the device is a switch. Ideally it remains off (voltage blocking state), or appears to have an infinite impedance until both the anode and gate terminals have suitable positive voltages with respect to the cathode terminal. The thyristor then *switches on* and current flows and continues to conduct without further gate signals. Ideally the thyristor has zero impedance in conduction state. For switching off or reverting to the blocking state, there must be no gate signal and the anode current must be reduced to zero. Current can flow only in one direction.

In absence of external bias voltages, the majority carriers in each layer diffuses until there is a built-in voltage that retards further diffusion. Some majority carriers have enough energy to cross the barrier caused by the retarding electric field at each junction. These carriers then become minority carriers and can recombine with majority carriers. Minority carriers in each layer can be accelerated across each junction by the fixed field, but because of absence of external circuit in this case the sum of majority and minority carrier currents must be zero.

A voltage bias, as shown in Fig. 8.3, and an external circuit to carry current allow internal currents which include the following terms:

The current I_1 is due to (*i*) majority carriers (holes) crossing junction J_1; (*ii*) minority carriers crossing junction J_1; (*iii*) holes injected at junction J_2 diffusing through the N-region and crossing junction J_1; and (*iv*) minority carriers from junction J_2 diffusing through the N-region and crossing junction J_1.

Similarly I_2 is due to six terms and I_3 is due to four terms.

The two simple analogues to explain the basic action for the thyristor are those of the diode and the two transistor models.

1. *Diode Model*. The thyristor is similar to three diodes in series as there are three P-N junctions. Without gate bias, there is always at least one reverse biased junction to prevent conduction irrespective of the polarity of an applied voltage between anode and cathode. If the anode is made positive and the gate is also biased positively with respect to cathode, the P-layer at the gate is flooded by the electrons from the cathode and loses its identity as a P-layer. Accordingly the thyristor becomes equivalent to a conducting diode.

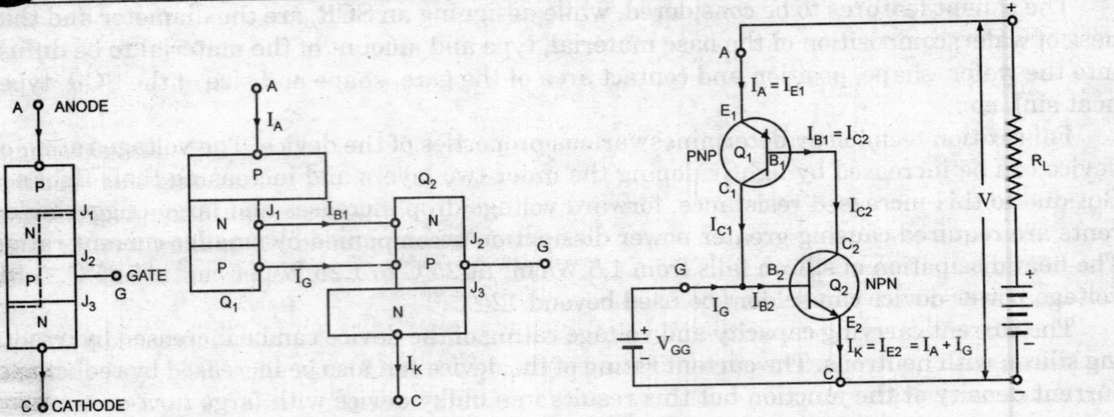

(a) (b) SCR Split-up into Two Transistors (c) Two Transistor Equivalent Circuit of An SCR

Fig. 8.4 *Two Transistor Model of An SCR*

2. *Two Transistor Model*. Imagine the SCR cut along the dotted line, as shown in Fig. 8.4 (*a*). Then we can have two devices, as shown in Fig. 8.4 (*b*). These two devices can be recognised as two transistors. The upper left one is P-N-P transistor and the lower right N-P-N type. Further it can be recognized that the base of the P-N-P transistor is joined to the collector of the N-P-N transistor while the collector of P-N-P is joined to the base of N-P-N transistor, as illustrated in Fig. 8.4 (*c*). The gate terminal is brought out from the base of the N-P-N material. This construction has been conceived merely to explain the working of SCR, otherwise in physical shape the SCR has four solid layers of P-N-P-N type only.

Now we can see that the two transistors are connected in such a manner that the collector of Q_1 is connected to the base of Q_2 *i.e.* the output collector current of Q_1 becomes the base current for Q_2. In the similar way the collector of Q_2 is joined to the base of Q_1 which shows that the output collector current of Q_2 is fed to Q_1, as input base current. These are back to back connections of transistors in such a way that the output of one goes into as input of other transistor and vice-versa. This gives net gain of loop circuit as $\beta_1 \times \beta_2$ where β_1 and β_2 are current gains of two transistors respectively.

When the gate current is zero or the gate terminal is open, the only current in circulation is the leakage current, which is very small in case of silicon device specially and the total current is a little higher than sum of individual leakage currents. Under these conditions P-N-P-N device is said to be in its forward blocking or high impedance 'off' state. As soon as a small amount of gate current is given to the base of transistor Q_2 by applying forward bias to its base-emitter junction, it generates the collector current as β_2 times the base current. This collector current of Q_2 is fed as input base current to Q_1 which is further multiplied by β_1 times as I_{C_1} which forms input base current of Q_2 and undergoes further amplification. In this way both transistors feedback each other and the collector current of each goes on multiplying. This process is very quick and soon both the transistors drive each other to saturation. Now the device is said to be in *on-state*. The current through the on-state SCR is controlled by external impedance only.

Analysis of Two Transistor Model. If I_{E_1} and I_{B_1} are the emitter current and base current respectively of the P-N-P transistor; I_A and I_K are the respective anode current and cathode current of the SCR; I_{C_1} and I_{C_2} are the collector currents of P-N-P and N-P-N transistors respectively; I_G is the gate current of the SCR; and α_1 and α_2 are the fraction of holes and electrons injected from anode and cathode on junction J_2 respectively, then neglecting leakage currents of transistors we have

$$I_{B_1} = I_A - I_{C_1} = I_A - \alpha_1 I_A = (1-\alpha_1) I_A \quad ...(8.1)$$

$$\text{Also } I_{C_2} = \alpha_2 I_K \text{ and } I_{B_1} = I_{C_2} \text{ or } (1-\alpha_1) I_A = \alpha_2 I_K \quad ...(8.2)$$

$$\text{Since } I_K = I_A + I_G,$$

$$\frac{(1-\alpha_1) I_A}{\alpha_2} = I_A + I_G$$

$$\text{or } (1-\alpha_1) I_A = \alpha_2 (I_A + I_G)$$

$$\text{or } I_A = \frac{\alpha_2 I_G}{1-(\alpha_1+\alpha_2)} \quad ...(8.3)$$

From above Eq. (8.3) it is obvious that if $(\alpha_1 + \alpha_2)$ becomes equal to unity, then anode current I_A attains infinite value. In other words, the device suddenly latches into conduction state from the original non-conducting state. The reason for such a behaviour is the regenerative manner in which the two constituent transistors Q_1 and Q_2 are interconnected. Thus if the gate current I_G is of such a magnitude that $\alpha_1 + \alpha_2 = 1$, the SCR will be triggered.

The value of $(\alpha_1 + \alpha_2)$ can be made almost unity by a positive gate current of proper magnitude and for short duration at the gate. Signal I_G when applied to the gate causes a flow of current in transistor Q_2. This gives rise to the collector current I_{C_2} in transistor Q_2. Now the transistor Q_1 base is energized because its base current is the collector current of transistor Q_2 and so first transistor is switched *on*. Thus each of the transistor supplies base current to the other and action is, therefore, regenerative. At this point even if the gate signal is removed, the device does not turn *off* as long as the magnitude of the current flowing does not fall below that of holding current.

It implies that when the device is forward biased it can be triggered by applying a small duration positive pulse at the gate. Once the device latches into the conducting state, the gate loses its control and the device remains in the conducting state, even if the gate is opened, till the current level is maintained to the minimum of that of the holding current.

When $(\alpha_1 + \alpha_2)$ becomes appreciably less than unity, the anode current I_A, according to Eq. (8.3), becomes extremely small and the device is said to be in the *'off'* or *non-conducting state*. On the other hand, with $(\alpha_1 + \alpha_2) = 1$, the current is extremely large and the device is said to be in the *on* or *conducting state*. In conducting state, the voltage drop across the device drops to a low value and a large current flows through it, limited only by the resistor R in the external circuit. Thus the existence of the device either in the *'on' state* or in the *'off' state* depends on the applied voltage and the gate current.

8.3.3. Volt-Ampere Characteristics

As already mentioned, the SCR is a four-layer device with three terminals, namely, the anode, the cathode and the gate. When the anode is made positive with respect to the cathode, junctions J_1 and J_3 are forward biased and junction J_2 is reverse-biased and only the leakage current will flow through the device. The SCR is then said to be in the *forward blocking state* or in the *forward mode* or *off state*. But when the cathode is made positive with respect to the anode, junctions J_1 and J_3 are reverse-baised, a small reverse leakage current will flow through the SCR and the SCR is said to be in the *reverse blocking state* or *in reverse mode*.

When the anode is positive with respect to cathode *i.e.* when the SCR is in forward mode, the SCR does not conduct unless the forward voltage exceeds certain value, called the *forward breakover voltage*, V_{FBO}. In non-conducting state, the current through the SCR is the leakage current which is very small and is negligible. If a positive gate current is supplied, the SCR can become conducting at a voltage much lesser than forward breakover voltage. The larger the gate current, lower the breakover voltage [Fig. 8.5(*c*)]. With sufficiently large gate current, the SCR behaves identical to PN rectifier. Once the SCR is switched *on*, the forward voltage drop across it is suddenly reduced to very small value, say about 1 volt. In the *conducting* or *on-state*, the current through the SCR is limited by the external impedance.

When the anode is negative with respect to cathode *i.e.* when the SCR is in reverse mode or in blocking state no current flows through the SCR except very small leakage current of the order of few microamperes. But if the reverse voltage is increased beyond a certain value, called the *reverse breakover voltage*, V_{RBO}, *avalanche breakdown* takes place. Forward breakover voltage V_{FBO} is usually higher than reverse breakover voltage, V_{RBO}.

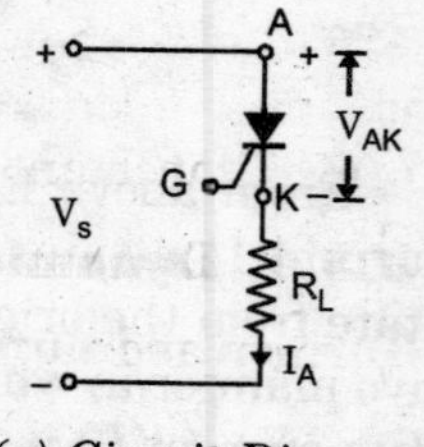

(a) Circuit Diagram

From the foregoing discussion, it can be seen that the SCR has two stable and reversible operating states. The change over from *off-state* to *on-state*, called *turn-on*, can be achieved by increasing the forward voltage beyond V_{FBO}. A more convenient and useful method of *turn-on* the device employs the gate drive. If the forward voltage is less than the forward breakover voltage, V_{FBO}, it can be *turned on* by applying a positive voltage between the gate and the cathode. This method is called the *gate control*. Another very important feature of the gate is that once the SCR is triggered to *on-state* the gate loses its control.

The switching action of gate takes place only when

(*i*) SCR is forward biased *i.e.* anode is positive with respect to cathode, and

(*ii*) suitable positive voltage is applied between the gate and the cathode.

Once the SCR has been switched *on*, it has no control on the amount of current flowing through it. The current through the SCR is entirely controlled by the external impedance connected in the circuit and the applied voltage. There is, however, a very small, about 1 V, potential drop across the SCR. The forward current through the SCR can be reduced by reducing the applied voltage or by increasing the circuit impedance. There is, however, a minimum forward current that must be maintained to keep the SCR in conducting state. This is called the *holding current* rating of SCR. If the current through the SCR is reduced below the level of holding current, the device returns to *off-state* or blocking state.

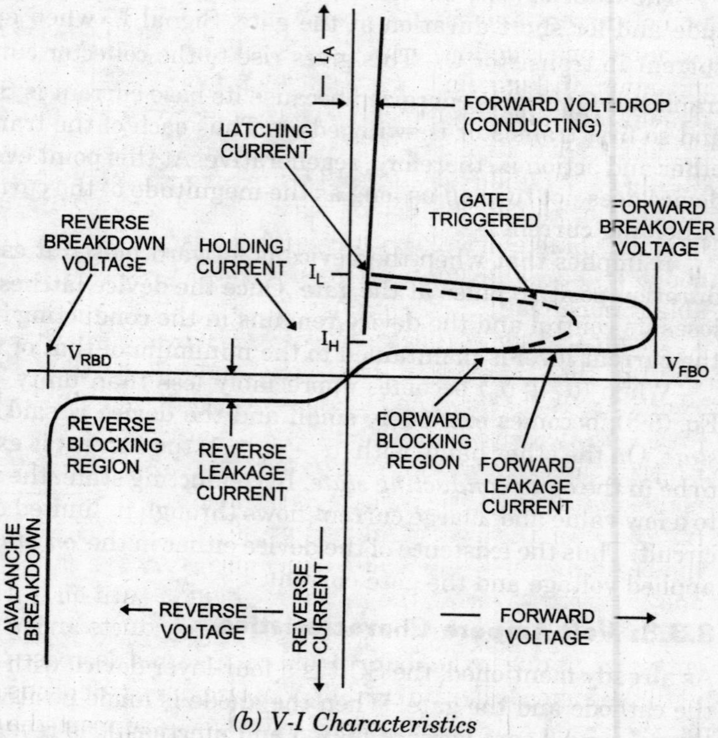

(b) V-I Characteristics

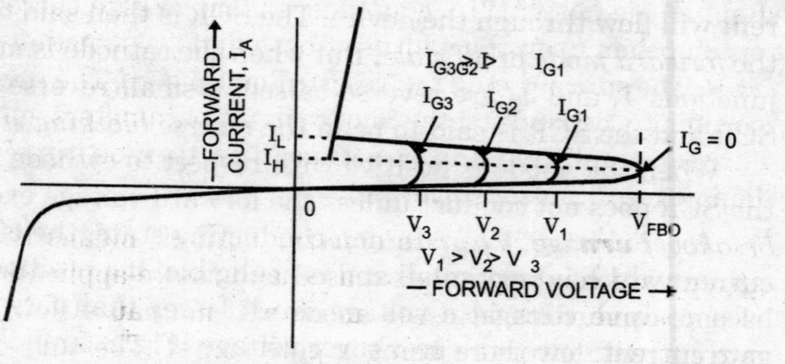

Fig. 8.5 *(c) Effects of Gate Current On Forward Blocking Voltage*

The SCR can be switched *off* by reducing the forward current below the level of holding current which may be done either by reducing the applied voltage or by increasing the circuit impedance.

Note : *The gate can only trigger or switch-on the SCR, it cannot switch off.*

Alternatively the SCR can be *switched off* by applying negative voltage to the anode (reverse mode), the SCR naturally will be *switched off*.

Here one point is worthmentioning, the SCR takes certain time to switch *off*. The time, called the *turn-off time,* must be allowed before forward voltage may be applied again otherwise the device will switch *on* with forward voltage without any gate pulse. The *turn-off* time is about 15 microseconds, which is immaterial when dealing with power frequency, but this becomes important in the inverter circuits, which are to operate at high frequency.

8.3.4. Dynamic Characteristics

Turn-on and turn-off characteristics of an SCR are called the *dynamic* characteristics of the SCR.

(a) Turn-On Characteristic. The turn-on time characteristic shows the variation of current and voltage during turn-on. Turn-on time is defined as the time from the initiation of triggering, when the SCR offers infinite impedance to the flow of anode current, to the time when an equilibrium charge distribution is established throughout the device together with a steady state forward voltage drop. The turn-on time is about 1 to 3 μs for the SCR readily available commercially.

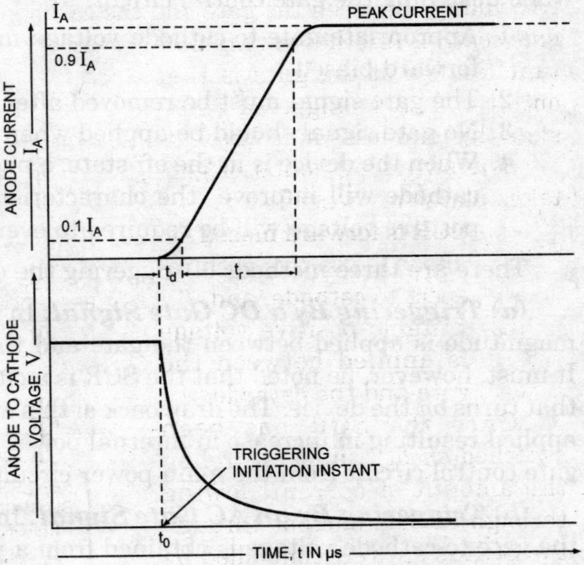

Fig. 8.6

Figure 8.6 shows the form of current rise during the transition from the non-conducting to the fully conducting state. The time t_0 indicates the initiation of turn-on which is caused by a step function of voltage applied to the gate. The period t_d is the delay time between the front of the gate pulse and the beginning of a rapid rate of increase of anode current. During the initial interval of turn-on (t_d) only a small area near the gate electrode conducts anode current. t_d is due to the charge build-up time and decreases with the increase in gate current. The rise time, t_r depends on the speed at which charges build up to support the final anode current, I_A. t_r is independent of gate current but depends upon the widths of base and construction of SCR. The sum of t_d and t_r gives turn-on time t_{on}. The turn-on time t_{on} depends on the anode circuit parameters, gate current amplitude and rise time. The turn-on time is important in pulse triggering.

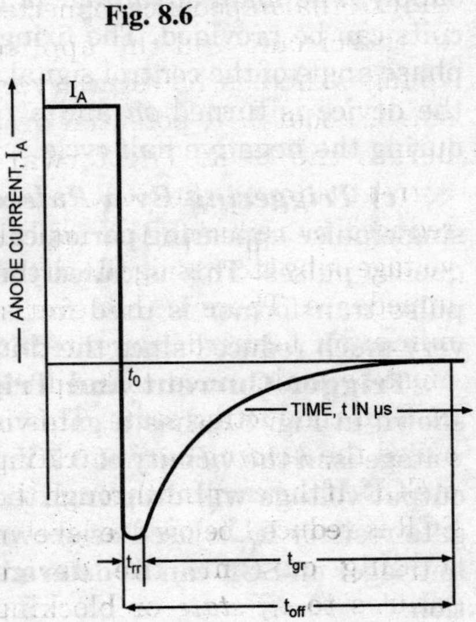

Fig. 8.7

(b) Turn-off Characteristic. Turn-off means that all forward conduction has ceased and the re-application of a positive voltage to the anode will not cause flow of current without there being a gate signal. The turn-off characteristic is shown in Fig. 8.7. The time t_0 indicates the instant of application of reverse voltage, t_{rr} is the duration for which the reverse recovery current flows after application of reverse voltage and t_{gr} is the time required for the recombination of all excess carriers in the inner two layers of the device. The sum of time duration t_{rr} and time duration t_{gr} gives the turn-off time t_{off} *i.e.* $t_{off} = t_{rr} + t_{gr}$. At the end of turn-off time, a depletion layer gets developed across junction J_2 and now SCR is capable of withstanding forward voltage. The turn-off time t_{off} depends upon anode current, magnitude of reverse voltage and the rate of application of forward voltage.

8.3.5. Turning-on Methods

As mentioned earlier, the SCR can be switched on either by increasing the forward voltage beyond forward breakover voltage V_{FBO} or by applying a positive gate signal when the device is forward biased. Of these two methods, the latter called the *gate control method*, is used as it is more efficient and easy to implement for power control. The following points have to be noted when designing the gate control circuit.

1. Appropriate gate-to-cathode voltage must be applied for turn-on when the device is forward biased.
2. The gate signal must be removed after the device is turned-on.
3. No gate signal should be applied when the device is reverse-biased.
4. When the device is in the off-state, a negative voltage applied between the gate and the cathode will improve the characteristics of the device. In such an instance, a large positive voltage will be required to overcome this negative bias for turn-on.

There are three methods of triggering the device by gate control.

(a) Triggering By a DC Gate Signal. In this method, a dc voltage of proper polarity and magnitude is applied between the gate and the cathode when the device is to be turned *on*. It must, however, be noted that the SCR is a current-operated device and it is the gate current that turns *on* the device. The drawback of this method is that gate signal has to be continuously applied resulting in increase in internal power dissipation and that there is no isolation of the gate control circuit from the main power circuit.

(b) Triggering By an AC Gate Signal. In many power control circuits that use ac input, the gate-to-cathode voltage is obtained from a phase-shifted ac voltage derived from the main supply. The main advantage of this method is that proper isolation of power and control circuits can be provided. The firing angle control is obtained very conveniently by changing the phase angle of the control signal. However, the gate drive is maintained for one half cycle after the device is turned *on* and a reverse voltage is applied between the gate and the cathode during the negative half cycle.

(c) Triggering By a Pulsed-Gate Signal. In this method, the gate drive consists of a single pulse appearing periodically, or a sequence of high-frequency pulses. This is called the carrier *frequency gating*. A pulse transformer is used for isolation. The gate losses are very much reduced since the drive is discontinuous.

Trigger Current and Trigger Voltage. An SCR, as shown in Fig. 8.8, has a gate voltage of V_G. When this gate voltage is in the vicinity of 0.7 V, the SCR will turn *on* and the output voltage will drop from + V_{CC} to a low value. When a gate resistor R_G is used as shown, the input voltage required to trigger an SCR can be determined from the following equation:

$$V_{IN} = V_T + I_T R_G \quad ...(8.4)$$

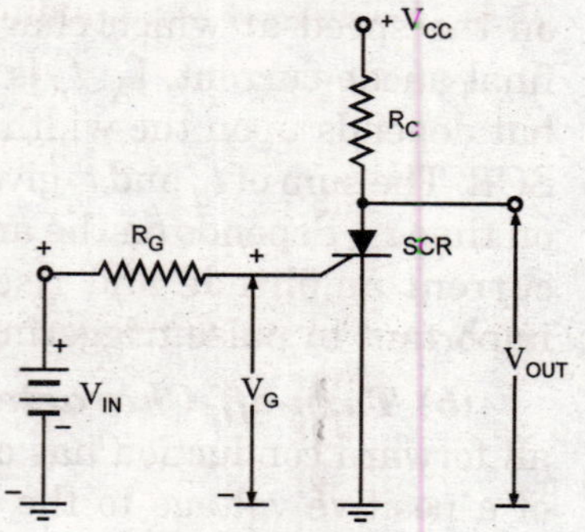

Fig. 8.8. ***An SCR Circuit***

where V_T and I_T are the trigger voltage and trigger current respectively, required for the device gate. This information is available on data sheets. For instance, the data sheet of a 2 N4441 gives $V_T = 0.75$ V and $I_T = 10$ mA. Sometimes a gate resistor is not used. In this case, R_G is the Thevenin's resistance of the circuit driving the gate. Unless Eq. (8.4) is satisfied, the SCR cannot turn *on*. After the SCR has turned *on*, it stays *on* even V_{IN} is reduced to zero. In this case, the output voltage remains low indefinitely.

8.3.6. Turning-off Methods

As already mentioned, once the SCR is fired, it remains *on* even when triggering pulse is removed. This ability of the SCR to remain *on* even when gate current is removed is referred to as *latching*. So SCR cannot be *turned off* by simply removing the gate pulse.

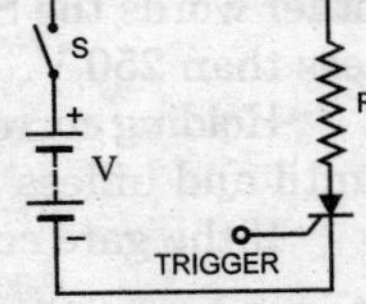

Fig. 8.9

There are three methods of switching *off* the SCR, namely *natural commutation*, *reverse bias turn-off*, and *gate turn-off*.

(a) *Natural Commutation*. When the anode current is reduced below the level of the holding current, the SCR turns *off*. However, it must be noted that rated anode current is usually larger than 1,000 times the holding value. Since the anode voltage remains positive w.r.t. the cathode in a dc circuit, the anode current can only be reduced by opening the line switch S (Fig. 8.9), increasing the load impedance R_L or shunting part of the load current through a circuit parallel to the SCR, *i.e.*, short-circuiting the device.

(b) *Reverse-bias Turn-off*. A reverse anode to cathode voltage (the cathode is positive with respect to the anode) will tend to interrupt the anode current. The voltage reverses every half cycle in an ac circuit, so that an SCR in the line would be reverse biased every negative cycle and would turn *off*. This is called *phase commutation* or ac *line commutation*. To create a reverse biased voltage across the SCR, which is in the line of a dc circuit, capacitors can be used. The method of discharging a capacitor in parallel with an SCR to turn *off* the SCR is called *forced commutation*.

In power electronic applications one advantage of using SCRs is that they are compact. The control equipment is also compact if integrated circuits are used. There has also been an attempt to miniaturize capacitors used for forced commutation and for filtering. The former use is important because the currents can be high and thermal dissipation takes high priority in design considerations. Small sizes of capacitors are at present being achieved by the use of matallized plastic film or a plastic film and aluminium foil.

(c) *Gate Turn Off*. In some specially designed SCRs the characteristics are such that a negative gate current increases the holding current so that it exceeds the load current and the device turns *off*. The current ratings are presently below 10 A and this type will not be considered further.

8.3.7. Normal Operating Conditions

For normal operation of an SCR, the points to be kept in view are :

1. The supply voltage is to be kept much less than breakover voltage, otherwise it will get damaged.
2. The SCR is turned *on* by passing an appropriate gate current and not by applying supply voltage more than breakover voltage.
3. In case of operation with ac the peak reverse voltage should not exceed the reverse breakdown voltage.
4. While turning *off* the SCR from on state, the anode current should be reduced to holding current.
5. For high value of gate current, the SCR will close at much reduced supply voltage.

Example 8.1. An SCR has a breakover voltage of 250 V, a trigger current of 12 mA and holding current of 12 m A. Explain the inferences you make from these datas. What will happen if the gate current is 16 mA ?

Solution : As defined above breakover voltage is the minimum forward voltage, with gate open, at which SCR starts conducting heavily. So, breakover voltage of 250 V means that if the gate is open the SCR will not start conducting until and unless the voltage applied becomes equal to or exceeds 250 V.

Trigger current of 12 mA means that the SCR will start conducting heavily only when a minimum gate current of 12 mA is passed in cases the applied voltage is less than breakover voltage of 250 V. In other words the SCR will not start conducting if trigger current is less than 12 mA and voltage applied is less than 250 V.

Holding current of 12 mA means the SCR will not stop conducting, even if triggering current is removed, until and unless the anode current is reduced to a value below 12 mA by reducing the supply voltage.

If the gate current is increased to 16 mA, the SCR will start conducting with lower supply voltage.

8.3.8. Typical SCR Parameters

Commercially available SCRs have a wide range of voltage and current ratings, turn-on and turn-off times and dynamic resistances in cut-off and in saturation regions. Typical values of these parameters for SCRs are given below in tabular form.

TABLE 8.1. Typical SCR Parameters

S.No.	Parameters	Typical Values
1.	Forward Breakover voltage, V_{FBO}	50 to 500 volts
2.	Maximum on-state voltage	About 1.5 V
3.	Peak reverse voltage, PRV	Upto 2.5 kV
4.	Holding voltage, V_H	0.5 to 20 volts
5.	Forward breakover current	Less than a few hundred μA
6.	Peak forward current	30 A to over 100 A
7.	Holding current	A few mA to few hundred mA
8.	Turn-on and turn-off times	A few tenths of μs for fast acting SCRs A few μs for slow acting SCRs
9.	Dynamic resistance in cut-off region	A few MΩ to a few hundred MΩ
10.	Dynamic resistance in saturation region	Lesser than 1 Ω for currents of several amperes Lesser than 10 Ω for large currents

8.4. SCR as a Switch

We have seen that SCR operates either in *on-state* or in *off-state* and no other state in between *i.e.* SCR behaves like a mechanical switch. As such it is called *electronic switch*.

An SCR has following advantages over a mechanical switch or electromechanical relay:

1. Noiseless operation owing to absence of moving parts.
2. Very high switching speed (say 10^9 operations per second).
3. High efficiency.
4. Low maintenance.
5. Small size and trouble free service for long period.
6. Large control current range (say from 30 A to 100 A) with small gate current of few mA.
7. Long life as no wear and tear is involved.

However, SCR suffers from the following drawbacks:

1. Cut-off current is not exactly zero.
2. There is some voltage drop across SCR when in *on-state*; hence there is some wastage of power.
3. It is more costly and need more care in handling.

8.5. P-N-P-N OR SHOCKLEY DIODE

The four-layer diode, also called the *Shockley diode* after its inventor Willian Shockley, is essentially a low-current SCR without a gate. It is classified as a diode because it has only two external terminals viz. anode and cathode. Because of its four doped regions it is often called a P-N-P-N diode. The basic structure, two transistor version, equivalent transistor circuit and schematic symbol of a Shockley diode are shown in Fig. 8.10.

The easiest way to understand how it operates is to visualize it to be formed of two transistors Q_1 and Q_2 placed back to back, as shown in Fig. 8.10 (*b*). Figure 8.10 shows the diode split-up into two parts, displaced mechanically but connected electrically. The left half is a P-N-P transistor while the right half is an N-P-N transistor. It may be seen that N-type base region of P-N-P transistor forms the collector of N-P-N transistor while P-type base of N-P-N transistor forms the collector of P-N-P transistor. Thus the four-layer diode is equivalent to the latch shown in Fig. 8.10.

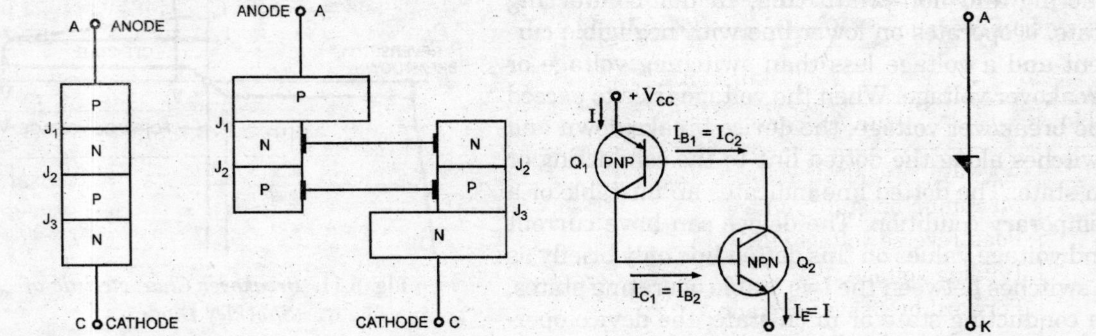

(*a*) *Basic Structure* (*b*) *Two Transistor Version* (*c*) *Equivalent Transistor Circuit* (*d*) *Schematic Symbol*

Fig. 8.10. *Shockley Diode*

Collector currents I_{C_1} are I_{C_2} for transistors Q_1 and Q_2 are indicated. In the active region the collector current is given by

$$I_C = \alpha I_E + I_{CO} \quad \text{...(8.5)}$$

where I_E is the emitter current, I_{CO} is the reverse saturation current, and α is the short-circuit common-base forward current gain.

Applying above Eq. (8.5), in turn, to transistors Q_1 and Q_2 we have

$$I_{C_1} = \alpha_1 I + I_{CO_1} \quad \text{...(8.6)} \because I_{E_1} = I$$

$$\text{and } I_{C_2} = \alpha_2 I + I_{CO_2} \quad \text{...(8.7)} \because I_{E_2} = I$$

But the sum of currents I_{C_1} and I_{C_2} is equal to the total current through the device *i.e.*

$$I_{C_1} + I_{C_2} = I \quad \text{...(8.8)}$$

Solving above equations we have

$$(\alpha_1 + \alpha_2)\, I + I_{CO_1} + I_{CO_2} = I$$

$$\text{or } I = \frac{I_{CO_1} + I_{CO_2}}{1 - (\alpha_1 + \alpha_2)} \quad \text{...(8.9)}$$

From above Eq. (8.9) it is obvious that the current I through the device is small (approximately the sum of collector saturation currents of the two equivalent transistors) as long as the sum $(\alpha_1 + \alpha_2)$ is small as compared to unity. It seems that the current I increases rapidly when the sum $(\alpha_1 + \alpha_2)$ approaches unity. The current does not increase beyond limit as Eq. (8.9) implies, however, because the derivation is no longer valid as $(\alpha_1 + \alpha_2)$ approaches unity. Since junction J_2 becomes forward biased in the forward-conducting state, both transistors become saturated after switching. The two transistors remain in saturation while the device is in the forward-conducting state, being held in saturation by the device current.

Because there are no trigger inputs, the only way to switch the device *on* is to increase the anode-to-cathode voltage V_{AK} to the forward switching voltage, and the only way to open it is by low current drop out. With a four layer diode it is not necessary to reduce the current all the way to zero to open the latch. The internal transistors of the device will come out of saturation when the current is reduced to a low value, called the *holding current*. The *forward switching voltage* V_s is equivalent of the SCR *forward breakover voltage*, and the minimum current at which device will switch *on* is the *switching current* I_s.

Breakover Characteristic. Voltage-current characteristic of a Shockley diode is shown in Fig. 8.11. The device has two operating states: conducting and non-conducting. In non-conducting state, it operates on lower line with negligible current and a voltage less than switching voltage or breakover voltage. When the voltage tries to exceed the breakover voltage, the device breaks down and switches along the dotted line to the conducting or on-state. The dotted line indicates an unstable or a temporary condition. The device can have current and voltage values on this dotted line only briefly as it switches between the two stable operating states. In conducting state or in on-state, the device operates on the upper line. As long as the current through the device is greater than the holding current I_H, then the voltage across it is slightly greater than knee voltage, V_K. When the current falls below the level of the holding current I_H, the device switches back along the dotted line to the non-conducting or off-state.

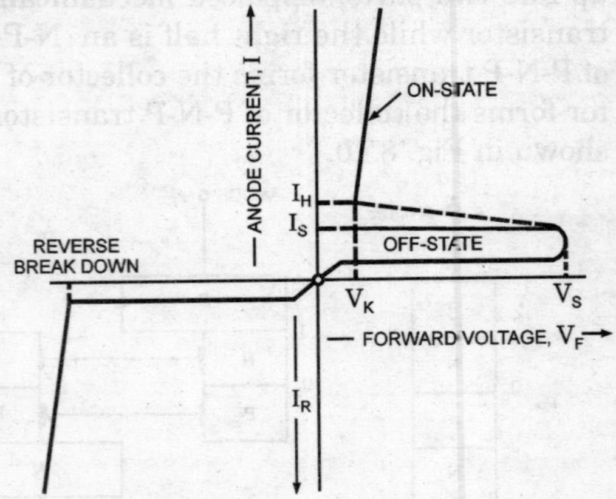

Fig. 8.11. *Breakover Characteristic of Shockley Diode*

One common application of the Shockley diode is as a trigger switch for an SCR. The circuit is shown in Fig. 8.12. When the circuit is energized, the capacitor will start getting charged and eventually, the voltage across the capacitor will be sufficiently high to first turn-on Shockley diode and then the SCR.

Another application of this diode is as a relaxation oscillator.

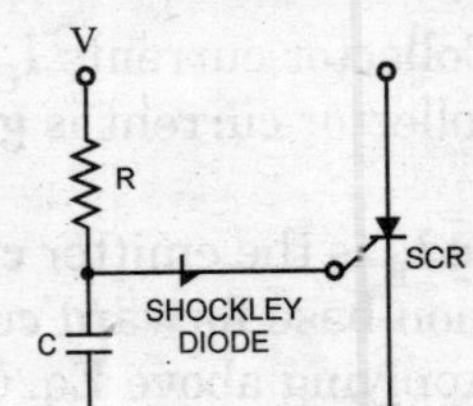

Fig. 8.12. *Shockley Diode as a Trigger Switch*

8.6. Silicon Controlled Switch (SCS)

Silicon controlled switch (SCS), like the SCR, is a unilateral, four layer, three junction P-N-P-N silicon device with four electrodes namely cathode C, cathode gate G_1, anode gate G_2 and the anode A, as shown in Fig. 8.13. Infact, the SCS is a low power device compared with the SCR. It handles currents in milliamperes rather than amperes. SCS differs from an SCR in the following aspects.

1. It has an additional gate—the anode gate.
2. It is physically smaller than SCR.
3. It has smaller leakage and holding currents than SCR.
4. It needs small triggering signals.
5. It gives more uniform triggering characteristics from sample to sample.

The basic structure and schematic symbol of SCS are shown in Figs. 8.13 (*a*) and 8.13 (*d*) respectively. It may be fabricated by using either the grown junction technique or the planar technique.

Operation. The easiest way to understand how it operates is to realize it to be formed of two transistors Q_1 and Q_2 placed back-to-back, as shown in Fig. 8.13 (*b*).

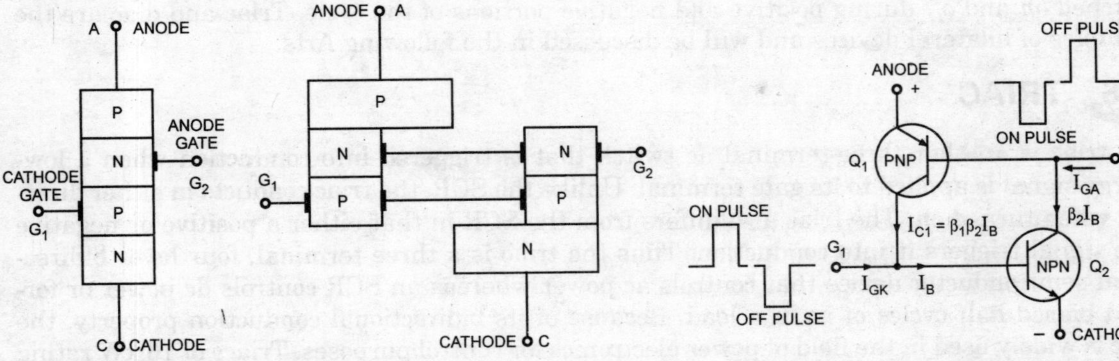

(a) Basic Structure *(b) Two Transistor Version* *(c) Equivalent Transistor Circuit*

In a two-transistor equivalent circuit shown in Fig. 8.13 (c), it is seen that a negative pulse at the anode gate G_2 causes transistor Q_1 to switch *on*. Transistor Q_1 supplies base current to transistor Q_2, and both transistors switch *on*. Similarly, a positive pulse at the cathode gate G_1 can switch the device *on*. Since only small currents are involved, the SCS may be switched *off* by an appropriate polarity pulse at one of the gates. At the cathode gate a negative pulse is required for switching *off* while at the anode gate a positive pulse is needed.

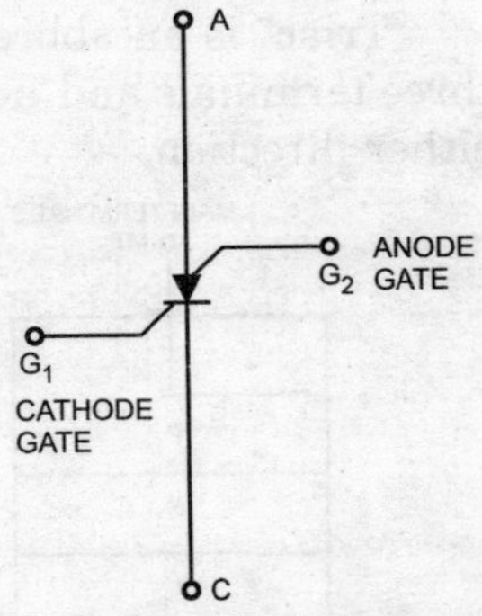

(d) Schematic Symbol

Fig. 8.13

Volt-Ampere Characteristic. The volt-ampere characteristic of an SCS is similar to that of an SCR and is shown in Fig. 8.14. With the increase in applied voltage, the current first increases slowly up to point A and then rapidly in the region AB, as shown in the figure. At point B, the product $\beta_1 \beta_2$ exceeds unity and the device is suddenly switched on. In the on-state, the current increases enormously and is limited by the external series resistor. SCS also exhibits negative differential resistance in the on region similar to SCR. SCS gets switched *on* accidentally if the anode voltage gets applied suddenly. This is known as *rate effect*, which is caused by inter-electrode capacitance between electrodes G_1 and G_2, known as transition capacitance.

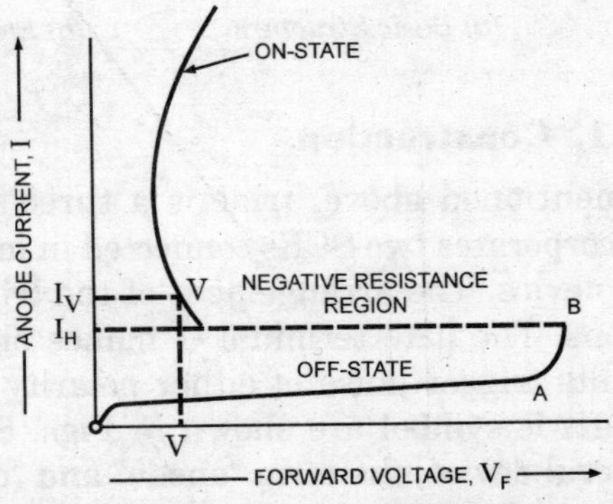

Fig. 8.14. ***Volt-Ampere Characteristic of SCS***

Advantages and Applications. An advantage of SCS over an SCR is the reduced turn-off time, typically within the range of 1 to 10 μs for the SCS and 5 to 30 μs for the SCR. Other advantages of the SCS over SCR are increased control and triggering sensitivity and a more predictable firing situation. However, the SCS is limited to low power, current, and voltage ratings (typical maximum anode currents range from 100 mA to 300 mA with dissipation rating of 100 to 500 mW).

A few of the more common areas of application of SCS include a variety of computer circuits (such as counters, registers, and timing circuits) voltage sensors, pulse generators, oscillators etc.

8.7. BILATERAL DEVICES

In many applications we need devices which switch symmetrically with forward and reverse bias. Such bilateral devices are particularly useful in ac cicruits in which sinusoidal signals are switched *on* and *off* during positive and negative portions of the cycle. Triac and diac are the examples of bilateral devices and will be discussed in the following Arts.

8.8. TRIAC

The triac is another three-terminal ac switch that is triggered into conduction when a low-energy signal is applied to its gate terminal. Unlike the SCR, the triac conducts in either direction when turned *on*. The triac also differs from the SCR in that either a positive or negative gate signal triggers it into conduction. Thus the triac is a three terminal, four layer bidirectional semiconductor device that controls ac power whereas an SCR controls dc power or forward biased half cycles of ac in a load. Because of its bidirectional conduction property, the triac is widely used in the field of power electronics for control purposes. Triacs of 16 kW rating are readily available in the market.

"Triac" is an abbreviation for three terminal ac switch. 'Tri'-indicates that the device has three terminals and 'ac' indicates that the device controls alternating current or can conduct in either direction.

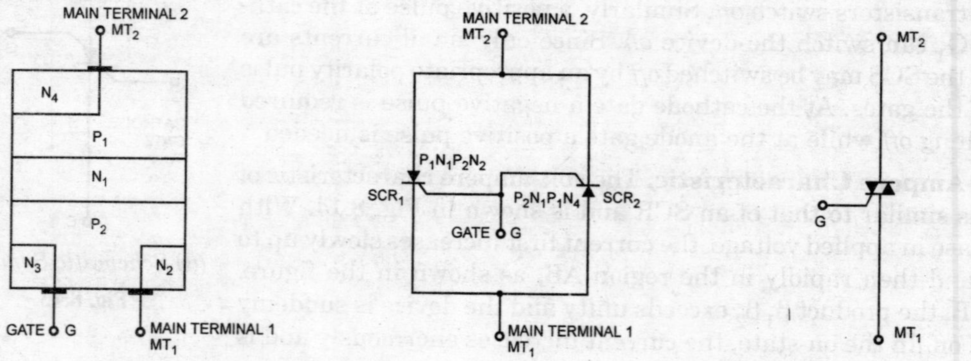

(a) Basic Structure (b) Electrical Equivalent Circuit (c) Schematic Symbol

Fig. 8.15. *Triac*

8.8.1. Construction

As mentioned above, triac is a three terminal, four layer bilateral semiconductor device. It incorporates two SCRs connected in inverse parallel with a common gate terminal in a single chip device. The arrangement of the triac is shown in Fig. 8.15 (*a*). As seen, it has six doped regions. The gate terminal G makes ohmic contacts with both the N and P materials. This permits trigger pulse of either polarity to start conduction. Electrical equivalent circuit and schematic symbol are shown in Figs. 8.15 (*b*) and 8.15 (*c*) respectively. Since the triac is a bilateral device, the term "anode" and "cathode" has no meaning and, therefore, terminals are designated as main terminal 1 (MT_1), main terminal 2 (MT_2) and gate G. To avoid confusion, it has become common practice to specify all voltages and currents using MT_1 as the reference.

8.8.2. Operation

Though the triac can be turned *on* without any gate current provided the supply voltage becomes equal to the breakover voltage of the triac but the normal way to turn *on* the triac is by applying a proper gate current. As in case of SCR, here too, the larger the gate current, the smaller the supply voltage at which the triac is turned *on*.

Triac can conduct current irrespective of the voltage polarity of terminals MT_1 and MT_2 with respect to each other and that of gate and terminal MT_2. Consequently four different possibilities of operation of triac exists. They are:

1. *Terminal MT_2 and gate are positive with respect to terminal MT_1.* When terminal MT_2 is positive w.r.t. terminal MT_1 current flows through path P_1-N_1-P_2-N_2. The two junctions P_1-N_1 and P_2-N_2 are forward biased whereas junction N_1 P_2 is blocked. The triac is now said to be positively biased.

A positive gate w.r.t. terminal MT_1 forward biases the junction P_2-N_2 and the breakdown occurs as in a normal SCR.

2. *Terminal MT_2 is positive but gate is negative w.r.t. terminal MT_1.* Though the flow path of current remains the same as in mode 1 but now junction P_2-N_3 is forward biased and current carriers injected into P_2 turn on the triac.

3. *Terminal MT_2 and gate are negative with respect to terminal MT_1.* When terminal MT_2 is negative w.r.t. terminal MT_1, the current flow path is P_2-N_1-P_1-N_4. The two junctions P_2-N_1 and P_1-N_4 are forward biased whereas junction N_1-P_1 is blocked. The triac is now said to be negatively biased.

A negative gate w.r.t. terminal MT_1 injects current carriers by forward biasing junction P_2-N_3 and thus initiates the conduction.

4. *Terminal MT_2 is negative but gate is positive with respect to terminal MT_1.* Though the flow path of current remains the same as in mode 3 but now junction P_2-N_2 is forward biased, current carriers are injected and, therefore, the triac is turned *on*.

Generally, trigger mode 4 should be avoided especially in circuits where high $\frac{di}{dt}$ may occur. The sensitivity of triggering modes 2 and 3 is high and in case of marginal triggering capability negative gate pulses should be used. Though the triggering mode 1 is more sensitive compared to modes 2 and 3, it requires a positive gate trigger. However, for bidirectional control and uniform gate trigger modes 2 and 3 are preferred.

8.8.3. Characteristics

Typical V-I characteristics of a triac are shown in Fig. 8.16. The triac has *on* and *off* state characteristics similar to SCR but now the characteristic is applicable to both positive and negative voltages. This is expected because triac consists of two SCRs connected in parallel but opposite in directions.

MT_2 is positive with respect to MT_1 in the first quadrant and it is negative in the third quadrant. As already said in Art. 8.8.2, the gate triggering may occur in any of the following four modes.

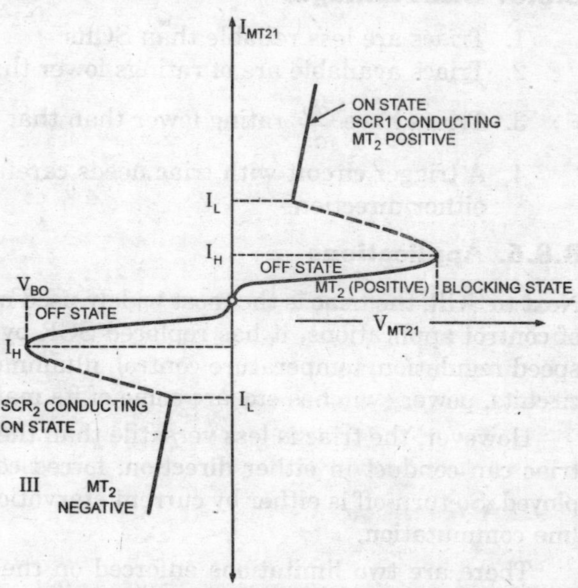

Fig. 8.16. *V-I Characteristic of a Triac*

Quadrant I operation : V_{MT21} positive; V_{G_1} positive
Quadrant II operation : V_{MT21} positive; V_{G_1} negative
Quadrant III operation : V_{MT21} negative; V_{G_1} negative
Quadrant IV operation : V_{MT21} negative; V_{G_1} positive

where V_{MT21} and V_{G_1} are the voltages of terminal MT_2 and gate with respect to terminal MT_1.

The device, when starts conduction, permits a very heavy amount of current to flow through it. This large inrush of current must be restricted by employing external resistance, otherwise the device may get damaged.

The gate is the control terminal of the device. By applying proper signal to the gate, the firing angle of the device can be controlled. The circuits used in the gate for triggering the device are called the *gate triggering circuits*. The gate triggering circuits for the triac are almost same like those used for SCRs. These triggering circuits usually generate trigger pulses for firing the device. The trigger pulse should be of sufficient magnitude and duration so that firing of the device is assured. Usually, a duration of 35 μs is sufficient for sustaining the firing of the device.

A typical triac has the following voltage/current values:

Instantaneous on-state voltage :	1.5 V
On-state current :	25 A
Holding current, I_H :	75 mA
Average triggering current, I_G :	5 mA

8.8.4. Advantages

1. Triacs can be triggered with positive or negative polarity voltages.
2. A Triac requires a single heat sink of slightly larger size, whereas antiparallel SCR pair requires two heat sinks of slightly smaller size, but due to the clearance total space required is more for SCRs.
3. A Triac requires a single fuse for protection, which also simplifies construction.

8.8.5. Disadvantages

1. Triacs are less reliable than SCRs.
2. Triacs available are of ratings lower than those of SCRs.
3. Triacs have $\frac{dv}{dt}$ rating lower than that of SCRs.
4. A trigger circuit with triac needs careful consideration because it can be triggered in either direction.

8.8.6. Applications

Next to SCR, the triac is the most widely used member of the thyristor family. In fact, in many of control applications, it has replaced SCR by virtue of its bidirectional conductivity. Motor speed regulation, temperature control, illumination control, liquid level control, phase control circuits, power switches etc. are some of its main applications.

However, the triac is less versatile than the SCR when turn-off is considered. Because the triac can conduct in either direction, forced commutation by reverse-biasing cannot be employed. So turn-off is either by current starvation, which is usually impracticable, or else by ac line commutation.

There are two limitations enforced on the use of triac at present state of commercially available devices (200 A and 1,000 PRV). The first is the frequency handling capability produced by the limiting $\frac{dv}{dt}$ at which the triac remains blocking when no gate signal is applied. This $\frac{dv}{dt}$ value is about 20 Vμs^{-1} compared with a general figure of 200 Vμs^{-1} for the SCR, so that the limitation of frequency is at the power level of 50 Hz. The same $\frac{dv}{dt}$ limitation means the load to be controlled is preferably a resistive one. When high frequencies and high $\frac{dv}{dt}$ are involved then the back-to-back SCRs cannot be replaced by the triac.

1. *High Power Lamp Switching.* Use of the triac as an ac *on/off* switch is shown in Fig. 8.17. When the switch S is in position 1, the triac is cut-off and so the lamp is dark. When the switch is put in position 2, a small gate current flowing through the gate turns the triac *on* and so the lamp is switched *on* to give rated output.

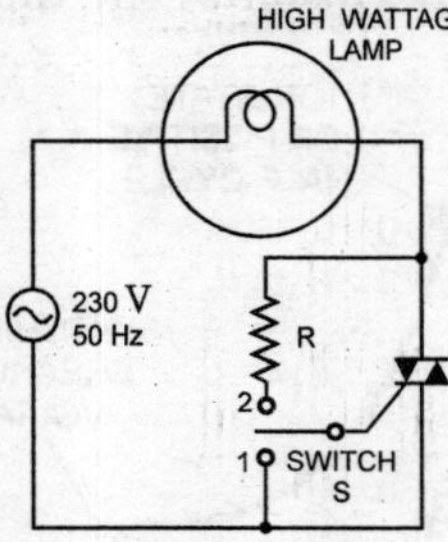

Fig. 8.17

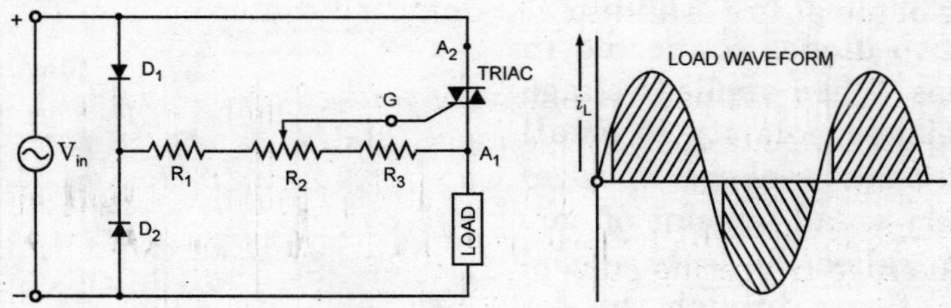

Fig. 8.18. ***Power Control By Triac***

2. *AC Power Control.* A triac control circuit is shown in Fig. 8.18. Here it is controlling ac power to load by switching *on* and *off* during the positive and negative half cycles of the input sinusoidal signal.

During the positive half cycle of the input voltage, diode D_1 is forward biased, D_2 is reverse-biased, and the gate terminal is positive with respect to A_1. During the negative half cycle, the diode D_2 is forward biased and diode D_1 is reverse-biased, so that the gate becomes positive with respect to terminal A_2. The point of commencement of conduction is controlled by adjusting the resistance R_2.

8.9. DIAC

A diac is an important member of the thyristor family and is usually employed for triggering triacs. A diac is a two-electrode bidirectional avalanche diode which can be switched from off-state to the on-state for either polarity of the applied voltage. This is just like a triac without gate terminal, as shown in Fig. 8.19 (*a*). Its equivalent circuit is a pair of inverted four layer diodes. Two schematic symbols are shown in Fig. 8.19 (*b*). Again the terminal designations are arbitrary since the diac, like triac, is also a bilateral device. The switching from off-state to on-state is achieved by simply exceeding the avalanche breakdown voltage in either direction.

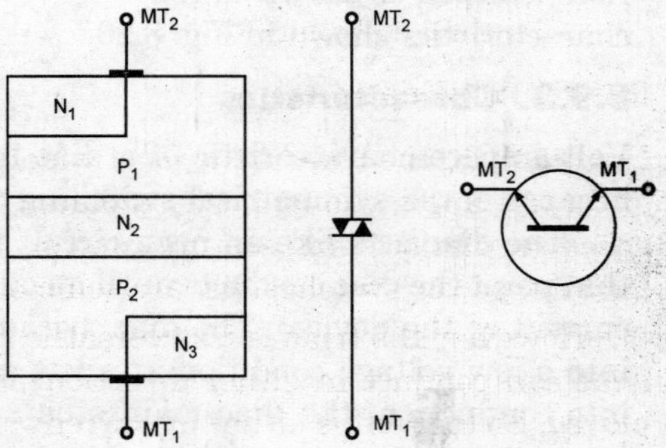

(*a*) *Basic Structure* (*b*) *Schematic Symbols*

Fig. 8.19. ***Diac***

8.9.1. Construction

A diac is a P-N-P-N structured four-layer, two-terminal semiconductor device, as shown in Fig. 8.19 (*a*). MT_2 and MT_1 are the two main terminals of the device. There is no control terminal in this device. From the diagram, a diac unlike a diode, resembles a bipolar junction transistor (BJT) but with the following exceptions.

(*i*) there is no terminal attached to the middle layer (base),

(*ii*) the three regions are nearly identical in size,

(*iii*) the doping level at the two end P-layers is the same so that the device gives symmetrical switching characteristics for either polarity of the applied voltage.

8.9.2. Operation

When the terminal MT_2 is positive, the current flow path is P_1-N_2-P_2-N_3 while for positive polarity of terminal MT_1 the current flow path is P_2-N_2-P_1-N_1. The operation of the diac can be explained by imagining it as two diodes connected in series. When applied voltage in either polarity is small (less than breakover voltage) a very small amount of current, called the *leakage current*, flows through the device. Leakage current caused due to the drift of electrons and holes in the depletion region, is not sufficient to cause conduction in the device. The device remains in non-conducting mode. However, when the magnitude of the applied voltage exeeds the avalanche breakdown voltage, breakdown takes place and the diac current rises sharply, as shown in the characteristics shown in Fig. 8.20.

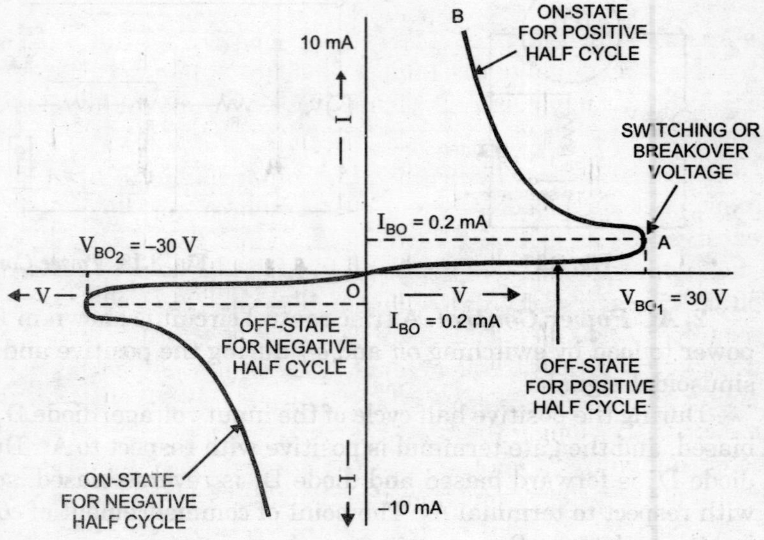

Fig. 8.20 ***V-I Characteristic of a Diac***

8.9.3. Characteristics

Volt-ampere characteristic of a diac is shown in Fig. 8.20. It resembles the English letter Z because of the symmetrical switching characteristics for either polarity of the applied voltage.

The diac acts like an open-circuit until its switching or breakover voltage is exceeded. At that point the diac conducts until its current reduces toward zero (below the level of the holding current of the device). The diac, because of its peculiar construction, does not switch sharply into a low voltage condition at a low current level like the SCR or triac. Instead, once it goes into conduction, the diac maintains an almost continuous negative resistance characteristic *i.e.*, voltage decreases with the increase in current. This means that, unlike the SCR and the triac, the diac cannot be expected to maintain a low (on) voltage drop until its current falls below a holding current level.

8.9.4. Applications

The diacs, because of their symmetrical bidirectional switching characteristics, are widely used as triggering devices in triac phase control circuits employed for lamp dimmer, heat control, universal motor speed control etc.

Although a triac may be fired into the conducting state by a simple resistive triggering circuit, but triggering devices are typically placed in series with the gates of SCRs and triacs as they give reliable and fast triggering. Diac is the most popular triggering device for the triac. This is illustrated in the following applications.

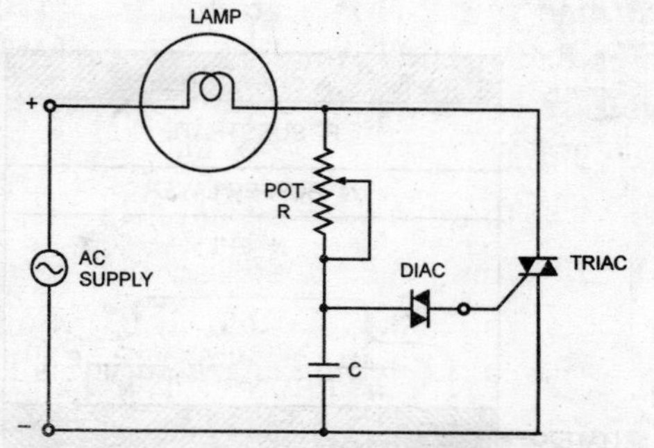

Fig. 8.21. ***Triac Lamp Dimmer Circuit***

1. ***Triac Lamp Dimmer Circuit.*** The circuit for a triac controlled by an R-C phase-shift network and a diac is given in Fig. 8.21. This circuit is an example of a simple lamp dimmer. The triac conduction angle is adjusted by adjusting the potentiometer R. The longer the triac conducts, the brighter the lamp will be. The diac acts like an open-circuit until the voltage across the capacitor exceeds its breakover or switching voltage (and the triac's required gate trigger voltage).

2. ***Heat Control Circuit.*** A typical diac-triac circuit used for smooth control of ac power to an heater is shown in Fig. 8.22. The capacitor C_1 in series with choke L across the triac slows-up the voltage rise across the device during off-state. The resistor R_4 across the diac ensures smooth control at all positions of potentiometer R_2. The triac conduction angle is adjusted by adjusting the potentiometer R_2. The longer the triac conducts, the larger the output will be from the heater. Thus a smooth control of the heat output from the heater is obtained.

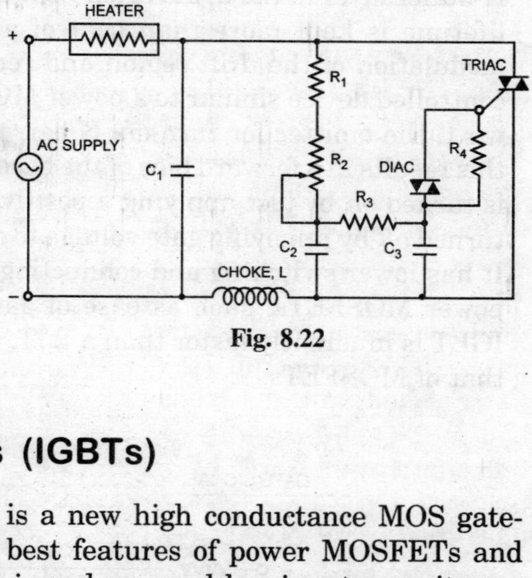

Fig. 8.22

8.10. Insulated Gate Bipolar Transistors (IGBTs)

An *insulated gate bipolar junction transistor* (IGBT) is a new high conductance MOS gate-controlled power switch. It incorporates some of the best features of power MOSFETs and power transistors. Like a MOSFET, it has high input impedance and low input capacitance. On the other hand, in the on-state, it has low resistance and high current handling capability, like a BJT or SCR. Because of these factors, and because it can turn *off* more easily than a SCR, the IGBT is gradually becoming the power device of choice, in place of the more traditional SCR. It has no second breakdown problem as with BJT. By chip design and structure, the equivalent drain-to-source resistance R_{DS} is controlled to behave like that of a BJT.

The silicon cross-section of an IGBT is depicted in Fig. 8.23(*a*), which is identical to that of an MOSFET except the P^+ substrate. However, the performance of an IGBT is closer to that of a BJT than an MOSFET. This is due to the P^+ substrate which is responsible for the minority carrier injection into the N region. The equivalent circuit is shown in Fig. 8.23 (*b*), which can be simplified to that shown in Fig. 8.23(c). An IGBT is made of four alternate PN PN layers, and could latch like a thyristor given the necessary condition $(\alpha_{NPN} + \alpha_{PNP}) > 1$. The N^+-buffer layer and the wide epi base reduce the gain of the NPN-terminal by internal design, thereby avoiding

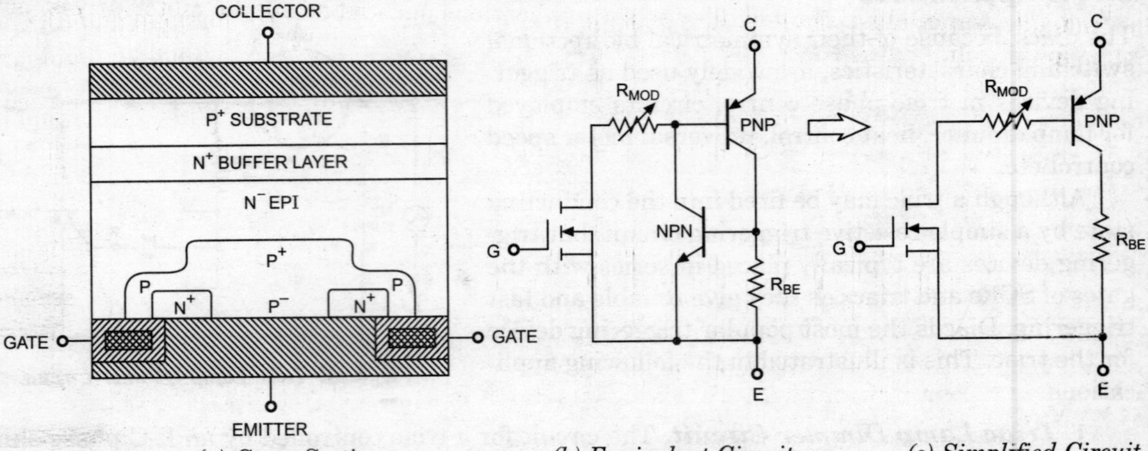

(a) Cross-Section *(b) Equivalent Circuit* *(c) Simplified Circuit*

Fig. 8.23. *IGBT*

latching. IGBTs have two structures of IGBTs: punch-through and non-punch through. In the punch-through IGBT structure, the switching time is reduced by using a heavily doped N-buffer layer in the drift region near the collector. In the non-punch through structure, carrier lifetime is kept more than that of a punch-through structure, which causes conductivity modulation of the drift region and reduces the on-state voltage drop. An IGBT is a voltage-controlled device similar to a power MOSFET. Like a MOSFET, when the gate is made positive w.r.t. the emitter for turn-on, N carriers are drawn into the P-channel near the gate region; this results in a forward bias of the base of the NPN transistor, which thereby turns *on*. An IGBT is turned *on* by just applying a positive gate voltage to open the channel for N carriers and is turned *off* by removing gate voltage to close the channel. It requires a very simple driver circuit. It has lower switching and conducting losses while sharing many of the appealing features of power MOSFETs, such as ease of gate drive, peak current, capability, and ruggedness. An IGBT is inherently faster than a BJT. However, the switching speed of the IGBTs is inferior to that of MOSFETs.

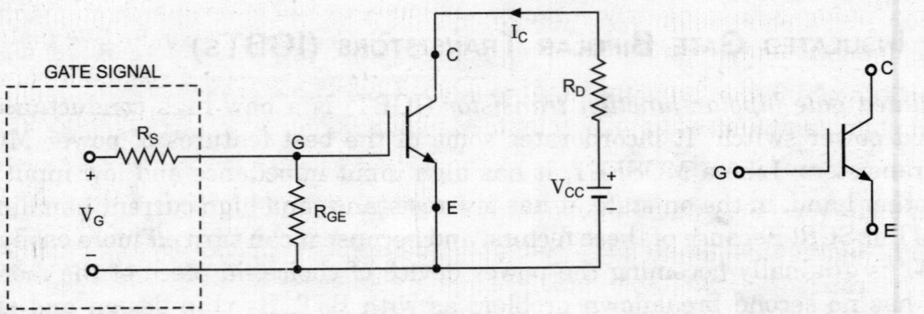

(a) Circuit *(b) Symbol*

Fig. 8.24. *IGBT*

The circuit and symbol of an IGBT are shown in Fig. 8.24. The three terminals are gate, collector, and emitter instead of gate, drain and source for a MOSFET. The typical output characteristics of i_C versus v_{CE} are shown in Fig. 8.25(*a*) for different gate-emitter voltage v_{GE}. The typical transfer characteristic of i_C versus v_{GE} is shown in Fig. 8.25(*b*). The parameters and their symbols are similar to that of MOSFETs, except that the subscripts for source and drain are changed to emitter E and collector C respectively. The current rating of a single IGBT can be up to 1200 V, 400 A, and the switching frequency can be up to 20 kHz. IGBTs are finding increasing applications in medium-power applications such as dc and ac motor drives, power supplies, solid-state relays, and contractors.

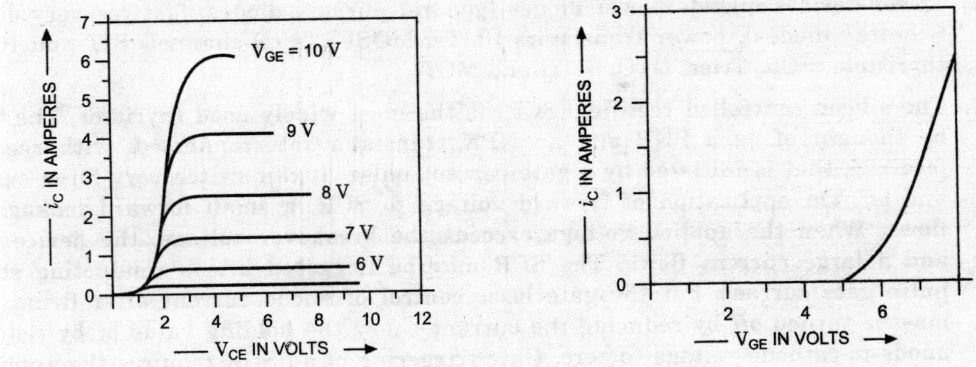

(a) Typical Output Characteristics *(b) Transfer Characteristic*

Fig. 8.25 ***IGBT***

As the upper limits of commercially available IGBT ratings are increasing (e.g., as high as 6500 V and 2400 A), the IGBTs are finding and replacing applications where BJTs and conventional MOSFETs were predominantly used for switching.

Handling Precautions For IGBTs. Insulated Gate Bipolar Transistors are susceptible to gate insulation damage by the electrostatic discharge of energy through the devices. When handling these devices, care should be exercised to assure that the static charge built in the handler's body capacitance is not discharged through the device. With proper handling and application procedures, however, IGBTs are currently being extensively used in production by numerous equipment manufacturers in military, industrial and consumer applications, with virtually no damage problems due to electrostatic discharge. IGBTs can be handled safely if the following basic precautions are taken:

1. Prior to assembly into a circuit, all leads should be kept shorted together either by the use of metal shorting springs or by the insertion into conductive material.
2. When devices are removed by hand from their carriers, the hand being used should be grounded by any suitable means, for example, with a metallic wristband.
3. Tips of soldering irons should be grounded.
4. Devices should never be inserted into or removed from circuits with power on.
5. **Gate Voltage Rating.** Never exceed the gate-voltage rating of V_{GEM}. Exceeding the rated V_{GE} can result in permanent damage to the oxide layer in the gate region.
6. **Gate Termination.** The gates of these devices are essentially capacitors. Circuits that leave the gate open-circuited or floating should be avoided. These conditions can result in turn-*on* of the device due to voltage buildup on the input capacitor due to leakage currents or pickup.
7. **Gate Protection.** These devices do not have an internal monolithic Zener Diode from gate to emitter. If gate protection is required an external Zener is recommended.

ECCOSORBD™ is a Trademark of Emerson and Cumming, Inc.

Highlights

1. Power devices include power diodes (general purpose diodes, fast recovery diodes and Schottky diodes), power transistors [BJTs, MOSFETs (N-channel) SIT and IGBT] and thyristors (SCR, Triac, GTO, SITH and MCT).
2. The silicon controlled rectifier (SCR) is the most widely used thyristor. The SCR may be thought of as a PNP and an NPN transistor interconnected, with regenerative feedback that is initiated by a gate current pulse. It can switch very large currents *on* and *off*. On application of forward voltage to SCR, a small forward leakage current flows. When the applied voltage exceeds the breakover voltage, the device turns *on* and a large current flows. The SCR may be triggered into a conducting state by a pulse gate current, but the gate loses control of anode current after firing. An SCR may be turned *off* by reducing the current below the holding value or by reducing the anode-to-cathode voltage to zero. Gate triggering of an SCR requires the application of gate voltage and current that must lie in the preferred area of the graphical gate characteristics is to ensure reliable switching. SCR has turn-on time of 1-4 μs and turn-off time of 10-300 μs. An SCR has a number of voltage, current and power ratings. These ratings should not be exceeded.

   The applications of SCRs include power control, switching, zero voltage switching, overvoltage protection, pulse circuits, battery charging regulator, heater control circuit, emergency lighting system etc.
3. The four-layer diode, also called a Shockley diode, is the simplest thyristor. It is essentially a low-current SCR without a gate.
4. Silicon controlled switch (SCS), like an SCR, is a unilateral, four-layer, three junction P-N-P-N silicon device, with four electrodes namely cathode, cathode gate, anode gate and anode. It is a low power device compared with the SCR. A few of the more common areas of application of SCS include a variety of computer circuits (such as counters, registers, and timing circuits) voltage sensors, pulse generators, oscillators etc.
5. A triac is a bidirectional device, conducting in either direction for either a positive or negative gate voltage. Because of its bidirectional conduction property, the triac is widely used in the field of power electronics for control purposes.
6. A diac is a two-electrode bidirectional avalanche diode which can be switched from off-state to the on-state for either polarity of the applied voltage. The diacs, because of their symmetrical bidirectional switching characteristics, are widely used as triggering devices in triac phase control circuits.
7. **Power transistors** are generally of three types: BJTs, MOSFETs and IGBTs. BJTs are current-controlled devices and their parameters are sensitive to junction temperature. BJTs suffer from second breakdown and need reverse base current during turn-off to reduce the storage time, but they have low on-state or saturation voltage.

   MOSFETs are voltage-controlled devices and require very small gating power and their parameters are less sensitive to junction temperature. There is no second breakdown problem and no need for negative gate voltage during turn-off.

   IGBTs, which combine the advantages of BJTs and MOSFETs, are voltage-controlled devices and have low on-state voltage similar to BJTs. They have no second breakdown phenomenon.

Exercises

1. Explain the operation of thyristor in detail with its V-I characteristics. Also discuss its two transistor model. **[M.D. Univ. Third Semester December 2007]**
2. What are the differences between a transistor and a thyristor ? Explain the operation of a thyristor with the help of its characteristics.
3. Discuss the construction and explain the characteristic curve of an SCR showing how forward voltage depends on the value of gate current.
4. Draw the block and symbolic representation for SCR. State its application. Sketch its V-I characteristics for different gate currents and indicate thereupon holding current, latching current, and breakover voltage.
5. Explain the basic construction of a silicon controlled rectifier (SCR). Explain its principle of operation using two transistor models. Explain the turn-off mechanism of SCR. Give a circuit to explain your answer.
6. Give the two transistor analogous circuit of an SCR and explain its operation. Sketch its V-I characteristics. Mark 'hold current' and 'latch current' on it and explain their importance in SCR power control circuits.
7. What are the normal methods of turn-on of an SCR ? Also state the schemes of commutation.
8. What are the advantages and disadvantages of an SCR switch over a mechanical switch ?
9. Describe the characteristics and applications of SCR and P-N-P-N diode. **[U.P. Technical Univ. Fundamentals of Electronic Devices 2009-10]**
10. Why a four-layer diode is called the Shockley diode ? Give its basic structure, equivalent transistor circuit and schematic symbol. Give its breakover characteristic and explain.
11. What is silicon controlled switch (SCS) ? How is it different from an SCR ? Give its basic structure, equivalent transistor circuit and schematic symbol. Give its advantages over an SCR and applications.
12. Why triac is called a double ended SCR ? Explain the construction and operation of a triac with the help of a neat diagram.
13. Enumerate the basic differences between a triac and thyristor. Draw and explain V-I characteristic of a triac. **[U.P.S.C. I.E.S. Electrical Engineering II, 2000]**
14. With the help of V-I characteristics show the four triggering modes of triac.
15. With neat diagram explain the constructional features of a triac. Sketch its V-I characteristics and explain its operation. How does it differ from an SCR ? Mention its applications.
16. Explain the working of TRIAC alongwith its V-I characteristics. **[M.D. Univ. Third Semester May 2008]**
17. Explain the construction of diac. Draw a circuit showing use of a diac to trigger a triac.
18. What is the origin of the name diac? Explain the construction, operation and characteristics of a diac. Discuss same applications of diac.
19. Explain the construction and working of (*i*) TRIAC and (*ii*) IGBT. **[M.D. Univ. Third Semester May 2006]**
20. Explain the characteristics of IGBT. **[M.D. Univ. Third Semester May 2005]**
21. What is IGBT ? Draw its equivalent circuit and enumerate its special features. **[U.P. Technical Univ. 2007-08]**
22. What is IGBT ? Give the operation and characteristics of the device. What precautions are taken while handling IGBTs ?
23. Draw the physical structure and equivalent circuit of the IGBT. Explain the special features. Draw its I-V characteristics.

24. Write short notes on the following:

(i) Thyristor. **[M.D. Univ. December 2005; May 2008]**

(ii) SCR. **[U.P. Technical Univ. 2006-07]**

(iii) TRIAC **[M.D. Univ. December 2006]**

(iv) DIAC and TRIAC. **[M.D. Univ. December 2005]**

(v) DIAC. **[M.D. UNIV. May 2007; 2008]**

(vi) IGBT. **[M.D. Univ. May 2007; 2008]**

Short Answer Type Questions With Answers

Q. 1. How is current limited in conducting state of an SCR ?

Ans. The current in conducting state of an SCR is controlled by external impedance.

Q. 2. What is the effect of negative gate current on a normal SCR ?

Ans. Negative gate current will increase the holding current of the SCR.

Q. 3. What is false triggering ?

Ans. False triggering is unintended turn-on of an SCR either through gate due to noise pick-up or excessive anode voltage.

Q. 4. What are the factors on which 'turn-off time' of an SCR depends ?

Ans. Turn-of time of an SCR depends upon doping densities, recombination time, junction temperature, on-state current, rate of decay of on-state current.

Q. 5. How the forced turn-off of an SCR is different from natural turn-off ?

Ans. In forced turn-off of an SCR current is brought below the holding current by an additional delicated circuit employing energy storage elements while in natural turn-off, it is the natural voltage/ current variation.

Q. 6. Why 'pulse triggering' is preferred and when does it fail ?

Ans. Pulse triggering is preferred as it is effective to turn *on* an SCR keeping the device gate dissipation low and ensures a fast turn *on* keeping the $\frac{di}{dt}$ stress on the device low.

Pulse triggering shall fail if the pulse is of short duration and the load circuit contains large inductance. With highly inductive load, the device current builds up slowly. If the pulse duration is small and the anode current in not able to build up to latching current, I_L, the device turns *off* on removal of gate pulse.

Q. 7. Why is it necessary to keep supply voltage much less than breakover voltage of an SCR ?

Ans. The supply voltage is kept much less than breakover voltage of an SCR otherwise the device will get damaged.

Q. 8. What is meant by breakover voltage of an SCR ?

Ans. The minimum forward voltage, with gate open, at which SCR starts conducting heavily (*i.e.* turns *on*) is called the breakover voltage.

Q. 9. What is maximum on-state voltage in reference to an SCR ?

Ans. Maximum on-state voltage is the maximum value of the voltage appearing across the SCR during conduction.

Q. 10. What is meant by forward current rating of an SCR ?

Ans. The maximum value of anode current, that an SCR can handle safely (without any damage) is called the forward current rating of the SCR.

Q. 11. What is holding current in an SCR ?

Ans. Holding current is the minimum on-state current required to keep the SCR in conducting state without any gate drive.

Q. 12. Define latching current of an SCR.

Ans. Latching current is the minimum device current, which must be attained by the device, before the gate drive is removed.

Q. 13. Give the typical values of current ratings of SCR.

Ans. Forward breakover current—less than a few hundred microamperes.

Peak forward current 30 A to over 100 A.

Holding current—a few mA to few hundred milliamperes.

Q. 14. How Shockley diode is switched *on* ?

Ans. Shockley diode has no trigger inputs, the only way to switch the device *on* is to increase the anode-to-cathode voltage to the switching voltage equivalent to SCR forward breakover voltage.

Q. 15. What is silicon controlled switch ?

Ans. Silicon controlled switch (SCS), like the SCR, is a unilateral, four-layer, three junction P-N-P-N silicon device with four electrodes namely cathode, cathode gate, anode gate and anode.

Q. 16. What are bilateral devices ?

Ans. Bilateral devices are those devices which switch symmetrically with forward and reverse bias. Triac and diac fall under this category.

Q. 17. What is triac ?

Ans. The triac is a three terminal, four layer bidirectional semiconductor device. It incorporates two SCRs connected in inverse parallel with a common gate terminal in a single chip. Triac is an abbreviation for a triode ac switch.

Q. 18. What is diac ?

Ans. A diac is a P-N-P-N structured four-layer, two-terminal semiconductor device. This is just like a triac but without gate terminal. Diac is an abbreviation for a diode ac switch.

Q. 19. What is IGBT ?

Ans. IGBT (insulated gate bipolar junction transistor) is a new high conductance MOS controlled power switch and incorporates some of the best features of power MOSFETs and power transistors.

Q. 20. Give the applications of IGBTs.

Ans. IGBTs are finding increasing applications in medium power applications such as dc and ac motor drives, power supplies, solid-state relays and contactors.

❑❑❑

Appendices

APPENDIX 1.1

1 N 4001 through 1 N 4007 Rectifiers

SURMETIC RECTIFIERS

Subminiature size, axial lead-mounted rectifiers for general purpose, low power applications

Designer Data for "Worst Case" Conditions

The Designers Data Sheets permit the design of most circuits entirely from the information presented. Limit curves-representing boundaries on device characteristics — are given to facilitate "worst case" design.

MAXIMUM RATINGS*

Rating	Symbol	1N 4001	1N 4002	1N 4003	1N 4004	1N 4005	1N 4006	1N 4007	Unit
Peak Repetitive Reverse Voltage Working Peak Reverse Voltage DC Blocking Voltage	V_{RRM} V_{RWM} V_R	50	100	200	400	600	800	1000	Volts
Non-Repetitive Peak Reverse Voltage (Half Wave Single Phase 60 Hz)	V_{RSM}	60	120	240	480	720	1000	1200	Volts
RMS Reverse Voltage	$V_{R\,(RMS)}$	35	70	140	280	420	560	700	Volts
Average Rectified Forward Current (Single Phase Resistive Load, 60 Hz, T_A = 75°C)	I_0	1.0							Amp
Non-Repetitive Peak Surge Current (Surge applied at rated load conditions	I_{FSM} or $I_{FM\,(surge)}$	30 (for 1 cycle)							Amp
Peak Repetitive Forward Current	$I_{FM\,(rep)}$	10							Amp
Operating and Storage Junction Temperature Range	T_j, T_{stg}	– 65 to + 175							°C

LEAD-MOUNTED
SILICON RECTIFIERS
50-100 VOLTS
DIFFUSED JUNCTION

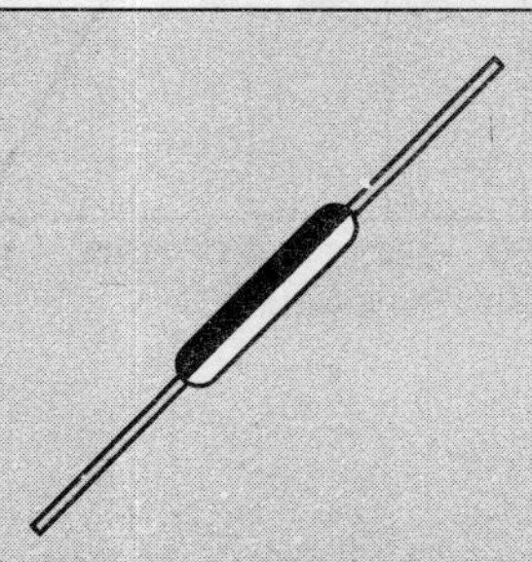

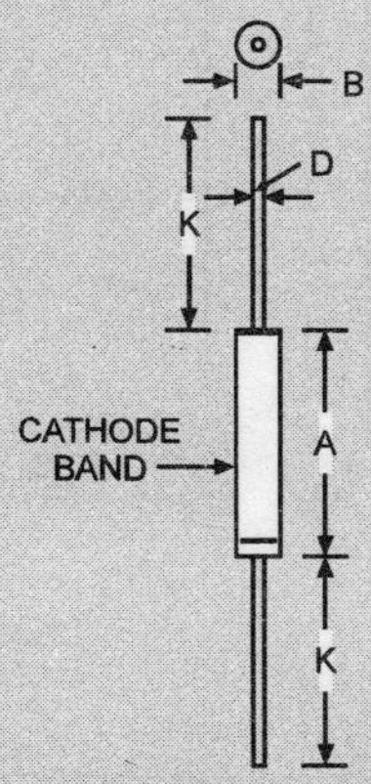

DIM	MILLIMETERS MIN	MILLIMETERS MAX	INCHES MIN	INCHES MAX
A	5.97	6.60	0.235	0.260
B	2.79	3.05	0.110	0.120
D	0.76	0.86	0.030	0.034
K	27.94	—	1.100	—

CASE 59-04
Does Not Conform to 00-41 outline

ELECTRICAL CHARACTERISTICS*

Characteristics and Conditions	Symbol	Typical	Max	Unit
Maximum Instantaneous Forward Voltage Drop (1.0 Amp Continuous DC, 25° C)	v_F	0.93	1.1	Volts
Maximum Full-Cycle Average Forward Voltage Drop (I_0 = 1.0 Amp, T_L = 75°C 1 inch leads)	$V_{F(AV)}$	—	0.8	Volts
Maximum Reverse Current (rated dc voltage) T_j = 25° C T_j = 100° C	I_R	0.05 1.0	10 50	µA
Maximum Full-Cycle Average Reverse Current (I_0 = 1.0 Amp, T_L = 75° C, 1 inch lead)	$I_{R(AV)}$	—	30	µA

* Indicates JEDEC Registered Data

MECHANICAL CHARACTERISTICS

CASE : Void free, transfer Molded

MAXIMUM LEAD TEMPERATURE FOR SOLDERING PURPOSES :

350° C, $\frac{3}{8}"$ from case for 10 seconds at 5 lbs tension

FINISH : All external surfaces are corrosion-resistant, leads are readily solderable

POLARITY : Cathode indicated by color band

WEIGHT : 0.40 grams (approximately)

TYPICAL FORWARD CHARACTERISTICS

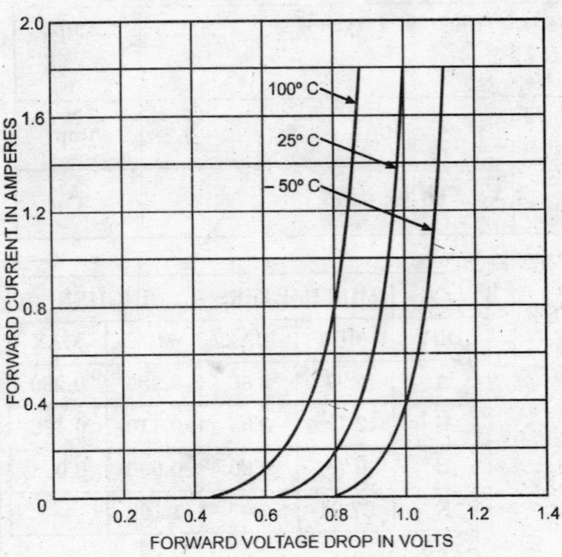

MAXIMUM DC OUTPUT

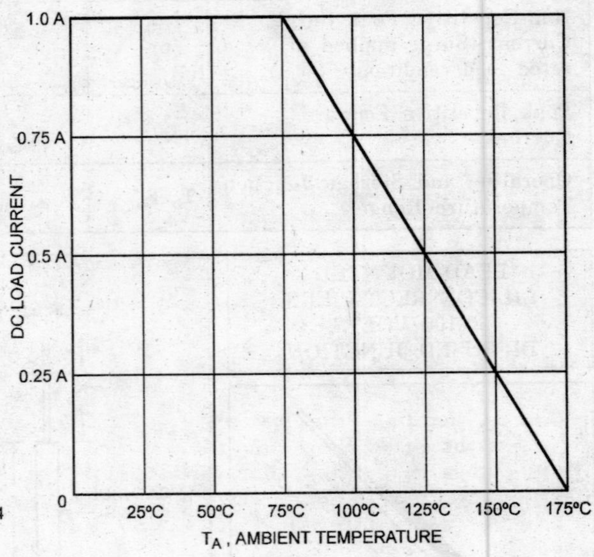

Courtesy of Motorola, Inc.

APPENDIX 1.2

1 N 746 through 1 N 759 Silicon

Voltage Regulator Diodes

3.3 To 12 V; 400 mW

Guaranteed Dynamic Zener Impedance

Available in 5 % and 10 % tolerances
– 65 to 175° C operation and storage

* Electrical Characteristics at 25° C free-air temperature (unless otherwise stated)

Parameter	Zener Breakdown Voltage, V_Z					Temperature Coefficient of Breakdown Voltage, α_Z	Small Signal Breakdown Impedance r_Z	Static Reverse Current, I_R	
Test Conditions	I_{ZT} = 20 mA					I_{ZT} = 20 mA	I_{ZT} = 20 mA I_{ZT} = 1 mA	V_R = 1 V	V_R = 1 V T_A = 150°C
Limit	NOM	1 N 746 – 1 N 759 MIN	MAX	1 N 746A – 1 N 759A MIN	MAX	TYP	MAX	MAX	MAX
UNIT	V	V	V	V	V	% / °C	Ω	μ A	μ A
1 N 746	3.3	2.97	3.63	3.135	3.465	– 0.62	28	10	30
1 N 747	3.6	3.24	3.96	3.420	3.780	– 0.055	24	10	30
1 N 748	3.9	3.51	4.29	3.705	4.095	– 0.049	23	10	30
1 N 749	4.3	3.87	4.73	4.085	4.515	– 0.036	22	2	30
1 N 750	4.7	4.23	5.17	4.465	4.935	– 0.018	19	2	30
1 N 751	5.1	4.59	5.61	4.845	5.355	– 0.008	17	1	20
1 N 752	5.6	5.04	6.16	5.320	5.880	+ 0.006	11	1	20
1 N 753	6.2	5.58	6.82	5.890	6.510	+ 0.022	7	0.1	20
1 N 754	6.8	6.12	7.48	6.460	7.140	+ 0.035	5	0.1	20
1 N 755	7.5	6.75	8.25	7.125	7.875	+ 0.045	6	0.1	20
1 N 756	8.2	7.38	9.02	7.790	8.610	+ 0.052	8	0.1	20
1 N 757	9.1	8.19	10.01	8.645	9.555	+ 0.056	10	0.1	20
1 N 758	10.0	9.00	11.00	9.50	10.500	+ 0.060	17	0.1	20
1 N 759	12.0	10.80	13.20	11.400	12.000	+ 0.060	30	0.1	20

* **Absolute Maximum Ratings**

Average Rectified Forward Current at (or below) 25° C Free-Air Temperature .... 230 mA
Average Rectified Forward Current at 150° C Free-Air Temperature .... 85 mA
Continuous Power Dissipation at (or below) 50° C Free-Air Temperature .... 400 mW
Continuous Power Dissipation at 150° C Free-Air Temperature .... 100 mW
Operating Free-Air Temperature Range .... – 65° C to 175° C
Storage Temperature Range .... – 65° C to 175° C

* Indicates JEDEC Registered Data

Courtesy of Texas Instruments, Incorporated

APPENDIX 1.3

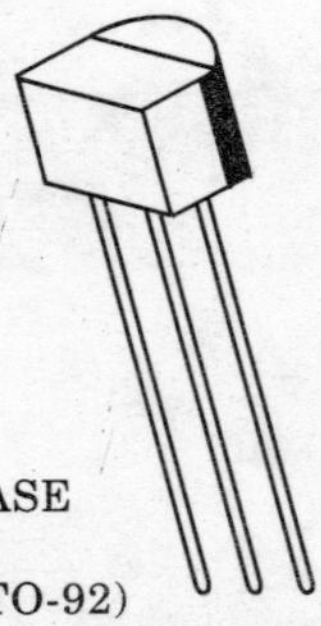

CASE
29
(TO-92)

2N3903, 2N3904 (Silicon) NPN Transistors

NPN silicon annular transistors, designed for general purpose switching and amplifier applications, features one-piece, injection-molded plastic package for high reliability. The 2N3903 and 2N3904 are complementary with types 2N3905 and 2N3906, respectively.

MAXIMUM RATINGS (T_A = 25 °C unless otherwise stated)

Characteristic	Symbol	Rating	Unit
Collector-Base Voltage	V_{CB}	60	Vdc
Collector-Emitter Voltage	V_{CEO}	40	Vdc
Emitter-Base Voltage	V_{EB}	6	Vdc
Collector Current	I_C	200	mAdc
Total Device Dissipation @ T_A = 60 °C	P_D	210	mW
Total Device Dissipation @ T_A = 25 °C Derate above 25 °C	P_D	310 2.81	mW mW/°C
Thermal Resistance, Junction to Ambient	θ_{JA}	0.357	°C/mW
Junction Operating Temperature	T_J	135	°C
Storage Temperature Range	T_{stg}	− 55 to + 135	°C

ELECTRICAL CHARACTERISTICS (T_A = 25 °C unless otherwise noted)

Characteristic	Symbol	Min	Max	Unit
OFF CHARACTERISTICS				
Collector-Base Breakdown Voltage (I_C = 10 μA dc, I_E = 0)	BV_{CBO}	60	—	Vdc
Collector-Emitter Breakdown Voltage* (I_C = 1 mA dc)	BV_{CEO}*	40	—	Vdc
Emitter-Base Breakdown Voltage (I_E = 10 μA dc, I_C = 0)	BV_{EBO}	6	—	Vdc
Collector Cut-off Current (V_{CE} = 40 V dc, V_{OB} = 3 Vdc)	I_{CEX}	—	50	nAdc
Base Cut-off Current (V_{CE} = 40 V dc, V_{OB} = 3 Vdc)	I_{BL}	—	50	nAdc

* Pulse Test. Pulse Width = 300 μ sec, Duty Cycle = 2% V_{OB} = Base-Emitter Reverse Bias

* Courtesy of Motorola, Inc.

Appendix 1.3 (Continued)

Characteristic		Symbol	Min	Max	Unit
ON CHARACTERISTICS					
DC Current Gain*		h_{FE}*			—
(I_C = 0.1 m A dc, V_{CE} = 1 Vdc)	2N3903		20	—	
	2N3904		40	—	
(I_C = 1.0 m A dc, V_{CE} = 1 Vdc)	2N3903		35	—	
	2N3904		70	—	
(I_C = 10 m A dc, V_{CE} = 1 Vdc)	2N3903		50	150	
	2N3904		100	300	
(I_C = 50 m A dc, V_{CE} = 1 Vdc)	2N3903		30	—	
	2N3904		60	—	
(I_C = 100 m A dc, V_{CE} = 1 Vdc)	2N3903		15	—	
	2N3904		30	—	
Collector-Emitter Saturation Voltage*		$V_{CE\,(sat)}$*			Vdc
(I_C = 10 m A dc, I_B = 1 m A dc)			—	0.2	
(I_C = 50 m A dc, I_B = 5 m A dc)			—	0.3	
Base-Emitter Saturation Voltage*		$V_{BE\,(sat)}$*			Vdc
(I_C = 10 m A dc, I_B = 1 m A dc)			0.65	0.85	
(I_C = 50 m A dc, I_B = 5 m A dc)			—	0.95	
SMALL SIGNAL CHARACTERISTICS					
High Frequency Current Gain	2N3903	$\|h_{fe}\|$	2.5	—	—
(I_C = 10 m A, V_{CE} = 20 V, f = 100 mc)	2N3904		3.0	—	
Current Gain– Bandwidth Product	2N3903	f_T	250	—	mc
(I_C = 10 m A, V_{CE} = 20 V, f = 100 mc)	2N3904		300	—	
Output Capacitance		C_{ob}			pf
(V_{CB} = 5 Vdc, I_E = 0, f = 100 kc)			—	4	
Input Capacitance		C_{ib}			pf
(V_{OB} = 0.5 V dc, I_C = 0, f = 100 kc)			—	8	
Small Signal Current Gain	2N3903	h_{fe}	50	200	—
(I_C = 1.0 m A, V_{CE} = 10 V, f = 1 kc)	2N3904		100	400	
Voltage Feedback Ratio	2N3903	h_{re}	0.1	5.0	$\times 10^{-4}$
(I_C = 1.0 m A, V_{CE} = 10 V, f = 1 kc)	2N3904		0.5	8.0	
Input Impedance	2N3903	h_{ie}	0.5	8	Kohms
(I_C = 1.0 m A, V_{CE} = 10 V, f = 1 kc)	2N3904		1.0	10	
Output Admittance		h_{oe}			μ mhos
(I_C = 1.0 m A, V_{CE} = 10 V, f = 1 kc)	Both Types		1.0	40	
Noise Figure		NF			db
(I_C = 100 μ A, V_{CE} = 5 V, R_g = 1 Kohms),	2N3903		—	6	
Noise Bandwidth = 10 cps to 15.7 kc)	2N3904		—	5	
SWITCHING CHARACTERISTICS					
Delay Time V_{CC} = 3 V dc, V_{OB} = 0.5 V dc,		t_d	—	35	nsec
Rise Time I_C = 10 m A dc, I_{B1} = 1 m A		t_r	—	35	nsec
Storage Time V_{CC} = 3 V dc, I_C = 10 m A dc,	2N3903	t_E	—	175	nsec
I_{B1} = I_{B2} = 1 m A dc	2N3904		—	200	
Fall Time		t_f	—	50	nsec

* Pulse Test: Pulse Width = 300 μ sec, Duty Cycle = 2 % V_{OB} = Base-Emitter Reverse Bias

APPENDIX 1.4

Type 2N3055 NPN Single-Diffused Mesa Silicon Power Transistor
FOR POWER-AMPLIFIER APPLICATIONS

Mechanical Data

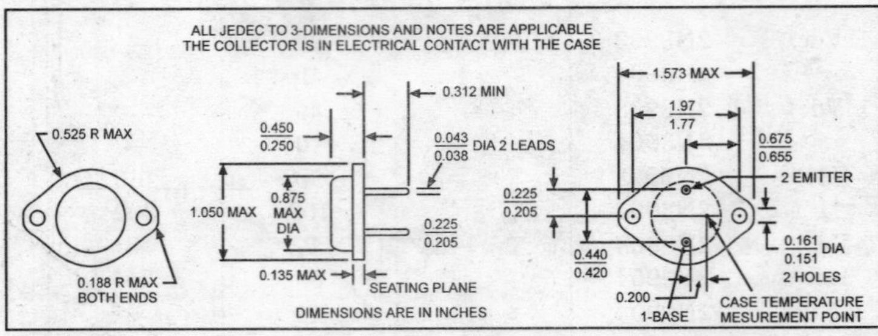

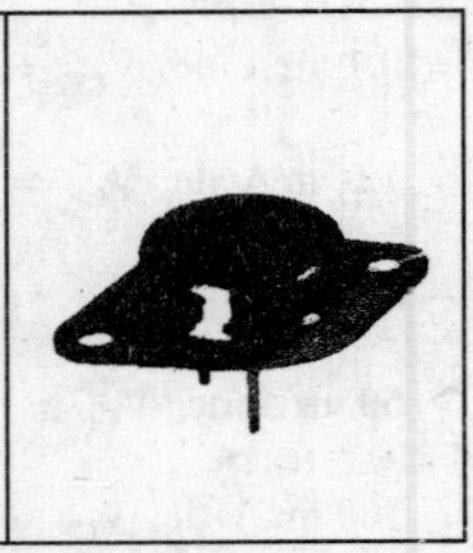

Absolute maximum ratings at 25 °C case temperature (unless otherwise noted)

Collector-Base Voltage .... 100 V
Collector-Emitter Voltage (See Note 1) .... 70 V
Emitter-Base Voltage .... 7 V
Continuous Collector Current .... 15 A
Continuous Base Current .... 7 A
Continuous Device Dissipation at (or below) 25 °C Case Temperature (See Note 2) .... 115 W
Operating Case Temperature Range .... – 65 °C to 200 °C
Storage Temperature Range .... – 65 °C to 200 °C
Lead Temperature 1/32 Inch from Case for 10 Seconds .... 235 °C

NOTES 1. This value applies when the base-emitter resistance R_{BE} = 100 Ω
2. Derate linearly to 200 °C case temperature at the rate of 0.66 W/deg.

*** Electrical characteristics at 25 °C case temperature (unless otherwise noted)**

PARAMETER		TEST CONDITIONS	MIN	MAX	UNIT
$V_{(BR)CEO}$	Collector-Emitter Breakdown Voltage	I_C = 200 mA, I_B = 0, See Note 4	60		V
$V_{(BR)CER}$	Collector-Emitter Breakdown Voltage	I_C = 200 mA, R_{BE} = 100 Ω	70		V
I_{CEO}	Collector Cutoff Current	V_{CE} = 30 V, I_B = 0		0.7	mA
I_{CEV}	Collector Cutoff Current	V_{CE} = 100 V, V_{BE} = – 1.5 V		5	mA
		V_{CE} = 100 V, V_{BE} = – 1.5 V, T_C = 150 °C		30	
I_{EBO}	Emitter Cutoff Current	V_{EB} = 7 V, I_C = 0		5	mA
h_{FE}	Static Forward Current Transfer Ratio	V_{CE} = 4 V, I_C = 4 A, See Notes 3 and 4	20	70	
		V_{CE} = 4 V, I_C = 10 A, See Notes 3 and 4	5		
V_{BE}	Base-Emitter Voltage	V_{CE} = 4 V, I_C = 4 A, See Notes 3 and 4		1.8	V
$V_{CE(sat)}$	Collector-Emitter Saturation Voltage	I_B = 400 mA, I_C = 4 A See Notes 3 and 4		1.1	V
		I_B = 3.3 A, I_C = 10 A See Notes 3 and 4		8	
h_{fe}	Small-Signal Common-Emitter Forward Current Transfer Ratio	V_{CE} = 4 V, I_C = 1 A, f = 1 kHz	15	120	
f_{hfe}	Small-Signal Common-Emitter Forward Current Transfer Ratio Cutoff Frequency	V_{CE} = 4 V, I_C = 1 A, See Notes 5	20		kHz

Thermal Characteristics

PARAMETER		MAX	UNIT
θ_{J-C}	Junction-to-Case Thermal Resistance	1.52	deg/W

NOTES 3. These parameters must be measured using pulse technique t_p = 300 μs, duty cycle ≤ 2 %.
4. These parameters are measured with voltage-sensing contacts separate from the current-carrying contacts.
5. f_{hfe} is the frequency at which the magnitude of the small-signal forward current transfer ratio is 0.707 of its low-frequency value. For this device, the reference measurement is made at 1 kHz.

* Courtesy of Texas Instruments, Incorporated

APPENDIX 1.5

2N5457, 2N5458, 2N5459 (Silicon) N-Channel FET

Silicon N-channel junction field-effect transistors depletion mode (Type A) designed for general-purpose audio and switching applications.

CASE 29 (5)
(TO-92)

Drain and source may be interchanged

MAXIMUM RATINGS

Rating	Symbol	Value	Unit
Drain-Source Voltage	V_{DS}	25	V dc
Drain-Gate Voltage	V_{DG}	25	V dc
Reverse Gate Source Voltage	$V_{GS(r)}$	25	V dc
Gate Current	I_G	10	mA dc
Total Device Dissipation @ T_A = 25 °C Derate above 25 °C	$P_D^{(2)}$	310 2.82	mW mW/°C
Operating Junction Temperature	$T_J^{(2)}$	135	°C
Storage Temperature Range	$T_{stg}^{(2)}$	− 65 to + 150	°C

ELECTRICAL CHARACTERISTICS (T_A = 25 °C unless otherwise noted)

Characteristic		Symbol	Min	Typ	Max	Unit
OFF CHARACTERISTICS						
Gate-Source Breakdown Voltage (I_G = − 10 μAdc, V_{DS} = 0)		BV_{GSS}	25	—	—	Vdc
Gate Reverse Current (V_{GS} = − 15 Vdc, V_{DS} = 0) (V_{GS} = − 15 Vdc, V_{DS} = 0, T_A = 100 °C)		I_{GSS}	— —	— —	1.0 200	n A dc
Gate-Source Cutoff Voltage (V_{DS} = 15 Vdc, I_D = 10 nAdc)	2N5457 2N5458 2N5459	$V_{GS(off)}$	0.5 1.0 2.0	— — —	6.0 7.0 8.0	V dc
Gate-Source Voltage (V_{DS} = 15 Vdc, I_D = 100 μAdc) (V_{DS} = 15 Vdc, I_D = 200 μAdc) (V_{DS} = 15 Vdc, I_D = 400 μAdc)	2N5457 2N5458 2N5459	V_{GS}	— — —	2.5 3.5 4.5	— — —	V dc
ON CHARACTERISTICS						
Zero-Gate-Voltage Drain Current (1) (V_{DS} = 15 Vdc, V_{GS} = 0)	2N5457 2N5458 2N5459	I_{DSS}	1.0 2.0 4.0	3.0 6.0 9.0	5.0 9.0 16	m A dc
DYNAMIC CHARACTERISTICS						
Forward Transfer Admittance (1) (V_{DS} = 15 Vdc, V_{GS} = 0, f = 1 kHz)	2N5457 2N5458 2N5459	$\|y_{fs}\|$	1000 1500 2000	3000 4000 4500	5000 5500 6000	μ mhos
Output Admittance (1) (V_{DS} = 15 Vdc, V_{GS} = 0, f = 1 kHz)		$\|y_{os}\|$	—	10	50	μ mhos
Input Capacitance (V_{DS} = 15 Vdc, V_{GS} = 0, f = 1 MHz)		C_{iss}	—	4.5	7.0	pF
Reverse Transfer Capacitance (V_{DS} = 15 Vdc, V_{GS} = 0, f = 1 MHz)		C_{rss}	—	1.5	3.0	pF

(1) Pulse Test : Pulse Width ≤ 630 ms, Duty Cycle ≤ 10 %
(2) Continuous package improvements have enhanced these guaranteed Maximum Ratings as follows P_D = 1.0 W @ T_C = 25 °C, Derate above 25 °C - 8.0 mW/°C, T_J = − 65 to + 150 °C, θ_{JC} = 125 °C/W

* Courtesy of Motorola, Inc.

Model Test Papers

MODEL TEST PAPER–1

Electronic Devices and Circuits

Time : 3 Hours *Maximum Marks : 100*

Note : Question No. 1 is ***compulsory***. Attempt ***One*** question from each of Sections – A, B, C and D.

1. (*a*) What is free electron theory ?

(*b*) What is superconductivity and how is it affected by frequency ?

(*c*) Define mobility of a carrier. Show that mobility constant of electron is larger than that of a hole.

(*d*) Explain the term doping and its need.

(*e*) Differentiate between Zener breakdown and Avalanche breakdown mechanisms.

(*f*) What is the significance of arrow head in the transistor symbol ?

(*g*) What is MESFET and how does it differ from P-N junction FET ?

(*h*) What is meant by LED ? What are its applications ?

(*i*) What is tunnel diode ? Give its few applications.

(*j*) Define holding current and latch current in reference to SCR and give their typical values.

SECTION–A

2. (*a*) Classify the various conducting materials. Also discuss their properties and applications.

(*b*) Prove that the heat developed per m^3 per second in a conductor carrying a current of density J as a result of an applied electric field E is given by $W = JE = \sigma E^2$ where σ is conductivity of conductor.

3. (*a*) Explain the terms "drift velocity", "mobility" and "collision time".

(*b*) Explain the phenomenon of superconductivity and also discuss its behaviour in magnetic field. Give some applications of this phenomenon.

SECTION–B

4. (*a*) Distinguish between intrinsic and extrinsic semiconductors. Discuss why intrinsic semiconductor is not used in semiconductor devices.

(*b*) Discuss V-I characteristics of P-N junction diode.

5. (*a*) What is metal-semiconductor junction ? How rectifying and non-rectifying contact type metal-semiconductor junctions are formed ? Explain.

(*b*) Discuss optical and electrical excitation in diodes.

SECTION–C

6. (*a*) Explain different components of currents flowing through the structure of an N-P-N transistor. How the emitter injection efficiency and base transport factor influences the amplification factor ?

   (*b*) Explain the working of transistor as an amplifier.

7. (*a*) What is MESFET ? Draw its structure and explain its working.

   (*b*) Explain the construction and working of DE-MOSFET. Give its applications too.

SECTION–D

8. (*a*) Explain the operation of solar cell and photodetectors in detail.

   (*b*) What are the basic constructional differences between a conventional P-N junction diode and a tunnel diode ? Explain briefly the operation of a tunnel diode.

9. (*a*) What is IGBT. Give its construction, working and applications.

   (*b*) Describe the characteristics and applications of TRIAC and DIAC.

MODEL TEST PAPER–2

Electronic Devices and Circuits

Time : 3 Hours *Maximum Marks : 100*

Note : Question No. 1 is ***compulsory.*** Attempt ***One*** question from each of Sections – A, B, C and D.

1. (*a*) What is crystal ? What is crystal lattice ?

(*b*) Describe the factors on which the conductivity of a material depends.

(*c*) Define diffusion current in a semiconductor.

(*d*) What are metal-semiconductor junctions and how are they fabricated ?

(*e*) What is zener diode and how it can be used as a voltage regulator ?

(*f*) What are 'emitter injection efficiency' and 'base transport factor' and how do they influence the transistor operation ?

(*g*) Give the advantages of HEMT over silicon MOSFET.

(*h*) What are photovoltaic or solar cells ?

(*i*) What are GUNN diodes ?

(*j*) What is IGBT ? Give its some applications.

SECTION–A

2. (*a*) Define and explain relaxation time, collision time and mean free path. Derive mathematical relation for each term.

(*b*) Discuss the factors affecting the conductivity of a conducting material.

3. (*a*) Explain thermal conductivity and how is it related to Wiedemann Franz law ?

(*b*) Explain the term 'superconductivity'. State various applications of superconductors.

SECTION–B

4. (*a*) Explain the concept of continuation equation in a semiconductor.

(*b*) Discuss the various capacitances of a diode.

5. (*a*) Explain in brief planar technology for device fabrication.

(*b*) Explain the breakdown mechanism in a diode.

SECTION–C

6. (*a*) Compare CE, CB and CC configurations of BJT. Also discuss the input and output characteristics of a transistor in CE configuration.

(*b*) Determine the values of emitter current and collector current of a transistor having $\alpha_{dc} = 0.98$ and collector-to-base leakage current $I_{CBO} = 4\ \mu A$. The base current is 50 μA.

7. (*a*) Explain the working of N-channel JFET. Also discuss its V-I characteristics.
(*b*) Explain the working of E-MOSFET in detail. Also discuss its V-I characteristics.

SECTION–D

8. (*a*) Define photonic devices and give their classification with categories in each category.
(*b*) What is LED ? Give its principle of working, construction, circuit symbol, merits, demerits and applications.
9. (*a*) Obtain V-I characteristics of a tunnel diode with special reference to its negative resistance starting from equilibrium condition to forward and reverse bias conditions.
(*b*) Explain the construction, working and applications of SCR.

❑❑❑

EXAMINATION PAPER

B.TECH.

THIRD SEMESTER (ELECTRICAL ENGINEERING) EXAMINATION, DECEMBER–2010

ELECTRONIC DEVICES AND CIRCUITS

Time : Three Hours] ***[Maximum Marks : 100***

Before answering the question, candidates should ensure that they have been supplied the correct and complete question paper. No complaint in this regard, will be entertained after examination.

Note : Question No. 1 is ***compulsory.*** Select ***one*** question from each Section. Attempt ***five*** questions out of **9** questions. All questions carry equal marks.

1. (*a*) Explain Widmann Franz Law. What is its physical significance ? **4**

(*b*) Explain the term optical excitation in diode. **4**

(*c*) Define the pinch off voltage and channel ohmic region. **4**

(*d*) Define "punch through" in Early effect. **4**

(*e*) Define the degenerate semiconductor concept. **4**

SECTION–A

2. (*a*) Explain energy band diagram of material considering conductor, insulator and semiconductor. **10**

(*b*) Derive the expression for mean free path. **10**

3. Explain the effect of magnetic field on conducting material, also give their application. **20**

SECTION–B

4. Derive an expression for continuity equation for P-N function. **20**

5. Explain the construction, working and application of LED. **20**

SECTION–C

6. (*a*) Derive the MOSFET Design Equations for all regions. **15**

(*b*) What do you mean by Channel length modulation in MOSFET ? **5**

7. (*a*) Explain the following : **10**

(*i*) Body effect.

(*ii*) Speed of response.

(*b*) Derive the expression for collector current in terms of current gain and reverse saturation current. **10**

SECTION–D

8. (*a*) Explain the working of IMPATT diode. **10**

(*b*) Discuss the spectral response of photodiode. **10**

9. Describe the mechanism by which light increases the conductivity of bulk type conductive cell. Give three characteristics of bulk type photocell. **20**

❑❑❑

B.TECH.

THIRD SEMESTER (ELECTRICAL ENGINEERING) EXAMINATION, DECEMBER–2011

ELECTRONIC DEVICES AND CIRCUITS

Time : Three Hours] ***[Maximum Marks : 100***

Before answering the question, candidates should ensure that they have been supplied the correct and complete question paper. No complaint in this regard, will be entertained after examination.

Note : Question No. 1 is *compulsory.* Select *one* question from each Section. Attempt *five* questions out of 9 questions. All questions carry equal marks.

1. (*a*) **Define the term mobility and scattering. Why scattering is less significant at higher temperature ? 4**

(*b*) **What is planar technology for device fabrication ? 4**

(*c*) **Why is collector wider than emitter and base explain ? 4**

(*d*) **Discuss the spectral response of LED. 4**

(*e*) **Explain briefly the working of IMPATT. 4**

SECTION–A

2. (*a*) **What are factors affecting the conductivity of materials—explain ? 10**

(*b*) **Explain and derive the expression for diffusion current density. 10**

3. Explain the following : 20

(*i*) **Drift velocity,**

(*ii*) **Mean free path,**

(*iii*) **Relaxation time,**

(*iv*) **Effect of magnetic field on conducting material.**

SECTION–B

4. (*a*) **Derive an expression for diffusion capacitance in a p-n junction diode. 10**

(*b*) **Explain the working and construction of solar cell. 10**

5. (*a*) **Explain the working of zener diode. 10**

(*b*) **Find the electron concentration in p-type semiconductor when the acceptor concentration is 5×10^{14} cm^{-3} and intrinsic value of germanium material at 300°K is 2.5×10^{13} cm^{-3}. 20**

SECTION–C

6. Explain the construction and working of MOSFET in both enhancement and depletion modes. **20**

7. (*a*) Draw the input and output characteristics of BJT in common-emitter configuration. **10**

   (*b*) Explain that how transistor used as an amplifier. **10**

SECTION–D

8. (*a*) Give the construction of a semiconductor photodiode. Draw and discuss its V-I characteristics. **10**

   (*b*) Explain construction and working of Tunnel Diode. **10**

9. (*a*) How does an avalanche photodiode obtains a higher current sensitivity than a PIN diode ? Does it have a fast response ? **5**

   (*b*) A LED has minimum drop of 1.5 V of the supply voltage be 10 V and R_S = 470 Ω. What will be the value of current. **5**

SOLUTION TO NUMERICAL PROBLEMS

5. (*b*)

Intrinsic carrier concentration, $n_i = 2.5 \times 10^{13}\ \text{cm}^{-3}$

Acceptor concentration, $N_A = 5 \times 10^{14}\ \text{cm}^{-3}$

Concentration of holes, $p \simeq N_A = 5 \times 10^{14}\ \text{cm}^{-3}$

$$\text{Electron concentration, } n = \frac{n_i^2}{p} = \frac{(2.5 \times 10^{13})^2}{5 \times 10^{14}} = 1.25 \times 10^{12}\ \text{cm}^{-3} \ \textbf{Ans.}$$

9. (*b*)

Supply voltage, $V_S = 10$ V

Forward bias voltage, $V = 1.5$ V

Serial resistor, $R_S = 470\ \Omega$

$$\text{Current, } I = \frac{V_S - V}{R_S} = \frac{10 - 1.5}{470} = 0.0181 \text{ A or } 18.1 \text{ mA} \ \textbf{Ans.}$$

❑❑❑

INDEX

INDEX

A

B

C

H

I

J

L

M

N

O

P

Q

R

S

T

U

V

W

Z

❑❑❑

OTHER DETAILS

(i) Who recommended you this book ? (Please tick (✓) in the box near the option relevant to you.

☐ Teacher ☐ Friends ☐ Bookseller

....

....

....

....

(ii) Name of the recommending teacher, his designation and address:

....

....

....

....

(iii) Name and address of the bookseller you purchased the book from :

....

....

....

....

(iv) Name and address of your institution (Please mention the University or Board as the case may be)

....

....

....

....

(v) Your name and complete address :

....

....

....

....

(vi) Course you are enrolled in :

....

....

....

....

The best assessment will be awarded each month : The award will be in the form of our publications as decided by the Editorial Board.

Please mail the filled up coupon at the vearliest to :

S.K. KATARIA & SONS®

4760-61/23, Ansari Road,

Daryaganj, New Delhi-110002

Phone: +91-11-23243489, 23269324

TeleFax : +91-11-23243489

email : katariabooks@yahoo.com